PROBABILITY THEORY AND
MATHEMATICAL STATISTICS

Volume I

PROBABILITY THEORY AND MATHEMATICAL STATISTICS

Proceedings of the Fourth Vilnius Conference

Vilnius, USSR, 24–29 June 1985

Edited by

Yu. V. Prohorov, V. A. Statulevicius, V. V. Sazonov and B. Grigelionis

VNU SCIENCE PRESS

Utrecht, The Netherlands

VNU Science Press BV
P.O. Box 2093
3500 GB Utrecht
The Netherlands

First published 1987

CIP-DATA KONINKLIJKE BIBLIOTHEEK, DEN HAAG
Probability theory and mathematical statistics:
proceedings of the Fourth Vilnius Conference: Vilnius,
USSR, 24–29 June 1985/ed. by Yu.V. Prohorov . . . [*et al.*] — Utrecht: VNU Science Press. — Ill.
ISBN 90-6764-067-0 (vol. I)
ISBN 90-6764-069-7 (set)
SISO 517 UDC 519.2(063)
Subject headings: probability theory/mathematical statistics.

Printed in Great Britain by J. W. Arrowsmith Ltd., Bristol

CONTENTS

PREFACE

The International Vilnius Conferences on Probability Theory and Mathematical Statistics have been held every four years since 1973 and have become traditional. However, their proceedings were never published. Only a limited number of copies of the abstracts of contributions were available, published locally.

The present publication makes it possible for a wide international scientific community to become familiar with the invited lectures delivered at the 4th Vilnius Conference which was held from 24–29 June, 1985. The main topics of the conference were (controlled) random processes and fields, limit theorems of probability theory, and asymptotic methods of statistics.

The organizing committee of the conference invited all the 45- and 30-minute speakers to present a paper to the Proceedings. We included in the Proceedings all of the papers received. There were 453 participants at the conference of whom 150 were invited speakers. We hope that the 4th Vilnius Conference and its Proceedings will give rise to many important and interesting investigations in the future.

Yu. Prohorov

Prob. Theory and Math. Stat., Vol.1, pp. 1-21
Prohorov et al. (eds)
1986 VNU Science Press

QUANTUM STOCHASTIC CALCULUS

Luigi Accardi
Princeton University
Department of Statistics
Princeton, New Jersey 08544

1.) INTRODUCTION

Hudson and Parthasarathy have introduced the Fock stochastic calculus over $L^2(R_+, dt)$ which is a beautiful generalization of the classical stochastic calculus that already found several applications in different questions both of mathematics and of physics . However , since there are uncountably many inequivalent representations of the CCR and infinitely many choices for the Hilbert spaces on which they are based , one might suspect that there are infinitely many "stochastic calculi" and that for each of these one should develop an ad hoc machinery based on the particular properties of the representation chosen . Since in the theory of classical stochastic processes there is a general theory of stochastic integration which does not depend on the particular features of the integrator process [11], [13] , [17], [19], [20], [22], it is natural to ask oneself if also in the theory of quantum stochastic processes it is possible to develop a general theory of stochastic integration which is in some sense canonical i.e. , independent of the particular features of the integrator processes . Such a theory has been developed recently [5] , [6], [7] and the goal of the present paper is to outline some of the main ideas of this representation free stochastic calculus.

It will be clear from what follows that the quantum stochastic calculus we are going to describe is the natural translation , in a quantum probabilistic context , of the classical stochastic calculus for general stochastic processes and all the basic constructions and techniques introduced in the representation free calculus reduce to the classical ones when the algebras considered are commutative i.e. the algebras generated by the random variables of some classical stochastic process (a property which is not shared by Hudson and Parthasarathy 's Fock stochastic calculus). Moreover the quantum Levy representation theorem (cf. Section (7)) shows the relation between the representation free stochastic calculus and the Fock stochastic calculus , or the "universally invariant" stochastic calculus introduced by Hudson and Lindsay [16] is roughly speaking the same as the relations between these calculi and the usual Wiener process : as the latter calculi can be looked at as "lumping together in a non-commutative way" a "bunch" of classical Wiener processes , the representation free stochastic calculus can be looked at as the result of the same procedure applied to a "bunch" of Husdson and Parthasarathy (or universally invariant) processes .

The results I am going to describe have been obtained in collaboration with A. Calzolari (Sect. (7)) , J.Quaegebeur (Sects. (5) ,(6),(10)) and K.R.Parthasarathy (Sections (3) , (9))

The present exposition is aimed at presenting to an audience of classical probabilists the main ideas of the quantum stochastic calculus with emphasis on the strict analogy between the classical and the quantum theory and on the computational and algorithmic aspects, rather than on detailed proofs (which would be impossible within the prescribed limitations of space

) . In particular, when I had to choose between a rigorous definition or notation and a clear and intuitive one I did not hesitate in favoring the latter. Therefore many technical details have been skipped in the exposition and for these the reader is referred to the bibliography at the end of the paper .

2.) CLASSICAL AND QUANTUM STOCHASTIC CALCULUS

A quick , intuitive idea of the basic facts about quantum stochastic processes can be obtained by translating in abstract Hilbert space language the theory of classical square integrable stochastic processes . Thus, for example , starting from a probability space $(\Omega, \mathcal{F}, P)$, the associated space $L^2(\Omega, \mathcal{F}, P)$ of complex valued , square integrable functions is replaced by an abstract complex separable Hilbert space $\mathcal{H}$:

$$L^2(\Omega, \mathcal{F}, P) \longrightarrow \mathcal{H} \tag{1}$$

The constant function 1 on Ω is replaced by a unit vector Φ in $\mathcal{H}$

$$1 \epsilon L^2(\Omega, \mathcal{F}, P) \longrightarrow \Phi \epsilon \mathcal{H}; \qquad ||\Phi|| = 1 \tag{2}$$

The algebra $L^\infty(\Omega, \mathcal{F}, P)$, which acts on $L^2(\Omega, \mathcal{F}, P)$ by multiplication

$$f \ : \ \xi \epsilon L^2(\Omega, \mathcal{F}, P) \rightarrow f\xi \ \epsilon L^\infty(\Omega, \mathcal{F}, P)$$

is replaced by an arbitrary sub-algebra $\mathcal{A}$ of the algebra of all bounded operators on $\mathcal{H}$:

$$L^\infty(\Omega, \mathcal{F}, P) \longrightarrow \mathcal{A} \subseteq \mathcal{B}(\mathcal{H}) \tag{3}$$

So that the condition that the set

$$L^\infty(\Omega, \mathcal{F}, P) \cdot 1 = \{f \ \epsilon L^\infty \cap L^2(\Omega, \mathcal{F}, P)\}$$

is dense in $L^2(\Omega, \mathcal{F}, P)$ becomes that the set

$$\mathcal{A} \cdot \Phi = \{a\Phi \ : \ a \epsilon \mathcal{A}\}$$

is dense in $\mathcal{H}$. This property is usually expressed by saying that Φ is cyclic for $\mathcal{A}$. If moreover

$$a\Phi = 0 \Longleftrightarrow a = 0$$

then we say that Φ is separating for $\mathcal{A}$.

The integral , associated to the probability measure P , is replaced by the expectation functional (or "state" in the quantum mechanical terminology) associated to the vector Φ through the formula :

$$\int_\Omega f dP \longrightarrow \varphi(a) = \langle \Phi, a\Phi \rangle \tag{4}$$

where $f \epsilon L^\infty(\Omega, \mathcal{F}, P)$ and a is in $\mathcal{A}$.

Since the bounded complex valued random variables , i.e. the elements of $L^\infty(\Omega, \mathcal{F}, P)$, are replaced by arbitrary bounded operators on $\mathcal{H}$, it is natural to expect that the unbounded random variables will be replaced by unbounded operator on $\mathcal{H}$. Usually it is required that the operators , corresponding to random variables , are pre-closed so that a minimum of functional calculus is allowed . In this language , the fact that X is an L^2-random variable is equivalent to the statement that the function 1 is in the domain of X (considered as a multiplication operator on $L^2(\Omega, \mathcal{F}, P)$) . Thus we have

$$X \ \epsilon L^2(\Omega, \mathcal{F}, P) \longrightarrow \Phi \epsilon D(X)$$

where, here and in the following , D(X) will denote the domain of the operator X in $\mathcal{H}$.

A classical stochastic process is a family of random variables (X_t). Similarly , in first approximation, a quantum stochastic process can be thought of simply as a family (X_t) of pre-closed operators on $\mathcal{H}$. A deeper analysis should involve in the definition the notion of "stochastic equivalence", but we skip this point here (cf. [3]) . For example, one could choose $\mathcal{H}$ to be the space $L^2(\Omega, \mathcal{F}, P; H)$ of square integrable H-valued functions on Ω , for some Hilbert space H , and $\mathcal{A}$ to be the algebra of all the operators on $\mathcal{H}$, identified with the tensor product

$$L^\infty(\Omega, \mathcal{F}, P) \bigotimes \mathcal{B}(\mathcal{H})$$

The quantum stochastic processes corresponding to this choices coincide with the classical operator valued processes which are studied in the quantum theory of measurement [9], [14], and they include the operator valued processes studied in [25] . A filtration $(\mathcal{F}_{t]})$ of sub-σ-algebras of $\mathcal{F}$ is uniquely determined by the corresponding filtration of algebras $L^\infty(\Omega, \mathcal{F}, P)$ (t$\geq$ 0) ,and this will be replaced in the quantum model by a filtration $(\mathcal{A}_{t]})$ of sub-algebras of $\mathcal{A}$:

$$L^\infty(\Omega, \mathcal{F}, P) \longrightarrow \mathcal{A}_{t]} \subseteq \mathcal{B}(\mathcal{H})$$

$$s \leq t \Longrightarrow \mathcal{A}_{s]} \subseteq \mathcal{A}_{t]}$$

If $\mathcal{F}$(s,t) is the σ-algebra generated by the random variables X_r with r in the interval (s,t) then the algebras $L^\infty(\Omega, \mathcal{F}(s,t), P)$ are replaced by the so-called local algebras $\mathcal{A}$(s,t) $\subseteq \mathcal{A}$ which satisfy

$$(s,t) \subseteq (s',t') \Longrightarrow \mathcal{A}(s,t) \subseteq \mathcal{A}(s',t')$$

Similarly the Hilbert space filtration

$$L^2(\Omega, \mathcal{F}_{t]}, P) =: \mathcal{H}_{t]}$$

will be replaced by the filtration of Hilbert spaces

$$\mathcal{H}_{t]} =: \ closure\ of\ \mathcal{A}_{t]}\Phi = closure\ of\ \{a\Phi : a \in \mathcal{A}_{t]}\}$$

If (X_t) is a classical stochastic process , the property of being adapted to the filtration $\mathcal{F}_{t]}$ can be expressed in Hilbert space language by saying that for each t and for each vector ξ in $\mathcal{H}_{t]} \cap D(X_t)$ also the vector $X_t\xi$ lies in $\mathcal{H}_{t]}$. Phrased in this way the definition of adaptedness makes sense for any family (X_t) of pre-closed operators on $\mathcal{H}$ and therefore for any quantum stochastic process .

In the notations introduced above , for any Hilbert space $\mathcal{H}$ we denote

$$P_{t]} : \mathcal{H} \longrightarrow \mathcal{H}_{t]}$$

the orthogonal projection and for any filtration $(\mathcal{A}_{t]})$ in $\mathcal{B}(\mathcal{H})$ we denote

$$E_{t]} : \mathcal{A} \longrightarrow \mathcal{A}_{t]}$$

the φ-conditional expectation from $\mathcal{A}$ onto $\mathcal{A}_{t]}$ which is a linear operator characterized by the properties

$$E_{t]}(a_{t]}a) = a_{t]}E_{t]}(a); \qquad a_{t]} \in \mathcal{A}_{t]}; \qquad a \in \mathcal{A} \tag{5}$$

$$E_{t]}(1) = 1 \tag{6}$$

$$\varphi \cdot E_{t]} = \varphi \tag{7}$$

Warning : while conditional expectations in the sense of the above definition always exist in the classical case (i.e. $\mathcal{A} = L^\infty(\Omega, \mathcal{F}, P)$), they rarely exist in the quantum case . From the technical point of view this is the main difference between classical and quantum probability

, but in the present paper we will restrict ourselves to a particularly nice class of processes , for which the conditional expectations needed to develop a quantum stochastic calculus always exist . Of course this class does not exhaust the class of physically interesting quantum stochastic processes , but nevertheless it has many interesting physical applications .

One easily checks that in the classical case , i.e. when φ is defined through the correspondence (4) , the three properties (5) , (6), (7) characterize the P-conditional expectation on $\mathcal{F}_{t]}$ defined in the usual way through the Radon-Nikodym theorem . Having the notion of conditional expectation , one can define a martingale in the usual way , i.e. as an adapted process M such that for each $s \leq t$ one has

$$E_{t]}(M_t) = M_s \tag{8}$$

Under some regularity conditions , which are much weaker in the classical case , one can prove the Doob-Meyer decomposition for a quantum stochastic process (cf. [6]).

Theorem(1) Let F be a bounded adapted process and let $\mathcal{T}$ be a locally convex topology on $\mathcal{B}(\mathcal{H})$ such that for each t in R_+ : (i) the conditional expectation $E_{t]}$ is $\mathcal{T}$-continuous ; (ii) the limit

$$\lim_{\epsilon \to 0} E_{t]}\left(\frac{F(t+\epsilon) - F(t)}{\epsilon}\right) =: D_+F(t) \tag{9}$$

exists in the $\mathcal{T}$-topology and (iii) the map $t \mapsto D_+F(t)$ is $\mathcal{T}$-continuous . Then the process

$$M(t) = F(t) - \int_0^t D_+F(s)ds \tag{10}$$

is a martingale . proh2.tex proh3.tex

3.) STOCHASTIC INTEGRALS WITH RESPECT TO A DIFFERENCE MARTINGALE

In the notations of Section (2) let $\xi \in \mathcal{H}$ be a vector such that the filtration $\mathcal{A}_{t]}$ is expected for the state φ_ξ associated to ξ through the correspondence (4) of Sect.(2) with ξ substituted for Φ and let $E_{t]}$ denote the corresponding conditional expectations ($t \geq 0$). Let dM be a ξ-difference martingale such that :

$$E_{t]}(|dM(t)|^2) = E_{t]}(\int_0^{t+dt} \sigma(s)ds) \tag{1}$$

for some operator valued function $\sigma(s)$ (which is necessarily positive ds-a.e.), and let

$$F = \sum_j F(t_j)\chi_{[t_j,t_{j+1})} \tag{2}$$

be a simple adapted process. For every bounded interval [t, u] define

$$\int_t^u dN_s F_s = \sum_j dN(t_j)F(t_j) \tag{3}$$

One has

$$\| \int_t^u dMF\xi \|^2 = \| \sum_j dM(t_j)F(t_j)\xi \|^2 = \tag{4}$$

$$= \sum_{j,k} \langle dM(t_j)F(t_j)\xi \ , \ dM(t_k)F(t_k)\xi \rangle =$$

$$\sum_{j,k} \varphi_\xi(F(t_j)^+ dM^+(t_j)dM(t_k)F(t_k) \ )$$

and using

$$\varphi_\xi \cdot E_{t]} = \varphi_\xi, \qquad E_{t]}(dM(t)) = 0 \tag{5}$$

and condition (1) one finds that the right hand side of (4) is equal to

$$\sum_j \varphi_\xi(F^+(t_j)E_{t]}(|dM(t_j)|^2)F(t_j) \ ) = \sum_j \int_{t_j}^{t_{j+1}} \varphi_\xi(F^+(t_j)\sigma(s)F(t_j) \ ) =$$

$$= \int_t^u \varphi_\xi(F^+(s)\sigma(s)F(s)) \ = \ \int_t^u \langle F(s)\xi \ , \ \sigma(s)F(s)\xi \ \rangle = \int_t^u \|\sigma(s)^{1/2}F(s)\xi\|^2$$

The identity

$$\| \int_t^u dMF\xi \|^2 \ = \ \int_t^u \| \ \sigma(s)^{1/2}F(s)\xi\|^2 ds \tag{6}$$

will be referred to as the Wiener isometry. With standard arguments one extends the Wiener isometry to each adapted process F such that ξ is in the domain of F(s) for each s ≥ 0 and for every bounded interval [t, u]

$$\int_t^u \|\sigma(s)^{1/2}F(s)\xi\|^2 ds \ \leq \ \infty \tag{7}$$

Sometimes the following, more general, criterion is useful to establish the N-integrability of some adapted process F :

Proposition (1) Let N be a ξ-martingale such that for each bounded interval [t, u]

$$\limsup_{|P|\to 0} \sum_{t_j \in P} \|E_{t_j]}(|dN(t_j)|^2)\|_\infty < \infty \tag{8}$$

Then for any adapted process F such that the map

$$t\epsilon R_+ \longrightarrow F(t)\xi$$

is right continuous in norm , the limit

$$\lim_{|P|\to\infty} \sum_{t_j \epsilon P} dN(t_j)F(t_j)\xi \tag{9}$$

exists in norm in H .

Proof. Let [t,u] be a bounded interval and let P , P' be two partitions of [t, u] such that P' is a refinement of P. We can assume that P and P' have the form :

$$P = P(t,u) \equiv t = t_1 < ... < t_n = u$$

$$P' = P'(t,u) \equiv t = t_1 \leq t'_1 < t_2 \leq t'_2 < ... < t_j \leq t'_j < t_{j+1} < ... < t_n = t$$

In this case

$$\sum_{t_j \epsilon P} dN(t_j)F(t_j) - \sum_{t'_j \epsilon P'} dN(t'_j)F(t_j) = \sum_{j=1}^{n} N(t'_j, t_j)[F(t_j) - F(t'_j)]$$

and therefore , using the conditions (5) above and denoting Δ_j the difference $[F(t_j) - F(t'_j)]$ one finds

$$||\sum_j N(t'_j, t_{j+1})[F(t_j) - F(t'_j)]\xi||^2 = \sum_j \varphi_\xi(\Delta_j^+ E_{t'_j]}(|N(t'_j, t_{j+1})|^2)\Delta_j) =$$

$$\leq \sum_j ||E_{t'_j]}(|N(t'_j, t_{j+1})|^2)||_\infty ||[F(t_j) - F(t'_j)]\xi||^2$$

from which the thesis follows.

The criteria discussed above allow to define the stochastic integrals

$$\int_t^u dN_s F_s \xi \tag{10}$$

for a fixed vector ξ in $\mathcal{H}$, but we want to define the stochastic integral

$$\int_t^u dN_s F_s \tag{11}$$

as a well behaved operator (densely defined, pre-closed,...) on $\mathcal{H}$. In order to achieve this goal, the set of vectors ξ for which the stochastic integral (10) is defined must be sufficiently large. In the examples considered up to now this property is guaranteed by one of the following circumstances :

i) the vector ξ is separating for $\mathcal{A}$

ii) there is a dense set D of vectors in $\mathcal{H}$ such that, for each ξ in D , N has a forward derivative with respect to ξ (such an N was called in [5] a regular semi-martingale).

In case (i) the inequality

$$||\sum_P dNFa'\xi - \sum_{P'} dNFa'\xi|| \leq ||a'||_\infty ||\sum dNF\xi - \sum dNF\xi||$$

holds for any a' in the commutant $\mathcal{A}'$ of $\mathcal{A}$ and implies the existence of the stochastic integral

$$\int_t^u dNFa'\xi$$

for every $a'\xi$ in the dense set $\mathcal{A}'\xi$.

In case (ii) , from Theorem (1) in Sect.(2) it follows that for every ξ in D we can decompose dN as

$$dN = b_\xi dt + dM_\xi \tag{13}$$

where b_ξ is the forward derivative of N with respect to ξ and M_ξ is a ξ-martingale. Using the decomposition (13) we can now define

$$\int dNF\xi = \int b_\xi F\xi dt + \int dM_\xi F\xi \tag{14}$$

where on the right hand side of (14) there is a usual vector valued integral plus a stochastic integral with respect to the ξ-martingale M_ξ to which the criteria discussed at the beginning of this section can be applied.

The typical case when the situation (i) above is realized is with the "universally invariant quantum Brownian motion" , first discussed by Hudson and Lindsay [16] . The situation (ii) is realized in the Fock quantum Brownian motion, originally considered by Hudson and Parthasarathy [15] . The technique used here has the advantage of being representation independent and has the technical advantage that , even when particularized to the Fock case, the stochastic integrals converge in the strong rather than the weak topology .

4.) STOCHASTIC DIFFERENTIAL EQUATIONS

Let $M_1, ..., M_n$ $(n \epsilon N)$ be martingales , satisfying the condition of Proposition (1) of Section (3) . An adapted process is said to satisfy the stochastic differential equation

$$dX_t = B_t dt + dM_\alpha F_\alpha(X_t); \qquad t\epsilon[s,u] \tag{1}$$

with the initial condition

$$X(s) = X \tag{2}$$

if for every t in [s ,u] one has

$$X_t = X + \int_s^t B(r)dr + \int_s^t dM_\alpha F_\alpha(X_s) \tag{3}$$

where both integrals exist in a topology $\mathcal{T}$ In the above formulas the F_α' s are suitably defined operator functions (e.g. polynomials , linear maps , Borel functions in case the X_t ' s are self-adjoint , ...) and, here and in the following, when two repeated greek indices appear in a formula it is understood that these indices must be summed over their range.

Remark(1) When confusion might arise one should speak of a $\mathcal{T}$-stochastic differential equation.

Remark(2) In our definition , " solution of a stochastic differential equation" means adapted solution .

Remark(3) The development of an abstract theory of stochastic differential equations of the form ( ) for general functions F_α 's can be based on estimates of X_t with respect to semi-norms of the form

$$|||X_t - X_s||| \leq \left(\int_s^t ||B_r\xi||dr\right)^{1/2} + c_\sigma\left(\int_s^t ||F_\alpha(X_r)\xi||_\sigma^2 dr\right)^{1/2}$$

where ξ denotes a vector in $\mathcal{H}$ and c_σ is a constant depending on the covariance of the martingales M_α (cf. [5]). However, at the moment, the only explicit examples concern the case in which the F_α 's have the form

$$F_\alpha(X_t) = C_\alpha(t) \cdot X_t$$

for some operator valued functions $C_\alpha(t)$.

5.) MEYER BRACKETS AND ITO FORMULA

Let M and N be two adapted processes . Assume that for each bounded interval [s,t] the limits

$$\lim_{|P|\to 0} \sum_{t_k \epsilon P} M(t_k, t_{k+1})N(0, t_k) = \int_s^t dMN \tag{1}$$

$$\lim_{|P|\to 0} \sum_{t_k \epsilon P} M(0, t_k)N(t_k, t_{k+1}) = \int_s^t MdN \tag{2}$$

exist in some topology $\mathcal{T}$ and that the product M(t)N(t) is well defined as an adapted process . Then, summing the algebraic identity

$$d(MN) = dM \cdot N + M \cdot dN + dMdN \tag{3}$$

over the intervals [t , t+ dt] of the partition P of [s, t] , one finds

$$M(t)N(t) - M(s)N(s) = \sum_k M(t_k, t_{k+1})N(0, t_k) +$$

$$+ \sum_k M(0, t_k)N(t_k, t_{k+1}) + \sum_k M(t_k, t_{k+1})N(t_k, t_{k+1})$$

and since the limits (1) and (2) are supposed to exist, it follows that also the limit

$$\lim_{|P|\to 0} \sum_k M(t_k, t_{k+1})N(t_k, t_{k+1}) \tag{4}$$

exists in the topology $\mathcal{T}$. This limit will be called the $\mathcal{T}$-Meyer bracket of the processes M and N and will be denoted

$$\langle M, N\rangle(s, t) \tag{5}$$

By construction one has

$$M(t)N(t) - M(s)N(s) = \int_s^t dMN + \int_s^t MdN + \int_s^t d\langle M, N\rangle$$

that we can write symbolically in differential form :

$$d(MN) = (dM)N + M(dN) + d\langle M, N\rangle \tag{6}$$

Equation (6) above expresses the essence of Ito formula in the present context.

Sometimes even if the operators (dM)N and M(dN) are well defined , one might have difficulties in controlling the domain of dMdN . In these cases it is convenient to use a weaker form of the identity (3) namely

$$d\langle M\xi, N\eta\rangle = \langle dM\xi, N\eta\rangle + \langle M\xi, dN\eta\rangle + \langle dM\xi, dN\eta\rangle \tag{7}$$

where ξ and η are vectors which range over a suitable domain in $\mathcal{H}$. In this case arguing as above and under similar assumptions, one arrives to establish the existence of the limit

$$\lim_{|P|\to 0} \sum_j \langle dM(t_j)\xi, dN(t_j)\eta\rangle \tag{8}$$

which defines a measure on R_+ with values in the sesquilinear forms defined on a suitable domain D in $\mathcal{H}$. Denoting

$$\langle M, N\rangle_w(s, t)(\xi, \eta) \tag{9}$$

the sesquilinear form defined by the limit (8) , in some cases this form can be associated to an operator, still denoted $\langle M, N\rangle$ which we call the weak Meyer bracket of M and N . When both methods are applicable , the relation between the weak and the (strong) Meyer bracket of M and N is given ,by

$$\langle M^+, N\rangle_w(s,t)(\xi,\eta) = \langle \xi, \langle M, N\rangle(s,t)\eta\rangle \tag{10}$$

The notation in (10) is particularly unhappy for the several meanings of the brackets $\langle\, , \cdot\rangle$ however in the present context the notation [. , .] is traditionally reserved for the commutator.

If N is a bounded variation process, in the sense that

$$dN(t) = \int_t^{t+dt} F_s ds$$

for some operator valued function F , then under quite general conditions on M and F one has

$$d\langle M, N\rangle = 0 \tag{11}$$

therefore in general

$$\langle dM + Fdt, dN + Gdt\rangle = \langle dM, dN\rangle \tag{12}$$

As a simple application of (11), consider a process Y satisfying the stochastic differential equation

$$Y = dM_\alpha F_\alpha(Y) \tag{12}$$

and assume that Z is an operator valued process satisfying the ordinary differential equation

$$\frac{dZ}{dt} = ZB \tag{13}$$

Then the adapted process $X = YZ$ satisfies the equation

$$dX = (dY)Z + Y(dZ) + d\langle Y, Z\rangle = dM_\alpha F_\alpha(Y)Z + YZBdt \tag{14}$$

So that if either the F_α 's have the form

$$F_\alpha(X) = C_\alpha X$$

or if they are arbitrary linear functions and the process B (hence Z) is a scalar process , then X will satisfy the equation

$$dX = dM_\alpha F_\alpha(X) + XBdt \tag{15}$$

This remark allows to simplify considerably the proof of the existence theorems in [4] and [8] .

In [7] the theory of quantum Meyer brackets has been developed for arbitrary quantum stochastic fields (i.e., processes in which the index t is multidimensional). This theory has then been applied to construct "quantum birth and death fields " and to obtain a characterization of the Fock representation uniquely in terms of Meyer brackets .

If the Meyer brackets of M and N is a certain measure dA , we shall use the notation

$$dMdN \equiv dA \tag{16}$$

which is better suited for computations (even if less precise) than the usual notation d$\langle M, N\rangle$.

6.) CONTINUITY OF THE TRAJECTORIES

Lemma Let M and N be adapted processes and let σ be their conditional covariance i.e. by definition

$$E_{t]}(dM(t)dN(t)) = \sigma(t, t+dt) = d\sigma(t) \tag{1}$$

Assume that for each bounded interval [s ,t] one has

$$\lim_{|P|\to 0} \sum_{t \epsilon P} ||dM(t)dN(t)\xi||^2 = 0 \tag{2}$$

$$\lim_{|P|\to 0} \sum_{dt \epsilon P} ||d\sigma(t)\xi||^2 = 0 \tag{3}$$

Then

$$\lim_{|P|\to 0} \sum_{dt \epsilon P} dM(t)dN(t)\xi = \sigma([s,t])\xi \tag{4}$$

Proof Let P $\equiv t_1 < \dots < t_n$ be a partition of the interval [s, t] and denote Δ the quantity

$$\Delta = ||\sum_j dM(t_j)dN(t_j)\xi - \sigma(s,t)\xi||^2 = ||\sum_j dM_j dN_j \xi - \sum_k d\sigma_k \xi||^2 = \tag{5}$$

$$= ||\sum_j dM_j dN_j \xi||^2 + ||\sum_k d\sigma_k \xi||^2 - 2Re\langle \sum_j dM_j dN_j \xi, \sum_k d\sigma_k \xi\rangle$$

$$= \sum_{jk} \langle dM_j dN_j \xi, dM_k dN_k \xi\rangle + \sum_{jk} \langle d\sigma_j \xi, d\sigma_k \xi +$$

$$-2Re \sum_{jk} \langle dM_j dN_j, d\sigma_k \xi\rangle$$

$$= \sum_j ||dM_j dN_j \xi||^2 + 2Re \sum_{\substack{jk \\ t_j < t_k}} \langle dM_j dN_j \xi, d\sigma_k \xi\rangle +$$

$$+ \sum_j ||d\sigma_j \xi||^2 + \sum_{j \neq k} \langle d\sigma_j, d\sigma_k \xi\rangle - 2Re \sum_{jk} \langle dM_j dN_j \xi, d\sigma_k \xi\rangle$$

But, because of (1)

$$2Re \sum_{t_j > t_k} \langle dM_j dN_j \xi, d\sigma_k \xi\rangle = 2Re \sum_{t_j > t_k} \langle d\sigma_j \xi, d\sigma_k \xi\rangle$$

and from this one easily deduces that

$$\Delta = \sum_j ||dM_j dN_j \xi||^2 - 2Re \sum_j \langle dM_j dN_j \xi, d\sigma \xi\rangle + \sum_j ||d\sigma_j \xi||^2 \leq$$

$$\leq 2Re \sum_j ||dM_j dN_j \xi||^2 + 2\sum_j ||d\sigma_j \xi||^2$$

To understand the meaning of the conditions (2) and (3) of the above Lemma consider the case in which σ has the form

$$d\sigma(t) = E_{t]}\left(\int_t^{t+dt} \sigma(s)ds\right) \tag{6}$$

where σ is an operator valued function such that the map

$$r \epsilon [s,t] \longrightarrow ||\sigma(r)\xi|| \tag{7}$$

is locally in $L^1(R,dt)$ In this case condition (3) is surely satisfied. While condition (2) is surely satisfied if

$$||dM(t)dN(t)\xi||^2 = o(dt) \tag{8}$$

In the classical case a process M satisfying (8) with M = N has continuous trajectories because of the Kolmogorov criterion. For this reason in [6] and [7] a process M satisfying (8) has been called a process with continuous trajectories . The above lemma implies in particular that the Meyer bracket of a process with continuous trajectories with itself coincides with its conditional covariance. The creation and annihilation processes have continuous trajectories in the above sense while the gauge or the birth and death processes , which are the analogue of the Poisson processes have not.

7.) THE HERMITE-WIENER MARTINGALES.

As a first application of the representation free quantum stochastic calculus , let us extend to the quantum stochastic processes a known and useful result of classical stochastic calculus .

Theorem Let A and M be adapted processes with Ito table

$$dMdM \equiv dA; \qquad dAdM \equiv 0 \tag{1}$$

(which implies $dAdA \equiv 0$) and such that for each t both M(t) and A(t) commute with dM(t) = M(t + dt) - M(t) . Let

$$H_n(s,x) = \frac{(-s)}{n!}\exp\frac{x^2}{2s}\frac{d^n}{dx^n}\exp -\frac{x^2}{2s} \tag{2}$$

denote the n-th Hermite polynomial with parameter s . Then if for each n≥ 0 the stochastic integral

$$\int_0^t dM_s H_n(A_s, M_s)$$

is well defined , it follows that the adapted processes $H_n(A,M)$ satisfy the recurrence relation

$$H_{n+1}(A,M) = \int_0^t dM_s H_n(A_s,M_s); \qquad n \geq 0 \tag{3}$$

Proof. Since

$$H_0(A,M) = 1; \qquad H_1(A,M) = M$$

it follows that (3) is true for n = 0 ,1 . Assume that (3) holds for $1 \leq k \leq n$ $(n \geq 2)$ then by the recurrence relation for the Hermite polynomials one has

$$H_{n+1}(A,M) = \frac{M}{n+1}H_n(A,M) - \frac{A}{n+1}H_{n-1}(A,M)$$

therefore

$$dH_{n+1} = \frac{dM}{n+1}H_n + \frac{M}{n+1}dH_n + \frac{dMdH_n}{n+1} - \frac{dA}{n+1}H_{n-1} - \frac{A}{n+1}dH_{n-1} - \frac{dAdH_{n-1}}{n+1}$$

Thus , using the induction assumption

$$dH_{n+1} = \frac{dM}{n+1}H_n + \frac{M}{n+1}dMH_{n-1} + \frac{dMdM}{n+1}H_{n-1} - \frac{dA}{n+1}H_{n-1} - \frac{A}{n+1}dH_{n-1} - \frac{dAdM}{n+1}H_{n-2}$$

Therefore, using the Ito table for dA and dM, the commutativity of A and M with dM and again the recurrence relation for H_n , we obtain

$$dH_{n+1} = \frac{dM}{n+1}H_n + \frac{n}{n+1}dM[\frac{M}{n}H_{n-1} - \frac{A}{n}H_{n-2}] =$$

$$= \frac{dM}{n+1}H_n + \frac{n}{n+1}dMH_n = dMH_n$$

hence (3) holds for each $n \geq 0$.

Remark. If M is such that the pair M, M^+ is a Levy pair in the sense of [6] then the conditions of the theorem above are certainly satisfied with

$$dA(t) = \sigma(t)dt$$

where σ is the conditional variance of M .

8.)REPRESENTATIONS OF THE CCR OVER A COMPLEX HILBERT SPACE.

The main source of examples of quantum stochastic processes is provided by the representations of the canonical commutations (or anticommutation) relations (CCR or CAR). In this section we briefly review the main notions concerning the representations of the CCR over a complex Hilbert space which should be considered as a real Hilbert space with an involution (given by the multiplication by the imaginary unit i). One can convince oneself that this is not a mathematical snobbery , but an essential intrinsic feature of the problem considered , by looking at the proof of the quantum Levy representation theorem [6] where in fact the covariance of the Levy pair defines canonically a space of complex valued functions with complex valued scalar product which is however a real and not a complex pre-Hilbert space .

Let be a complex Hilbert space with scalar product denoted, $\langle\cdot,\cdot\rangle$, antilinear in the first component. Denote σ the symplectic form on H defined by the imaginary part of $\langle\cdot,\cdot\rangle$ i.e.

$$\sigma(\cdot,\cdot) = Im\langle\cdot,\cdot\rangle \tag{1}$$

Denote $W_o(H)$ the algebra generated by the set

$$\{W_f : f\epsilon H\}$$

with the relations

$$W_f W_g = \exp -i\sigma(f,g) W_{f+g}; \qquad f,g\epsilon H \tag{2}$$

$$W(0) = 1 \tag{3}$$

Because of the relations (2) , (3) , $W_o(H)$ coincides with the set of all (finite) linear combinations of the W_f 's $(f\epsilon H)$.

It can be proved that there exists a unique C^*-norm on $W_o(H)$ (i.e. a norm $||.||$ satisfying the relation $||x^*x|| = ||x||^2$). By definition the Weyl algebra over H is the closure of $W_o(H)$ with respect to this C^*-norm . It will be denoted W(H).

A representation of the W(H) is defined by :

- a complex separable Hilbert space $\mathcal{H}$
- a *-homomorphism

$$\pi : W(H) \longrightarrow \mathcal{B}(\mathcal{H}) = \{\text{the bounded linear operators on } \mathcal{H}\}$$

(i.e. an algebraic homomorphism satisfying $\pi(x^*) = \pi(x)^*$) such that for each f in H the map

$$t\epsilon R \longrightarrow \pi(W(tf))\,\epsilon\mathcal{B}(\mathcal{H}) \tag{4}$$

is continuous in the strong operator topology.

Note that we include in the definition of representation of W(H) the conditions that $\mathcal{H}$ is separable and that the map (4) is strongly continuous.

Given a representation $\mathcal{H},\pi$ of W(H) the field operators are defined , through Stone' s theorem, by the condition

$$\pi(W(tf)) = \exp itB(f) \tag{5}$$

and the creation and annihilation operators are defined respectively by

$$a^+(f) = \frac{i}{2}[B(if) - iB(f)] \tag{6}$$

$$a(f) = \frac{1}{2}[B(if) + iB(f)] \tag{7}$$

The relations (2) for the W(f)' s imply that

$$[B(f), B(g)] = 2iIm\langle f,g\rangle \tag{8}$$

or equivalently

$$[a(f), a(g)] = \langle f, g\rangle \tag{9}$$

where [. , .] denotes the commutator.

A state on W(H) is a linear functional φ which is positive on positive elements and normalized (i.e. $\varphi(x^*x) \geq 0$; $\varphi(1) = 1$). A state φ on W(H), such that for each f in H the map

$$t\epsilon R \mapsto \varphi(W(tf))$$

is continuous , defines a representation $\mathcal{H}, \pi$ of W(H) in which the space $\mathcal{H}$ is the completion of W(H) under the scalar product

$$x, y\epsilon W(H) \mapsto \phi(x^* y)$$

(or of a suitable quotient of W(H) , in case some norms are zero) and the representation π is the one induced by the action of W(H) on itself by left multiplication. This multiplication is called the Gelfand Naimark Segal (GNS) representation and has the additional feature that there exists a vector Φ in $\mathcal{H}$ such that the sub-space

$$\pi(W(H))\Phi = \{\pi(W_f)\Phi : f\epsilon H\}$$

is dense in $\mathcal{H}$ and moreover

$$\varphi(W_f) = \langle \Phi, \pi(W_f)\Phi\rangle \tag{10}$$

The triple $\mathcal{H}, \pi, \Phi$ is called the GNS triple associated to the state φ . The representation π is uniquely determined up to isomorphism by the functional

$$f\epsilon H \mapsto \phi(W_f)$$

The state φ is called quasi-free , or gaussian , if for each f in H

$$\varphi(W_f) = \langle \Phi, \pi(W_f)\Phi\rangle = \exp -\frac{1}{2}\langle f, Qf\rangle \tag{11}$$

where Q is a densely defined operator on H such that

$$Q \geq 1 \tag{12}$$

The quasi-free state corresponding to the choice Q = 1 is called the Fock state . The states corresponding to the choice Q = μ with $1 \leq \mu < \infty$ are called universally invariant states.

Remark1.) The above definition of quasi-free states is not the most general one and does not includes some interesting examples. However it will be sufficient for our purposes since in the following we will only consider quasi-free states for which the associated operator Q is bounded .

Remark2.) If the scalar product on H is real then $\sigma \equiv 0$ and W(H) is an abelian algebra. In this case the definition of quasi-free states reduces to the usual definition of gaussian process over the Hilbert space H . In particular the Fock quasi-free state defines the unit gaussian process on H.

Finally if H is chosen to be the space $L^2(R_+)$ with scalar product

$$\langle f, g\rangle = Re \int fgdt$$

then the Fock state on H coincides with the classical Wiener measure and the field operators

$$B(\chi_{[0.t]}) = w_t$$

define the classical Wiener process .

If ϕ is a quasi-free state on W(H) and $\mathcal{H}, \pi, \Phi$ is the associated GNS triple , then for any strongly continuous 1-parameter group (V_t) on H such that

$$V_t Q = Q V_t, \qquad t \epsilon R$$

there exists a unique strongly continuous 1-parameter group (U_t) on $\mathcal{H}$ characterized by the property

$$U_t \pi(W_f)\Phi = W(V_t f)\Phi; \qquad f \epsilon H$$

For U_t one also uses the notations

$$U_t = \pi(W(V_t)) = \Gamma(V_t)$$

the last notation especially if π is the Fock representation. If T is the generator of (V_t) then the infinitesimal generator of (U_t) is denoted Λ_T or $\Lambda(T)$. Thus by definition

$$\exp it\Lambda_T = \pi(W(\exp itT)) = \Gamma(\exp itT)$$

9.) THE QUANTUM LEVY MARTINGALE REPRESENTATION THEOREM

In the following the Hilbert spaces $\mathcal{H}, \mathcal{H}_{t]}$, the algebras $\mathcal{A}$, $\mathcal{A}_{t]}$, the vector Φ and the state $\varphi(\cdot) = \langle \Phi, (\cdot)\Phi \rangle$, will be as in Section (2) and we will moreover assume that

$$\mathcal{A}'\Phi \qquad \text{is dense in} \mathcal{H}$$

i.e. that Φ is separating for $\mathcal{A}$. It is also assumed that for each t ≥ 0 there exist a conditional expectation $E_{t]} : \mathcal{A} \longrightarrow \mathcal{A}_{t]}$ compatible with the state φ i.e.

$$\varphi \cdot E_{t]} = \varphi$$

If M is a martingale for the above structure, we denote

$$M_1 = M; \qquad M_2 = M^+ \tag{1}$$

and for a bounded interval [s ,t] , we define $\mathcal{H}_{[s,t]} = \mathcal{H}_{[s,t]}(M)$ to be the closed subspace of $\mathcal{H}$ generated by the vectors of the form

$$M_{\alpha_n}(t_{n-1}, t_n) M_{\alpha_{n-1}}(t_{n-2}, t_{n-1}) ... M_{\alpha_1}(t_o, t_1) U_o \Phi$$

where $t_o < t_1 < ... < t_n$ is a partition of [s ,t] and U_o is a bounded operator in the initial space $\mathcal{H}_{0]}$.

For a martingale M , the pair (M, M^+) will be called a Levy pair if :

I.) There exists a scalar function $\sigma_{\alpha,\alpha}$ (α =1,2) on R such that

$$E_{t]}(|dM_\alpha(t)|^2) = E_{t]}(dM_\alpha(t)^+ dM_\alpha(t)) = \int_t^{t+dt} \sigma_{\alpha,\alpha}(s) ds$$

II.) dM_α commutes with the past in the sense that

$$dM_\alpha(t) F_t = F_t dM_\alpha(t)$$

for any adapted process F and for any t and dt .

III.) dM_α is well defined on the past Hilbert space and on this space one can form the products $dM_\alpha dM_\beta$ i.e.

$$D(dM_\alpha(t)) \supseteq \mathcal{H}_{t]} \cup dM_\beta(t) \mathcal{H}_{t]}$$

IV.) dM_α is affiliated to $\mathcal{A}$ i.e. , dM_α commutes with $\mathcal{A}'$ and on the domain $\mathcal{A}'\Phi$ the two operators dM_α and dM_α^+ are adjoint of each other.

V.) M and M^+ have continuous trajectories in the sense of [5] , [6] i.e. for any vector ξ in $\mathcal{H}_{t]}$ one has

$$||dM_\alpha(t) dM_\beta(t)\xi||^2 = o(dt); \qquad \alpha, \beta = 1, 2$$

where o(dt) denotes a number , depending on t , dt , and ξ such that

$$\lim_{dt \to 0} \frac{o(dt)}{dt} = 0$$

uniformly for t in a bounded interval .

From now on , in this Section , M will denote a martingale such that the pair (M, M^+) is a Levy pair .

It is not difficult to show that there exists a scalar valued function $\sigma_{1,2}(t)$ such that the identity

$$E_{t]}(dM_\alpha(t) dM_\beta(t)) = \int_t^{t+dt} \sigma_{\alpha,\beta}(s) \tag{2}$$

holds with $\sigma_{2.1} = \sigma_{1.2}^{+}$. Moreover for each t the 2×2-matrix $(\sigma_{\alpha,\beta})$ is positive semi-definite . In the following the 2×2-matrix valued function σ will be called the covariance of the pair (M,M^+) .

We use this 2×2-matrices to define a (possibly degenerate) scalar product on the space of the compact support , complex valued functions on R_+ with at most a finite number of points of discontinuity in which they admit a finite limit on the right and on the left. Namely, if f is such a function , we denote

$$f_1 = f^+; \qquad f_2 = f \tag{3}$$

and define

$$\langle f, g\rangle = \frac{1}{2}\int f_\alpha^+(s)\sigma_{\alpha,\beta}(s)g_\beta(s)ds \tag{4}$$

The resulting pre-Hilbert space will be denoted $H(\sigma)$. As anticipated in Section (8) , $H(\sigma)$ is a real and not a complex pre-Hilbert space . We also introduce the notations :

$$(f,g)_\sigma(s) = f_\alpha^+(s)\sigma_{\alpha,\beta}(s)g_\beta(s) \tag{5}$$

$$f_\sigma(s) = (f,g)_\sigma(s) \tag{6}$$

$$e_{t]}f(s) = \chi_{[0,t]}(s)f(s) \tag{7}$$

i.e. $e_{t]}$ denotes the operator (projector) on $H(\sigma)$ defined by the multiplication by the characteristic function of the interval [0,t] .

The following theorem shows that any Levy pair (M,M^+) can be realized as the pair of creation and annihilation operators in a representation of the CCR which is uniquely determined by the covariance of the pair .

Theorem For each f in $H(\sigma)$ there exists a unique continuous adapted process $U_f(t)$ with values in the unitary operators on $\mathcal{H}$ satisfying the stochastic differential equation

$$dU_f(t) = (if_\alpha(t)dM_\alpha(t) - \frac{1}{2}f_\sigma(t)dt)U_f(t) \tag{8}$$

with initial condition

$$U_f(0) = 1$$

Moreover if , for each vector ξ in $\mathcal{H}_{0]}$, $\Gamma(H(\sigma),\xi)$ denotes the closed sub-space of $\mathcal{H}$ generated by the vectors of the form $U_f(t)\xi$ for some t≥0 and some f in $H(\sigma)$, then there exists a family W(f) (f $\epsilon H(\sigma)$) of unitary operators on $\Gamma(H(\sigma),\xi)$ such that :

i) the triple $\Gamma(H(\sigma),\xi), W, \xi$ is isomorphic to the Fock representation of the CCR over $H(\sigma)$.

ii) for each t and each f in $H(\sigma)$ one has

$$W(e_{t]}f) = U_f(t) \tag{9}$$

iii) denoting $a(\cdot)$ the annihilation operator associated to the representation W for every t, dt ≥ 0 and for every f in $H(\sigma)$, the identity

$$a(\chi_{[t,t+dt]}f) = \frac{1}{\sqrt{2}}\int_t^{t+dt} f^+(s)dM(s) \tag{10}$$

takes place on the subspace of $\Gamma(H(\sigma),\xi)$ generated by the vectors W(f)ξ with support of f contained in [0,t] .

The idea of the proof of the above theorem is the following : first we use the continuity of the trajectories of M and the results of Sections (5),(6), to show that (in the notation (16) of Section (5))

$$dM_\alpha dM_\beta \equiv \sigma_{\alpha^+,\beta}dt \tag{11}$$

Thus , defining for any f in $H(\sigma)$ the martingale

$$dN_f = i f_\alpha dM_\alpha \tag{12}$$

we obtain

$$dN_f dN_g = -f_\alpha g_\beta dM_\alpha dM_\beta \equiv -f_\alpha \sigma_{\alpha^+.\beta} g_\beta dt = -f_\alpha^+ \sigma_{\alpha.\beta} g_\beta dt \tag{13}$$

that is , in view of (12)

$$dN_f dN_g \equiv -(f,g)_\sigma dt \tag{14}$$

The results of Section (7) then imply that the adapted process

$$V_f = \sum_{n=0}^{\infty} H_n(N_f(t), \int_0^t f_\sigma(s)ds)$$

where the $H_n(x,s)$'s are the Wiener-Hermite polynomials with parameter s , are the unique solutions of the stochastic differential equation

$$V_f = dN_f V_f; \qquad V_f(0) = 1 \tag{15}$$

Using the fact that the V_f 's commute with the past and have continuous trajectories one then shows that

$$dV_f dV_g \equiv dN_f V_f dN_g V_g \equiv -(f,g)_\sigma dt V_f V_g \tag{16}$$

which implies that the adapted process $V_f V_g$ satisfies the stochastic differential equation

$$d(V_f V_g) = (dN_{f+g} - (f,g)_\sigma dt) V_f V_g$$

Thus, again using the results of Section (4), it follows that the adapted process

$$U_f = V_f \exp -\frac{1}{2}\int_0^t f_\sigma(s)ds \tag{17}$$

is the unique solution of

$$dU_f = (dN_f - \frac{1}{2} f_\sigma dt) U_f$$

which , because of (12) is the same as (8) . Using this and (14) it follows that

$$d(U_f U_g) = (dU_f)U_g + U_f(dU_g) + dU_f dU_g =$$

$$= (dN_{f+g} - \frac{1}{2}[f_\sigma + g_\sigma + 2(f,g)_\sigma]dt) U_f U_g$$

but since $(f,g)_\sigma^* = (g,f)_\sigma$, one finds

$$f_\sigma + g_\sigma + 2(f,g)_\sigma = (f+g)_\sigma + 2i Im(f,g)_\sigma$$

and therefore

$$d(U_f U_g) = (dN_{f+g} - \frac{1}{2}(f,g)_\sigma dt - i Im(f,g)_\sigma dt) U_f U_g$$

But this equation , again by the arguments in Section (4) , is also satisfied by the process

$$U_{f+g} \exp -i Im \int_0^t (f,g)_\sigma(s)ds$$

therefore by uniqueness

$$U_f(t) U_g(t) = U_{f+g}(t) \exp -i Im \int_0^t (f,g)_\sigma(s)ds \tag{18}$$

Thus for fixed t the $U_f(t)$'s satisfy the CCR . Finally , since each f has compact support , for t large enough $U_f(t)$ does not depend on t , hence the limit

$$\lim_{t\to\infty} U_f(t) =: W(f)$$

exists in a trivial way and , because of (18) defines a representation of the CCR over the pre-Hilbert space $H(\sigma)$. Starting from (8) and (9) , the identification (10) is a relatively simple corollary. Therefore the theorem is proved.

In the classical case under general regularity conditions on the covariance, a continuous trajectory martingale can be transformed, with a change of scale, into the Wiener process . In the present context it has been shown [6] that there exist exactly four canonical forms whose covariance has "constant type " (more precisely : three and a one parameter family) which correspond respectively to : (i) the classical real Wiener process ; (ii) the complex classical Wiener process ; (iii) the Fock Wiener process introduced by Hudson and Parthasarathy ; (iv) the universal invariant Wiener process with parameter λ with $1 \le \lambda \le \infty$, introduced by Hudson and Lindsay. While the quantum Wiener processes whose covariance has a non constant type correspond to previously non investigated forms of quantum Wiener processes .

The fermion version of the Levy martingale representation theorem has been obtained in [7] .

10.) THE STRUCTURE OF UNITARY MULTIPLICATIVE FUNCTIONALS

In the notations of Section (2) let U(s,t) be a two parameter family ($0 \le s \le t$) of unitary operators satisfying

$$U(s,t)\in\mathcal{A}(s,t)$$

$$U(r,s)U(s,t) = U(r,t); \qquad r < s < t$$

$$U(s,s) = 1$$

such a family will be called a unitary multiplicative functional by obvious analogy with the classical probabilistic terminology . Assume that the forward derivative (cf. Section (2)) of U(s,t) exists

$$D_+U(s,t) = \lim_{\epsilon\to 0} E_{t]}\left(\frac{U(s,t+\epsilon) - U(s,t)}{\epsilon}\right) = U(s,t)b(t)$$

If b(.) is continuous , then according to Theorem (1) of Section (2) ,

$$U(0,t) = 1 + \int_0^t U(s,r)b(r)dr + M(t) \tag{1}$$

where M is a martingale . We will denote the relation (1) symbolically by

$$dU = Ubdt + dM \tag{2}$$

Now assume that the martingale M is absolutely continuous with respect to a Levy pair (A, A^+) in the sense that there exist adapted processes λ and μ satisfying

$$dM = \lambda^+ dA + \mu dA^+$$

or equivalently

$$M(t) - M(s) = \int_s^t [dA(r)\lambda^+(r) + dA^+(r)\mu(r)]$$

Then the equation (2) becomes

$$dU = Ubdt + dA\lambda^+ + dA^+\mu \tag{3}$$

and , denoting $\sigma = (\sigma_{i,j})$ the covariance of the Levy pair (A, A^+) , the unitarity of U implies

$$0 = d(UU^+) = (dU)U^+ + U(dU^+) + dUdU^+ =$$

$$= (Ubdt + dA\lambda^+ + dA^+\mu)U^+U(dtb^+U^+ + \lambda dA^+ + \mu^+dA) +$$

$$+(Ubdt + \lambda^+dA + \mu dA^+)(b^+U^+dt + dA^+\lambda + dA\mu^+)$$

$$= U(b+b^+)U^+dt + (dA\lambda^+U^+ + U^+\mu^+dA) + (da^+\mu U^+ + U\lambda dA^+) + (\mu,\lambda^+)\begin{pmatrix}\sigma_{11} & \sigma_{12}\\ \sigma_{21} & \sigma_{22}\end{pmatrix}\begin{pmatrix}\mu^+\\ \lambda\end{pmatrix}dt$$

From this , assuming the independence of dA ,dA^+ and dt , one finds two conditions :

$$U(b+b^+)U^+ = -(\mu,\lambda^+)_\sigma \tag{4}$$

$$\lambda^+U^+ = -U\mu^+ \tag{5}$$

where we have introduced the notation

$$(\mu,\lambda)_\sigma = (\mu,\lambda^+)\begin{pmatrix}\sigma_{11} & \sigma_{12}\\ \sigma_{21} & \sigma_{22}\end{pmatrix}\begin{pmatrix}\mu^+\\ \lambda\end{pmatrix}$$

Writing , as it is always possible

$$b = K + iH; \qquad K = K^+; \qquad H = H^+$$

condition (4) is equivalent to

$$K = -\frac{1}{2}U^+(\mu,\lambda^+)_\sigma U$$

and writing

$$\mu = U\mu_o$$

condition (5) implies that

$$\lambda^+ = -U\mu_o^+$$

Thus equation (3) becomes

$$dU = U[(iH - K)dt + dA^+\mu_o - dA\mu_o^+]$$

and applying again to this the unitarity condition we find

$$K = |\mu_o|^2_\sigma$$

where by definition

$$|\mu_o|^2_\sigma = (\mu_o, \mu_o)$$

Therefore we conclude that the unitary multiplicative functional U must satisfy the equation

$$dU = U[(iH - \frac{1}{2}|\mu_o|^2_\sigma)dt + dA^+\mu_o - dA\mu_o^+] \tag{6}$$

Keeping in mind the quantum Levy theorem we conclude that every unitary multiplicative functional whose associated martingale is absolutely continuous with respect to a Levy pair should be obtained (modulo regularity conditions) as the solution of a quantum stochastic differential equation of the form (6) where the dA , dA^+ are annihilation and creation operators in a suitable quasi-free representation of the CCR and in the right hand side of (6) one should add an additional term coming from the gauge or the birth and death processes. In the universal invariant case this has been proved by Hudson and Lindsay [16] in the boson case and under the assumption that the multiplicative functional U(s,t) is continuous in norm and satisfies a covariance condition with respect to the shift (i.e. , in the terminology of [1] , U(s,t) is a markovian cocycle). The more difficult problem of a strongly continuous markovian cocycle has been solved by Journet [18] in the Fock case . In this case the operators μ_0 and H in equation (6) can be unbounded and some qualitatively new phenomena appear . Also the results by Lindsay [21] and Parthasarathy and Sinha [24] give support to this conjecture .

A different technique for constructing markovian cocycles , entirely based on the classical Wiener process , has been developed by Alicki and Fannes [8] . They show that with this technique it is possible to generate enough markovian cocycles to dilate any norm continuous quantum markovian semi-group . This result solves positively a problem that has been open for some years after the work of Kummerer [19] (cf. also Frigerio 's paper [12]) i.e. : is it possible to construct a Feynman-Kac dilation of a (norm continuous) quantum markovian semi-group by letting a quantum system interact with an appropriately chosen classical white noise ? On the other hand , since it is reasonable to expect uniqueness in the solution of the stochastic differential equation (6) under quite general conditions, in order to describe the structure of a general unitary markovian cocycle the classical stochastic calculus will not be in general sufficient , and we must use the quantum stochastic calculus . *ACKNOWLEDGEMENTS* Support from the Office of Naval Research under the contract No. 00014-84-K-4421 is gratefully acknowledged .

BIBLIOGRAPHY

[1] Accardi L. On the quantum Feynman-Kac formula. Rend. Sem. Mat. Fis. Univ. Polit. Milano, 48(1978)135-179

[2] Accardi L. , Calzolari A. Linear quantum filtering . to appear

[3] Accardi L., Cecchini C. Conditional expecations in von Neumann algebras and a theorem of Takesaki. J. Funct. Anal. 45(1982)245-273

[4] Accardi L., A.Frigerio , J.T.Lewis Quantum stochastic processes. Publications RIMS 18(1982)97-133

[5] Accardi L., Parthasarathy K.R. Stochastic Calculus on local algebras. in: Quantum Probability and Applications; L.Accardi, W. von Waldenfels (eds.) ; Springer LNM N.1136

[6] Accardi L., Parthasarathy K.R. The quantum Levy martingale representation theorem. Preprint 1986

[7] Accardi L., Quaegebeur J. Quantum birth and death processes. to appear

[8] Alicki R., Fannes M. Dilations of quantum dynamical semigroups with classical Brownian motion . Preprint 1985

[9] Barchielli A. Measurement theory and stochastic differential equations in quantum mechanics . Preprint 1985

[10] Barnett C., Streater R.F., Wilde I.F. The Ito-Clifford integral. J. Funct. Anal. 48(1982)172-212

[11] Dellacherie M., Meyer P.A. Probabilites et potentiel. Hermann (1975)

[12] Frigerio A. Construction of stationary quantum Markov processes through quantum stochastic calculus . in : Quantum Probability and applications II . L.Accardi , W. von Waldenfels (eds.) Springer LNM N 1136

[13] Gihman I.I., Shorokhod A.V. Stochastic differential equations. Springer 1972

[14] Holevo A. Probabilistic and statistical aspects of quantum theory. North Holland 1982

[15] Hudson R.L., Parthasarathy K.R. Quantum Ito' s formula and stochastic evolutions. Comm. Math. Phys. 93(1984)301-323

[16] Hudson R.L., Lindsay M. Use of non-Fock quantum Brownian motion and a quantum martingale representation theorem. in : Quantum Probability and applications II L.Accardi , W. von Waldenfels (eds.) Springer LNM N 1136

[17] Ikeda N., Watanabe S. Stochastic differential equations and diffusion processes. North-Holland 1981

[18] Journet J.L. Structure des cocycles markoviens sur l' espace de Fock . Preprint 1986

[19] Kummerer B., Schroder W. A new construction of unitary dilations : singular coupling to white noise . in : Quantum Probability and applications II . Accardi L., von Waldenfels W. (eds.) Springer LNM N 1136

[20] Letta G. Martingales et Integration stochastique. Quaderni Scuola Normale Superiore Pisa 1984

[21] Lindsay J.M. Fermion martingales . Prob. Theory 71(1986)307-320

[22] Metivier M., Pellaumail J. Stochastic integration. Academic Press 1980

[23] Nelson E. Dynamical Theories of Brownian motion. Princeton University Press 1972

[24] Parthasarathy K.R., Sinha K. Boson-Fermion relations in several dimensions. preprint 1985

[25] Skhorohod A.V. Random Linear Operators. Reidel 1984

[26] Stroock D. Topics in stochastic differential equations. Lectures Tata Institute of Fundamental Research Bombay 1982

Prob. Theory and Math. Stat., Vol.1, pp. 23-30
Prohorov et al. (eds)
1986 VNU Science Press

GLIVENKO-CANTELLI CONVERGENCE FOR WEIGHTED EMPIRICAL AND QUANTILE PROCESSES OF U-STATISTIC STRUCTURE

M. Aerts, Limburgs Universitair Centrum, Diepenbeek, Belgium

P. Janssen, Limburgs Universitair Centrum, Diepenbeek, Belgium

D. Mason, Math. Institut der Univ. München, München, W. Germany

1. INTRODUCTION

Let $X_1, X_2, \ldots$ be a sequence of i.i.d. random elements of some space S and let h be a kernel of degree m : i.e. a symmetric function from S^m to $\mathbb{R}$, with $m \in \mathbb{N}_0$. Denote the c.d.f. of $h(X_1, \ldots, X_m)$ by H_F. A natural empirical d.f. for estimation of H_F is given by

$$H_n(x) = n_m^{-1} \Sigma_{C_{nm}} I[h(X_{i_1}, \ldots, X_{i_m}) \leq x], \; x \in \mathbb{R};$$

where $n_m = \binom{n}{m}$ (a shorthand notation used throughout) and where C_{nm} is the set of all m-tuples $(i_1, \ldots, i_m)$ with $1 \leq i_1 < i_2 < \ldots < i_m \leq n$. Notice that $H_n(x)$ is a U-statistic for every fixed x. The corresponding quantile function H_F^{-1} and sample quantile function H_n^{-1} are defined as :

$$H_F^{-1}(t) = \inf\{x : H_F(x) \geq t\}, \; 0<t<1;$$

$$H_n^{-1}(t) = W_{n,i} \text{ for } \frac{i-1}{n_m} < t \leq \frac{i}{n_m}, \; i=1, \ldots, n_m;$$

where $W_{n,1} \leq \ldots \leq W_{n,n_m}$ denote the ordered values of the outcomes $h(X_{i_1}, \ldots, X_{i_m})$, $(i_1, \ldots, i_m) \in C_{nm}$.

The main object of this paper is to study the almost sure behaviour of the Kolmogorov distance of the weighted empirical and quantile process of U-statistic structure. The case m=1 corresponds to the weighted empirical and quantile process of i.i.d. structure for which results on the almost sure behaviour have been obtained by Wellner (1977,1978), Gaenssler (1983,p.18), Mason (1982) and in addition an excellent account is given in Csörgő (1983). For related work on the a.s. behaviour of processes with U-statistic structure see Nolen and Pollard (1985) and Schneemeier (1985).

In Serfling (1984) and Janssen et al. (1984) it is described how a wide variety of non-parametric statistics can be expressed as functionals of the empirical d.f. H_n. This provides a strong statistical motivation for the study of processes of U-statistic structure. Since for $m \geq 2$ H_n is an empirical d.f. of identically distributed but dependent random variables for which the standard theory of empirical processes is not directly applicable, the study of the behaviour of these U-statistic processes presents a number of interesting and challenging problems.

2. GLIVENKO-CANTELLI CONVERGENCE FOR EMPIRICAL PROCESSES OF U-STATISTIC STRUCTURE

Let Q be the class of weight functions defined as :
$Q=\{q:[0,1]\to\mathbb{R}$, q continuous, $q(0)=q(1)=0$, $q(t)>0$ for all $t \in (0,1)$, q monotone increasing on $[0,\delta_1]$ and monotone decreasing on $[1-\delta_2,1]$ for appropriate $\delta_i = \delta_i(q)$, $i=1,2$ with $0<\delta_1<1-\delta_2<1\}$.
For a given $q \in Q$ and with H_n and H_F as in the introduction the Kolmogorov distance of the weighted empirical process of U-statistic structure is :

$$D_n(q) = \sup_{x \in \mathbb{R}} |(H_n(x) - H_F(x))/q(H_F(x))| .$$

Remark that an interesting subclass of Q is given by

$$Q_0=\{q:[0,1]\to\mathbb{R} : q(t) = t^{\nu_1}(1-t)^{\nu_2} \text{ where } \nu_i > 0,\ i = 1,2\}.$$

Finally for $q \in Q$ put

$$\psi_1(t) = 1/q(t); \quad \psi_m(t) = (1/q(t))^{1/m}$$

and let

$$\gamma_E(q) = \limsup_{n\to\infty} D_n(q).$$

<u>Theorem 1</u>. Let $q \in Q$ and assume H_F continuous.

(i) If $\int_0^1 \psi_1(t)dt < \infty$, then $\gamma_E(q) = 0$ a.s.

(ii) If $\int_0^1 \psi_m(t)dt = \infty$, then $\gamma_E(q) = \infty$ a.s.

<u>Proof</u>. See Helmers et al. (1985) for (i). For (ii) either

(a) $\int_0^{\delta_1} \psi_m(t)dt = \infty$ or (b) $\int_{1-\delta_2}^1 \psi_m(t)dt = \infty$. Let us assume that (a) holds (case (b) follows by a similar argument).

Let k>0, then (see e.g. Theorem 3.2.1 in Chung (1974))

$$\infty = E\left[\frac{\psi_m(H_F(h(X_1,\dots,X_m))\wedge\delta_1)}{k^{1/m}}\right]$$

$$\le (m+1) + \sum_{i=m}^{\infty} P\{\psi_1(H_F(h(X_1,\dots,X_m))\wedge\delta_1)>i^m k\}$$

$$= (m+1) + \sum_{j=1}^{\infty}\sum_{\ell=0}^{m-1} P\{\psi_1(H_F(h(X_{(j-1)m+1},\dots,X_{jm}))\wedge\delta_1)>(jm+\ell)^m k\}$$

$$\le (m+1) + m\sum_{j=1}^{\infty} P\{\psi_1(H_F(h(X_{(j-1)m+1},\dots,X_{jm}))\wedge\delta_1)>(jm)^m k\}.$$

An application of Theorem 4.4.1 in Galambos (1978) and the monotonicity of ψ_1 give

$$\limsup_{n\to\infty} \frac{\psi_1(H_F(W_{n,1}))}{n^m} = \infty \text{ a.s.}$$

The result now follows from the inequality

$$D_n(q) \ge \frac{1}{2}\,\psi_1(H_F(W_{n,1}))\,|H_n(W_{n,1}) - H_n(W_{n,1}-0)|$$

obtained from the remark on p.21 in Gaenssler (1983) and the continuity of H_F. □

For $q \in Q_o \subset Q$ (i.e. $q(t) = t^{\nu_1}(1-t)^{\nu_2}$ for some ν_1, $\nu_2 > 0$) and using for this subclass of weight functions the notation $\gamma_E(\nu_1,\nu_2)$ rather than $\gamma_E(q)$, Theorem 1 rephrases as follows.

Corollary 1. Let $q \in Q_o$ and assume H_F continuous

(i) If $\nu_1,\nu_2 \in [0,1)$, then $\gamma_E(\nu_1,\nu_2) = 0$ a.s.

(ii) If $\nu_1 \ge m$ or $\nu_2 \ge m$, then $\gamma_E(\nu_1,\nu_2) = \infty$ a.s. □

The a.s. behaviour of weighted empirical processes of U-statistic structure with more general weight functions (see Mason (1982) for weighted empiricals of i.i.d. structure) can be studied in a similar way.

For m=1, Theorem 1 and Corollary 1 coincide with Theorem 1 in Gaenssler (1983), which provides (for continuous d.f.) a characterization of the a.s. behaviour of the weighted empirical process (viz. $\gamma_E(q) = 0$ or ∞ according to $\int_0^1 \psi_1(t)dt < \infty$ or $= \infty$). Since in part (ii) of the theorem and the corollary we have to assume $\int_0^1 \psi_m(t)dt = \infty$ which is a stronger assumption than $\int_0^1 \psi_1(t)dt = \infty$, our present theorem, holding for general m, fails to provide that characterization. As a consequence, there is a subclass of weight functions q belonging to Q for which the theorem does not provide a result on the a.s. behaviour of the corresponding weighted empirical process $D_n(q)$. We refer to Section 4 for a further discussion.

3. GLIVENKO-CANTELLI CONVERGENCE FOR QUANTILE PROCESSES OF U-STATISTIC STRUCTURE

As in Mason (1982) we use weight functions belonging to Q_o. Therefore, with H_n^{-1} and H_F^{-1} as in Section 1, we define the Kolmogorov distance of the weighted quantile process of U-statistic structure as :

$$Q_n(\nu_1,\nu_2) = \sup_{0<t<1} t^{\nu_1}(1-t)^{\nu_2} \,|\, H_n^{-1}(t) - H_F^{-1}(t) \,|$$

an we put

$$\gamma_Q(\nu_1,\nu_2) = \limsup_{n\to\infty} Q_n(\nu_1,\nu_2).$$

The main result of this section reads as follows.

Theorem 2.

(i) Assume H_F^{-1} continuous. If $E\,|\,h(X_1,\dots,X_m)\wedge 0\,|^{1/\nu_1} < \infty$ and $E[h(X_1,\dots,X_m)\vee 0]^{1/\nu_2} < \infty$, then $\gamma_Q(\nu_1,\nu_2) = 0$ a.s.

(ii) If $E\,|\,h(X_1,\dots,X_m)\wedge 0\,|^{1/m\nu_1} = \infty$ or $E[h(X_1,\dots,X_m)\vee 0]^{1/m\nu_2} = \infty$, then $\gamma_Q(\nu_1,\nu_2) = \infty$ a.s.

The proof of the theorem uses the following three lemmas.

Lemma 1. Let $0<t<1$. If $H_F^{-1}(t)$ is the unique solution x of $H_F(x-0)\le t\le H_F(x)$, then $H_n^{-1}(t) \to H_F^{-1}(t)$ a.s.

Proof. The argument is similar to that of Theorem 2.3.1 in Serfling (1980), but now using the strong law for U-statistics.□

The proof of the next lemma follows from a classical Glivenko-Cantelli argument.

Lemma 2. Assume H_F^{-1} continuous. Let $0<t_1\le t_2<1$, then

$$\sup_{t_1\le t\le t_2} |\, H_n^{-1}(t) - H_F^{-1}(t) \,| \to 0 \text{ a.s. } \square$$

The final lemma can be obtained by a slight modification of the strong law of large numbers for generalized L-statistics as obtained in Helmers et al. (1985). See also van Zwet (1980).

Lemma 3.
If $\int_0^d |H_F^{-1}(t)|^{1/\nu_1} dt < \infty$ for $0<d<1$, then for all $c \in (0,d)$

$$n_m^{-1} \Sigma_{k=1}^{[cn_m]+1} |W_{n,k}|^{1/\nu_1} \to \int_0^c |H_F^{-1}(t)|^{1/\nu_1} dt \text{ a.s.} \quad \square$$

Proof of Theorem 2.
(i) For any $0<c<1/2$,

$$\gamma_Q(\nu_1,\nu_2) \le \sup_{0<t<c} t^{\nu_1} |H_n^{-1}(t)| + \sup_{0<t<c} t^{\nu_1} |H_F^{-1}(t)|$$

$$+ \sup_{c\le t\le 1-c} t^{\nu_1}(1-t)^{\nu_2} |H_n^{-1}(t)-H_F^{-1}(t)| + \sup_{1-c<t<1} (1-t)^{\nu_2} |H_n^{-1}(t)|$$

$$+ \sup_{1-c<t<1} (1-t)^{\nu_2} |H_F^{-1}(t)| \equiv J_{1n}(c)+J_2(c)+J_{3n}(c)+J_{4n}(c)+J_5(c).$$

First remark that $J_{3n}(c)\to 0$ a.s. by Lemma 2. Next we show that $J_1 \equiv \lim_{c\downarrow 0} \limsup_{n\to\infty} J_{1n}(c) = 0$ a.s. and $\lim_{c\downarrow 0} J_2(c) = 0$. We have to consider two cases.

Case 1 : $H_F^{-1}(t) \ge 0$ for all $0<t<1$. Then, using Lemma 1,

$$0 \le J_1 \le \lim_{c\downarrow 0} \limsup_{n\to\infty} c^{\nu_1} H_n^{-1}(c) \overset{a.s.}{=} \lim_{c\downarrow 0} c^{\nu_1} H_F^{-1}(c) = 0.$$

Moreover $0 \le \lim_{c\downarrow 0} J_2(c) \le \lim_{c\downarrow 0} c^{\nu_1} H_F^{-1}(c) = 0.$

Case 2 : $H_F^{-1}(t) < 0$ for all $0<t<c$ (c sufficiently small). Since

$$(J_{1n}(c))^{1/\nu_1} \le \max_{1\le k\le [cn_m]+1} \frac{k}{n_m} |W_{n,k}|^{1/\nu_1} \le n_m^{-1} \sum_{k=1}^{[cn_m]+1} |W_{n,k}|^{1/\nu_1}$$

an application of Lemma 3 (with $d = \sup\{u:H_F^{-1}(u)<0\}$) gives

$$\limsup_{n\to\infty} (J_{1n}(c))^{1/\nu_1} \le \int_0^c |H_F^{-1}(t)|^{1/\nu_1} dt \text{ a.s.}$$

Hence $J_1 = 0$ a.s.
The fact that $\lim_{c\downarrow 0} J_2(c) = 0$ follows from the inequality

$$(J_2(c))^{1/\nu_1} \le \int_0^c |H_F^{-1}(t)|^{1/\nu_1} dt.$$

Analogous arguments hold for $J_{4n}(c)$ and $J_5(c)$.

(ii) Assume for instance $E|h(X_1,\ldots,X_m)\wedge 0|^{1/m\nu_1}=\infty$ and put

$\alpha = \limsup_{t\downarrow 0} t^{\nu_1}|H_F^{-1}(t)|$.

First assume $\alpha < \infty$. Similar as in the proof of Theorem 1 one obtains from the moment condition and Theorem 4.4.1 in Galambos (1978) that $\limsup_{n\to\infty} \frac{|W_{n,1}|}{n^{m\nu_1}} = \infty$ a.s.

Therefore the result follows from the inequality $\gamma_Q(\nu_1,\nu_2) \geq$

$$\limsup_{n\to\infty} n_m^{-\nu_1}(1-n_m^{-1})^{\nu_2}|H_n^{-1}(n_m^{-1})-H_F^{-1}(n_m^{-1})| \geq \limsup_{n\to\infty} \frac{|W_{n,1}|}{n^{m\nu_1}} - \alpha.$$

In the case that $\alpha = \infty$, $\gamma_Q(\nu_1,\nu_2) = \infty$ a.s. follows immediately from the inequality

$$Q_n(\nu_1,\nu_2) \geq \sup_{0<t<1/n_m} t^{\nu_1}(1-t)^{\nu_2}|W_{n,1} - H_F^{-1}(t)|$$

$$\geq (1-1/n_m)^{\nu_2} \sup_{0<t<1/n_m} t^{\nu_1}|H_F^{-1}(t)| - |W_{n,1}| \sup_{0<t<1/n_m} t^{\nu_1}(1-t)^{\nu_2}. \square$$

For m=1, our Theorem 2 coincides with Theorem 3 in Mason (1982). However, as in the case for weighted empirical processes, our present theorem on weighted quantile processes does not provide the complete characterization of the a.s. behaviour for $m\geq 2$. A discussion on this problem is given in the next section.

4. INDICATING EXAMPLES AND A CONJECTURE

At the end of Sections 2 and 3 we already noticed that Theorem 1 and Theorem 2 do not provide a complete characterization of the a.s. behaviour of the Kolmogorov distance of the weighted empirical and quantile process of U-statistic structure. This section gives a discussion on this problem, which will be based on the following simplified version of Theorem 2.

Assume $h\geq 0$ and H_F^{-1} continuous. Put

$$\tilde{\gamma}_Q = \limsup_{n\to\infty} \sup_{0<t<1} (1-t)|H_n^{-1}(t) - H_F^{-1}(t)|; \text{ then}$$

(i) $E[h(X_1,\ldots,X_m)] < \infty$ implies $\tilde{\gamma}_Q = 0$ a.s.

(ii) $E[h^{1/m}(X_1,\ldots,X_m)] = \infty$ implies $\tilde{\gamma}_Q = \infty$ a.s.

Further we consider the following specific example. Let $U_1, U_2, \ldots$ be independent r.v. uniformly distributed on (0,1) and define $h(U_1,U_2) = (U_1 \wedge U_2)^{-1}$. As before $W_{n,n_2} = \max\{h(U_i,U_j);(i,j) \in C_{n2}\}$. As further notation we use $U_{n,1} \leq \ldots \leq U_{n,n}$ to denote the order statistics of the first n of the U_i's.

Property 1. For this specific example, with m=2 and a positive valued kernel, it holds that :

(1) $H_F^{-1}(t) = 1/(1-\sqrt{t})$, $0<t<1$; continuous.

(2) $E(U_1 \wedge U_2)^{-1} = \infty$ and $E(U_1 \wedge U_2)^{-1/2} < \infty$.

(3) $n^{-2} W_{n,n_2} \to 0$ a.s.

(4) $\tilde{\gamma}_Q = \infty$ a.s.

Proof. (1) and (2) are immediate. Since (see e.g. Ex.4 on p.44 in Chung (1974)) $E[U_1^{-1/2}]<\infty$ is equivalent to $\sum_{n=1}^{\infty} P\{U_1^{-1/2}>cn\} = \sum_{n=1}^{\infty} P\{U_1^{-1}>c^2n^2\}<\infty$ for all $c>0$; it follows from Theorem 4.4.1 in Galambos (1978) that $\limsup_{n\to\infty} \frac{1}{n^2} \max\left(\frac{1}{U_1},\ldots,\frac{1}{U_n}\right) = \limsup_{n\to\infty} \frac{1}{n^2 U_{n,1}} = 0$ a.s.

Hence (3) follows since $n^{-2}\, W_{n,n_2} = n^{-2} U_{n,1}^{-1}$ a.s.

To prove (4) remark that the inequality $\sup_{0<t<1} (1-t)H_n^{-1}(t) \geq \frac{2}{nU_{n,1}}$ a.s. and the fact that $\limsup_{n\to\infty} \frac{1}{nU_{n,1}} = \infty$ a.s. imply that

$$\tilde{\gamma}_Q \geq \limsup_{n\to\infty} \left| \sup_{0<t<1} (1-t)H_n^{-1}(t)-2 \right| = \infty \text{ a.s.} \quad \square$$

Remark 1. In our example we have $Eh^{1/2}(U_1,U_2)<\infty$ and $\tilde{\gamma}_Q = \infty$ a.s. Therefore in part (i) of the simplified Theorem 2 the condition $E[h(X_1,\ldots,X_m)]<\infty$ cannot be replaced by $E[h^{1/m}(X_1,\ldots,X_m)]<\infty$.

Remark 2. The proof of part (ii) of the simplified Theorem 2 is, for $\alpha<\infty$, based on the fact that $\limsup_{n\to\infty} n^{-m} W_{n,n_m} = \infty$ a.s.. From (2) and (3) it follows that at least the method of proof will fail if we replace $E[h^{1/m}(X_1,\ldots,X_m)] = \infty$ by $E[h(X_1,\ldots,X_m)] = \infty$.

Remark 3. The above discussion suggests that a complete characterization of when $\tilde{\gamma}_Q = 0$ a.s. for any $m \geq 2$, analogous to that obtained for the case m=1, is presumably hard or even impossible to get.

However, by looking at specific examples, we conjecture that the following statement holds :

If $h \geq 0$ and H_F^{-1} is continuous on (0,1); then

$\tilde{\gamma}_Q = 0$ a.s. iff $E[h(X_1,\ldots,X_m)] < \infty$.

Remark 4. A similar discussion based on the above example can be given for Theorem 1.

REFERENCES

Chung, K.L. (1974). A Course in Probability Theory 2nd ed. Academic Press, New York.

Csörgö,M. (1983). Quantile Processes with Statistical Applications. SIAM, Philadelphia.

Galambos, J. (1978). The Asymptotic Theory of Extreme Order Statistics, J. Wiley, New York.

Gaenssler, P. (1983). Empirical Processes. IMS Lecture Notes 3, Hayward.

Helmers, R., Janssen, P. and Serfling, R. (1985). Asymptotic theory of the e.d.f. of U-statistic structure and applications to generalized L-statistics. Technical report, Johns Hopkins University.

Janssen, P., Serfling, R. and Veraverbeke, N. (1984). Asymptotic normality for a general class of statistical functions and applications to measures of spread. Ann. Statist. 12, 1369-1379.

Mason, D. (1982). Some characterizations of a.s. bounds for weighted multidimensional empirical distributions and a Glivenko-Cantelli theorem for sample quantiles. Z. Wahrsch. Verw. Gebiete 59, 505-513.

Nolen, D. and Pollard, D. (1985). U-processes : rates of convergence. Technical report, Yale University.

Serfling, R. (1980). Approximation Theorems of Mathematical Statistics. J. Wiley, New York.

Serfling, R. (1984). Generalized L-, M- and R-statistics. Ann. Statist. 12, 76-86.

Schneemeier, W. (1985). Unpublished Ph. D. Dissertation. University of Munich. (in German).

Wellner, J. (1977). A Glivenko-Cantelli theorem and a strong law of large numbers for functions of order statistics. Ann. Statist. 5, 473-480; Correction, ibid. 6, 1394.

van Zwet, W. (1980). A strong law for linear functions of order statistics. Ann. Probab. 8, 986-990.

Prob. Theory and Math. Stat., Vol.1, pp. 31-38
Prohorov et al. (eds)
1986 VNU Science Press

ON SYMMETRY PROPERTIES AND NONPARAMETRIC ESTIMATES OF THE ν-TH ORDER SPECTRAL DENSITY OF A STATIONARY RANDOM PROCESS

A.G.Alekseev

Institute of Physics of the Atmosphere, Moscow,USSR

1. Let $\{\xi(k), k\in Z\}$ be a real-valued random process with zero mean, which is stationary up to the eighth order (i.e., all its moments up to the eighth order exist and are invariant under time translations), and let $r_j(\ell_1,\dots,\ell_{j-1}) = S_j(k, k+\ell_1,\dots,k+\ell_{j-1})$, where $S_j(k_1,\dots,k_j)$ is the simple j-th cumulant of the random vector $\{\xi(k_1),\dots,\xi(k_j)\}$. Assume that the cumulants $r_j(\ell_1,\dots,\ell_{j-1})$, $j=2,3,\dots,8$, decrease so rapidly (in absolute value) as $\ell_1^2+\dots+\ell_{j-1}^2\to\infty$, that the spectral densities

$$f^{(\nu)}(\lambda_1,\dots,\lambda_{\nu-1}) = \frac{1}{(2\pi)^{\nu-1}}\sum_{\ell_1,\dots,\ell_{\nu-1}\in Z} r_\nu(\ell_1,\dots,\ell_{\nu-1}) \times$$
$$\times \exp[-i(\ell_1\lambda_1+\dots+\ell_{\nu-1}\lambda_{\nu-1})], \quad \nu=2,3,\dots,8 \tag{1}$$

exist. The densities $f^{(\nu)}(\lambda_1,\dots,\lambda_{\nu-1}) = f^{(\nu)}(\lambda)$ can be considered only for $\lambda\in\Pi^{\nu-1}$, where $\Pi=(-\pi,\pi]$, but it is more convenient to assume that $f^{(\nu)}(\lambda)$ is defined for all $\lambda\in R^{\nu-1}$ and is 2π-periodic in each of its arguments.

The spectral densities $f^{(\nu)}(\lambda)$, $\lambda\in R^{\nu-1}$ for $\nu>2$ were introduced first by Shiryaev [1] and later they were studied (jointly with their estimates) by Brillinger and Rosenblatt [2, 3], Van Ness [4], Alekseev [5], Kalmykov [6], Zurbenko [7] and others.

The second order spectral density $f(\lambda) = f^{(2)}(\lambda)$ is clearly an even function, i.e. it is symmetric with respect to the point $\lambda = 0$. The full description of the symmetry properties for function $f^{(3)}(\lambda_1, \lambda_2)$ is given in [4] (see also [8, 1.4]).

A convex set $\Lambda \subset R^{\nu-1}$ is called a basic set for spectral density $f^{(\nu)}(\lambda)$, if

(I) the values of $f^{(\nu)}(\lambda)$ in the points of Λ uniquely determine (by the symmetry and periodicity properties) its values in the whole space $R^{\nu-1}$ and

(ii) no proper subset $\Lambda_1 \subset \Lambda$ posesses the property (i).

A system of the basic sets of the spectral density $f^{(\nu)}(\lambda)$, $\lambda \in R^{\nu-1}$ is called a full system, if it covers (taking into account the periodicity of $f^{(\nu)}(\lambda)$) the whole cube $\Pi^{\nu-1}$. The full system of the basic sets if formed by two segments $[0, \pi]$ and $[-\pi, 0]$, if $\nu = 2$, and by 12 closed triangles given by Van Ness [4], if $\nu = 3$.

An attempt to describe all the symmetry properties of the spectral density $f^{(4)}(\lambda)$, $\lambda \in R^3$, was made in [3], where a hypothesis on the form of a basic set of the function $f^{(4)}(\lambda)$ was formulated. According to the stated hypothesis, this set is the sum of two pyramids with total volume $5\pi^3/27$. However, in fact, the description of all symmetry properties for the function $f^{(\nu)}(\lambda)$, $\lambda \in R^{\nu-1}$ where $\nu \geq 4$, is rather complicated, and solution of this problem given in [3] for the case $\nu = 4$ is incorrect. As it will be shown below, the volume of the basic set for the spectral density $f^{(4)}(\lambda)$ if equal to $\pi^3/6$, i.e. it is slightly smaller than $5\pi^3/27$.

It is easy to see that the definition (1) of the spectral density $f^{(4)}(\lambda_1, \lambda_2, \lambda_3)$ and the stationary of

$\xi(k)$ imply the symmetry conditions

$$f^{(4)}(\lambda_1,\lambda_2,\lambda_3) = f^{(4)}(\lambda_2,\lambda_1,\lambda_3) =$$
$$= f^{(4)}(\lambda_1,\lambda_3,\lambda_2) = f^{(4)}(-\lambda_1,-\lambda_2-\lambda_3,\lambda_2,\lambda_3) =$$
$$= \overline{f^{(4)}(-\lambda_1,-\lambda_2,-\lambda_3)}. \tag{2}$$

The detailed analysis of conditions (2) leads to the conclusion that the full system of the basic sets for $\nu = 4$ consists of 48 closed pyramids (tetrahedrons) with the same volume $\pi^3/6$. All the pyramids can be obtained by cutting the dodecahdron

$$\begin{cases} -2\pi \le \lambda_j - \lambda_k \le 2\pi, & j,k = 1,2,3, \quad j \ne k, \\ -2\pi \le \lambda_j + \lambda_k + 2\lambda_l \le 2\pi, & j \ne k \ne l \ne j, \end{cases}$$

by nine planes $\lambda_j = \pm\lambda_k$, $j,k = 1,2,3$, $j \ne k$ and $\lambda_j + \lambda_k + 2\lambda_l = 0$, $j \ne k \ne l \ne j$.

The symmetry conditions for the spectral density $f^{(5)}(\lambda)$, $\lambda \in R^4$, are analogous to (2). Their detailed analysis permits constructing in the case $\nu = 5$ three different full systems of basic sets. Each of the systems consists of 240 related closed polyhedrons with the same four-dimensional volume $\pi^4/15$. One system of inequalities determining 6 basic sets (2 basic sets from each of three systems) is given below.

To obtain the sought-for basic sets we cut the four-dimensional icosahedron

$$\begin{cases} -2\pi \leq \lambda_j - \lambda_k \leq 2\pi, & j,k=1,2,3,4, \quad j \neq k, \\ -2\pi \leq \lambda_j + \lambda_k + \lambda_l + 2\lambda_m \leq 2\pi, & j<k<l, \\ & j,k,l \neq m, \end{cases}$$

by the hyperplanes $\lambda_j = \lambda_k$, $j,k=1,2,3,4$, $j \neq k$, and $\lambda_j + \lambda_k + \lambda_l + 2\lambda_m = 0$, $j<k<l$, $j,k,l \neq m$. This procedure generates 120 pentahedrons Γ_j. Each of them is the set-theoretical sum' of two basic sets. The inequalities

$$\begin{cases} -(\lambda_2+\lambda_3+\lambda_4)/2 \leq \lambda_1 \leq \lambda_2 \leq \lambda_3 \leq \lambda_4, \\ \lambda_1 + \lambda_2 + \lambda_3 + 2\lambda_4 \leq 2\pi, \end{cases} \qquad (3)$$

describe one of the pentahedrons Γ_j. By cutting the pentahedron (3) by one of three hyperplanes $\lambda_1 - 2\lambda_2 + \lambda_3 = 0$, $2\lambda_1 + \lambda_2 + 2\lambda_3 = 0$ and $\lambda_1 + 3\lambda_2 + \lambda_3 = 0$, we receive 2 polyhedrons (pentahedrons or hexahedrons), which belong to one of the three dsystems of basic sets.

Finally, in case $\nu = 6$ at least one system consisting of 1440 basic sets can be constructed. Two of the basic sets are closed heptahedrons, obtained by cutting the five-dimensional hexahedron $-\frac{1}{2}(\lambda_2+\lambda_3+\lambda_4+\lambda_5) \leq \lambda_1 \leq \lambda_2 \leq \lambda_3 \leq \lambda_4 \leq \lambda_5 \leq \pi - (\lambda_1+\lambda_2+\lambda_3+\lambda_4)/2$ by the hyperplane $\lambda_1 + 2\lambda_2 + 2\lambda_3 + \lambda_4 = 0$. The five-dimensional volume of each basic set is equal to $\pi^5/45$.

Some preliminary considerations allow us to presume also that in the case $\nu = 7$ ($\nu = 8$) the full system of basic sets consists of 10 080 (corresp. 80 640) closed polyhedrons with the same $(\nu-1)$-dimensional volume $2\pi^6/315$ (corresp. $\pi^7/630$).

In the cases where $\nu = 2,\dots,6$ (and apparently for $\nu > 6$ too) the point $\lambda = (0,\dots,0)$ is the common vertex of all the basic sets (polyhedrons).

2. Let us now consider the statistical estimation problem for the ν -th spectral densities, $\nu > 2$. Consistent estimates $f_{2n}^{(\nu)}(\lambda)$, $\lambda \in R^{\nu-1}$ of the spectral densities of the orders $\nu = 3$ and 4 can be obtained by the convolution of the ν -th order periodogram $I_{2n}^{(\nu)}(\lambda)$, $\lambda \in R^{\nu-1}$, $\nu = 3$ and 4, with the spectral windows of a special kind. However, there is an essential difference between the cases $\nu = 3$ and $\nu = 4$. While the periodogram $I_{2n}^{3}(\lambda)$ can be evaluated from a single realization $\{\xi(k), k \in Z_n\}$ of the process $\xi(k)$, the evaluation of the periodogram $I_{2n}^{(4)}(\lambda)$ requires two independent realizations $\{\xi_\ell(k), k \in Z_n\}$, $\ell = 1,2$. In both cases the set Z_n is assumed to be determined by the relation $Z_n = \{k : k \in Z_n, |k| < n\}$.

The periodograms of orders $\nu = 3$ and $\nu = 4$ can be determined as

$$I_{2n}^{(3)}(\lambda_1,\lambda_2) = \frac{n}{2\pi^2(n^2+1)} \prod_{j=1}^{3} \sum_{k_j \in Z_n} a(k_j)\xi(k_j)\exp(-ik_j\lambda_j),$$

where $(\lambda_1,\lambda_2) \in R^2$ and $\lambda_1+\lambda_2+\lambda_3 \equiv 0 \pmod{2\pi}$, and

$$I_{2n}^{(4)}(\lambda_1,\lambda_2,\lambda_3) = [15n^3/16\pi^3(6n^4+10n^2-1)] \times$$

$$\times \Big[\sum_{\ell=1}^{2} \prod_{j=1}^{4} \sum_{k_j \in Z_n} a(k_j)\xi_\ell(k_j)\exp(-ik_j\lambda_j) -$$

$$- \sum_{q=1}^{6} \prod_{j=1}^{4} \sum_{k_j \in Z_n} a(k_j)\xi_{\ell(q,j)}(k_j)\exp(-ik_j\lambda_j)\Big],$$

where $(\lambda_1,\lambda_2,\lambda_3) \in R^3$, $\lambda_1+\lambda_2+\lambda_3+\lambda_4 \equiv 0$, and the values of the index $\ell(q,j)$ are given in Table 1. The data window $a(k)$ in both formulas can be defined as $a(k) = 1 - |k|/n$, $k \in Z_n$. In both cases this data window essentially simplifies the investigation of the periodogram expectation.

Using the periodograms $I_{2n}^{(\nu)}(\lambda)$, $\lambda \in R^{\nu-1}$, $\nu = 3$ and $\nu = 4$, we define the estimates of the corresponding spectral densities by the formula

$$f_{2n}^{(\nu)}(\lambda) = \frac{1}{h^{\nu-1}} \int_{R^{\nu-1}} w\left(\frac{\omega}{h}\right) I_{2n}^{(\nu)}(\lambda+\omega)\,d\omega,$$

where $\lambda \in R^{\nu-1}$, $h = h(n)$ is a sequence of positive numbers which goes to zero (but not too rapidly:) as $w(\omega)$, and $\omega \in R^{\nu-1}$, $n \to \infty$, as a weight function, which must satisfy some special conditions indicated later.

The asymptotical (for $n \to \infty$) behaviour of the mean square estimation error Δ^2_{2n} of the estimates $f_{2n}^{(\nu)}(\lambda)$, $\nu = 3$ and $\nu = 4$, depends essentially on the smoothness of the corresponding spectral density and on the rate of decreasing of the cumulants $r_j(\ell_1, \ldots, \ell_{j-1})$, $j = 2, \ldots, 2\nu$, as $\ell_1^2 + \ldots + \ell_{j-1}^2 \to \infty$.

<u>Theorem</u>. Assume that

(i) All partial derivatives up to the r-th order (where $r \geq 1$) of the real and imaginary parts of the estimated spectral density are bounded, and

(ii) $$\sum_{\ell_1, \ldots, \ell_{j-1} \in Z} |r_j(\ell_1, \ldots, \ell_{j-1})| < \infty$$

for $j = 2, \ldots, 2\nu - 2$ and

$$\sum_{\ell_1, \ldots, \ell_{2\nu-1} \in Z} |r_{2\nu}(\ell_1, \ldots, \ell_{2\nu-1})|^m < \infty,$$

where $m = 1$ or 2.

If the weight function $w(\omega) = w(\omega_1, \ldots, \omega_{\nu-1})$ can be represented in the form

$$w(\omega) = \prod_{j=1}^{\nu-1} u(\omega_j) ,$$

where function $u(\omega)$ is even, bounded and satisfies conditions

$$u(\omega) = 0 \quad \text{for} \quad |\omega| \geqslant 1 , \quad \int_{-1}^{1} u(\omega)\, d\omega = 1$$

and (in case where $r > 2$)

$$\int_{-1}^{1} \omega^j u(\omega)\, d\omega = 0 , \quad j = 1, 2, \ldots, r-1 ,$$

while the sequence $h = h(n)$ is chosen to be propertional to $n^{-1/m(2r+\nu-1)}$ as $n \to \infty$, then for every $\lambda \in R^{\nu-1}$:

$$\Delta_{2n}^2 = O\left(n^{-2r/m(2r+\nu-1)}\right) .$$

For the case $\nu = 3$ the theorem is proved in [5]. Its proof for the case $\nu = 4$ is analogous, but more cumbersome.

Table 1

q	j			
	1	2	3	4
1	1	1	2	2
2	1	2	1	2
3	1	2	2	1
4	2	2	1	1
5	2	1	2	1
6	2	1	1	2

REFERENCES

1. Shiryaev A.N. Some problems in the spectral theory of higher-order moments. I. - Theory of Probab.and Applic., 1960, v.5, N 3, p.293-313 (in Russian).
2. Brillinger D., Rosenblatt M. Asymptotic theory of estimates of k-th order spectra. - In: Advanced seminar on spectral analysis of time series (ed.by B. Harris). New York: John Wiley, 1967, p.153-188.
3. Brillinger D., Rosenblatt M. Computation and interpretation of k-th order spectra. - Ibid., p.189-232.
4. Van Ness J.W. Asymptotic normality of bispectral estimates. Ann.Math.Statist., 1966, v.37, N 5, p.1257-1272.
5. Alekseev V.G. Some problems of the estimation of bispectral density function of stationary random process. - Probl.Inform.Transmission, 1983, v.19, N 3, p.38-51 (in Russian).
6. Kalmykov V.A. On computation of the bispectral densities of stationary random processes. - Probl.Inform. Transmission. 1983, v.19, N 4, p.100-103 (in Russian).
7. Zurbenko I.G. Statistical estimation of higher order spectra. - Theory of Probab.and Applic., 1985, v.30, N 1, p.66-77 (in Russian).
8. Subba Rao T., Gabr M.M. An introduction to bispectral analysis and bilinear time series models. (Lecture Notes in Statist., N 24). New York-Berlin-Heidelberg-Tokyo: Springer-Verlag, 1984, 280 p.

Prob. Theory and Math. Stat., Vol.1, pp. 39-67
Prohorov et al. (eds)
1986 VNU Science Press

LARGE DEVIATIONS FOR FIRST-PASSAGE TIMES

A.K.Aleškevičienė
Institute of Mathematics and Cybernetics of the Academy of Sciences of the Lithuanian SSR, 232021 Vilnius, Akademijos 4, USSR

Let ξ_ℓ, $\ell = 1,2,\ldots$ be a sequence of independent identically distributed random variables with a non-degenerate distribution function $F(x)$. Assume

$$a = E\xi_1, \sigma^2 = D\xi_1, \beta_3 = E|\xi_1 - a|^3, \quad S_n = \sum_{\ell=1}^{n} \xi_\ell, \quad n = 1,2,\ldots,$$

$$F_n(x) = P\{S_n < x\}, \quad \bar{S}_n = \max_{1 \le k \le n} S_k .$$

In the given paper we shall investigate the asymptotic behaviour of large deviations of the first-passage time

$$T_x = \min\{n : S_n > x\}$$

of random walk $\{S_n, n=1,2,\ldots\}$ through the given level $x>0$.

Beginning with an obvious relation

$$P\{T_x > n\} = P\{\bar{S}_n \le x\} \qquad (1)$$

and using the well-known results on the asymptotic distribution of the maximum of partial sums $\bar{S}_n$ one can easily prove the convergence of distributions of random variables T_x as $x \to \infty$ to the limiting law. But certain diffuculties have already arisen obtaining the rate of convergence in integral limit theorems for T_x. It is obvious from papers [1]-[3], devoted namely to this problem. We shall present here the most definitive results obtained by Sh.A.Mirakhmedov [3]: if $E|\xi_1|^3 < \infty$ then for $a > 0$

$$\sup_y \left| P\left\{ \frac{T_x - x/a}{\sigma x^{1/2} a^{-3/2}} < y \right\} - \Phi(y) \right| \le C\left[\frac{\beta_3}{\sigma^3 \sqrt{x/a}} + \frac{\sigma}{\sqrt{ax}} \right], \quad y \in R_1, \quad (2)$$

$$\Phi(y) = \frac{1}{\sqrt{2\pi}} \int_{-\infty}^{y} e^{-\frac{u^2}{2}} du,$$

and for $a = 0$

$$\sup_y \left| P\left\{ \frac{\sigma^2 T_x}{x^2} < y \right\} - 2\left(1 - \Phi(y^{-1/2})\right) \right| \le \frac{C\beta_3}{\sigma^2 x}, \quad y > 0. \quad (3)$$

Here and further C will denote an absolute constant but not always one and the same. (In paper [3] the rate of convergence was obtained in a more general case, namely, when $E|\xi_1|^p < \infty$, $p > 2$.)

In the case when $a > 0$ and when it is possible introduce conjugate random variables (i.e. when the Cramer condition (4) is valid, see below), D.Siegmund [4] established the first correction term in the integral limit theorem for T_x. In the same case S.P.Lalley [5] set the asymptotical behaviour of probabilities $P\{T_x < x/a_h\}$ up to equivalence as $x \to \infty$, a_h is a mean of the conjugate distribution. In the case of non-negative random variables the theorems on large deviations for T_x were proved by the author [6].

Let us formulate our new results.

We shall say that the Cramer condition is fulfilled if there exists such $A > 0$ that

$$\int_{-\infty}^{\infty} e^{hx} dF(x+a) < \infty, \quad |h| \le A. \quad (4)$$

<u>Theorem 1</u>. If $a > 0$ and the Cramer condition (4) is satisfied, then for $1 \le y = o(\sqrt{x})$ and $x \to \infty$ holds the relation

$$P\left\{ \frac{T_x - x/a}{\sigma x^{1/2} a^{-3/2}} \ge y \right\} = \left[1 - \Phi(y)\right] \exp\left\{ -\frac{y^3}{\sqrt{x/a}} \lambda\left(-\frac{y}{\sqrt{x/a}}\right) \right\} \left[1 + O\left(\frac{y}{\sqrt{x/a}}\right)\right],$$

where $\lambda(x)$ is a power series converging for all sufficiently small $|x|$. Symbol $\mathcal{O}(\cdot)$ is used to denote a variable bounded in y and x.

Theorem 2. If $\alpha=0$ and $\beta_3<\infty$ then for all $y\geq 1$ and $x\to\infty$ the relation

$$P\left\{\frac{\sigma^2 T_x}{x^2}\geq y\right\}=\frac{1}{\sqrt{2\pi}}\int_y^\infty \frac{1}{u^{3/2}}e^{-\frac{1}{2u}}du\left[1+\theta\frac{c\beta_3}{\sigma^2 x^2}\right],\ |\theta|\leq 1. \tag{5}$$

is fulfilled.

Denote that the function

$$G(y):=\frac{1}{\sqrt{2\pi}}\int_0^y u^{-3/2}e^{-1/2u}du \tag{6}$$

is a stable distribution function, whose canonical representation of the characteristic function is defined by the parameters $\alpha=\frac{1}{2}$, $\lambda=1$, $\beta=-1$ (cf. e.g. [16], §34).

Let us pass over the local limit theorems for a variable T_x. The asymptotical behaviour of the probabilities $P\{T_x=n\}$ with moment restrictions for random variables ξ_i was studied in [9], [11], e.g., in [9] it was shown that if $\alpha>0$ and $\beta_3<\infty$ then as $x\to\infty$ and $n\to\infty$

$$P\{T_x=n\}=\frac{\alpha}{\sigma\sqrt{2\pi n}}e^{-\frac{(x-\alpha n)^2}{2\sigma^2 n}}+\mathcal{O}\left(\frac{1}{n}+\frac{1}{x\sqrt{n}}\right). \tag{7}$$

If $\alpha=0$ and $\sigma^2<\infty$ then there exists an increasing sequence k_n such that

$$P\{T_x=n\}=\frac{x}{\sigma\sqrt{2\pi}\,n^{3/2}}e^{-\frac{x^2}{2\sigma^2 n}}+o\left(\frac{x}{n^{3/2}}\right) \tag{8}$$

uniformly in $x>k_n$ (cf. [11]). Having rewritten relation (8) in terms of Theorem 2, when $x\geqslant k_n$ we have

$$P\{T_x=n\}=\frac{x}{\sigma\sqrt{2\pi}\,n^{3/2}}e^{-\frac{x^2}{2\sigma^2 n}}(1+o(1))=\frac{\sigma^2}{x^2}\,\frac{1}{\sqrt{2\pi}\,y_n^{3/2}}e^{-\frac{1}{2y_n}}(1+o(1)), \tag{9}$$

where $y_n = \sigma^2 n / x^2$. Consequently, as $y_n \to \infty$ (it is possible for $k_n \le x \le \varepsilon_n \sqrt{n}$, $\varepsilon_n \to 0$ as $n \to \infty$) relation (9) gives probability estimates of large deviations in the local form for the random variable T_x with respect to its values.

If $a=0$ and $\beta_3 < \infty$ then some information on the behaviour of probabilities $P\{T_x = n\}$ for $\varepsilon_n \sqrt{n} < x < \varepsilon_n^{-1} \sqrt{n}$ is given by the estimate (cf. [9])

$$P\{T_x = n\} = \frac{x}{\sigma\sqrt{2\pi}\, n^{3/2}} e^{-\frac{x^2}{2\sigma^2 n}} + O\left(\frac{x}{n^{3/2}}\left[\frac{1}{x} + \frac{n^{3/2}}{x^4}\right]\right).$$

Under stronger restrictions, namely, under the fulfilled Cramer condition (4) in the case $a=0$ the behaviour of probabilities $P\{T_x = n\}$ for $\sqrt{n} \le x \le \varepsilon_n n$ is determined by Theorem 4 given below.

When $a > 0$ relation (7) defines the behaviour of probabilities $P\{T_x = n\}$ for $\varepsilon_n \sqrt{n} \le |x - an| \le \varepsilon_n^{-1} \sqrt{n}$. In a more wide zone $\sqrt{n} \le |x - an| \le \varepsilon_n n$ Theorem 3 gives the answer concerning the behaviour of probabilities $P\{T_x = n\}$.

It should be noted that the asymptotics for probabilities $P\{T_x = n\}$ with the estimates for arbitrary deviations x was obtained by Borovkov [11] in 1962. Since the results of Borovkov are expressed in terms of factorized components of the moment generating function we assume that it is worthwhile to prove Theorems 3 and 4.

Theorem 3. If $a > 0$ the Cramer condition (4) is fulfilled and if random variables ξ_ℓ have the bounded density $p(x) \le C$ or they are lattice with the maximal distribution step $h=1$, then for $1 \le |x_n| = o(\sqrt{n})$, $x_n = (x - an)/\sigma\sqrt{n}$ the relation

$$P\{T_x = n\} = \frac{a}{\sigma\sqrt{2\pi n}} e^{-\frac{x_n^2}{2}} \exp\left\{\frac{x_n^3}{\sqrt{n}} \lambda_1\left(\frac{x_n}{\sqrt{n}}\right)\right\}\left[1 + O\left(\frac{x_n}{\sqrt{n}}\right)\right],$$

holds, where $\lambda_1(x)$ is the Cramer power series converging for sufficiently small values $|x|$.

<u>Theorem 4</u>. If $a=0$ the Cramer condition (4) satisfied and if random variables ξ_l have the bounded density $p(x)\leq C$ or are lattice with $h=1$ then for $\sqrt{n}\leq x\leq \leq \varepsilon_n n$ and $n\to\infty$

$$P\{T_x=n+1\}=\frac{x}{\sigma\sqrt{2\pi}\,n^{3/2}}e^{-\frac{x^2}{2\sigma^2 n}}\exp\left\{\frac{x^3}{n^2}\lambda_1\left(\frac{x}{n}\right)\right\}\left[1+O\left(\frac{1}{\sqrt{n}}+e^{-x^2/16n}\right)\right].$$

Theorems 3 and 4 are the theorems on large deviations for probabilities $P\{T_x=n\}$ with respect to the given level x.

<u>Proof of Theorem 1</u>. Let us denote

$$T_x^*=\frac{T_x-\frac{x}{a}}{\sigma x^{1/2}a^{-3/2}}\,,\quad \bar{S}_n^*=\frac{\bar{S}_n-an}{\sigma\sqrt{n}}\,,\qquad \sigma_a=\frac{\sigma}{a}\,,\quad x_a=\frac{x}{a}\,.$$

By virtue of (1)

$$P\{T_x^*\geq y\}=P\{T_x\geq y\sigma_a x_a^{1/2}+x_a\}=P\left\{\bar{S}_{[x_a+y\sigma_a x_a^{1/2}]}\leq x\right\}=$$

$$=P\left\{\bar{S}^*_{[x_a+y\sigma_a x_a^{1/2}]}\leq-\frac{y x_a^{1/2}-\delta/\sigma_a}{\sqrt{[x_a+y\sigma_a x_a^{1/2}]}}\right\}, \tag{10}$$

where $[x]$ is an integer part of x and δ is the fractional part of number $x_a+y\sigma_a x_a^{1/2}$, $0\leq\delta<1$.

In further calculation of the right side of relation (10) we shall use the following statement of Theorem 2 from [7].

<u>Theorem</u>. If $a>0$ and the Cramer condition (4) is satisfied, then for $y\geq 1$, $y=o(\sqrt{n})$ and $n\to\infty$ the relation

$$P\{\bar{S}_n^*<-y\}=\Phi(-y)\exp\left\{-\frac{y^3}{\sqrt{n}}\lambda_1\left(-\frac{y}{\sqrt{n}}\right)\right\}\left[1+O\left(\frac{y}{\sqrt{n}}\right)\right],$$

is valid, where $\lambda_1(\cdot)$ is the Cramer power series, defined in [17].

From (1), taking into consideration the theorem cited, we shall find out for all y, satisfying the inequalities

$$1 \leq \frac{y x_a^{1/2} - \delta/\sigma_a}{\sqrt{x_a + y\sigma_a x_a^{1/2}}} \leq \varepsilon_x \sqrt{x_a + y\sigma_a x_a^{1/2}}, \tag{11}$$

where $\varepsilon_x \to 0$ as $x \to \infty$, the relation

$$P\left\{ \bar{S}_{[x_a + y\sigma_a x_a^{1/2}]} \leq - \frac{y x_a^{1/2} - \delta/\sigma_a}{\sqrt{[x_a + y\sigma_a x_a^{1/2}]}} \right\} =$$

$$= \Phi\left(- \frac{y\sqrt{x_a} - \delta/\sigma_a}{\sqrt{[x_a + y\sigma_a x_a^{1/2}]}} \right) \exp\left\{ - \frac{(y\sqrt{x_a} - a\delta)^3}{[x_a + y\sigma_a\sqrt{x_a}]^2} \lambda_1\left(- \frac{y\sqrt{x_a} - a\delta}{[x_a + y\sigma_a\sqrt{x_a}]} \right) \right\} \times$$

$$\times \left[1 + O\left(\frac{y\sqrt{x_a}}{x_a + y\sigma_a\sqrt{x_a}} \right) \right]. \tag{12}$$

holds.

Let us denote $y_0 = \dfrac{y}{\sqrt{1 + y\sigma_a x_a^{-1/2}}}$. Then it follows from (10)-(12) that for $1 + y\sigma_a x_a^{-1/2} \leq y = o(\sqrt{x_a})$ and $x_a \to \infty$

$$P\{T_x^* \geq y\} = \Phi(-y_0) \exp\left\{ - \frac{y^3}{\sqrt{x_a}\,(1 + y\sigma_a/x_a)^2} \lambda_1\left(- \frac{y_0}{\sqrt{x_a}} \right) \right\} \left[1 + O\left(\frac{y}{\sqrt{x}} \right) \right]. \tag{13}$$

Using the expansions of functions $(1 + y\sigma_a/\sqrt{x_a})^{-1}$ and $(1 + y\sigma_a/\sqrt{x_a})^{-2}$ in the power series of the variable $\frac{\sigma_a y}{\sqrt{x_a}}$ and the definition of the variable y_0 we obtain

$$\exp\left\{ - \frac{y^3}{\sqrt{x_a}(1 + y\sigma_a/x_a)^2} \lambda_1\left(- \frac{y_0}{\sqrt{x_a}} \right) \right\} = \exp\left\{ - \frac{y^3}{\sqrt{x_a}} \lambda_2\left(\frac{y}{\sqrt{x_a}} \right) \right\}, \tag{14}$$

where $\lambda_2(y)$ is a power series of y converging for sufficiently small values $|y|$. Here the multipliers $\sigma^k_a{}^k$ by which the variables $(\sigma_a y/\sqrt{x_a})^k$ differ from $(y/\sqrt{x_a})^k$ were attributed to the coefficients of the power series $\lambda_2(y/\sqrt{x_a})$.

Further, having expanded the function

$$\Psi(y_0) := y_0\sqrt{2\pi}\, e^{\frac{1}{2}y_0^2}[1-\Phi(y)] \tag{15}$$

into the Teylor series in the neighbourhood of y we have

$$\Psi(y_0) = y\sqrt{2\pi}\, e^{\frac{1}{2}y^2}[1-\Phi(y)] + \Psi'(y_\theta)(y_0-y), \quad y_0 < y_\theta < y. \tag{16}$$

One can easily get convinced that

$$\Psi'(y_\theta) = -y_\theta + [1+y_\theta^2]\sqrt{2\pi}\, e^{\frac{1}{2}y_\theta^2}[1-\Phi(y_\theta)].$$

Hence, making use of expansion (cf. [18], Chap.7, §6 and [19], formula (7.1.24))

$$1-\Phi(y) = \frac{1}{y\sqrt{2\pi}} e^{-\frac{1}{2}y^2}\left[1-\frac{1}{y^2}+\frac{3}{y^4}+R(y)\right], y>0, |R(y)| \leq \frac{15}{y^6},$$

we get

$$\Psi'(y_\theta) = \frac{2}{y_\theta^3} + R_1(y_\theta), \qquad |R_1(y_\theta)| \leq \frac{C}{y_\theta^5}.$$

Consequently,

$$\Psi'(y_\theta)|y_0-y| = \Psi'(y_\theta)\left|\left[y\left(\sqrt{1+\sigma_a y/x_a^{1/2}}\right)^{-1}-y\right]\right| \leq \frac{C\sigma_a}{y^2 x_a^{1/2}} = O\left(\frac{1}{y^2\sqrt{x}}\right). \tag{17}$$

It follows from (16) and (17) that

$$\Psi(y_0) = y\sqrt{2\pi}\, e^{\frac{1}{2}y^2}[1-\Phi(y)] + O\left(\frac{1}{y^2\sqrt{x}}\right). \tag{18}$$

According to (15) and (18),

$$1-\Phi(y_0) = \frac{1}{y_0} e^{-\frac{1}{2}y_0^2}\, y e^{\frac{1}{2}y^2}[1-\Phi(y)] + O\left(\frac{1}{y^2\sqrt{x}}\right).$$

Hence, since

$$\frac{1}{y_0}e^{-\frac{1}{2}y_0^2}=\frac{1}{y}e^{-\frac{1}{2}y^2}\exp\left\{\frac{y^2}{2}\sum_{k=1}^{\infty}(-1)^{k+1}\left(\frac{\sigma_a y}{\sqrt{x_a}}\right)^k\right\}\left[1+O\left(\frac{y}{\sqrt{x}}\right)\right]=$$

$$=\frac{1}{y}e^{-\frac{1}{2}y^2}\exp\left\{\frac{y^3}{\sqrt{x_a}}\lambda_3\left(\frac{y}{\sqrt{x_a}}\right)\right\}\left[1+O\left(\frac{y}{\sqrt{x}}\right)\right],$$

then

$$1-\Phi(y_0)=[1-\Phi(y)]\exp\left\{\frac{y^3}{\sqrt{x_a}}\lambda_3\left(\frac{y}{\sqrt{x_a}}\right)\right\}\left[1+O\left(\frac{y}{\sqrt{x}}\right)\right]. \qquad (19)$$

Finally, from relations (13), (14) and (19) we get

$$P\{T_n^*\geq y\}=[1-\Phi(y)]\exp\left\{\frac{y^3}{\sqrt{x_a}}\left[\lambda_2\left(\frac{y}{\sqrt{x_a}}\right)+\lambda_3\left(\frac{y}{\sqrt{x_a}}\right)\right]\right\}\left[1+O\left(\frac{y}{\sqrt{x}}\right)\right]=$$

$$=[1-\Phi(y)]\exp\left\{\frac{y^3}{\sqrt{x/a}}\lambda\left(\frac{y}{\sqrt{x/a}}\right)\right\}\left[1+O\left(\frac{y}{\sqrt{x}}\right)\right].$$

This proves Theorem 1.

Proof of Theorem 2. Taking into consideration the equality

$$2\left(1-\Phi(-x^{-1/2})\right)=\frac{1}{\sqrt{2\pi}}\int_0^x\frac{1}{u^{3/2}}e^{-\frac{1}{2u}}du=G(x)$$

and repeating the proof of relation (2) from [3], we obtain

$$\left|P\left\{\frac{\sigma^2T_x}{x^2}<y\right\}-G(y)\right|\leq\frac{C\beta_3}{\sigma^2x\sqrt{y}}, \quad y\geq 1.$$

Hence and from the obvious estimate

$$1-G(y)=\frac{1}{\sqrt{2\pi}}\int_y^{\infty}\frac{1}{u^{3/2}}e^{-\frac{1}{2u}}du=O\left(\frac{1}{\sqrt{y}}\right), \quad y\geq 1,$$

there follows the statement of the theorem

$$P\left\{\frac{\sigma^2 T_x}{x^2} \geqslant y\right\} = 1 - G(y) + \theta \frac{c\beta_3}{\sigma^2 x \sqrt{y}} = \left[1 - G(y)\right]\left[1 + \theta \frac{c\beta_3}{\sigma^2 \tau}\right], \ |\theta| \leq 1.$$

Proof of Theorem 3. It should be noted that while proving Theorems 3 and 4 we shall make use of certain arguments from the author's paper [8].

From (1) we have

$$P\{T_x = n+1\} = P\{\bar{S}_n < x\} - P\{\bar{S}_{n+1} \leq x\} = \bar{F}_n(x) - \bar{F}_{n+1}(x). \qquad (20)$$

Further, if

$$f(t) = \int_{-\infty}^{\infty} e^{itx}\, dF(x), \quad f_n(t) = \int_{-\infty}^{\infty} e^{itx}\, d\bar{F}_n(x), \quad \bar{\varphi}_n(t) = \int_{-\infty}^{0-} (1 - e^{itx})\, d\bar{F}_n(x)$$

then (cf. [12])

$$f_n(t) = f^n(t) + \sum_{k=1}^{n-1} f^k(t)\bar{\varphi}_{n-k}(t). \qquad (21)$$

Continuous case (when the bounded density exists). From (21) we have

$$\chi_n(t) := \int_{-\infty}^{\infty} e^{itx}\left[\bar{F}_n(x) - \bar{F}_{n+1}(x)\right] dx =$$

$$= \frac{f(t) - 1}{it}\left[f^n(t) + \sum_{k=1}^{n-1} f^k(t)\bar{\varphi}_{n-k}(t)\right] + \frac{f(t)\bar{\varphi}_n(t)}{it} \qquad (22)$$

Note

$$\bar{\varphi}_k(t) = it\tilde{\varphi}_k(t), \quad \tilde{\varphi}_k(t) = \int_{-\infty}^{0-} e^{itx}\bar{F}_k(x)\, dx, \quad |\bar{\varphi}_k(t)| \leq 2P\{\bar{S}_k < 0\}. \qquad (23)$$

Since the distribution $F(x)$ has the bounded density then according to (22) and (23) we have

$$P\{T_x = n+1\} = \frac{1}{2\pi} \int_{-\infty}^{\infty} e^{itx} \chi_n(t)\, dt =$$

$$=\frac{1}{2\pi}\int_{-\infty}^{\infty} e^{-itx}\left\{\frac{f(t)-1}{it}\left[f^n(t)+\sum_{k=1}^{n-1} f^k(t)\bar{\varphi}_{n-k}(t)\right]+f(t)\tilde{\varphi}_n(t)\right\}dt. \qquad (24)$$

First of all let us prove the statement of the theorem for x for which $x_n=(x-an)/\sigma\sqrt{n} \geqslant 2\sqrt{\ln n}$. Assume

$$P\{T_x=n+1\}=J_1+\sum\nolimits_1+\sum\nolimits_2, \qquad (25)$$

where

$$J_1=\frac{1}{2\pi}\int_{-\infty}^{\infty} e^{-itx}\frac{f(t)-1}{it}\left[f^n(t)+\sum_{k=n_1}^{n-1} f^k(t)\bar{\varphi}_{n-k}(t)\right]dt,$$

$$\sum\nolimits_1=\sum_{k=1}^{n_1-1}\bar{F}_{n-k}(0)\left[F_{k+1}(x)-F_k(x)\right],$$

$$\sum\nolimits_2=\sum_{k=1}^{n_1-1}\int_{-\infty}^{0}\left[F_{k+1}(x-y)-F_k(x-y)\right]d\bar{F}_{n-k}(y)+\int_{-\infty}^{0}p(x-y)\bar{F}_n(y)dy,$$

$F_k(x)=P\{S_k<x\}$ and the number n_1 will be chosen later (regard that $n_1\leqslant\frac{1}{6}n$).

Let us estimate the sums $\sum_1$ and $\sum_2$. Obviously

$$|\sum\nolimits_2|\leqslant\sum_{k=1}^{n_1-1}\bar{F}_{n-k}(0)\sup_{y\geqslant x}\left[F_{k+1}(y)-F_k(y)\right]+\left[1-F(x)\right]\bar{F}_n(0)\leqslant$$

$$\leqslant\sum_{k=1}^{n_1-1}\bar{F}_{n-k}(0)\left\{\left[1-F_{k+1}(x)\right]+\left[1-F_k(x)\right]\right\}+\left[1-F(x)\right]\bar{F}_n(0). \qquad (26)$$

In order to estimate the functions $1-F_k(x)$ when $1\leqslant k\leqslant n_1-1$ let us make use of the following lemma from [14].

<u>Lemma</u>. Let ξ be a random variable and $\gamma_\xi(h)=$ $=\ln Ee^{h\xi}$, $h\in[-\Delta,\Delta]$, $\Delta>0$. If $\gamma_\xi(h)\leqslant\frac{1}{2}h^2H$ for all $0<h\leqslant\Delta$, $H>0$, then the estimate

$$P\{\xi \geq x\} \leq \begin{cases} e^{-x^2/2H} & \text{for} \quad 0 \leq x \leq H\Delta, \\ e^{-x\Delta/2} & \text{for} \quad x > H\Delta, \end{cases} \tag{27}$$

is valid.

It follows from (4) that there exists a constant C_1 such that

$$\left|\ln E e^{z(\xi_1 - a)}\right| \leq C_1 \quad \text{for} \quad |z| \leq A_1 < A$$

(the main value of the logarithm is taken). Let $\xi = \frac{S_K - aK}{\sigma\sqrt{K}}$. Then for $|z| \leq A_1 \sigma \sqrt{K}$

$$\left|\ln E e^{z\xi}\right| = k \left|\ln E \exp\left\{z \frac{\xi_1 - a}{\sigma\sqrt{K}}\right\}\right| \leq k C_1. \tag{28}$$

Assume $\Delta = A_1 \sigma \sqrt{K}$ and $H_1 = C_1/\sigma^2 A_1^2$. Then from (28) we obtain

$$\ln E \exp\left\{h \frac{S_K - ak}{\sigma\sqrt{k}}\right\} \leq H_1 h^2, \qquad 0 < h \leq \Delta.$$

Consequently, for $\xi = (S_K - ak)/\sigma\sqrt{k}$ (27) is satisfied when

$$H = 2H_1 = \frac{2C_1}{\sigma^2 A_1^2}; \quad \Delta = A_1 \sigma \sqrt{K}.$$

Making use of (27), we get

$$1 - F_k(x) = P\left\{\frac{S_K - ak}{\sigma\sqrt{k}} > \frac{x - ak}{\sigma\sqrt{k}}\right\} \leq \begin{cases} \exp\left\{-\frac{\sigma^2 A_1^2}{4C_1}\left(\frac{x-ak}{\sigma\sqrt{k}}\right)^2\right\}, & 0 \leq \frac{x-ak}{\sigma\sqrt{k}} \leq \frac{2C_1\sqrt{k}}{\sigma A_1}, \\ \exp\left\{-\frac{A_1\sigma\sqrt{k}}{2}\left(\frac{x-ak}{\sigma\sqrt{k}}\right)\right\}, & \frac{x-ak}{\sigma\sqrt{k}} > \frac{2C_1\sqrt{k}}{\sigma A_1}. \end{cases} \tag{29}$$

Assume $n_1 = \left[\min\left(\frac{\sigma^2 A_1^2}{8C_1}, \frac{1}{6}\right) n\right]$. Then from (25), taking into consideration the inequalities

$$\frac{x - aK}{\sigma\sqrt{K}} = x_n \sqrt{\frac{n}{K}} + \frac{a(n-k)}{\sigma\sqrt{n}} \sqrt{\frac{n}{K}} > x_n \sqrt{\frac{n}{n_1}},$$

which holds for $x_n \geqslant 2\sqrt{\ln n}$, $x_n < \varepsilon_n \sqrt{n}$ and sufficiently large n, we get

$$1 - F_k(x) \leqslant e^{-2x_n^2}, \quad x_n = \frac{x - an}{\sigma\sqrt{n}}, \quad 1 \leqslant k \leqslant n_1. \qquad (30)$$

Finally, from (26) taking into consideration estimations (30) and the inequality

$$P\{\bar{S}_k < 0\} \leqslant P\{S_k < 0\} = P\left\{\frac{S_k - ak}{\sigma\sqrt{k}} < -\frac{a\sqrt{n}}{\sigma}\right\} \leqslant \frac{\sigma^2}{a^2 k} E\left(\frac{S_k - ak}{\sigma\sqrt{k}}\right)^2 = O\left(\frac{1}{k}\right) \qquad (31)$$

for $x_n \geqslant 2\sqrt{\ln n}$, $x_n < \varepsilon_n \sqrt{n}$ we get

$$|\Sigma_2| = O\left(\sum_{k=1}^{n_1} \frac{1}{n-k} e^{-2x_n^2} + \frac{1}{n} e^{-2x_n^2}\right) = o\left(\frac{1}{n} e^{-x_n^2}\right). \qquad (32)$$

Analogously

$$|\Sigma_1| = o\left(\frac{1}{n} e^{-x_n^2}\right). \qquad (33)$$

Let us pass over to the estimation of the integral $\mathcal{J}_1$ in relation (25). Since $p(x) \leqslant C$ then $f(t) \in L_2$, $|f(t)| < 1$ for $t \neq 0$ and $|f(t)| \to 0$ as $t \to \infty$. Having in mind all this and taking into account inequalities (23) we get that for any $\varepsilon > 0$

$$\int\limits_{|t|>\varepsilon} \left| \frac{f(t) - 1}{it} \left[f^n(t) + \sum_{k=n_1}^{n-1} f^k(t) \bar{\varphi}_{n-k}(t) \right] \right| dt \leqslant$$

$$\leqslant 2\left[\exp\{-\eta_1(\varepsilon)(n-1)\} + B\exp\{-\eta_1(\varepsilon)(n_1-1)\} \sum_{k=0}^{n-n_1} \frac{1}{n} \exp\{-\eta_1(\varepsilon)k\}\right] \int\limits_{|t|>\varepsilon} \frac{f(t)}{|t|} dt =$$

$$= B \exp\{-\eta_2(\varepsilon) n\}, \qquad (34)$$

where $\eta_1(\varepsilon)$ and $\eta_2(\varepsilon)$ are positive quantity, $\eta_2(\varepsilon) < \eta_1(\varepsilon)$ (here and further) the bounded function is denoted by B, not always one and the same, i.e. $B = O(1)$.

According to (34),

$$J_1 = \frac{1}{2\pi} \int\limits_{|t|\leq\varepsilon} e^{-itx} \frac{f(t)-1}{it} \left[f^n(t) + \sum_{k=n_1}^{n-1} f^k(t) \bar{\varphi}_{n-k}(t) \right] dt + B e^{-\eta_2(\varepsilon) n}. \tag{35}$$

Let $M(it) = f(t) = E e^{it\xi_1}$. By virtue of (4) an analytical extension $M(it)$ from the segment of the imaginary axis $|t| \leq A_1$ to the circle $|z| \leq \frac{1}{2} A = A_1$ is possible. The same can be said about the function $\bar{\varphi}_k(t)$ because

$$\bar{\varphi}_k(t) = \int\limits_{-\infty}^{0-} (1 - e^{itx}) d\bar{F}_k(x) = it \int\limits_{-\infty}^{0-} e^{itx} \bar{F}_k(x) dx \tag{36}$$

and for any $x < 0$

$$\bar{F}_k(x) \leq F(x).$$

We shall regard that in (9) ε is chosen smaller than A_1 and, in addition, such that in the circle $|z| \leq \varepsilon$ $|M(z)| > \frac{1}{2}$. Let

$$K(z) = \ln M(z),\ \bar{K}(z) = \ln \bar{M}(z) = \ln e^{-az} M(z),\ |z| \leq \varepsilon,\ K(0) = 0.$$

Then for $|z| \leq \varepsilon$, $K(z)$ is the analytical function of the complex variable z in this circle and is presented in it by the MacLorren series

$$\bar{K}(z) = \sum_{k=2}^{\infty} \frac{\gamma_k z^k}{k!},$$

where γ_k are the cumulants of the variable ξ_1, $\gamma_2 = \sigma^2$.

Further, let z_0 be a solution of the equation

$$\bar{K}'(z) - \sigma\tau = 0, \quad \tau = \frac{x_n}{\sqrt{n}}, \quad x_n = \frac{x - an}{\sigma\sqrt{n}}, \quad 1 \leq x_n \leq \varepsilon_n \sqrt{n}. \tag{37}$$

For sufficiently small x_n according to the rules of inversion of the power series we get

$$z_0 = Z_0(x_n) = \frac{\tau}{\sigma} - \frac{\gamma_3 \tau^2}{2\sigma^4} + \frac{3\gamma_3^2 - \gamma_4 \sigma^2}{6\sigma^7} \tau^3 + \dots \tag{38}$$

Consequently, for sufficiently small x_n series being on the right side of relation (38), will converge whereas point z_0 will remain within the circle $|z| \le \varepsilon/2 = \varepsilon_1$ on the positive semi-axis.

Replacing ε by ε_1 in [35] and using analogous proving methods as those for formulas (25)-(30) in [8] we obtain

$$J_1 = \frac{1}{2\pi i} \int_{z_0 - i\varepsilon_1}^{z_0 + i\varepsilon_1} \exp\{-n\sigma\tau z - anz\} \frac{M(z)-1}{z} \Big[M^n(z) + \sum_{k=n_1}^{n-1} M^k(z) \overline{\varphi}_{n-k}(-iz) \Big] dz +$$

$$+ B \exp\{-\eta_3(\varepsilon_1) n\}, \quad \eta_3 \le \eta_2, \quad z = z_0 + it. \qquad (39)$$

When $z = z_0 + it$ in the neighbourhood of point z_0

$$\overline{K}(z) - \sigma\tau z = \overline{K}(z_0) - \sigma\tau z_0 + \sum_{j=2}^{\infty} \frac{K^{(j)}(z_0)(it)^j}{j!}. \qquad (40)$$

By virtue of (37)

$$\overline{K}(z_0) - \sigma\tau z_0 = \overline{K}(z_0) - z_0 \overline{K}'(z_0) = -\sum_{m=2}^{\infty} \frac{m-1}{m!} \gamma_m z_0^m$$

or

$$\overline{K}(z_0) - \sigma\tau z_0 = -\frac{\tau^2}{2} + \tau^3 \lambda(\tau), \qquad (41)$$

where

$$\lambda(\tau) = \frac{\gamma_3}{6\sigma^3} + \frac{\gamma_4 \sigma^2 - 3\gamma_3^2}{24\sigma^6} \tau + \dots$$

On the other hand, when $|t| \le \varepsilon_1$ we have

$$\sum_{j=2}^{\infty} \frac{K^{(j)}(z_0)(it)^j}{j!} = -\frac{1}{2} K''(z_0) t^2 + O(|t|^3). \qquad (41')$$

If ε_1 is chosen sufficiently small, then for $|t| \le \varepsilon_1$

$$\mathrm{Re}\Big(\sum_{j=2}^{\infty} \frac{1}{j!} K^{(j)}(z_0)(it)^j \Big) < -\frac{K''(z_0) t^2}{3},$$

$$K(z_0)=\frac{1}{2}\sigma^2 z_0^2+Bz_0^3, \quad K''(z_0)=\sigma^2+Bz_0. \tag{42}$$

Further, according to (36)

$$|\bar{\varphi}_k(-iz)|=\left|z\int_{-\infty}^{0-}e^{zy}\bar{F}_{n-k}(y)dy\right|\le|z|\int_{-\infty}^{0-}e^{z_0 y}\bar{F}_{n-k}(y)dy\le$$

$$\le|z|\int_{-\infty}^{0-}\bar{F}_{n-k}(y)dy=|z||\bar{a}_{n-k}| \tag{43}$$

and for $a>0$ (cf. [10], formula (16))

$$|\bar{a}_k|=Bk^{-3/2}.$$

Hence taking into account (41) and (43) we get

$$\left|\frac{1}{2\pi i}\int_{z_0-i\varepsilon_1}^{z_0+i\varepsilon_1}\exp\{-n\sigma\tau z-anz\}\frac{M(z)-1}{z}\sum_{k=n_1}^{n-1}M^k(z)\bar{\varphi}_{n-k}(-iz)dz\right|\le$$

$$\le\frac{a}{2\pi}\exp\{n[\bar{K}(z_0)-\sigma\tau z_0]\}\int_{-\varepsilon_1}^{\varepsilon_1}\sum_{k=n_1}^{n-1}\exp\{-\frac{1}{3}\bar{K}''(z_0)t^2\}(z_0+|t|)|\bar{a}_{n-k}|dt=$$

$$=B\exp\{n[\bar{K}(z_0)-\sigma\tau z_0]\}\sum_{k=n_1}^{n-1}\left(\frac{z_0}{\sqrt{K''(z_0)k}}+\frac{1}{K''(z_0)k}\right)\frac{1}{(n-k)^{3/2}}=$$

$$=B\exp\{n[\bar{K}(z_0)-\sigma\tau z_0]\}\left(\frac{z_0}{\sqrt{n}}+\frac{1}{n}\right). \tag{44}$$

On the other hand, according to (40) and (42) when $1\le x_n\le\varepsilon_n\sqrt{n}$, $z=z_0+it$, $t\in D_1$, $D_1:=\left[\frac{\ln^2 n}{\sqrt{n}},\varepsilon_1\right)$ we have

$$\left|\frac{1}{2\pi i}\int_{|t|\in D_1}\exp\{-n\sigma\tau z-anz\}\frac{M(z)-1}{z}M^n(z)dz\right|\le a\int_{|t|\in D_1}\exp\{n[\bar{K}(z_0)-\sigma\tau z_0]\}dt\le$$

$$\leq \alpha \exp\{n[\bar{K}(z_0)-\sigma\tau z_0]\} \int\limits_{|t|\in D_1} \exp\{-\tfrac{1}{3} n K''(z_0) t^2\} dt =$$

$$= \frac{B}{\sqrt{n}} \exp\{n[\bar{K}(z_0)-\sigma\tau z_0]\} \exp\{-\tfrac{1}{4}\sigma^2 \ln^4 n\}. \tag{45}$$

By virtue of (44) and (45) relation (39) can be expressed as follows

$$J_1 = \frac{1}{2\pi i} \int\limits_{z_0-id_1}^{z_0+id_1} \exp\{-n\sigma\tau z - anz\} \frac{M(z)-1}{z} M^n(z) dz +$$

$$+ B \exp\{n[\bar{K}(z_0)-\sigma\tau z_0]\} \left(\frac{z_0}{\sqrt{n}} + \frac{1}{n}\right) + B \exp\{-\eta_3(\varepsilon_1) n\}, \tag{46}$$

$$z = z_0 + it, \quad d_1 = \frac{\ln^2 n}{\sqrt{n}}.$$

Further, according to the expansion

$$\frac{M(z)-1}{z} = \alpha + \frac{1}{2}\mu_2 z + \frac{1}{6} z^2 \int e^{\theta z y} dF(y) =$$

$$= (\alpha + \tfrac{1}{2}\mu_2 z_0) + \tfrac{1}{2}\mu_2 it + \tfrac{1}{6} z^2 \int e^{\theta z y} dF(y), \quad \mu_k = \int y^k dF(y),$$

let us assume

$$\frac{1}{2\pi i} \int\limits_{z_0-id_1}^{z_0+id_1} \exp\{-n\sigma\tau z - anz\} \frac{M(z)-1}{z} M^n(z) dz =$$

$$= \frac{1}{2\pi i} (\alpha + \tfrac{1}{2}\mu_2 z_0) \int\limits_{z_0-id_1}^{z_0+id_1} \exp\{n[\bar{K}(z) - \sigma\tau z]\} dz +$$

$$+\frac{M_2}{4\pi}\int_{-id_1}^{id_1} t\exp\{n[\bar{K}(z_0+it)-\sigma\tau(z_0+it)]\}\,dt\,+$$

$$+\frac{1}{2\pi i}\int_{-id_1}^{id_1}(z_0+it)^2\exp\{n[\bar{K}(z_0+it)-\sigma\tau(z_0+it)]\}\int e^{\theta(z_0+it)y}dF(y)\,dt=$$

$$=\mathcal{J}_{1,1}+\mathcal{J}_{1,2}+\mathcal{J}_{1,3}\,. \qquad (47)$$

By using the same methods as in calculating the corresponding integral in the proof of the W.Richter's local theorem (cf., e.g. [15]) we get

$$\mathcal{J}_{1,1}=\frac{a}{\sqrt{2\pi K''(z_0)\,n}}\exp\{n[\bar{K}(z_0)-\sigma\tau z_0]\}\left[1+B\frac{\ln^2 n}{n}\right]. \qquad (48)$$

Similarly

$$|\mathcal{J}_{1,2}|\le\frac{M_2}{4\pi}\exp\{n[\bar{K}(z_0)-\sigma\tau z_0]\}\int_{-id_1}^{id_1}|t|\exp\{-\tfrac{1}{2}n\bar{K}''(z_0)t^2\}dt\left(1+B\frac{\ln^6 n}{\sqrt{n}}\right)=$$

$$=\frac{B}{n}\exp\{n[\bar{K}(z_0)-\sigma\tau z_0]\}. \qquad (49)$$

By virtue of (4), (42)

$$|\mathcal{J}_{1,3}|\le B\left(z_0^2+\frac{\ln^4 n}{n}\right)\exp\{n[\bar{K}(z_0)-\sigma\tau z_0]\}\int_{|t|\le d_1}\exp\{-\tfrac{1}{3}\bar{K}''(z_0)t^2\}dt=$$

$$=B\left(\frac{z_0^2}{\sqrt{n}}+\frac{\ln^4 n}{n^{3/2}}\right)\exp\{n[\bar{K}(z_0)-\sigma\tau z_0]\}. \qquad (50)$$

It follows from (39) and (46)-(50) that

$$J_1 = \frac{a}{\sigma\sqrt{2\pi n}} \exp\{n[\bar{K}(z_0) - \sigma\tau z_0]\}\left[1 + B\left(z_0 + \frac{1}{\sqrt{n}}\right)\right] + B\exp\{-\eta_3(\varepsilon_1)n\}.$$

Finally, according to (25), (32), (33) and (50)

$$P\{T_x = n+1\} = \frac{a}{\sigma\sqrt{2\pi n}} \exp\{n[\bar{K}(z_0) - \sigma\tau z_0]\}\left[1 + B\left(z_0 + \frac{1}{\sqrt{n}}\right)\right] + B\exp\{-\eta_3(\varepsilon_1)n\}.$$

Hence, taking into consideration (38), (41) and the inequalities

$$\exp\left\{-\frac{x_n^2}{2} + \frac{x_n^3}{\sqrt{n}}\lambda\left(\frac{x_n}{\sqrt{n}}\right)\right\} > e^{-x_n^2} > \exp\left\{-\frac{1}{2}\eta_3(\varepsilon_1)n\right\}, \tag{51}$$

$$1 \leq x_n \leq \varepsilon_n\sqrt{n}, \quad n \to \infty,$$

we finally find that for $x_n > 2\sqrt{\ln n}$, $x_n = o(\sqrt{n})$ and sufficiently large n

$$P\{T_x = n+1\} = \frac{a}{\sigma\sqrt{2\pi n}} e^{-\frac{x_n^2}{2}} \exp\left\{\frac{x_n^3}{\sqrt{n}}\lambda\left(\frac{x_n}{\sqrt{n}}\right)\right\}\left[1 + O\left(\frac{x_n}{\sqrt{n}}\right)\right]. \tag{52}$$

Theorem 3 is proved in a continuous case when $x_n > 2\sqrt{\ln n}$. In order to prove Theorem 3 for $1 \leq x_n \leq 2\sqrt{\ln n}$ in relation (25) we have to substitute $n_2 = \left[\frac{5\ln^2 n}{\min(1, \eta_3(\varepsilon_1))}\right]$ for n_1 and use similar arguments as in the proof of relation (52).

Let us pass over the lattice case. Since for the lattice case of the variable ξ_1

$$\sum_{k=-\infty}^{\infty} e^{zk} P\{\xi_1 \leq k\} = \frac{M(z)}{1-e^{-z}}, \quad M(z) = Ee^{z\xi_1} = \sum_{m=-\infty}^{\infty} e^{mz} P\{\xi_1 = m\},$$

taking into consideration (21), similarly as (22) we have

$$\bar{\chi}_n(z) := \sum_{k=-\infty}^{\infty} e^{zk}\left[P\{\bar{S}_n \le k\} - P\{\bar{S}_{n+1} \le k\}\right] =$$

$$= \frac{M(z)-1}{1-e^{-z}}\left[M^n(z) + \sum_{k=1}^{n-1} M^k(z)\bar{\varphi}_{n-k}(-iz)\right] + \frac{M(z)\,\bar{\varphi}_n(-iz)}{1-e^{-z}}. \qquad (53)$$

Let, as before, z_0 be a saddle point, defined by relation (38). Let us multiply (53) by e^{-zx} and integrate over the vertical straight line from $z_0 - i\pi$ to $z_0 + i\pi$. We shall obtain

$$P\{T_x = n+1\} =$$

$$= \frac{1}{2\pi i}\int_{z_0-i\pi}^{z_0+i\pi} e^{-zx}\left\{\frac{M(z)-1}{1-e^{-z}}\left[M^n(z) + \sum_{k=1}^{n-1} M^k(z)\bar{\varphi}_{n-k}(-iz)\right] + \frac{M(z)\bar{\varphi}_n(-iz)}{1-e^{-z}}\right\}dz. \qquad (54)$$

It is known ([15], Chap. 7, §4) that for any $c \in [-\frac{1}{2}A, \frac{1}{2}A]$ and $|t| \neq 0$

$$|M(c+it)| < M(c).$$

Hence and from the continuity $M(z)$ it follows that for $\varepsilon_1 \le |t| \le \pi$

$$|M(z_0+it)| < M(z_0)\exp\{-\eta_3(\varepsilon_1)\}, \qquad \eta_3(\varepsilon_1) > 0. \qquad (55)$$

In view of (36), (43) and (55) for $z = z_0 + it$

$$\left|\left(\int_{z_0+i\varepsilon_1}^{z_0+i\pi} + \int_{z_0-i\pi}^{z_0-i\varepsilon_1}\right) e^{-zx}\,\frac{M(z)-1}{1-e^{-z}}\left[M^n(z) + \sum_{k=n_2}^{n-1} M^k(z)\bar{\varphi}_{n-k}(-iz)\right]dz\right| =$$

$$= B\exp\{-n\sigma\tau z_0 - anz_0\}\, M^n(z_0)\left[\exp\{-n\eta_3(\varepsilon_1)\} +\right.$$

$$\left. + \sum_{k=n_2}^{n-1} M^{-(n-k)}(z_0)\exp\{-k\eta_3(\varepsilon_1)\}\,|\bar{a}_{n-k}||z|\right]dz =$$

$$= B \exp\{n[\bar{K}(z_0) - \sigma\tau z_0]\}\Big[\exp\{-n\eta_3(\varepsilon_1)\} +$$

$$+ \sum_{k=n_2}^{n-1} (n-k)^{-3/2} \exp\{-m_3\,\eta_3(\varepsilon_1)\}\Big] = \frac{B}{n^{3/2}} \exp\{n[\bar{K}(z_0) - \sigma\tau z_0]\}, \quad (56)$$

$$n_2 = \left[\frac{5\ln^2 n}{\min(1,\ \eta_3(\varepsilon_1))}\right] .$$

Further, similarly as in the proofs of (26), (32) and (33) and regarding that x is an integer, we find

$$\frac{1}{2\pi i} \int_{z_0 - i\pi}^{z_0 + i\pi} e^{-zx} \left\{ \frac{M(z)-1}{1-e^{-z}} \sum_{k=1}^{n_2-1} M^k(z)\,\bar{\varphi}_{n-k}(-iz) + \frac{M(z)\,\bar{\varphi}_n(-iz)}{1-e^{-z}} \right\} dz =$$

$$= \sum_{k=1}^{n_2-1} \left[P\{S_{k+1} \leq n+1\} - P\{S_k \leq n+1\}\right] \bar{F}_{n-k}(0) -$$

$$- \sum_{y=-\infty}^{-1} \left[P\{S_{k+1} \leq n+1-y\} - P\{S_k \leq n+1-y\}\right] P\{\bar{S}_{n-k} = y\} +$$

$$+ P\{\xi_1 > n+1\} \bar{F}_n(0) - \sum_{y=-\infty}^{-1} P\{\xi_1 > n+1-y\} P\{\bar{S}_n = y\} = o\left(\frac{1}{n} e^{-x_n^2}\right). \quad (57)$$

Inserting (56) and (57) in to relation (54), we get

$$P\{T_x = n+1\} =$$

$$= \frac{1}{2\pi i} \int_{z_0 - i\varepsilon_1}^{z_0 + i\varepsilon_1} \exp\{-n\sigma\tau z - anz\} \frac{M(z)-1}{1-e^{-z}} \times$$

$$\times\Big[M^{n}(z)+\sum_{k=n_2}^{n-1}M^{k}(z)\,\overline{\varphi}_{n-k}(-iz)\Big]dz+$$

$$+B\exp\{n[\bar{K}(z_0)-\sigma\tau z_0]\}\frac{1}{n^{3/2}}+o\Big(\frac{1}{n}e^{-x_n^2}\Big). \qquad (58)$$

Since the integrals in relations (58) and (39) differ only in the multipliers $1/(1-e^{-z})$ and $1/z$ the integral from (58) is calculated similarly to that in (39). The changes which should be done here are obvious. Theorem 3 is proved for $1\leqslant x_n\leqslant \varepsilon_n\sqrt{n}$. The proof of the theorem for $-\varepsilon_n\sqrt{n}\leqslant x_n\leqslant -1$ is analogous, therefore we shall not give it here.

<u>Proof of Theorem 4.</u> The same as in the proof of Theorem 3 in the continuous case (see the derivation of the relations (32), (33) and (39)), for

$$x_n\geqslant 2\sqrt{\ln n},\quad x_n=o(\sqrt{n}),\quad x_n=\frac{x}{\sigma\sqrt{n}},\quad \tau=\frac{x_n}{\sqrt{n}}$$

we obtain

$$P\{T_x=n+1\}=J_1+\sum\nolimits_1+\sum\nolimits_2, \qquad (59)$$

where

$$J_1=\frac{1}{2\pi}\int_{z_0-i\varepsilon_1}^{z_0+i\varepsilon_1}e^{-n\sigma\tau z}\,\frac{M(z)-1}{z}\Big[M^{n}(z)+$$

$$+\sum_{k=n_1}^{n-1}M^{k}(z)\,\overline{\varphi}_{n-k}(-iz)\Big]dz+B\exp\{-\eta_3(\varepsilon_1)n\}, \qquad (60)$$

$$\sum\nolimits_2=B\sum_{k=1}^{n-1}\frac{1}{\sqrt{n-k}}e^{-2x_n^2}+\frac{B}{\sqrt{n}}e^{-2x_n^2}=o\Big(\frac{1}{n}e^{-x_n^2}\Big),$$

$$\sum\nolimits_1 = o\left(\frac{1}{n} e^{-x_n^2}\right), \tag{61}$$

since for $a=0$ $\bar{F}_n(0) = \frac{B}{\sqrt{n}}$ (cf. [13]).

Further let $t \in D$ where $D = \left(\min\left(\frac{3}{4} z_0, \frac{\ln^2 n}{\sqrt{n}}\right), \varepsilon_1\right)$. Then according to (42) for $z = z_0 + it$

$$\left| \int\limits_{|t| \in D} e^{-n\sigma\tau z} \frac{M(z)-1}{z} \sum_{k=n_1}^{n-1} M^k(z) \bar{\varphi}_{n-k}(-iz) dt \right| \leq$$

$$\leq \exp\{n[K(z_0) - \sigma\tau z_0]\} \int\limits_{|t| \in D} |z|^2 \sum_{k=n_1}^{n-1} |\bar{a}_{n-k}| \exp\{-(n-k)K(z_0) - \tfrac{1}{3} k K''(z_0) t^2\} dt. \tag{62}$$

Let us calculate the integral on the right side of relation (62). First let $\frac{3}{4} z_0 < \ln^2 n / \sqrt{n}$, $\frac{3}{4} z_0 \leq |t| \leq 6 z_0$. Then, taking into consideration (42), (43), the estimate

$$|\bar{a}_k| = \frac{B}{\sqrt{k}} \tag{63}$$

valid for $a=0$ (cf. [13]), and the inequality

$$x^{\alpha} e^{-x} < e^{-\alpha} \alpha^{\alpha}, \quad x > 0, \ \alpha > 0, \tag{64}$$

we get that for $z = z_0 + it$ and $n \to \infty$

$$\int\limits_{\frac{3}{4} z_0 \leq |t| \leq 6 z_0} |z|^2 \sum_{k=n_1}^{n-1} |\bar{a}_{n-k}| \exp\{-(n-k)K(z_0) - \tfrac{1}{3} k K''(z_0) t^2\} dt =$$

$$= B z_0^3 \exp\{-\tfrac{1}{8} n \sigma^2 z_0^2\} \sum_{k=n_1}^{n-1} \frac{1}{\sqrt{n-k}} =$$

$$= B z_0^3 \sqrt{n} \exp\{-\tfrac{1}{8} n \sigma^2 z_0^2\} = \frac{B z_0}{\sqrt{n}} \exp\left\{-\frac{1}{12} n \sigma^2 z_0^2\right\}. \tag{65}$$

Now let $\frac{3}{4} z_0 < \ln^2 n / \sqrt{n}$ and $6 z_0 < |t| \leq \varepsilon_1$ By virtue of (42), (43) and (63)

$$\int\limits_{6z_0<|t|\le\varepsilon_1}|z|^2\sum_{k=n_1}^{n-1}|\bar a_{n-k}|\exp\left\{-(n-k)K(z_0)-\frac{1}{3}kK''(z_0)t^2\right\}dt=$$

$$=B\exp\left\{-\frac{1}{4}n\sigma^2z_0^2\right\}\sum_{k=n_1}^{n-1}\frac{1}{\sqrt{n-k}}\int\limits_{6z_0}^{\varepsilon_1}t^2\exp\left\{-\frac{1}{8}k\sigma^2t^2\right\}dt=$$

$$=B\exp\left\{-\frac{1}{4}n\sigma^2z_0^2\right\}\sum_{k=n_1}^{n-1}\frac{1}{k^3\sqrt{n-k}}=\frac{B}{n}\exp\left\{-\frac{1}{4}n\sigma^2z_0^2\right\}. \tag{66}$$

Finally, let $\frac{3}{4}z_0\geqslant\frac{\ln^2n}{\sqrt{n}}$, $\frac{\ln^2n}{\sqrt{n}}<|t|\le\varepsilon_1$. Then having in mind (42), (43) and (63) for $z=z_0+it$ we obtain

$$\int\limits_{\frac{\ln^2n}{\sqrt{n}}\le|t|\le\varepsilon_1}|z|^2\sum_{k=n_1}^{n-1}|\bar a_{n-k}|\exp\left\{-(n-k)K(z_0)-\frac{1}{3}kK''(z_0)t^2\right\}dt=$$

$$=B\sum_{k=n_1}^{n-1}\frac{1}{\sqrt{n-k_1}}\int\limits_{\ln^2n/\sqrt{n}}^{\varepsilon_1}t^2\exp\left\{-\frac{1}{4}k\sigma^2t^2\right\}dt=$$

$$=B\sum_{k=n_1}^{n-1}\frac{1}{k^{3/2}\sqrt{n-k}}\exp\left\{-\frac{\sigma^2}{24}\ln^4n\right\}=\frac{B}{n}\exp\left\{-\frac{\sigma^2}{24}\ln^4n\right\}. \tag{67}$$

It follows from (62), (65)-(67) that for $z=z_0+it$

$$\left|\int\limits_{|t|\in D}e^{-n\sigma\tau z}\frac{M(z)-1}{z}\sum_{k=n_1}^{n-1}M^k(z)\,\overline{\varphi}_{n-k}(-iz)dt\right|=$$

$$=B\exp\{n[K(z_0)-\sigma\tau z_0]\}\left[\frac{z_0}{\sqrt{n}}+\frac{1}{n}\right]\left[\exp\left\{-\frac{1}{12}n\sigma^2z_0^2\right\}+\exp\left\{-\frac{1}{24}\sigma^2\ln^4n\right\}\right]. \tag{68}$$

Analogously

$$\int\limits_{|t|\in D} e^{-n\tau\sigma z_0}\left|\frac{M(z_0+it)-1}{z_0+it}\right||M(z_0+it)|^n dt =$$

$$=B\exp\{n[K(z_0)-\sigma\tau z_0]\}\frac{1}{n}\exp\left\{-\frac{1}{5}\sigma^2 n\left[\min\left(\frac{3}{4}z_0, \frac{\ln^2 n}{\sqrt{n}}\right)\right]^2\right\}. \qquad (69)$$

Inserting estimates (68)-(69) into (61) and taking into account (38) we get that

$$J_1=\int\limits_{z_0-id}^{z_0+id} e^{-n\sigma\tau z}\frac{M(z)-1}{z}\left[M^n(z)+\sum_{k=n_1}^{n-1}M^k(z)\bar{\varphi}_{n-k}(-iz)\right]dz+$$

$$+B\exp\{n[K(z_0)-\sigma\tau z_0]\}\left[\frac{z_0}{\sqrt{n}}+\frac{1}{n}\right]\left[\exp\left\{-\frac{1}{16}\sigma^2 x_n^2\right\}+\exp\left\{-\frac{\sigma^2}{24}\ln^2 n\right\}\right]. \qquad (70)$$

Further the following lemma will be necessary.

<u>Lemma</u> (cf. [8], lemma 3). If $a=0$ and the random variables ξ_ℓ satisfy condition (4), then for $z=z_0+it$, $|t|\leq\min\left(\frac{3}{4}z_0, \frac{\ln^2 n}{\sqrt{n}}\right)$, any n_0, $1\leq n_0<\frac{n}{2}$ and sufficiently large n

$$M^n(z)+\sum_{k=n_0}^{n-1}M^k(z)\bar{\varphi}_{n-k}(-iz)=2M^n(z)\left[1+B\left(z_0+\frac{1}{z_0\sqrt{n}}\exp\left\{-\frac{1}{14}n\sigma^2 z_0^2\right\}\right)\right].$$

Now in order to finish the proof of Theorem 4 for $x_n\geq 2\sqrt{\ln n}$, $x_n=o(\sqrt{n})$ it remains only to make use of the above cited Lemma and of the analogous arguments employed in the proof of Theorem 3.

While proving Theorem 4 for $1\leq x_n\leq 2\sqrt{\ln n}$ it is necessary to replace n_1 by n_2 in (59) and to employ the arguments analogous to those in the proof of Theorem 4 for $x\geq 2\sqrt{\ln n}$. Here we shall give only the calculation of the integral

$$J_0 := \int_{D_0} e^{-n\sigma\tau z} \frac{M(z)-1}{z} \sum_{k=n_2}^{n-1} M^k(z) \bar{\varphi}_{n-k}(-iz)dt,$$

$$z = z_0 + it, \quad \frac{3}{4} z_0 < \frac{\ln^2 n}{\sqrt{n}}, \quad D_0 = [-\varepsilon_1, -6z_0) \cup (6z_0, \varepsilon_1],$$

since it differs from the calculation of the analogous integral for $x_n \geqslant 2\sqrt{\ln n}$.

Integrating by parts, we obtain that for $z = z_0 + it$

$$J_0 = \frac{i}{\sigma n\tau} \exp\{-n\sigma\tau(z_0+it)\}[M(z_0+it)-1] \sum_{k=n_2}^{n-1} M^k(z_0+it)\tilde{\varphi}_{n-k}(t-iz_0)\Big|_{|t|=6z_0}^{\varepsilon_1} -$$

$$-\frac{i}{\sigma n\tau} \int_{D_0} e^{-n\sigma\tau z} \Big\{ \frac{d}{dt}[M(z_0+it)-1] \sum_{k=n_2}^{n-1} M^k(z) \tilde{\varphi}_{n-k}(-iz) +$$

$$+ [M(z)-1] \sum_{k=n_2}^{n-1} k M^{k-1}(z) \frac{d}{dt}[M(z_0+it)] \tilde{\varphi}_{n-k}(-iz) +$$

$$+ [M(z)-1] \sum_{k=n_2}^{n-1} M^k(z) \frac{d}{dt} \tilde{\varphi}_{n-k}(t-iz_0) \Big\} dt = J_{01} + J_{02} + J_{03} + J_{04}. \quad (71)$$

In virtue (38), (40), (42) and (63)

$$|J_{01}| \leqslant \frac{1}{\sigma n\tau} \exp\{n[K(z_0) - \sigma\tau z_0]\} \times$$

$$\times \sum_{k=n_2}^{n-1} |\bar{a}_{n-k}| \, t^2 \exp\Big\{-(n-k)K(z_0) - \frac{1}{3} k K''(z_0) t^2\Big\}\Big|_{|t|=6z_0}^{\varepsilon_1} =$$

$$= \frac{B}{x_n\sqrt{n}} \exp\{n[K(z_0) - \sigma\tau z_0]\} \times$$

$$\times\sum_{k=n_2}^{n-1}\left[\exp\left\{-\frac{1}{4}n\sigma^2 z_0^2\right\}\frac{z_0^2}{\sqrt{n-k}}+\exp\left\{-\frac{1}{4}n\sigma^2\varepsilon_1^2\right\}\frac{\varepsilon_1^2}{\sqrt{n-k}}\right]=$$

$$=\frac{Bz_0^2}{x_n}\exp\{n[K(z_0)-\sigma\tau z_0]\}\exp\left\{-\frac{1}{4}n\sigma^2 z_0^2\right\}=$$

$$=\frac{Bx}{n^2}\exp\{n[K(z_0)-\sigma\tau z_0]\}. \qquad (72)$$

Further when $z=z_0+it,\ 6z_0<|t|\leq\varepsilon_1$ we have

$$\frac{d}{dt}M(z_0+it)=i\int_{-\infty}^{\infty}y\,e^{(z_0+it)y}dF(y)=iz\sigma^2+Bz^2,$$

$$\tilde{\varphi}_k(t-iz_0)=\int_{-\infty}^{0-}e^{(z_0+it)y}\bar{F}_k(y)dy=|\bar{a}_k|+Bz\bar{b}_k,\quad \bar{b}_k=\int_{-\infty}^{0-}y^2 d\bar{F}_k(y),$$

$$\frac{d}{dt}\tilde{\varphi}_k(t-iz_0)=i\int_{-\infty}^{0-}e^{(z_0+it)y}y\,\bar{F}_k(y)dy=-\frac{1}{2}\bar{b}_k i+Bz. \qquad (73)$$

Taking into account (38), (63), (73) and the estimate $\bar{b}_k=B/\sqrt{k}$ (cf. [13]) we get

$$J_{02}=-\frac{i}{\sigma n\tau}\int_{D_0}e^{-n\sigma\tau z}(iz\sigma^2+Bz^2)\sum_{k=n_2}^{n-1}M^k(z)\left[|\bar{a}_{n-k}|+Bzb_{n-k}\right]dt=$$

$$=\frac{i\sigma}{x_n\sqrt{n}}\int_{D_0}e^{-n\sigma\tau z}it\sum_{k=n_2}^{n-1}M^k(z)|\bar{a}_{n-k}|dt+$$

$$+\frac{B}{x_n\sqrt{n}}\int_{D_0}e^{-n\sigma\tau z_0}(z_0+t^2)\sum_{k=n_2}^{n-1}|M(z)|^k\frac{1}{\sqrt{n-k}}dt=J'+J''. \qquad (74)$$

It is easy to see that

$$\mathcal{J}'' = B\exp\{n[K(z_0)-\sigma\tau z_0]\}\sum_{k=n_2}^{n-1}\frac{1}{\sqrt{n-k}}e^{-(n-k)K(z_0)}\int_{D_0}(z_0+t^2)e^{-\frac{1}{3}kK''(z_0)t^2}dt=$$

$$=\frac{B}{x_n\sqrt{n}}\exp\{n[K(z_0)-\sigma\tau z_0]\}z_0e^{-\frac{1}{6}n\sigma^2z_0^2}+\sum_{k=n_2}^{n-1}\frac{1}{k^{3/2}\sqrt{n-k}}\exp\{-(n-k)K(z_0)\}=$$

$$=B\exp\{n[K(z_0)-\sigma\tau z_0]\}\left[\frac{1}{n}e^{-\frac{1}{6}n\sigma^2z_0^2}+\frac{1}{n^{3/2}}\right]. \tag{75}$$

Having integrated by parts once more we get analogously

$$|\mathcal{J}'| = B\exp\{n[K(z_0)-\sigma\tau z_0]\}\left[\frac{1}{n}e^{-\frac{1}{6}n\sigma^2z_0^2}+\frac{1}{n^{3/2}}\right]. \tag{76}$$

It follows from (74)-(76) that

$$|\mathcal{J}_{02}| = B\exp\{n[K(z_0)-\sigma\tau z_0]\}\left[\frac{1}{n}e^{-\frac{1}{6}n\sigma^2z_0^2}+\frac{1}{n^{3/2}}\right]. \tag{77}$$

We get such estimates for the integrals $\mathcal{J}_{03}$ and $\mathcal{J}_{04}$, too. Hence, and from the relations (71)-(73) we get that for $z=z_0+it$, $\frac{3}{4}z_0 < \ln^2 n/\sqrt{n}$, $6z_0 < |t| \leq \varepsilon_1$

$$\int_{D_0}e^{-n\sigma\tau z}\frac{M(z)-1}{z}\sum_{k=n_2}^{n-1}M^k(z)\bar{\varphi}_{n-k}(-iz)dt=$$

$$=B\exp\{n[K(z_0)-\sigma\tau z_0]\}\left[\frac{1}{n}e^{-\frac{1}{6}n\sigma^2z_0^2}+\frac{1}{n^{3/2}}\right]. \tag{78}$$

Theorem 4 is proved in continuous case. The proof of Theorem 4 in the lattice case is similar and is not presented here.

REFERENCES

1. Kennedy, D. (1972). Estimates for the rates of convergence in limit theorems for the first passage time of random walks. Ann.Math.Statist. 43, 2090-2094.
2. Migai, N.I., Nevzorov, V.B. (1976). Limit theorems for the first passage times of a certain level. Theory Probab.Appl. 21, 417-420.
3. Mirakhmedov, Sh.A. (1978). On the estimates of convergence in limit theorems for the first passage time. Matem.Zametki 23, 487-495.
4. Siegmund, D. (1979). Corrected diffusion approximations in certain random walk problems. Adv.Appl. Probab. 11, 701-719.
5. Lalley, S.P. (1984). Limit theorems for first-passage times in linear and non-linear renewal theory. Adv. Appl.Probab. 16, 766-803.
6. Aleškevičienė, A.K. (1967). Large deviations for the numbers of occurrence of recurrent events. Litovskij Matem.Sbornik 7, 187-193.
7. Aleškevičienė, A.K. (1979). On the probabilities of large deviations for the maximum of sums of independent random variables. Theory Probab.Appl. 24, 18-33; 318-331.
8. Aleškevičienė, A.K. (1978). Local limit theorems of large deviations for the maximum of sums of random variables. Litovskij Matem.Sbornik 18, 5-32.
9. Aleškevičienė, A.K. (1975). On the local limit theorem for the first passage time. Litovskij Matem.Sbornik 15, 23-66.
10. Aleškevičienė, A.K. (1974). On the rate convergence of distributions of the maxima of sums of independent random variables. Litovskij Matem.Sbornik 14, 19-46.
11. Borovkov, A.A. (1962). New limit theorems in boundary problems for sums of independent terms. Sibirskij Matem.Zurn. 3, 654-694 (in Russian); Selected Transl. Math.Statist.Probab. 5, 315-372 (English translation).
12. Eppel, M.S. (1979). A local limit theorem for the first passage time. Sibirskij Matem.Zurn. 20, 181-191.
13. Nagaev, S.V. (1969). Estimation of the convergence rate of the distribution of the maximum of sums of independent random variables. Sibirskij Matem.Zurn. 10, 614-633.
14. Statulevičius, V. (1970). On the limit theorems for the random functions. Litovskij Matem.Sbornik 10, 583-592.
15. Ibragimov, I.A., Linnik, Y.V. (1971). Independent and Stationary sequences of random variables. Wolters-Norrdhoff, Groningen.
16. Gnedenko, B.V., Kolmogorov, A.N. (1954). Limit distributions for sums of independent random variables. Addison-Wesley Publishing Co., Inc., Mass.

17. Cramér, H. (1938). Sur un nouveau théorème-limite de la théorie des probabilités. Actualités Scientifiques et industrielles. Paris 736, 23.
18. Feller, W. (1957). An introduction to probability theory and its applications.V.1. Wiley, New York.
19. Abramowitz, E.M., Stegun, I. (1964). Handbook of Mathematical Functions. National Bureau of Standards, Appl.Math., ser.55.

Prob. Theory and Math. Stat., Vol.1, pp. 69-84
Prohorov et al. (eds)
1986 VNU Science Press

QUASI-AVERAGE APPROACH TO THE DESCRIPTION OF THE LIMIT STATES OF THE n-VECTOR CURIE-WEISS FERROMAGNET

N.Angelescu, V.A.Zagrebnov

Central Institute of Physics, Bucharest, Romania
Joint Institute for Nuclear Research, USSR

1. Introduction

Bogolubov's quasi-average method [1] has appeared as a tool of removing the degeneracy (i.e. of separating the pure phases) in statistical mechanical models exhibiting phase transitions accompanied by a spontaneous symmetry breaking. It consists in switching on a source term (external field), in the Hamiltonian which reduces its symmetry to that of the corresponding pure phase below the transition point; after the thermodynamic limit of the correlation functions has been calculated, the source is switched off. This approach allowed the construction of the exact solution of the BCS model [2] and of some other mean-field models, see e.g. [3,4]. It is now an important tool in the study of models with a spontaneous symmetry breaking.

At present, the description of the infinite-volume equilibrium states of models in the classical statistical mechanics makes use of the general notion of limit Gibbs measure, as first introduced by Minlos [5]. These were characterised in the whole generality for systems with "bona-fide" interactions as measures with prescribed conditional expectations by Dobrushin [6] and Lanford and Ruelle [7] (DLR equations). For each systems the whole set of limit Gibbs measures coincides with the set of all weak limits of finite-volume Gibbs measures subject to various boundary conditions. On the other hand, for models of the mean-field type, where the interaction depends on volume and where there is no notion such as interaction in the infinite system, one cannot define limit Gibbs measures via DLR equations. The infinite-volume equilibrium states should be obtained as weak limits of the finite-volume Gibbs states [5].

In this note we construct for the n-vector Curie-Weiss model the infinite-volume equilibrium states as limits of the finite-volume Gibbs states. Though the Curie-Weiss model has been extensively studied from very different points of view, see e.g. [8-11], as far as we know, no complete constructive description of all its equilibrium states exists in the literature. To this end, we propose in Sec. 2 a generalization of Bogolubov's quasi-average method consisting in the following: the small external symmetry-breaking fields switched on in a finite volume are allowed to be inhomogeneous and to depend on the volume; also (as we are interested in zero-field states where multiple phases arise) they are allowed to go to zero simultaneously with the volume going to infinity.

It turns out that for the n-vector Curie-Weiss model: (i) all limit states obtained in this way possess the whole symmetry of the zero-field Hamiltonian under the permutation of spins; (ii) they are convex combinations of pure phases (as constructed by the usual quasi-average method). More precisely, the limit states are completely specified by the average magnetisation, and thus they correspond to the points of the ball in $\mathbb{R}^n$ of radius equal to the spontaneous magnetisation. The pure states are the points on the sphere, whereas the mixed states correspond to the interior of the ball. Thus, the set of equilibrium states we obtain, though in principle larger, coincides with that given in [10], where the permutational symmetry is presupposed and the above simplicial structure is derived via general theorems. The advantage of our procedure consists in its constructive character, which allows us to give an explicit expression for the mixture of pure phases in terms of physical parameters related to the preparation of the state. Sec. 3 is devoted to the limit states in non-zero inhomogeneous external fields. It is shown that for almost all (with respect to a product probability measure) external field configurations, the limit states are convex combinations of product states associated to the solutions of one "average" self-consistency equation.

2. Limit states in the case of zero external field

We start with defining the n-vector Curie-Weiss ferromagnetic model in an external field for N spins. Let $\underset{\sim}{h}_N^{(N)} = (\vec{h}_1^{(N)}, \dots, \vec{h}_N^{(N)}) \in \mathbb{R}^{Nn}$ and $\beta > 0$ be given. To every $j = 1, 2, \dots, N$, we assign a spin, i.e. a random vector in $\mathbb{R}^n$ of unit length, $\vec{\sigma} \in S^{n-1}$; the unit sphere S^{n-1} is endowed with the potationally invariant probability measure $d\vec{\sigma}$. The joint distribution of $\underset{\sim}{\sigma}_N = (\vec{\sigma}_1, \dots, \vec{\sigma}_N)$ corresponding to the external fields $\underset{\sim}{h}_N^{(N)}$ and temperature β^{-1} is given according to the Gibbs prescription by the following density with respect to the free product measure $d\underset{\sim}{\sigma}_N = \prod_{1 \le j \le N} d\vec{\sigma}_j$:

$$\rho_N(\underset{\sim}{\sigma}_N; \beta, \underset{\sim}{h}_N^{(N)}) = \left[Z_N(\beta, \underset{\sim}{h}_N^{(N)})\right]^{-1} \exp\{-\beta \mathcal{H}_N(\underset{\sim}{\sigma}_N; \underset{\sim}{h}_N^{(N)})\} \quad (1)$$

where $Z_N(\beta, \underset{\sim}{h}_N^{(N)})$ is the partition function and the Hamiltonian is:

$$\mathcal{H}_N(\underset{\sim}{\sigma}_N, \underset{\sim}{h}_N^{(N)}) = -\frac{1}{2N}\left(\sum_{j=1}^{N} \vec{\sigma}_j\right)^2 - \sum_{j=1}^{N} \vec{\sigma}_j \vec{h}_j^{(N)} \quad (2)$$

From (1) one obtains the distribution of k spins, say $\vec{\sigma}_1, \dots, \vec{\sigma}_k$, by integrating over the remaining spins:

$$P_K^{(N)}(d\vec{\sigma}_1, \dots, d\vec{\sigma}_K; \beta, \underset{\sim}{h}_N^{(N)}) = \left\{ \int_{(S^{n-1})^{N-K}} \prod_{j=K+1}^{N} d\vec{\sigma}_j \, P_N(\underset{\sim}{\sigma}_N; \beta, h_N^{(N)}) \right\} d\underset{\sim}{\sigma}_K \quad (3)$$

Suppose now that a sequence $N_m \to \infty$ has been chosen such that, for every fixed k, the following limit exists:

$$P_K(d\vec{\sigma}_1, \dots, d\vec{\sigma}_K; \beta, \underset{\sim}{h}) = \lim_{m \to \infty} P_K^{(N_m)}(d\vec{\sigma}_1, \dots, d\vec{\sigma}_K; \beta, \underset{\sim}{h}_{N_m}^{(N_m)}) \quad (4)$$

It is easy to check that $\{P_K\}_{K=1}^{\infty}$ is a consistent family of probability distributions on the configuration space $\mathcal{S} = (S^{n-1})^{\mathbb{N}}$ (with the Borel Σ -algebra of the product topology). Then, by Kolmogorov's theorem (see e.g. [12]), there exists a unique probability measure P on $\mathcal{S}$, whose finite-dimensional distributions on $\prod_{1 \le j \le K} S^{n-1}$ are P_K.

A probability measure on $\mathcal{S}$ defined in this way from the Hamiltonian (2) will be called (in accordance with [5]) a Gibbs state of the infinite-volume Curie-Weiss n-vector model, or simply a limit state. Of course, P will depend, in general, on the sequences $\{\underset{\sim}{h}_{N_m}^{(N_m)}\}_{m=1}^{\infty}$. Whenever $\lim_{m\to\infty} \vec{h}_K^{(N_m)} \doteq \vec{h}_K$ exists for all $k \in \mathbb{N}$ we associate P with the external field configuration $\underset{\sim}{h} = \{\vec{h}_K\}_{K \in \mathbb{N}}$. In particular, we are interested in limit states corresponding to zero external field ($\vec{h}_K = 0$ for all K), because for $\underset{\sim}{h}_N = 0$ the Hamiltonian $\mathcal{H}_N$ is invariant under the permutation of spins and under simultaneous rotation of all spins, while the limit states are not necessarily such. Before stating the result, we fix some notation and briefly remind the calculation of the free-energy for the zero-field case, i.e. of $\lim_{N\to\infty} (N\beta)^{-1} \ln Z_N(\beta, 0)$.

Let us define the function $\varphi_n : [0,\infty) \to [0,\infty)$ by:

$$\varphi_N(y) = \beta^{-1} \ln \int_{S^{n-1}} d\vec{\sigma}\, \exp(\beta \vec{\sigma}\cdot\vec{y}),\ y = |\vec{y}| \tag{5}$$

This is possible because the r.h.s. is obviously rotation invariant. It is clear that $\varphi_n(y)$ is increasing and convex, while $\varphi_n'(y)$ is concave and increases monotonically from zero to one. Moreover, $\varphi_n''(y) = \beta/n$. Therefore, the function $f_n : [0,\infty) \to R^1$ defined by:

$$f_n(y) = \frac{1}{2} y^2 - \varphi_n(y). \tag{6}$$

has exactly one minimum point $y_0(\beta) \in [0,\infty)$, which equals zero if $\beta \le n$ and equals the nonzero solution of the equation

$$y = \varphi_n'(y) \tag{7}$$

if $\beta > n$. Now, using the identity:

$$\exp\left[\frac{\beta}{2N} \vec{S}^2\right] = \left(\frac{\beta N}{2\pi}\right)^{n/2} \int_{R^n} d\vec{y}\, \exp\left[-\beta\left(\frac{N}{2} y^2 + \vec{S}\cdot\vec{y}\right)\right] \tag{8}$$

and the definitions (5), (6), we have for Z_N :

$$Z_N(\beta, 0) = \int_{(S^{n-1})^N} d\underset{\sim}{\sigma}_N \exp\left[\frac{\beta}{2N}\left(\sum_{j=1}^{N} \vec{\sigma}_j\right)^2\right] = \tag{9}$$

$$\left(\frac{\beta N}{2\pi}\right)^{n/2} \int_{R^n} d\vec{y}\, \exp[-N\beta f_n(y)]$$

Integrating over the angular variables and applying the Laplace method, one obtains:

$$F_n(\beta) = \lim_{N\to\infty} (N\beta)^{-1} \ln Z_N(\beta,0) = -f_n(y_0(\beta)) \quad (10)$$

Thus, $\beta_c^{-1} = n^{-1}$ is the critical temperature and $y_0(\beta)$ is the spontaneous magnetisation.

<u>Proposition 1.</u> Suppose $\|\underset{\sim}{h}_N^{(N)}\| = \left\{\sum_{k=1}^{N} (\vec{h}_k^{(N)})^2\right\} \to 0$ as $N \to \infty$ and denote $\vec{H}_N = \sum_{1\le k\le N} \vec{h}_k^{(N)}$ and $\vec{\rho}_N = \vec{H}_N / H_N$. Then:

(i) If $H_N \to \infty$ and $\vec{\rho}_N \to \rho$ as $N \to \infty$, all the limits (4) exist and are equal to

$$P_K(d\vec{\sigma}_1, \dots, d\vec{\sigma}_K; \vec{\rho}) = \prod_{j=1}^{K} p(\vec{\sigma}_j; \vec{\rho})\, d\vec{\sigma}_j \quad (11)$$

where

$$p(\vec{\sigma}; \vec{\rho}) = \exp\left[\beta\left\{y_0(\beta)\vec{\sigma}\cdot\vec{\rho} - \varphi_n(y_0(\beta))\right\}\right] \quad (12)$$

(ii) If $\vec{H}_N \to \vec{H}$ as $N \to \infty$, the limits (4) exist and are equal to $\mathcal{P}_K(d\vec{\sigma}_1, \dots, d\vec{\sigma}_K; \vec{H}) =$

$$= \int_{S^{n-1}} d\vec{\rho}\; P_K(d\vec{\sigma}_1, \dots, d\vec{\sigma}_K; \vec{\rho}) \exp\left\{-\beta\left[y_0(\beta)\vec{H}\cdot\vec{\rho} + \varphi_n(H y_0(\beta))\right]\right\} \quad (13)$$

<u>Proof.</u> We consider firstly the case $\beta > \beta_c$, when $y_0(\beta) > 0$, and prove assertion (ii). Using again the identity (8) and the definition (6), one gets for the k-spin distribution of the finite-volume system, Eq.(3), the following expression

$$\rho_K^{(N)}(d\vec{\sigma}_1, \dots, d\vec{\sigma}_K; \beta, \underline{h}_N^{(N)}) =$$

$$\frac{\int_{R^n} d\vec{y} \exp\{-\beta[\frac{N}{2}y^2 + \vec{s}_K\cdot\vec{y} - \sum_{1\le j\le K} \vec{\sigma}_j \cdot \vec{h}_j^{(N)} - \sum_{K+1\le j\le N} \varphi_n(|\vec{y}-\vec{h}_j^{(N)}|)]\}}{\int_{R^n} d\vec{y} \exp\{-\beta[\frac{N}{2}y^2 - \sum_{1\le j\le N} \varphi_n(|\vec{y}-\vec{h}_j^{(N)}|)]\}} \, d\underset{\sim}{\sigma}_K$$

where $\vec{s}_K = \sum_{1\le j\le K} \vec{\sigma}_j$. Developing $\varphi_n(|\vec{y}-\vec{h}|)$ to the second order in $\vec{h}$: $\varphi_n(|\vec{y}-\vec{h}|) = \varphi_n(y) - \vec{h}\cdot\vec{y}\,\varphi_n'(y)/y + R_n(\vec{y};h)$ one obtains:

$$\rho_K^{(N)}(d\vec{\sigma}_1, \dots, d\vec{\sigma}_K; \beta, \underline{h}_N^{(N)}) = \Big(\int_{R^n} d\vec{y} \exp\{-\beta[N f_n(y) + \varphi_N'(y)\vec{H}_N\cdot\vec{y}/y]\} \times$$

$$\times \exp\{\beta \sum_{j=1}^{N} R_n(\vec{y};\vec{h}_j^{(N)})\}\Big)^{-1} \Big(\int_{R^n} d\vec{y} \exp\{-\beta[N f_n(y) + \varphi_n'(y)\vec{H}_N\cdot\vec{y}/y + \tag{14}$$

$$+ \vec{s}_K\cdot\vec{y} + K\varphi_n(y)]\} \exp\{\beta[\sum_{1\le j\le K} \vec{\sigma}_j\cdot\vec{h}_j^{(N)} + \sum_{K+1\le j\le N} R_n(\vec{y};\vec{h}_j^{(N)})]\} d\underset{\sim}{\sigma}_K\Big)$$

Now the assumption $\vec{H}_N \to \vec{H}$ (see (ii)) implies that $\varphi_n(y)\vec{H}_N\cdot\vec{y}/y \to \varphi_n(y)\vec{H}\cdot\vec{y}/y$ uniformly in $\vec{y}$ as $N \to \infty$. On the other hand, using the concavity of $\varphi_n'(y)$, we have

$$0 \le R_n(\vec{y};\vec{h}) = h^2 \varphi_n''(|\vec{y}_\theta|) + \Big[h^2 - \Big(\frac{\vec{h}\cdot\vec{y}_\theta}{|\vec{y}_\theta|}\Big)^2\Big]\frac{\varphi_n'(|\vec{y}_\theta|)}{|\vec{y}_\theta|} \le$$

$$\le 2h^2 \varphi_n''(0) = \frac{2\beta}{n} h^2 ,$$

where $\vec{y}_\theta = \vec{y} - \theta h$ for some $\theta \in (0,1)$. So, $\| \underset{\sim}{h}_N^{(N)} \| \to 0$ implies that the last exponentials in the numerator and denorminator of Eq. (14) converge uniformly to 1 as $N \to \infty$. Defining, for $g(\vec{y})$ bounded and continuous, the average

$$\langle g \rangle_1^{(N)} = \int_{R^n} d\vec{y}\, g(\vec{y}) \exp[-N\beta f_n(y)] \Big/ \int_{R^n} d\vec{y} \exp[-N\beta f_n(y)]$$

one obtains, by the usual Laplace method applied after integrating out the angular variables, that

$$\lim_{N\to\infty} \langle g \rangle_1^{(N)} = \int_{S^{n-1}} d\vec{\rho}\; g(y_0(\beta)\vec{\rho})$$

Taking into account that $|\langle g \rangle_1^{(N)}| \le \sup_{\vec{y}} |g(\vec{y})|$ (so that $\langle g_N - g \rangle_1^N \to 0$ whenever $g_N \to g$ uniformly), we obtain from Eq.(14)):

$$\lim_{N\to\infty} \rho_K^{(N)}(d\vec{\sigma}_1, \ldots, d\vec{\sigma}_K; \beta, \underset{\sim}{h}_N^{(N)}) =$$

$$= \lim_{N\to\infty} \left\langle \exp\left\{-\beta\left[\frac{\vec{H}\cdot\vec{y}}{y}\varphi_n'(y) + \vec{S}_K\cdot\vec{y} + K\varphi_n(y)\right]\right\}\right\rangle_1^{(N)}$$

$$\times \left\{\left\langle \exp\left\{-\beta\left[\frac{\vec{H}\cdot\vec{y}}{y}\varphi_n'(y)\right]\right\}\right\rangle_1^{(N)}\right\}^{-1} d\underset{\sim}{\sigma}_K =$$

$$= \frac{\int_{S^{n-1}} d\vec{\rho} \exp\{-\beta[\varphi_n'(y_0(\beta))\vec{H}\cdot\vec{\rho} + y_0(\beta)\vec{S}_K\cdot\vec{\rho} + K\varphi_n(y_0(\beta))]\}}{\int_{S^{n-1}} d\vec{\rho} \exp\{-\beta\varphi_n'(y_0(\beta))\vec{H}\cdot\vec{\rho}\}} \qquad (15)$$

which is identical with Eq.(13) (if we remember that $y_0(\beta)$ is the solution of Eq.(7)). Thus, assertion (ii) is proved.

More care is needed to prove (i), because angle-dependent part in the exponents in Eq.(14) has also the divergent factor H_N. Defining again an auxiliary average $\langle - \rangle_2^{(N)}$ by the density (Q_N^{-1} is a normalizing factor)

$$Q_N^{-1} \exp\left\{-N\beta\left[f_n(y) + \frac{\vec{H}_N \cdot \vec{y}}{Ny} \cdot \varphi_n'(y)\right]\right\}$$

we can discard, as above, the last exponentioals in Eq. (14). Thus, we need to evaluate only the limit

$$\lim_{N\to\infty} \left\langle \exp\left\{-\beta\left[\vec{S}_k^1 \cdot \vec{y} + K\varphi_n(y)\right]\right\}\right\rangle_2^{(N)} = \rho_K$$

Taking (with Eq.(5)) the integral over angles, we get:

$$\rho_k = \lim_{N\to\infty} \frac{\int_0^\infty dy\, y^{n-1} \exp\left\{-\beta\left[Nf_n(y) - \varphi_n\left(|\vec{H}_N \varphi_n'(y) + y\vec{S}_k^1|\right) + k\varphi_n(y)\right]\right\}}{\int_0^\infty dy\, y^{n-1} \exp\left\{-\beta\left[Nf_n(y) - \varphi_n\left(|\vec{H}_N| \varphi_n'(y)\right)\right]\right\}} \quad (16)$$

Clearly, because $N^{-1}H_N \to 0$ and the function $\varphi_n(y)$ has linear growth at infinity, only some compact interval $I \subset (0, \infty)$ containing the minimum of $f_n(y)$ will contribute to the limit (16). Let us expand $\varphi_n\left(|\vec{H}_N \varphi_N'(y) + y\vec{S}_K|\right)$ for $y \in I$ to the second order in $y\vec{S}_k$:

$$\varphi_n\left(|\vec{H}_N \varphi_n'(y) + y\vec{S}_K|\right) = \varphi_n\left(H_N \varphi_n'(y)\right) + y\vec{P}_N \cdot \vec{S}_k^1 \varphi_n'\left(H_N \varphi_N'(y)\right) + \Lambda \quad (17)$$

Then, because the derivative $\varphi_N'(y)$ is bounded away from zero on the compact I, one gets for the remainder Λ (with some $\theta_N(y) \in (0,1)$ and uniformly in

$y \in I$):

$$|\Lambda| \leq \frac{1}{2} |\vec{S}_K|^2 \sup_{y \in I} \{ y^2 \varphi_n''(|\vec{H}_N \varphi_N'(y) + \theta_N(y) y \vec{S}_K|) \} \xrightarrow[N \to \infty]{} 0 \quad (18)$$

As $\varphi_n'(y) \to 1$ for $y \to \infty$, we can again apply a uniform convergence argument to obtain from Eqs.(16)-(18):

$$P_K(d\vec{\sigma}_1, \dots, d\vec{\sigma}_K; \vec{P}) = \exp\{\beta[y_0(\beta)\vec{S}_K \cdot \vec{P} - K\varphi_n(y_0(\beta))]\} d\vec{\sigma}_K$$

what proves (11).

For $\beta \leq \beta_c$, $y_0(\beta) = 0$. One can see that, provided $\varepsilon_N > 0$ are chosen such that $\varepsilon_N \to 0$ and $\varepsilon_N > (H_N/N)^{1/3}$ then only a neighbourhood of the origin of the form $|\vec{y}|^2 < \varepsilon_N$ ($|\vec{y}|^4 < \varepsilon_N$, if $\beta = \beta_c$) will contribute to the $N \to \infty$ limit of Eq.(14). On the other hand, in this neighbourhood, $\exp\{-\beta[\vec{S}_K \cdot \vec{y} + K\varphi_n(y)]\}$ approaches 1 uniformly in the angular variables, implying that the limit coincides with Eqs.(11), (12) (and with Eq.(13) too) for $y_0(\beta) = 0$. □

We shall denote by $dP_{\beta}(\underline{\sigma})$, respectively by $dP_{H}^{+}(\underline{\sigma})$, the measures on $\mathcal{Y}$ associated with the finite-dimensional distributions (11)-(12), respectively (13). The picture of the set of zero-field limit states given in the Proposition is by no means unexpected. If $\beta > \beta_c$, then case (i) describes the pure states (pure phases), which correspond to the identical independent distribution of the spins according to the noninvariant distribution $\rho(\vec{\sigma}; \vec{P}) d\vec{\sigma}$. In other words, the rotation invariance, but not the permutation invariance, is broken in these states. On the other hand, case (ii) describes the mind states (mixed phases): the measure $d\mathcal{P}_H^+$ is explicitly decomposed into its ergodic components $dP_{\vec{\beta}}$, which are the extreme points of the convex set of the limit states:

$$d\,\mathcal{P}_{\vec{H}} = \int_{S^{n-1}} \mu_{\vec{H}}(d\vec{\rho})\, d\,P_{\vec{\rho}} \tag{19}$$

The probability measure $\mu_{\vec{H}}(d\vec{\rho})$ on S^{n-1} defining the mixture depends only on the "total magnetic field" $\vec{H}$, see Eq.(13). If $\beta \le \beta_c$, Eq.(12) becomes, as expected, the uniform distribution on S^{n-1}, and the distributions (11) and (13) coincide: one has uniqueness of the limit state above the critical temperature.

Remark 1. The following remarks concerning the assumption $\|\underset{\sim}{h}_N^{(N)}\| \to 0$ for $N \to \infty$ are in order here. This is a regularity condition on the convergence to zero of the external fields. Technically, it allows a simple proof of the Proposition. The main result specifying the set of pure states as well as the fact that all other limit states are superpositions of the latter should hold under the weaker assumption: $\max_{1\le j\le N} |\vec{h}_j^{(N)}| \to 0$ for $N \to \infty$. E.g. there is no difficulty to verify this for homogeneous fields: $\vec{h}_j^{(N)} = \vec{h}^{(N)}$, $j = 2, \dots, N$; $|\vec{h}^{(N)}| \le C N^{-\alpha}$; one obtains the statement (i) for all $0 < \alpha < 1$, though $\|\underset{\sim}{h}_N^{(N)}\| \to 0$ only for $\alpha > 1/2$. However, the explicit formula (13) for the mixed states may no longer hold if $\|\underset{\sim}{h}_N^{(N)}\| \to 0$. For instance, choose

$$\vec{h}_j^{(N)} = \begin{cases} N^{-1/6}\,\vec{\rho} & \text{for } 1 \le j < 2\sqrt{N} \\ (-2)\,N^{-1/6}\,\vec{\rho} & \text{for } 2\sqrt{N} \le j < 3\sqrt{N} \\ 0 & \text{for } 3\sqrt{N} \le j \le N \end{cases} \tag{20}$$

Then $H_N = 0$, but the 3rd order term in the expansion in powers of $\vec{h}_j^{(N)}$ in Eq.(14) will remove the degeneracy and lead to a limit state with nonzero magnetisation along $\vec{\rho}$.

Remark 2. As a particular case of certain physical significance, we point out the construction of limit states via "boundary conditions". If the configuration of the last $N-m_N$ spins is frozen: $\vec{\sigma}_j = \vec{S}_j$, $j = m_N+1, \dots, N$ then the distribution of the first m_N spins is given by the Hamiltonian

$$\mathcal{H}_N(\underset{\sim}{\sigma}) = -\frac{1}{2N}\left(\sum_{j=1}^{m_N}\vec{\sigma}_j\right)^2 - \left(\frac{1}{N}\sum_{j=m_N+1}^{N}\vec{S}_j\right)\sum_{j=1}^{N}\vec{\sigma}_j \tag{21}$$

Therefore, boundary conditions are equivalent to the homogeneous external fields $\vec{h}^{(N)} = \left(\sum_{m_N+1\le j\le N}\vec{S}_j\right)\cdot N^{-1}$ and to a relative shift in β equal to $N^{-1}(N-m_N)$.

In the conventional quasi-average approach, one calculates the following limits:

$$\lim_{\varepsilon\to 0}\lim_{N\to\infty} P_K^{(N)}(d\vec{\sigma}_1,\dots,d\vec{\sigma}_K;\beta,\varepsilon\underset{\sim}{h}_N) = \tilde{P}_K(d\vec{\sigma}_1,\dots,d\vec{\sigma}_K;\beta,\vec{h}),$$

where $\underset{\sim}{h}$ is a homogeneous external field $\vec{h}$ independent of N. One can easily see that the projections $\{\tilde{P}_k\}_{k=1}^{\infty}$ coincide with P_k in Eq.(11) for $\vec{\rho} = \vec{h}/h$ i.e. the quasi-average method and its generalization give the same result as far as the pure states are concerned.

We view our result as a strong argument in favour of the fact that the Curie-Weiss model has no inhomogeneous states at zero external field. In order to obtain inhomogeneous states one has to modify the interactions so as to allow for several mean fields, as has been done e.g. in Kac-Helfand models [13]; thereby, interesting phenomena related to the nonhomogeneity do appear [14].

3. Limit states in the case of random external field

In this section we shall consider the limit states associated to nonzero inhomogeneous external field. We shall somewhat simplify the situation and choose, for a given sequence $\underset{\sim}{h}=\{\vec{h}_j\}_{j=1}^{\infty}$ of vectors in $\mathbb{R}^n$, the external fields for the model with N spins as the first N vectors in $\underset{\sim}{h}$: $\underset{\sim}{h}_N^{(N)}=\underset{\sim}{h}_N=(\vec{h}_1,\dots,\vec{h}_N)$. By compactness, one can always choose subsequences $\{N_m\}$ such that all limits (4) exist. Let $dP(\mathfrak{S};\underset{\sim}{h})$ denote a limit state defined in this way. We shall show that for a suitable measure $d\lambda(\underset{\sim}{h})$ on the space of sequences $\underset{\sim}{h}$ and for typical external field configurations (i.e. $d\lambda$ -a.e.), $dP(\mathfrak{S};\underset{\sim}{h})$ is a superposition of product states, which are indexed by the solutions of one average self-consistency equation.

Let $d\nu(\vec{h})$ be a probability measure on $\mathbb{R}^n$ such that

$$\int_{\mathbb{R}^n} d\nu(\vec{h})\,|\vec{h}| < \infty \tag{22}$$

Then, the function

$$\mathcal{F}_\nu(\vec{y})=\tfrac{1}{2}y^2-\int_{\mathbb{R}^n} d\nu(\vec{h})\,\varphi_n(|\vec{y}-\vec{h}|)\equiv\tfrac{1}{2}y^2-\Phi_\nu(\vec{y}) \tag{23}$$

is well defined, convex and differentiable. As the gradient of $\Phi_\nu(\vec{y})$ is bounded in magnitude by unity, $\mathcal{F}_\nu(\vec{y})\longrightarrow\infty$ as $|\vec{y}|\to\infty$ uniformly in the angular variables. Denote by $\mathcal{M}(\nu;\beta)\subset\mathbb{R}^n$ the set where $\mathcal{F}_\nu(\vec{y})$ attains its absolute minimum. The set $\mathcal{M}(\nu;\beta)$ is compact and all $\vec{y}\in\mathcal{M}(\nu;\beta)$ are solutions of the equation

$$\vec{y}=\int_{\mathbb{R}^n} d\nu(\vec{h})\,\frac{(\vec{y}-\vec{h})}{|\vec{y}-\vec{h}|}\,\varphi_n'(|\vec{y}-\vec{h}|) \tag{24}$$

<u>Proposition 2.</u> Consider on the product space $\Omega=(\mathbb{R}^n)^N$ the product measure $d\lambda(\underset{\sim}{h})=\prod_{k=1}^{\infty} d\nu(\vec{h}_k)$. Then, for $d\lambda$ -almost all h and every limit state $dP(\mathfrak{S};\underset{\sim}{h})$ associated with it by the procedure described

above, there exist a measure $\omega(d\vec{y})$, concentrated on $\mathcal{M}(\nu;\beta)$, such that:

$$dP(\underset{\sim}{\xi};\underset{\sim}{h}) = \int_{\mathcal{M}(\nu;\beta)} \omega(d\vec{y})\, dP_{\vec{y}}(\underset{\sim}{\xi};\underset{\sim}{h}) \,, \tag{25}$$

where $dP_{\vec{y}}(\underset{\sim}{\xi};\underset{\sim}{h})$ is the measure on $\mathcal{S}$ whose projections on $\prod_{1\le j\le K} S^{n-1}$ are given by the densities:

$$p_K(\underset{\sim}{\xi}_K;\underset{\sim}{h},\vec{y}) = \prod_{j=1}^{K} \exp\left\{\beta\left[\vec{\xi}_j(\vec{y}-\vec{h}_j) - \varphi_n(|\vec{y}-\vec{h}_j|)\right]\right\} \tag{26}$$

<u>Proof.</u> Let $\Phi_N(\vec{y};\underset{\sim}{h}) = N^{-1}\sum_{1\le j\le N}\varphi_n(|\vec{y}-\vec{h}_j|)$. As before, using the identity (8), one obtains

$$P_K^{(N)}(d\vec{\xi}_1,\dots,d\vec{\xi}_K;\beta,\underset{\sim}{h}) = \left[\int \omega_{\underset{\sim}{h}}^{(N)}(d\vec{y})\, p_K(\underset{\sim}{\xi}_K;\underset{\sim}{h},\vec{y})\right] d\underset{\sim}{\xi}_K \tag{27}$$

where $\omega_{\underset{\sim}{h}}^{(N)}(d\vec{y}) =$

$$= d\vec{y}\cdot \exp\left\{-\beta N\left[\tfrac{1}{2}y^2 - \Phi_N(\vec{y};\underset{\sim}{h})\right]\right\}\left(\int_{\mathbb{R}^n} d\vec{y} \exp\left\{-\beta N\left[\tfrac{1}{2}y^2 - \Phi_N(\vec{y};\underset{\sim}{h})\right]\right\}\right)$$

By the individual ergodic theorem [,5], for every fixed $\vec{y}$, $\Phi_N(\vec{y};\underset{\sim}{h})$ converges to $\Phi_\nu(\vec{y})$ for $d\lambda$ -almost all $\underset{\sim}{h}$. A countable union of the sets of the zero measure also has zero measure, so the same will hold for a countable dense set, e.g. for all $\vec{y}$ with rational coordinates. Using the convexity of $_N(\vec{y};\underset{\sim}{h})$, one obtains the $d\lambda$ -a.e. convergence for <u>all</u> $\vec{y}\in\mathbb{R}^n$ uniformly on compacts. Moreover, the asymptotics of Φ_N at infinity is uniform in N. Hence, one concludes that the $N \to \infty$ limit does not change if one restricts the integration in (27) to a certain large ball, $\mathcal{B}$. Moreover, using the uniform convergence of Φ_N to Φ_ν on $\mathcal{B}$, one has that, for every continuous g with compact support in $\mathcal{B} \setminus \mathcal{M}(\nu;\beta)$, $\int \omega_{\underset{\sim}{h}}^{(N)}(d\vec{y})g(\vec{y}) \to 0$ as $N\to\infty$. Thus, every limit point $\omega(d\vec{y})$ of $\{\omega_{\underset{\sim}{h}}^{(N)}(d\vec{y})\}_N$ has support in $\mathcal{M}(\nu;\beta)$, that concludes the proof.□

Remark 3. If we perturb as in Sec. 2 the field $\underset{\sim}{h}_N$ by a vector $\underset{\sim}{\xi}_N^{(N)}$ such that $\|\underset{\sim}{\xi}_N^{(N)}\| \to 0$, the same proof goes through and every limit state of this kind can be again represented as in Eq. (25). In other words, one can explore, as before, the simplex with extremal points $\{dP_{\vec{y}}(\underset{\sim}{\varepsilon};\underset{\sim}{h});\vec{y}\in\mathcal{M}(\nu;\beta)\}$. However, we are not able to give an explicit form of the mixture.

Remark 4. We do not consider here the description of $\mathcal{M}(\nu;\beta)$ in terms of $d\nu$. This question is of relevance in the study of the type of transition and of the critical properties of the system in a random field. Such questions are especially interesting in the case of short-range models and are now under active study (e.g. [16]). We only remark, in order to make a connection with the quasi-average method, that, if ν_t denotes the scaled measure: $\nu_t(A)=\nu(t^{-1}A)$ ($A\subset\mathbb{R}^n$ measurable), then $\mathcal{M}(\nu_t;\beta)$ approaches $\vec{y}_0(\beta)\in S^{n-1}$ when $t\to 0$, so one recovers the situation in zero field described in Sec. 2:

$$(Pr=1)-\lim_{t\to 0}\int_{\mathcal{M}(\nu_t;\beta)}\omega(d\vec{y})\,dP_{\vec{y}}(\underset{\sim}{\varepsilon};\underset{\sim}{h}) = dP(\underset{\sim}{\varepsilon}) \qquad (28)$$

References

1. N.N.Bogolubov, Lectures in Quantum Statistics. Vol. 2. Quasi-Averages. Gordon- & Breach, New York (1970).
2. N.N.Bogolubov (jr.), A Method for Studying Model Hamiltonians. Pergamon Press, Oxford-New York (1972).
3. N.N.Bogolubov (Jr.), J.C.Brankov, V.A.Zagrebnov, A.M.Kurbatov, and N.S.Tonchev, The Approximating Hamiltonian Method in Statistical Physics. Publ. House of the Bulgarian Acad. Sci., Sofia (1981).
4. V.A.Zagrebnov, Z.Phys., B55: 75-85 (1984).
5. R.A.Minlos, Funkcional. Anal.i Prilozen. 2: 60-73 (1967); 3: 40-53 (1967).
6. R.L.Dobrushin, Teor. Verojatnost. i Primenen. 13: 197-224 (1968); 15: 458-486 (1970).
7. O.E.Lanford and D.Ruelle, Comm. Math. Phys. 13: 194-215 (1969).
8. R.S.Ellis and Ch.M.Newman. J.Statist.Phys. 19: 149-161 (1978).
9. A.Rieckers, Physica A108: 107-134 (1981).
10. M.Fannes, H.Spohn and A.Verbeure, J. Math. Phys. 21: 355-358 (1980).
11. J.G.Brankov and V.A.Zagrebnov, J. Phys. A 16: 2217-2224 (1983).
12. P.Billingsley, Convergence of Probability Measure, John Wiley & Sons, New York (1968).
13. M.Kac and E.Helfand, J. Math. Phys. 4: 1078-1088 (1963).
14. N.Angelescu, M.Bundaru, G.Costache and G.Nenciu, J. Statist. Phys. 24: 529-552 (1981).
15. I.P.Cornfeld, Ya.G.Sinai and S.V.Fomin, Ergodic Theory. Springer, New York (1982).
16. D.S.Fisher, J. Fröhlich and T.Spencer, J. Statist. Phys. 34: 863-870 (1984).

Prob. Theory and Math. Stat., Vol.1, pp. 85-97
Prohorov et al. (eds)
1986 VNU Science Press

INEQUALITIES AT THE MARKOV APPROXIMATION OF LUMPED PROCESSES

V.V. Anisimov
Kiev State University, Kiev, U.S.S.R.

Now, in the theory of asymptotical states enlarging limit theorems of weak convergence to Markov processes (MP) have been principally elaborated. They have been developing in 2 directions. The first direction is based upon the theory of inversion of operators perturbed on the spectrum and enables us to investigate homogeneous semi-Markov processes. It was developed in [1]. The second one elaborated in [2,3] is based on the methods of probabilistically-constructive analysis, switching processes techniques and embraces non-homogeneous in time processes. In this report for the first time explicit estimates for the deviation in variation between finite dimensional distributions of lumped non-homogeneous MP's and MP's with averaged characteristics are given. The proof uses constructive representations of the lumped process in terms of switching processes [3].

Let $(\mathcal{X}, \mathcal{B}_{\mathcal{X}})$ and $(\mathcal{Y}, \mathcal{B}_{\mathcal{Y}})$ be measurable spaces, $K(\cdot)$ be a measurable mapping of $\mathcal{X}$ onto $\mathcal{Y}$, $x(\ell)$, $\ell \geq 0$ be a MP taking values in $\mathcal{X}$ with the transition probability on the ℓ-th step

$$p(x, \ell, A) = P\{x(\ell+1) \in A / x(\ell) = x\}, \ x \in \mathcal{X}, A \in \mathcal{B}_{\mathcal{X}}, \ \ell \geq 0.$$

We introduce the lumped process $\hat{x}(\ell) = K(x(\ell))$, $\ell \geq 0$, which is, generally speaking, non-Markovian and proceed to approximate it with some MP.

Denote $I_y = \{x : K(x) = y\}$, $\hat{x}^{(y)}(\ell), \ell \geq 0$

a MP on I_y with a certain initial distribution and transition probability

$$\tilde{p}^{(y)}(x,\ell,A)=p(x,\ell,A)\,p(x,\ell,I_y)^{-1},\quad x\in I_y,\ A\in\mathcal{B}_{I_y},\ \ell\geq 0$$

(in the case $p(x,\ell,I_y)=0$, $\tilde{p}^{(y)}(x,\ell,A)$ may be arbitrary).

Introduce a MP $y(\ell)$, $\ell\geq 0$ in Y such that

$$P\{y(0)\in B\}=P\{x(0)\in K^{-1}(B)\},\quad B\in\mathcal{B}_Y, \tag{1}$$

and the transition probability on the ℓ-th step is

$$\hat{p}(y,\ell,B)=\int_{I_y}p(x,\ell,K^{-1}(B))\,\pi^{(y)}(\ell,dx),\quad \ell\geq 0,$$

$$y\in Y,\ B\in\mathcal{B}_Y,\ y\bar{\in}B,\ (\hat{p}(y,\ell,\{y\})=1-\hat{p}(y,\ell,Y\setminus\{y\})), \tag{2}$$

where

$$\pi^{(y)}(\ell,A)=P\{\tilde{x}^{(y)}(\ell)\in A\}.$$

Put

$$\tilde{p}^{(y)}(i,x,j,A)=P\{\tilde{x}^{(y)}(j)\in A/\tilde{x}^{(y)}(i)=x\},$$

$$\varphi^{(y)}_{i,j}=\sup\{|\tilde{p}^{(y)}(i,x_1,j,A)-\tilde{p}^{(y)}(i,x_2,j,A)|:$$

$$x_1,x_2\in I_y,\ A\in\mathcal{B}_{I_y}\},\ i\leq j.$$

Here $\varphi^{(y)}_{ij}$ is uniform strong mixing coefficient of the MP $\tilde{x}^{(y)}(\cdot)$ on the interval $[i,j]$.

Denote for an arbitrary integer $n>0$

$$\varepsilon_n=\sup_{y\in Y}\ \max_{k<n}\ \sup_{x\in I_y}\ p(x,k,X\setminus I_y),$$

$$\varphi_n=\sup_{y\in Y}\ \max_{k\leq n}\ \sum_{i=k}^{n}\varphi^{(y)}_{ki},\quad \alpha_n=n\varepsilon_n.$$

Theorem 1. For any $n>0$, $r>0$, $0\leq k_1<\ldots<k_r\leq n$, $A_i\in\mathcal{B}_Y$, $i=\overline{1,r}$, $\varepsilon_n\leq 1/2$

the inequality

$$|P\{\hat{x}(k_i)\in A_i,\ i=\overline{1,r}\}-P\{y(k_i)\in A_i,\ i=\overline{1,r}\}|\leq\rho_n \tag{3}$$

where $\rho_n = C_1(\alpha_n)\varepsilon_n\psi_n + C_2(\alpha_n)\varepsilon_n + C_3(\alpha_n)\frac{1}{n}$.

$$C_1(a) = 2(e^a(12a^3 + 24a^2 + a + 1) + 24a + 1),$$

$$C_2(a) = a(2a+1)(\frac{1}{6}e^a a^2(a+3)+2),\ C_3(a) = 2e^a a^2(a+2)$$

holds.

Proof. First of all we state some preliminary results. Let a sequence of nonnegative functions $f(k,x)$, $x \in X$, $k \geq 0$ be given. Put $\hat{f}(k,y) = E f(k, \tilde{x}^{(y)}(k))$ and let $\tilde{x}_\mu^{(y)}(k)$ be the MP $\tilde{x}^{(y)}(k)$ with initial distribution μ.

Lemma 1. For any distribution μ on I_y

$$\sup_{y \in Y} \max_{0 \leq k \leq j \leq n} E_\mu (\sum_{i=k}^{j} (f(i, \tilde{x}_\mu^{(y)}(i)) - \hat{f}(i,y)))^2 \leq 6nC^2\psi_n, \quad (4)$$

where

$$C = \sup_{y \in Y} \max_{k \leq n} \sup_{x \in I_y} |f(k,x) - \hat{f}(k,y)|.$$

Indeed, using the well-known ineqality (see(5)), for any measurable function $f(x)$ and measures $P(A)$, $Q(A)$

$$|\int f(x)P(dx) - \int f(x)Q(dx)| \leq 2\sup|f(x)| \sup_A |P(A) - Q(A)| \quad (5)$$

(note that for $f(x) \geq 0$ the factor 2 in the right side of (5) can be omited), we receive for $i \leq l$

$$|E_\mu(f(i,\tilde{x}_\mu^{(y)}(i)) - \hat{f}(i,y))(f(l,\tilde{x}_\mu^{(y)}(l)) - \hat{f}(l,y))| \leq$$
$$\leq C^2(\varphi_{0i}^{(y)} \cdot \varphi_{0l}^{(y)} + 2\varphi_{il}^{(y)}),$$

which after simple transformations entails (4).

Lemma 2. Let $0 \leq p_i \leq 1/2$, $0 \leq q_i \leq 1/2$, $i = \overline{0,k}$. Then

$$|\prod_{i=0}^{k}(1-p_i) - \prod_{i=0}^{k}(1-q_i) - \prod_{i=0}^{k}(1-q_i)\sum_{i=0}^{k}(1-q_i)^{-1}(q_i-p_i)| \leq$$
$$\leq 2(\sum_{i=0}^{k}(1-q_1)^{-1}(q_i-p_i))^2 + 2k\varepsilon^2(1+4k\varepsilon^2), \quad (6)$$

where $\varepsilon = \max_{i \leq k}|q_i - p_i|$.

Indeed, one can easily show that for any $a>0$ $-\infty<z\le a$

$$|e^{z}-1-z|\le z^{2}Q(a), \tag{7}$$

where $Q(a)=a^{-2}(e^{a}-1-a)$. Hence for any $x\ge 0$ $y\ge 0$ we obtain

$$|e^{-x}-e^{-y}-e^{-y}(y-x)|\le e^{-y}|y-x|^{2}Q(y)\le |y-x|^{2},$$

because $e^{-y}Q(y)\le 1$. The second order Taylor's expansion with a remainder term in Lagrange form yields

$$|\ln(1-p_i)-\ln(1-q_i)-(1-q_i)^{-1}(q_i-p_i)|\le \tfrac{1}{2}(1-q_i)^{-2}(q_i-p_i)^{2}.$$

Using the identity $\prod_{i=0}^{k}(1-p_i)=\exp\{\sum_{i=0}^{k}\ln(1-p_i)\}$ and inegualities written above we eventually after transformations obtain (6).

We proceed now to the proof of the theorem. Denote $\hat{\mu}(k), k\ge 0$ successive moments of $\hat{x}(\cdot)$ jumps, i.e. $\hat{\mu}(0)=0$ and

$$\hat{\mu}(k+1)=\min\{\ell:\ell>\hat{\mu}(k),\ \hat{x}(\ell)\ne\hat{z}(k)\},$$

where $\hat{z}(k)=\hat{x}(\hat{\mu}(k)),\ k\ge 0.$

Let $\mu(k), k\ge 0$ be respectively successive moments of jumps of $y(\cdot)$ and $z(k)=y(\mu(k))$. Denote

$$\hat{p}(k_1,\dots,k_r,A_0,\dots,A_r)=P\{\hat{\mu}(i)=k_i,\ i=\overline{1,r},\ \hat{\mu}(r+1)>n,$$

$$\hat{z}(k_j)\in A_j,\ j=\overline{0,r}\},\ 0=k_0<k_1<\dots<k_r\le n,\ A_j\in\mathcal{B}_y,\ j=\overline{1,r},\ r\ge 0.$$

(the value $r=0$ is corresponded to $\hat{p}(A_0)=P\{\hat{z}(0)\in A_0,$ $\hat{\mu}(1)>n\}$) and analogously define $p(k_1,\dots,k_r,A_0,\dots,A_r)$ for $y(\cdot)$. Further, put

$$q^{(1)}(k_1,\dots,k_r)=\sup\{(\hat{p}(k_1,\dots,k_r,A_0,\dots,A_r)-$$

$$-p(k_1,\dots,k_r,A_0,\dots,A_r)):A_i\in\mathcal{B}_y,\ i=\overline{0,r}\},$$

$$q^{(2)}(k_1,\dots,k_r) = \sup\{(p(k_1,\dots,k_r,A_0,\dots A_r)-$$

$$-\hat{p}(k_1,\dots,k_r,A_0,\dots,A_r) : A_i \in \mathcal{B}_y,\ i=\overline{0,r}\},$$

$$q^{(s)} = \max_{U_0} \sum\nolimits_{r\in U_0} \max_{U_i, i=\overline{1,r}} \sum\nolimits_{k_i \in U_i, i=\overline{1,r}} q^{(s)}(k_1,\dots,k_r), s=1,2.$$

Here U_i $i=\overline{0,r}$ are the sets of integer points of the interval $[0,n]$. We can write $P\{y(n_i)\in A_i, i=\overline{1,k}\}=$

$$=\sum\nolimits_{r=0}^{n} \sum\nolimits_{0<k_1<\dots<k_r\le n} \int_{Y_0}\cdots\int_{Y_r} \chi(y(n_i,\omega)\in A_i,$$

$$i=\overline{1,k})\, p(k_1,\dots,k_r,dy_0,\dots,dy_r),$$

where $n_i \in n$, $i=\overline{1,k}$, $Y_0 = Y$, $Y_j = Y\setminus\{y_{j-1}\}$, $j=\overline{1,r}$,

and $y(\cdot,\omega)$ is a fixed trajectory of the process $y(\cdot)$ related to the event $\mu(i)=k_i$, $i=\overline{1,r}$, $z(k_j)=y_j$, $j=\overline{0,r}$, $\chi(A)$ is the indicator of the event A . Then the analogous representation for the distributions of $\hat{x}(\cdot)$ and the inequality (5) imply that for any A_i, $i=\overline{1,k}$, $k>0$

$$|P\{\hat{x}(n_i)\in A_i, i=\overline{1,k}\} - P\{y(n_i)\in A_i, i=\overline{1,k}\}| \le q^{(1)} \vee q^{(2)}.$$

Therefore to prove (3) it is sufficient to show that $q^{(1)}\vee q^{(2)} < \rho_n$.

It is easy to prove now that if $x(0)=x\in I_y$, then by the construction of $\hat{x}^{(y)}(i)$ $(\hat{x}^{(y)}(0)=x)$ a representation $P\{\hat{\mu}(1)=i, \hat{z}(1)\in A\}=$

$$=E_x \prod\nolimits_{\ell=0}^{i-2} p(\hat{x}^{(y)}(\ell),\ell,I_y)\, p(\hat{x}^{(y)}(i-1), i-1, K^{-1}(A)) \qquad (8)$$

where $y \bar{\in} A$ holds (see [3]). Respectively

$$P\{\mu(1)=i, z(1)\in A\} = \prod\nolimits_{\ell=0}^{i-2} \hat{p}(y,\ell,\{y\})\,\hat{p}(y,i-1,A).$$

(here $\prod_{k}^{m}=1$ $k>m$). We introduce the notations

$$\hat{\alpha}_i=\prod_{\ell=k_i}^{k_{i+1}-2} p(\hat{x}^{(y_i)}(\ell),\ell,I_{y_i}),\ i=\overline{0,r},$$

where $y_i=K(x_i)$, $k_0=0$, $k_{r+1}=n+1$,

$$\alpha_i=\prod_{\ell=k_i} \hat{p}(y_i,\ell,\{y_i\}),\ i=\overline{0,r}.$$

Further on, let $\hat{\beta}_i=p(\hat{x}^{(y_i)}(k_{i+1}-1),k_{i+1}-1,dx_{i+1})$,

$$\beta_i=\hat{p}(y_i,k_{i+1}-1,dx_{i+1}),\ i=\overline{0,r-1},$$

$$\hat{p}(y,k,B)=\int_{I_y} p(x,k,B)\,\pi^{(y)}(k,dx),\ y\in K(B).$$

Then similarly to (8) we obtain

$$\hat{p}(k_1,\dots,k_r,A_0,\dots,A_r)=\int_{B_0} p(dx_0)\cdot\int_{B_1}\cdots\int_{B_r} E_{x_0}(\hat{\alpha}_0\hat{\beta}_0)\cdot$$
$$E_{x_1}(\hat{\alpha}_1\hat{\beta}_1)\dots E_{x_{r-1}}(\hat{\alpha}_{r-1}\hat{\beta}_{r-1})E_{x_r}\hat{\alpha}_r,$$

where in our case $z(i)=x_i$, i.e. $\hat{x}^{(y_i)}(k_i)=x_i$,

$$B_0=K^{-1}(A_0),\ B_i=K^{-1}(A_i\setminus\{y_{i-1}\}),\ i=\overline{1,r},$$

$$p(B)=p\{x(0)\in B\}.$$ Respectively

$$p(k_1,\dots,k_r,A_0,\dots,A_r)=\int_{B_0} p(dx_0)\int_{B_1}\cdots\int_{B_r}\alpha_0\beta_0\cdots$$
$$\cdots\alpha_{r-1}\beta_{r-1}\alpha_r,\ r\geq 0.$$

We write a representation

$$\hat{p}(k_1,\dots,k_r,A_0,\dots,A_r)-p(k_1,\dots,k_r,A_0,\dots,A_r)=\sum_{j=0}^{r}\delta_j+\sum_{j=0}^{r-1}\gamma_j,$$

where

$$\delta_j=\int_{B_0}\cdots\int_{B_r} p(dx_0)\prod_{i=0}^{j-1}E_{x_i}(\hat{\alpha}_i\hat{\beta}_i)E_{x_j}(\hat{\alpha}_j-\alpha_j)\beta_j\cdot$$
$$\prod_{i=j+1}^{r-1}(\alpha_i\beta_i)\alpha_r,\ j=\overline{0,r-1},$$

$$\delta_r = \int_{B_0}\cdots\int_{B_r} p(dx_0)\prod_{i=0}^{r-1} E_{x_i}(\hat{\alpha}_i\hat{\beta}_i)\cdot E_{x_r}(\hat{\alpha}_r - \alpha_r),$$

and

$$\gamma_j = \int_{B_0}\cdots\int_{B_r} p(dx_0)\prod_{i=0}^{j-1} E_{x_i}(\hat{\alpha}_i\hat{\beta}_i)\hat{\alpha}_j E_{x_j}(\hat{\beta}_j - \beta_j)\cdot$$

$$\prod_{i=j+1}^{r-1}(\alpha_i\beta_i)\cdot\alpha_r,\ j=\overline{0,r-1}.$$

Note that for any $A\in\mathcal{B}_{I_{y_i}}$, $K_i<K$

$$|E_{x_i}\chi(\tilde{x}^{(y_i)}(K)\in A)-\pi^{(y_i)}(K,A)|\le\varphi^{(y_i)}_{K_i K}$$ takes place.

Then the inequality (5) easily implies

$$\left|\int_{B_{j+1}} E_{x_j}(\hat{\beta}_j-\beta_j)\right|\le 2\varepsilon n\,\varphi^{(y_i)}_{K_j, K_{j+1}-1}\,.$$

for any B_{j+1}.

Owing to $|\alpha_i|\le 1$, $|\beta_i|\le\varepsilon_n$, $i=\overline{0,r}$, and applying (3) again we receive

$$\left|\sum_{K_i\in U_i,\ i=\overline{1,r}}\sum_{j=0}^{r-1}\gamma_j\right|\le 2r\varepsilon_n^r C_n^{r-1}\psi_n,\ r>0. \quad (10)$$

Taking into account that $|p(x,\ell,I_y)-\hat{p}(y,\ell,\{y\})|\le\varepsilon_n$,

as $x\in I_y$ and using the inequalities (4), (5) and (6) one can see that

$$|E_{x_j}(\hat{\alpha}_j-\alpha_j)|\le 2\varepsilon_n\sum_{i=K_j}^{K_{j+1}-1}\varphi^{(y_i)}_{K_j i}+48\varepsilon_n^2(K_{j+1}-K_j-1)\cdot$$

$$\sum_{i=K_j}^{K_{j+1}-1}\varphi^{(y_i)}_{K_j i}+2\varepsilon_n^2(K_{j+1}-K_j-1)(1+4n\varepsilon_n^2). \quad (11)$$

Hence estimating C_n^K by $n^K/K!$ we obtain

$$\left|\sum_{K_j\in U_i,\ i=\overline{1,r}}\sum_{j=0}^{r}\delta_j\right|\le\frac{(r+1)}{(r-1)!}\varepsilon_n^{r+1}n^{r+1}\Big(24\varepsilon_n\psi_n+\frac{1}{3}\varepsilon_n^2(r-1)\cdot$$

$$(1+4n\varepsilon_n^2)+\frac{2}{n^2}\psi_n\Big).$$

Using analogous correlation for $\hat{p}(A_0)-p(A_0)$ and summing by all $r>0$ we shall have after transformations $q_1\vee q_2\le\rho_n$ which proves (3).

We'll give now estimates for the distribution of the 1-st jump instance which in reliability theory is usually interpreted as the 1-st failure instance. One can easy derive from (7) that for any $y \geq b, x \geq b \geq 0$

$$|e^{-x} - e^{-y} - e^{-y}(y-x)| \leq e^{-b}|y-x|^2 G(y-b), \tag{12}$$

where $G(a) = \min(1, a^{-2})$.

We put $q_k^{(y)} = 1 - \hat{p}(y, k, \{y\}), \ a_k^{(y)} = \inf_{x \in I_y} p(x, k, X \setminus I_y)$,

$$Q_n^{(y)} = \sum_{k=0}^{n} q_k^{(y)}, \ A_n^{(y)} = \sum_{k=0}^{n} a_k^{(y)},$$

$$\delta_n^{(y)} = \max_{k \leq n} \sup_{x \in I_y} |p(x, k, I_y) - \hat{p}(y, k, \{y\})|.$$

Afterwards (12) similarly entails (11).

Corollary 1. If $q_k^{(y)} \leq 1/2$, $k \leq n$ and

$$P\{x(0) \in I_y\} = 1,$$

then $|P\{\hat{\mu}(1) < n\} - P\{\mu(1) < n\}| \leq$

$$\leq \exp\{-Q_n^{(y)}\} 2\delta_n^{(y)} \psi_n + \exp\{-A_n^{(y)}\} G(Q_n^{(y)} - A_n^{(y)}) \cdot$$

$$\cdot 2n(\delta_n^{(y)})^2 (24\psi_n + 1 + 4n(\delta_n^{(y)})^2).$$

<u>Remark</u>. If $x(\cdot)$ is a homogeneous MP and $\psi_n \leq \psi$, $n > 0$ then $\rho_n \leq \frac{1}{n} C(T, \psi)$ as $\varepsilon_n \leq 1/2, \alpha_n \leq T$, where for $C(T, \psi)$ an explicit expression is available. Note also that

$$|P\{\mu(1) > n\} - \exp\{-nq_1^{(y)}\}| \leq n(q_1^{(y)})^2.$$

Thus under our assumptions $|P\{\hat{\mu}(1) > nt\} - \exp\{-ntq_1^{(y)}\}| \leq$

$\leq \frac{1}{n} C_1(T, \psi)$ where one can easily compute $C_1(T, \psi)$.

Furthermore, one can give examples of homogeneous MP even with finite set of states for which $\psi_n \leq \psi$, $\alpha_n \leq \alpha$, but $|P\{\hat{\mu}(1) < n\} - P\{\mu(1) < n\}| > \frac{1}{2}\varepsilon_n$.

So, the order of estimates found in Theorem 1 and Corollary 1 is unimprovable in general case.

The above estimates were constructed in terms of auxiliary MP's $\tilde{x}^{(y)}(\cdot)$, which naturally corresponds to the structure of the process $\hat{x}(\cdot)$. It is constructed here as a switching process defined by the family $\tilde{x}^{(y)}(\cdot)$ and probabilities of switches $p(x,k,X\setminus I_y)$, $x \in I_y$ ([3]). Incidentally the problem of approximation of characteristics of the $\tilde{x}^{(y)}(\cdot)$ by those of more convenient processes or their stationary measures can arise.

Let the representation hold

$p(x,k,A) = p_0(x,k,A) + b(x,k,A)$, $x \in X$, $A \in \mathcal{B}_X$, $k \geq 0$,

(herewith $p_0(x,k,A) = 0$ if $x \in I_y$, $A \cap I_y = \emptyset$, $y \in Y$, $k \geq 0$).

Denote

$$\varepsilon = \max_{k \geq 0} \sup_{x \in X, A \in \mathcal{B}_X} |b(x,k,A)|,$$

$\psi_{ij}^{(y)}$ the coefficient of uniformly strong mixing of the MP $x_0^{(y)}(\cdot)$ (defined on I_y) with the transition probabilities $p_0(x,k,A)$ and some fixed initial distribution. We define the MP $y(\cdot)$ in Y according to (1),(2) where in (2)

$$\pi^{(y)}(l,A) = P\{x_0^{(y)}(l) \in A\}.$$

Theorem 2. Let for some integer $\tau > 0$

$$\sup_{i \geq 0, y \in Y} \psi_{i,i+\tau}^{(y)} \leq q_1, \quad q_1 + 8\tau\varepsilon \leq q < 1. \tag{13}$$

Then in (3)

$$\rho_n \leq \varepsilon C(T,\tau,q),$$

as $\varepsilon \leq 1/2$ and $n \leq \varepsilon^{-1} T$ where $C(\cdot)$ can be found explicitly.

Proof. We estimate at first the deviation of transition characteristics of the processes $x_0^{(y)}(\cdot)$ and $\tilde{x}^{(y)}(\cdot)$. It follows immediatelly from the construction that

$$\sup\{|p_0(x,k,A) - \tilde{p}^{(y)}(x,k,A)| : x \in I_y, A \in \mathcal{B}_{I_y}, k \geq 0\} \leq \leq \frac{2\varepsilon}{1-\varepsilon} \leq 4\varepsilon. \tag{14}$$

Denote $p_0(k,x,j,A)$, $x \in I_y$, $A \in \mathcal{B}_{I_y}$ the transition probability of the $x_0^{(y)}(\cdot)$ on the interval $[k,j]$. The inequality (5) together with Chapman-Kolmogorov equation imply for $k \le j$,

$$|p_0(k,x,j+1,A)-\tilde{p}^{(y)}(k,x,j+1,A)|=|\int_X p_0(k,x,j,dz)(p_0(z,j,A)-$$

$$-\tilde{p}^{(y)}(z,j,A))+\int_X(p_0(k,x,j,dz)-\tilde{p}^{(y)}(k,x,j,dz))p^{(y)}(z,k,A)|\le$$

$$\le \sup_{z\in I_y}|p_0(z,j,A)-\tilde{p}^{(y)}(z,j,A)|+\sup_{B\in \mathcal{B}_{I_y}}|p_0(k,x,j,B)-\tilde{p}^{(y)}(k,x,j,B)|.$$

Hence for all $x\in I_y$, $A\in\mathcal{B}_{I_y}$, $k\ge 0$

$$|p_0(k,x,k+\tau,A)-\tilde{p}^{(y)}(k,x,k+\tau,A)|\le 4\varepsilon\tau.$$

Then it is easy to count that under the condition

(13) $\varphi^{(y)}_{i,i+\tau}\le q<1$.

Using coefficient properties [4] one can prove that for every $i\le j$

$$\varphi^{(y)}_{ij}\le q^{[\tau^{-1}(j-i)]} \tag{15}$$

Extending the estimates in [5] to the case of discrete time and arbitrary state space one can prove in a similar way that (13), (14) imply

$$|\tilde{p}^{(y)}(i,x,j,A)-\pi^{(y)}(j,A)|\le 4\varepsilon\tau(1-q)^{-1}+\psi^{(y)}_{ij},\ i<j. \tag{16}$$

Now performing the same constructions as in the Theorem 1 we obtain that in respective estimates the values $\varphi^{(y)}_{ij}$ should be replaced by $\psi^{(y)}_{ij}+4\varepsilon\tau(1-q)^{-1}$. Herein ψ_n with account of (15) will be estimated by $\tau(1-q)^{-1}(1+n\varepsilon)$. So (3) implies as $n\le\varepsilon^{-1}T$ that $\rho_n\le\varepsilon C(T,\tau,q)$, where $C(\cdot)$ can be easily computed by virtue of $C_e(a), \ell=\overline{1,3}$, which proves the theorem.

As an illustration consider now homogeneous in time processes. In this case the characteristics of $\hat{y}(\cdot)$ can be explicitly expressed through the stationary measures of the classes I_y.

Let $x(\ell), \ell\ge 0$ be a homogeneous MP in X with transition probability for 1 step

$$p(x,A)=p_0(x,A)+b(x,A),$$

herewith $p_0(x,A)\equiv 0,\ x\in I_y,\ A\cap I_y=\emptyset,\ y\in Y$.

Put $\varepsilon=\sup_{x,A}|b(x,A)|$. Denote $\psi_k^{(y)}$ the coefficient of uniform strong mixing of the process $x_0^{(y)}(\cdot)$ for k steps. Suppose that for some integer $\tau>0$ for all $y\in Y$

$$\psi_\tau^{(y)}\le q_1 \text{ and } q_1+8\tau\varepsilon\le q<1. \tag{17}$$

Under this condition the processes $x_0^{(y)}(\cdot)$ are ergodic. Denote corresponding stationary measures $\pi^{(y)}(A),A\in\mathcal{B}_{I_y}$, $y\in Y$ and define a homogeneous MP $y(\cdot)$ with the transition probability for 1 step

$$\hat{p}(y,B)=\int_{I_y} b(x,K^{-1}(B))\pi^{(y)}(dx),\ y\bar{\in} B. \tag{18}$$

Since under our assumptions (16) holds with $\pi^{(y)}(j,A)=$ $=\pi^{(y)}(A)$ we receive

Corollary 2. If the condition (17) holds and $\varepsilon<1/2$ then for $n\le\varepsilon^{-1}T$ in (3)

$$\rho_n\le\varepsilon C(T,\tau,q).$$

The results obtained immediately entail limit theorems on the convergence of the $\hat{x}(\cdot)$ distributions to those of a MP. Consider a sequence of (for clearness homogeneous) MP's $x_n(k),\ k\ge 0$ with one step transition probabilities

$$p_n(x,A)=p_0(x,A)+\frac{1}{n}b(x,A),\ n>0$$

herewith the probabilities $p_0(x,A)$ are concordant with the fragmentation I_y and for some integer τ and $0<q<1$ $\psi_\tau^{(y)}\le q,\ y\in Y$. Denote $z(t),\ t\ge 0$ a MP with continuous time and transition intensities

$$\lambda(y,B)=\int_{I_y} b(x,K^{-1}(B))\pi^{(y)}(dx),\ y\bar{\in} B.$$

<u>Theorem 3.</u> Under the assumptions stated above for every $T>0,\ r>0$,

$$t_1<t_2<\ldots<t_r<T,\ A_i\in\mathcal{B}_Y,\ i=\overline{1,r}$$

$$\lim_{n\to\infty} P\{\hat{x}_n([nt_i])\in A_i,\ i=\overline{1,r}\}=P\{z(t_i)\in A_i,\ i=\overline{1,r}\}.$$

(it is assumed that A_i are the sets of continuity of the right side).

The proof is based on the fact that the approximating MP $y_n([nt])$ is approaching $z(t)$. It can be proved immediately in terms of constructive representations in Theorem 1 (the case of discrete state space was studied in [3]).

Thus we see that under quite general assumptions about smallness of the transition probabilities between the domains I_y $(\sim\varepsilon)$ and good mixing of the backup (unperturbed) process in each domain the transition probabilities of the lumped process are uniformly approximated up to ε by the characteristics of the MP with avaraged transition probabilities. In applied models these results in a number of cases allow us to decrease the dimension of the initial model and to obtain simple and asymptotically precise formulae. It should be noted that with the aid of the same technique analogous results can be obtained for processes with continuous time and extended to estimates uniform on the whole time axis.

References

1. V.S.Korolyuk and A.F.Turbin, Mathematical foundations of phase consolidation of complex systems, "Naukova Dumka", Kiev,1978. (Russian).
2. V.V.Anisimov, Limit theorems for switching processes and the r applications, Kybernetika,Kiev,No.6 (1978), 108-118; English transl. in Cybernetics, 14(1978).
3. V.V.Anisimov, Limit theorems for processes admitting the asymptotic consolidation of one component. Teor. Veroyatnost. i Mat.Statist. Vyp.22 (1980), 3-15; English transl. in Theor.Probab. and Math.Statist. No.22(1981), 1-13.

4. R.L.Dobrushin, The central limit theorem for nonhomogeneous Markov chains, I,II,Teor.Veroyatnost. i Primenen., I, Vyp.1,4,(1956), 72-89, 365-425; English transl. in Theor.Probab.Appl., I,(1956)

5. V.V.Anisimov, M.F.Tairov, Proximity estimates for transition characteristics of Markov chains, Abstract of Communications of the 4th International Vilnius conference on probability theory and mathematical statistics. Vol.1, Vilnius, 1985. (Russian), 28-29

Prob. Theory and Math. Stat., Vol.1, pp. 99-113
Prohorov et al. (eds)
1986 VNU Science Press

REMARKS ON LIMIT THEOREMS FOR SEQUENCES OF RANDOM VARIABLES WITH RANDOM INDEX

T.A.Azlarov, A.A.Dzamirzaev, A.G.Abdullaev
Tashkent State University, Tashkent, USSR

There are many papers (see e.g. [2 - 11]) concerning limit theorems for sequences of random variables (r.v.) with random index (r.i.). But in most of these papers in proving the so-called transfer theorems it is supposed, that the sequence of r.i. under some normalization converges to a positive r.v. This, obviously, means that the considered sequences of r.v. have the same order of growth at all elementary events of probability space. But it is not dificult to construct a sequence of r.v. which dos not satisfy this condition (see Example 1 below). Similar but more general situation occurs in practice too, e.g. in problems of reliability theory, queueing theory, sequential analysis and so on. Such an example from reliability theory can be found at the end of this paper. Therefore it is a natural problem to extend the transfer theorems to more general index sequences.

In this paper the authors introduce two classes of sequences of r.i. for which the transfer theorem are proved. The first part of this paper is divoted to the case when the weak convergence of the normalized sequence of r.i. is required and the second part is devoted to the case of convergence in probability.

Let us suppose that all r.v. considered below are defined on the same probability space $(\Omega, \mathcal{F}, P)$.

1. Let $\{\xi_n\}$ be a sequence of independent identically distributed r.v. with $M\xi_1 = m \neq 0$ and R be a class of sequences of positive integer-values r.v. $\{\nu_n\}$ which satisfy the following condition:

(R) For every $\varepsilon > 0$ there exist numbers $0<a<b<\infty$ and a sequence of numbers $\{k_n\}$ such that as $n \to \infty$ we have $k_n \to \infty$ and

$$P\{ak_n \leq \nu_n < bk_n\} \geq 1-\varepsilon$$

In [1] it is proved that

$$\frac{1}{m \cdot \nu_n} \sum_{i=1}^{\nu_n} \xi_i \xrightarrow{P} 1 \quad , \quad n \to \infty$$

for $\{\nu_n\} \in R$. In [2] using this fact the following theorem is proved.

<u>Theorem</u> ([2]). Let the following condition be satisfied:

(E) There exists a sequence of numbers $\{k_n\}$ and a distribution function (d.f.) $G(x)$ such that $k_n \to \infty$ as $n \to \infty$ and

$$P\left\{\frac{\nu_n}{k_n} < x\right\} \longrightarrow G(x) \, , \quad G(+0)=0,$$

at all continuity points of d.f. $G(x)$. Then

$$P\left\{\frac{1}{mk_n} \sum_{i=1}^{\nu_n} \xi_i < x\right\} \longrightarrow G(x), \, n \to \infty$$

at all continuity points of d.f. $G(x)$.

We note that for the following simple example the conditions (E) and (R) are not satisfied.

<u>Example 1</u>. For $A \in \mathcal{F}$, $P(A)>0$ we define ν_n as follows

$$\nu_n = \begin{cases} n_2, & \text{if } \omega \in A \\ n^2, & \text{if } \omega \in \bar{A} \end{cases}$$

We introduce a wider class of r.i. sequences, which contains the class $\mathcal{R}$.

Let M_r $(r=1,2,\dots)$ be the class of sequences of positive integer-values r.v. $\{\nu_n\}$, for which there exist a sequence of measurable partitions $B_{n1}, B_{n2}, \dots, B_{nr}$ and sequences of numbers $\{k_{ni}\}$, $i=\overline{1,r}$, such that

a) $\lim\limits_{n\to\infty} k_{ni}=\infty$, $\lim\limits_{n\to\infty} P(B_{ni})=p_i$, $i=\overline{1,r}$

b) for every $\varepsilon>0$ and for each i $(i=\overline{1,r})$ there exist numbers $0<a_i<b_i<\infty$ such that for large n we have

$$P\{a_i k_{ni}\leq \nu_n < b_i k_{ni} / B_{ni}\}\geq 1-\varepsilon,\ i=\overline{1,r} \quad (1)$$

Let $M=\bigcup\limits_{r=1}^{\infty} M_r$. If $\mathcal{E}$ is a class of r.i. sequences satisfying the condition (E), then

$$\mathcal{E}\subset\mathcal{R}\equiv M_1\subset M$$

The following theorem holds.

__Theorem 1__. Let $\{\nu_n\}\in M$ and the following condition is satisfied:

(E_1) There exist a sequence of numbers $\{k_n\}$ and a d.f. $G(x)$ such that $k_n\to\infty$ as $n\to\infty$ and

$$P\left\{\frac{\nu_n}{k_n}<x\right\}\to G(x),\ n\to\infty$$

at all continuity points of the d.f. $G(x)$.

Then

$$\underline{P}\left\{\frac{\xi_1+\dots+\xi_{\nu_n}}{m\cdot k_n}<x\right\}\to G(x)$$

at all continuity points of the d.f. $G(x)$.

At first we shall prove the following assertion which generalizes P.Revesz results [1].

Lemma 1. Let $\{\nu_n\} \in M$. Then as $n \to \infty$

$$\frac{1}{m\nu_n} \sum_{i=1}^{\nu_n} \xi_i \xrightarrow{P} 1 \tag{2}$$

Proof of the Lemma. Let us denote

$$S_n = \sum_{i=1}^{n} (\xi_i - m), \quad \eta_n = S_n/n,$$

$$N_i = [a_i k_n], \quad M_i = [b_i k_{ni}]$$

It is not hard to see that

$$\eta_{\nu_n} = \eta_{N_i} + \frac{N_i}{\nu_n}\left(\frac{S_{\nu_n} - S_{N_i}}{N_i}\right) + \eta_{N_i}\left(\frac{N_i}{\nu_n} - 1\right) \tag{3}$$

Consider events $A_{ni} = \{a_i k_{ni} \le \nu_n < b_i k_{ni}\}$

$$C_{ni} = \left\{\left|\frac{N_i}{\nu_n}\left(\frac{S_{\nu_n} - S_{N_i}}{N_i}\right) + \eta_{N_i}\left(\frac{N_i}{\nu_n} - 1\right)\right| < \frac{\varepsilon}{2}\right\} \tag{4}$$

Relation (2) is equivalent to

$$\eta_{\nu_n} \xrightarrow{P} 0 \quad \text{as} \quad n \to \infty \tag{5}$$

1) Let $\{\nu_n\} \in M_r$, where r is finite. We have

$$P\{|\eta_{\nu_n}| > \varepsilon\} = \sum_{i=1}^{r} P_{B_{ni}}(|\eta_{\nu_n}| > \varepsilon) P(B_{ni}) \tag{6}$$

were $P_B(\cdot) = P(\cdot / B)$. By a simple probability reasoning and (1), (3), (4) we have

$$P_{B_{ni}}\{|\eta_{\nu_n}| > \varepsilon\} \le P_{B_{ni}}\{|\eta_{N_i}| > \frac{\varepsilon}{2}\} + \\ + P_{B_{ni}}\{\overline{C}_{ni} \cdot A_{ni}\} + \varepsilon \tag{7}$$

According to the law of large numbers we have for large

$$P^{n}_{B_{ni}}\{|\eta_{N_i}|>\frac{\varepsilon}{2}\}\leq\frac{\varepsilon}{r\underline{P}(B_{ni})} \tag{8}$$

It is not hard to show, as in the proof of Theorem 1 [2], that

$$\underline{P}_{B_{ni}}\{\overline{C}_{ni}\cdot A_{ni}\}\leq\frac{\varepsilon}{r\underline{P}(B_{ni})} \tag{9}$$

From (6)-(9) we obtain (5).

2) Now let $\{\nu_n\}\in M_2$, where r may be infinite. For an arbitrary $\varepsilon>0$

$$\underline{P}\{|\eta_{\nu_n}|>\varepsilon\}=\sum_{i=1}^{\infty}\underline{P}_{B_{ni}}\{|\eta_{\nu_n}|>\varepsilon\}\underline{P}(B_{ni}),$$

For a given $\varepsilon>0$ we choose $N=N(\varepsilon)$, that

$$P\{|\eta_{\nu_n}|>\varepsilon\}=\sum_{i=1}^{N}P_{B_{ni}}\{|\eta_{\nu_n}|>\varepsilon\}\underline{P}(B_{ni})+\varepsilon$$

Convergence to zero of the sum in the eight hend side can be proved in the same way as in 1). Therefore (5) holds. Lemma 1 is proved.

Proof of the Theorem 1. Since

$$\frac{\xi_1+\ldots+\xi_{\nu_n}}{mk_n}=\frac{\xi_1+\ldots+\xi_{\nu_n}}{m\nu_n}\cdot\frac{\nu_n}{k_n}$$

using Lemma 1 and Kramer's lemma we obtain Theorem 1.

We note that if $\{\nu_n\}\in M$ then it can be proved that all similar results, which are proved under condition (E) (for example, Theorem 2 of [3], Theorem 1 and 2 of [4]), are also true under condition (E_1) instead of (E).

2. Let a sequence of r.v. $\{\eta_n\}$ and a sequence of positive integer-valued r.v. $\{\nu_n\}$ be given on the proba-

bility space $(\Omega, \mathcal{F}, P)$.

Definition (12). A sequence of r.v. $\{\eta_n\}$ is said to be stable if for any $B \in \mathcal{F}$ with $P(B)>0$

$$P\{\eta_n < x \mid B\} \to F_B(x)$$

holds as $n \to \infty$ at all continuity points of the d.f. $F_B(x)$.

If $F_B(x) = F(x)$ for all $B \in \mathcal{F}$, $P(B)>0$, then it is said that the sequence $\{\eta_n\}$ is a mixing sequence with the limit d.f. $F(x)$.

Now we state a well-known theorem of Guiasy [5].

Theorem (5). Let the following conditions be satisfied:

(A_1) The sequence $\{\eta_n\}$ is a mixing sequence with the limit d.f. $F(x)$;

(A_2) For every $\varepsilon > 0$ there exist $\delta > 0$ and $n_0 = n_0(\varepsilon, \delta)$ such, that for each $B \in \mathcal{F}$, $P(B)>0$, and $n > n_0$.

$$P\{\max_{|i-n| \le \delta n} |\eta_i - \eta_n| \ge \varepsilon / B\} \le \varepsilon ;$$

(C_1) There exists a sequence of numbers $\{k_n\}$ such that

$$k_n \to \infty \quad \text{and} \quad \frac{\nu_n}{k_n} \xrightarrow{P} \nu_0 \quad \text{as} \quad n \to \infty$$

where $\nu_0 > 0$ is a r.v.

Then

(D_1) $$P\{\eta_{\nu_n} < x\} \to F(x)$$

as $n \to \infty$ at all continuity points of the d.f. $F(x)$.

In a series of papers (see [6 - 10]), where limit theorems are proved for concrete sequences of r.v. with r.i., condition (C_1) are supposed to be satisfied. Obviously in Example 1 condition (C_1) are not satisfied.

Let $\mathcal{N}$ be a class of sequences of positive integer-valued r.v. for which condition (C_1) is satisfied.

We introduce a wider class of sequences of r.i. which

contains $\mathcal{N}$.

Let $\mathcal{L}_r, r=1,2,\ldots$ be a class of sequences of positive integer-valued r.v. $\{\nu_n\}$, for which there exist a partition $B_1, B_2, \ldots, B_r$ and sequences of numbers $\{k_{ni}\}$, $i=\overline{1,r}$ such that as $n \to \infty$

a) $k_{ni} \to \infty \quad , \quad i=\overline{1,r}$

b) $P\{ | \frac{\nu_n}{k_{ni}} - \nu_{0i} | > \varepsilon / B_i \} \to 0$

where $\nu_{01}, \nu_{02}, \ldots, \nu_{0r}$ are positive r.v.'s.

Let $\mathcal{L} = \bigcup_{r=1}^{\infty} \mathcal{L}_r$. Obviously $\mathcal{N} \equiv \mathcal{L}_1 \subset \mathcal{L}$. We also note that the r.v. considered in Example 1 belongs to $\mathcal{L}_2$.

We introduce the conditions:

(A_3) For an arbittary $B \in \mathcal{F}$, $P(B) > 0$,

$$\lim_{n \to \infty} P\{ \eta_n < x / B \} = F_B(x)$$

at all continuity points of the d.f. $F_B(x)$;

(C_2) $\{\nu_n\} \in \mathcal{L}$.

We shall prove the following theorems.

Theorem 2. Let condition (A_3), (A_2) and $\{C_2\}$ be satisfied. Then the following holds:

(D_2) For every $B \in \mathcal{F}$, $P(B) > 0$

$$\lim_{n \to \infty} P\{ \eta_{\nu_n} < x / B \} = F_B(x)$$

at all continuity points of the d.f. $F_B(x)$.

Theorem 3. Let conditions (A_1), (A_2) and (C_2) be satisfied. Then the sequence of r.v. $\{\eta_n\}$ is a mixing sequence with the limit d.f. .

In the same way it can be proved that all similar theorems which were proved under condition (C_1) (e.g. the main results of [6 - 10]) are also true under condition (C_2) instead of (C_1).

Before proving the theorems, we note that a weaker

variant of condition (C_1) was introduced in a paper by Aldous [11]: in [11] the condition was introduced instead of (C_1):

(C_3) For every $\varepsilon > 0$, $\delta > 0$ there exists a partition $B_1, \dots, B_N$ and a sequence of numbers $\{k_{ni}\}$, $i = \overline{1, N}$ such that

a) $\lim_{n \to \infty} k_{ni} = \infty$, $i = \overline{1, N}$

b) $\limsup_{n} \sum_{i=1}^{N} P\{|\frac{\nu_n}{k_{ni}} - 1| > \delta, B_i\} \leq \varepsilon$

In [11] the following was proved.

<u>Theorem</u> ([11]). If conditions (A_3), (A_2) and (C_3) are satisfied then (D_1) holds.

Our condition (C_2) is more general then condition (C_3) since in the case of (C_2) in each subset B_i a sequence of normalized r.i. behaves as some positive r.v. but in the case of (C_3) it behaves as a constant. In our theorem the stability of sequences of r.v. $\{\eta_{\nu_n}\}$ is proved but in the Aldous theorem in [11] the ordinary weak convergence of $\{\eta_{\nu_n}\}$ is proved. Furthermore, using the methods of this paper the Aldous theorm can be generalized as follows:

<u>Theorem 4</u>. Let (A_3), (A_2) and the following condition

(C_4) for each $\varepsilon > 0$ and $\delta > 0$ there exists a partition $B_1, \dots, B_N$ and sequences of numbers $\{k_{ni}\}$, $i = \overline{1, N}$ such, that

a) $\lim_{n \to \infty} k_{ni} = \infty$, $i = \overline{1, N}$;

b) $\limsup_{n} \sum_{i=1}^{N} P\{|\frac{\nu_n}{k_{ni}} - \nu_{0i}| > \delta, B_i\} \leq \varepsilon$,

where $\nu_{01}, \dots, \nu_{0N}$ are positive r.v.'s.

Then (D_2) holds.

Theorem 2, 3 and 4 can be proved in the same way. Therefore we prove only Theorem 2. First we state a Mogyorodi's lemma 6 in a more general form:

Lemma. Let $\{\tau_n\}$ be a sequence of r.v., τ_0 be some r.v. and let $a_1 < a_2$ be arbitrary points of continuity of the d.f. $\underline{P}\{\tau_0 < x\}$. Let $A_n = \{a_1 \le \tau_n < a_2\}$, $A_0 = \{a_1 \le \tau_0 < a_2\}$. If for some $B \in \mathcal{F}$, $P(B) > 0$ and for each $\varepsilon > 0$

$$\underline{P}\{|\tau_n - \tau_0| > \varepsilon / B\} \to 0 \quad \text{as } n \to \infty,$$

then

$$\underline{P}\{A_n \Delta A_0 / B\} \to 0 \quad \text{as } n \to \infty$$

where

$$A_n \Delta A_0 = A_n \bar{A}_0 + \bar{A}_n A_0$$

Proof. of Theorem 2. Let $G_i(x) = P\{\nu_{0i} < x\}$. From the set on points of continuity of the d.f. $G_i(x)$ we choose

$$0 < a_i = a_{i0} < a_{i1} < \ldots < a_{im} = b_i < \infty$$

such that

$$\underline{P}\{a_i \le \nu_{0i} < b_i\} > 1 - \varepsilon\, \underline{P}(A_i) \tag{10}$$

$$\max_{1 \le j \le m} (a_{i,j} - a_{i,j-1}) = \delta_{im}, \; \delta_{im} \to 0 \quad \text{as } n \to \infty \tag{11}$$

Consider $E_i = \{a_i \le \nu_{0i} < b_i\}$, $E_i^{(n)} = \{a_i k_{ni} \le \nu_n < b_i k_{ni}\}$

$$E_{ij} = \{a_{i,j-1} \le \nu_{0i} < a_{i,j}\}, \; E_{ij}^{(n)} = \{k_{ni} a_{i,j-1} \le \nu_n < k_{ni} a_{i,j}\}$$

Put for $B \in \mathcal{F}$

$$Q(x, B) = \begin{cases} 0 & \text{if } \underline{P}(B) = 0 \\ F_B(x) / \underline{P}(B) & \text{if } \underline{P}(B) > 0 \end{cases}$$

Then the stability of $\{\eta_n\}$ means

$$\underline{P}\{\eta_n < x, B\} \to Q(x, B) \quad \text{as } n \to \infty$$

at all continuity points of $Q(x, B)$.

Note that if $A \in \mathcal{F}$, $B \in \mathcal{F}$ and $AB = \emptyset$, then

$$Q(x, A+B) = Q(x, A) + Q(x, B) \tag{12}$$

1) Let $\{\nu_n\} \in \mathcal{L}_r$ and let r be a finite number. Obviously we have

$$\underline{P}\{\eta_{\nu_n} < x, B\} = \sum_{i=1}^{r} P\{\eta_{\nu_n} < x, BB_i\} \tag{13}$$

For each summand in (13) we can write

$$\underline{P}\{\eta_{\nu_n} < x, BB_i\} = \sum_{j=1}^{m} \underline{P}\{\eta_{\nu_n} < x, BB_iE_{ij}^{(n)}\} + \underline{P}\{\eta_{\nu_n} < x, BB_i\overline{E}_i^{(n)}\} \tag{14}$$

According to the definition of $\mathcal{L}_r$ and by (10) we obtain

$$\underline{P}\{\overline{E}_i^{(n)} / B_i\} \leq 2\varepsilon \tag{15}$$

Denoting

$$R_{ij}^{(n)} = \{\max_{a_{i,j-1} k_{ni} \leq k \leq a_i k_{ni}} |\eta_k - \eta_{n_{ij}}| \leq \varepsilon\}$$

one can see easily that

$$\underline{P}\{\eta_{\nu_n} < x, BB_iE_{ij}^{(n)}\} \leq \underline{P}\{\eta_{n_{ij}} < x+\varepsilon, BB_iE_{ij}^{(n)}\} + P\{B_iE_{ij}^{(n)}\overline{R}_{ij}^{(n)}\} \tag{16}$$

and

$$\underline{P}\{\eta_{\nu_n} < x, BB_iE_{ij}^{(n)}\} \geq \underline{P}\{\eta_{n_{ij}} < x+\varepsilon, BB_iE_{ij}^{(n)}\} - \underline{P}\{B_iE_{ij}^{(n)}\overline{R}_{ij}^{(n)}\}, \tag{17}$$

where $n_{ij} = [k_{ni} a_{i,j-1}]$.

Now by (14)-(17) we have

$$U_{in}^{-} - 2\varepsilon P(B_i) \le P\{\eta_{\nu_n} < x, BB_i\} \le U_{in}^{+} + 2\varepsilon P(B_i) \quad (18)$$

where

$$U_{in}^{\pm} = \sum_{j=1}^{m} \left[P\{\eta_{n_{ij}} < x \pm \varepsilon, BB_i E_{ij}^{(n)}\} \pm P(B_i E_{ij}^{(n)} \bar{R}_{ij}) \right]$$

Using the Lemma for large n we obtain

$$\left| \sum_{j=1}^{m} P(B_i E_i^{(n)} \bar{R}_{ij}^{(n)}) - \sum_{j=1}^{m} P(B_i E_{ij} \bar{R}_{ij}^{(n)}) \right| \le \varepsilon P(B_i) \quad (19)$$

It is not difficult to show that for large n and m

$$[k_{ni} a_{i,j-1}, k_{ni} a_{i,j}] \subset [n_{ij} - \delta n_{ij}, n_{ij} + \delta n_{ij}]$$

Consequently using condition (A_2) we have

$$\sum_{j=1}^{m} P(B_i E_{ij}^{(n)} R_{ij}^{(n)}) \le$$

$$\le \sum_{j=1}^{m} P\{\max_{|k-n_{ij}| \le \delta n_{ij}} |\eta_k - \eta_{n_{ij}}| > \varepsilon, B_i E_{ij}\} \le \varepsilon P(B_i) \quad (20)$$

From (19) and (20) we obtain

$$\sum_{j=1}^{m} P(B_i E_{ij}^{(n)} R_{ij}^{(n)}) \le 2\varepsilon P(B_i) \quad (21)$$

Again, using the Lemma we have

$$\left| \sum_{j=1}^{m} [P\{\eta_{n_{ij}} < x \pm \varepsilon, BB_i E_{ij}^{(n)}\} - \right.$$

$$\left. - P\{\eta_{n_{ij}} < x \pm \varepsilon, BB_i E_{ij}^{(n)}\}] \right| \le \varepsilon P(B_i) \quad (22)$$

According to condition (A_3) for suficiently large n

$$\left|\underline{P}\{\eta_{n_{ij}} < x \pm \varepsilon, BB_iE_{ij}\} - Q(x\pm\varepsilon, BB_iE_{ij})\right| \leq \varepsilon \underline{P}(B_iE_{ij}) \tag{23}$$

Using the equality (12) it is easy to show, that

$$\left|\sum_{j=1}^{m} Q(x\pm\varepsilon, BB_iE_{ij}) - Q(x\pm\varepsilon, BB_i)\right| \leq \varepsilon P(B_i) \tag{24}$$

From (22)-(24) for large n we obtain

$$\left|\sum_{j=1}^{m} P\{\eta_{ij} < x\pm\varepsilon, BB_iE_{ij}^{(n)}\} - Q(x\pm\varepsilon, BB_i)\right| < 3\varepsilon \underline{P}(B_i) \tag{25}$$

Now from (18), (22) and (25) we have

$$Q(x-\varepsilon, BB_i) - 7\varepsilon \underline{P}(B_i) \leq \underline{P}\{\eta_{\nu_n} < x, BB_i\} \leq$$

$$\leq Q(x\pm\varepsilon, BB_i) + 7\varepsilon \underline{P}(B_i)$$

Summation from $m=1$ to $i=r$ gives us the eneguality

$$Q(x-\varepsilon, B) - 7\varepsilon \leq \underline{P}\{\eta_{\nu_n} < x, B\} \leq Q(x+\varepsilon, B) + 7\varepsilon$$

Since $\varepsilon > 0$ is arbitrary the Theorem is proved in the case of finite r.

2) Let $\{\nu_n\} \in \mathcal{L}_r$, where r can be infinite. Obviously

$$\underline{P}\{\eta_{\nu_n} < x, B\} = \sum_{i=1}^{\infty} P\{\eta_{\nu_n} < x, BB_i\}$$

For an arbitrary $\varepsilon > 0$ we choose $N = N(\varepsilon)$ such that

$$\sum_{i \geq N} \underline{P}\{\eta_{\nu_n} < x, BB_i\} \leq \varepsilon$$

Then $$\sum_{i=1}^{N} \underline{P}\{\eta_{\nu_n} < x, BB_i\} \leq \underline{P}\{\eta_{\nu_n} < x, B\} \leq$$

$$\leq \sum_{i=1}^{N} \underline{P}\{\eta_{\nu_n} < x, BB_i\} + \varepsilon$$

In the same way as in 1) it can be proved that

$$\left| \sum_{i=1}^{N} P\{\eta_{\nu_n} < x, BB_i\} - Q(x,B) \right| \leq \varepsilon$$

Theorem 2 is proved.

3. Example 2. As is well known to solve statistical problems in reliability theory one-first make experiments to check the reliability of goods. According to the set of rules under which the experiment is made one says about one or another design of the experiment. Different designs of experiments are given in [13].

Consider the plan defined by the following conditions:

1) at time $t=0$ one piece of equipment is checked;

2) the piece of equipment which does not work is exchanged by identic new one;

3) if the r-th ristroy happened before time T then the experiment continues until time T. If until time T the r-th distroy have not happened then the experiment continues until $t=T^2$. Denote by ν_T the number pieces of equipment which does not work at the end of the experiment and by μ_T the number of pieces of equipment which failed before time T. Now suppose that the duration of time of work without a failure has exponential distribution with parameter λ. Then has the Puasson distribution with parameter λT (see [13]). We choose $r=[\lambda T]$. Then

$$\nu_T = \begin{cases} \mu_T & \text{if } S'_{[\lambda T]} < T \\ \mu_{T^2} & \text{if } S_{[\lambda T]} \geq T \end{cases}$$

where $S = \sum_{i=1}^{[\lambda T]} \xi_i$, $\xi_1, \xi_2, \ldots$ is the sequence of times of working of a piece of equipment without a failure. According to the Central limit theorem we have

$$\underline{P}\{ S_{[\lambda T]} < \tfrac{1}{2} \} \longrightarrow \tfrac{1}{2} \qquad \text{as} \quad T \to \infty$$

Choosing as normalization $\lambda^{-1}T^2$ we obtain

$$\underline{P}\left\{ \frac{\nu_T}{\lambda^{-1}T^2} < x \right\} = P\left\{ \frac{\mu_T}{\lambda^{-1}T^2} < x,\ S_{[\lambda T]} < T \right\} +$$

$$+ \underline{P}\left\{ \frac{\mu_{T^2}}{\lambda^{-1}T^2} < x,\ S_{[\lambda T]} \geq T \right\} \to \Psi(x) \qquad \text{as} \quad T \to \infty$$

where

$$\Psi(x) = \begin{cases} 0 & \text{if } x \leq 0 \\ \frac{1}{2} & \text{if } 0 < x \leq 1 \\ 1 & \text{if } x > 1 \end{cases}$$

It is not hard to see that $\nu_T \in \mathcal{E}$. But as it was shown $\nu_T \in M_2$ and the condition (E_1) is satisfyed. Therefore by Theorem 1

$$\lim_{T \to \infty} P\left\{ \frac{\lambda^2}{T^2} \sum_{i=1}^{\nu_T} \xi_i < x \right\} = \Psi(x)$$

REFERENCES

1. Revesz P. The law of large numbers, Akademiai Kiado, Budapest, 1967.
2. Modyorodi J., A remark on limiting distributions for sums of a random number of independent random variables, Rev.Roumaine Math.Pures.Appl. 1971, 16, 552-557.
3. Джамирзаев А.А. "Сбобщение теоремы А.Я. Хингана об относительной устойчивости", Научные труды Таш.ГУ, вып. 402 (1972), Ташкент, "Фан", с.39-44.
4. Азларов Т.А., Джамирзаев А.А. "Об относительной устойчивости для сумм случайного числа случайных вели-

чин", Изв.АН Уз.ССР, серия физ.-мат.наук, 1971, 112, с.7-14.

5. Guiasu S., On the asumptotic distribution of the sequences of random variables with random indices, Ann., Math.Stat., 1971, 42, p.2018-2028.
6. Moguorodi J., A central limit theorem for the sum of a random number of independent random variables, Publ.Math.Inst.Hung.Acad.Sci., 1962, 7, p.409-424.
7. Csörgö S. On limit distributions of sequences of random variables with random indices, Acta Math. Acad.Sci.Hung., 1974, 25, p.227-232.
8. Sreehari M., A limit theorem for the maximum of cumulative sums., Acta Math.Acad.Sci.Hung., 1968, 19, N 1-2, p.117-120.
9. Джамирзаев А.А., О свойстве перемешивания в смысле А.Реньи для положительных сумм., Acta Sci.Math., 1979, 41, с.47-53.
10. Джамирзаев А.А., Абдуллаев А.Г., Свойства перемешивания в смысле А.Реньи для статистики Пирсона., Сб. "Асимптотические задачи для вероятностных распределений", Ташкент, "Фан", 1984, с.72-78.
11. Alsous D.J., Weak convergence of randomly indexed sequences of random variables, Math.Proc.Camb.Phil., Soc., 1978, 83, p.117-126.
12. Renyi A. On stable sequence of events, Sankyha, 1963, 25, p.293-302.
13. Гнеденко Б.В., Беляев Ю.К., Соловьев А.Д., Математические методы в теории надежности., М., Наука, 1965 1965, с.524.

Prob. Theory and Math. Stat., Vol.1, pp. 115-118
Prohorov et al. (eds)
1986 VNU Science Press

CONTINUITY OF LOCAL TIME FOR MARKOV PROCESSES WITH STATIONARY INDEPENDENT INCREMENTS

M.T. Barlow
Statistical Laboratory
16 Mill Lane
Cambridge CB2 1SB
Great Britain

Let X_t be a one-dimensional Markov process with stationary independent increments. Let P^x denote the law of X starting at x ; we write P for P^0. The characteristic function of X is given by

$$E\, e^{i\theta X_t} = e^{-t\psi(\theta)},$$

where

$$\psi(\theta) = -ia\theta + \tfrac{1}{2}\sigma^2\theta^2 - \int_{-\infty}^{\infty} \left(e^{i\theta y} - 1 - \frac{i\theta y}{1+y^2}\right) \nu(dy) .$$

In Kesten (1969), and Bretagnolle (1971) it is shown that if

(1) either $\sigma^2 > 0$ or $\int (|x|\wedge 1)\, \nu(dx) = \infty$

and

(2) $$\int_{-\infty}^{\infty} \mathrm{Re}\, \frac{d\lambda}{1+\psi(\lambda)} < \infty$$

then 0 is regular for $\{0\}$; it then follows by the results of Blumenthal and Getoor (1964) that X has a jointly measurable local $L(a,t)$, $a \in \mathbb{R}$, $t \geq 0$. For each fixed a , the map $t \to L(a,t)$ is a.s. continuous, and L satisfies the density of occupation formula:

(3) for all bounded measurable f ,

$$\int_0^t f(X_s)ds = \int_{-\infty}^{\infty} f(a)\, L(a,t)da .$$

In Getoor and Kesten (1972) it is shown that there exist processes (the asymmetric Cauchy process for example) for which no jointly continuous version of $L(a,t)$ exists; the authors also pose the problem of finding necessary and sufficient conditions for the existence of a jointly continuous version of $(a,t) \to L(a,t)$.

Let

$u^1(x)$ be the 1-potential kernel density of X,

$$\phi(x)^2 = 2u^1(0) - u^1(x) - u^1(-x)$$

$$= \frac{1}{\pi}\int_{-\infty}^{\infty} (1 - \cos\theta x)\, \mathrm{Re}\, \frac{1}{1+\psi(\theta)}\, d\theta .$$

The monotone rearrangement $\bar{\phi}$ of ϕ is defined by

$$m(y) = |\{x: \phi(x) < y\}| , \quad \bar{\phi}(x) = \inf\{y: m(y) > x\} .$$

The function $\bar{\phi}$ is continuous and satisfies $\bar{\phi}(0) = 0$, $\bar{\phi}(x) > 0$ for $x \neq 0$. Set

$$I(\bar{\phi}) = \int_0^{1/e} \frac{\bar{\phi}(u)\, du}{u(\log 1/u)^{\frac{1}{2}}} .$$

Hawkes (1983) made the following conjecture:

(4) $I(\phi) < \infty$ is a necessary and sufficient condition for $L(a,t)$ to be continuous.

<u>Lemma 1</u> For each $\lambda > 0$, a, b,

(5) $$P(\sup_{s\geq 0} |\lambda\wedge L(a,s) - \lambda\wedge L(b,s)| > x) \leq 2 \exp\left[\frac{-x^2}{16\lambda\phi^2(b-a)}\right] .$$

<u>Proof</u>

(See Barlow, 1985). Let a, b be fixed, and let $\tau_t = \inf\{s: L(a,s) > t\}$. By considering the excursion process from a, it is easily seen that $L(b,\tau_t)$ is of the form

(6) $$L(b,\tau_t) = \sum_{i=1}^{N} V_i ,$$

where N is Poisson with mean $h^{-1}t$, the V_i are independent exponential r.v. with mean h, and $h = E^b L(b,T_a)$, where

$T_a = \inf\{t: X_t = a\}$.

A standard large deviation estimate now gives

$$P(\sup_{s\le\tau_t} (L(a,s) - L(b,s)) > x) \le \exp(-x^2/4th) .$$

Computations with the potential kernel density function $u^1(x)$ provide the estimate $h \le 4\phi^2(b-a)$, and the proof of the lemma is concluded by noting that

$$\{\omega: \sup_{s\le\tau_\lambda} (L(a,s) - L(b,s)) > x\} \subseteq \{\omega: \sup_{s\ge 0} (\lambda\wedge L(a,s) - \lambda\wedge L(b,s)) > x\}.$$

Feeding the estimate (5) into Dudley's theorem one obtains:

Theorem 2 (Barlow and Hawkes 1985) If $I(\bar{\phi}) < \infty$ then a continuous version of $L(a,t)$ exists.

Remark

Let $\Gamma(a,b) = u^1(a-b) + u^1(b-a)$, and Y_t be the stationary gaussian process with covariance Γ . By the Dudley-Fernique theorem (Dudley (1973), Fernique (1974)) Y has a continuous version if and only if $I(\bar{\phi}) < \infty$. Thus Hawkes' conjecture corresponds to the statement that $L(a,t)$ has a continuous version if and only if the gaussian process with the same covariance structure is continuous. See Hawkes (1986).

It is also possible to derive modulus of continuity results from (5).

Theorem 3 Let X be stable, of index $\alpha > 1$, and let

$$\rho_\alpha(x) = (|x|^{\alpha-1} \log 1/|x|)^{\frac{1}{2}}$$

$$L^*(t) = \sup_a L(a,t) .$$

Then $L^*(t) < \infty$, a.s., and there exist constants c_α , C_α , such

$$\text{(7)} \qquad c_\alpha \le \lim_{\delta\downarrow 0} \sup_{\substack{|a-b|<\delta \\ 0\le s\le t}} \frac{|L(a,s) - L(b,s)|}{L^*(t)\,\rho_\alpha(b-a)} \le C_\alpha .$$

Proof

The right hand inequality follows easily by applying Garsia's lemma to (5) (see Barlow (1985) for details). The left hand inequality is a consequence of a generalisation of Lemma 1, which will appear in Barlow (1986).

Note The author has recently established the 'necessity' side of Hawkes' conjecture (4): if $I(\bar{\phi}) = \infty$ then no continuous version of exists. The proof will appear in Barlow (1986).

References

Barlow, M.T. (1985). Continuity of local times for Lévy processes. Z. für Wahrs. 69, 23-35.

Barlow, M.T. (1986). Necessary and sufficient conditions for the continuity of local times of Lévy processes. To appear.

Barlow, M.T., and Hawkes, J. (1985). Application de l'entropie métrique à la continuité des temps locaux des processus de Lévy. C.R.A.S. Paris, 301, 237-239.

Blumenthal, R.M., and Getoor, R.K. (1964). Local times for Markov processes. Z. für Wahrs. 3, 50-74.

Bretagnolle, J. (1971). Resultats de Kesten sur les processus à accroisements indépendants. Sem de Prob. V, Lect. Notes Math. 191, 21-36. Springer, Berlin.

Dudley, R.M. (1973). Sample functions of the gaussian process. Ann. Prob. 1, 66-103.

Fernique, X. (1974). Régularité des trajectoires des fonctions aléatoires gaussiennes. Ecole d'été de probabilités de Saint-Fleur IV, Lect. Notes Math. 480, 1-96. Springer, Berlin.

Getoor, R.K., and Keston, H. (1972). Continuity of local times for Markov processes. Comp. Math. 24, 277-303.

Hawkes, J. (1983). Private communication.

Hawkes, J. (1986). Local times as stationary processes. To appear.

Kesten, H. (1969). Hitting probabilities for single points for processes with stationary independent increments. Memoir no. 93, Amer. Math. Soc.

Prob. Theory and Math. Stat., Vol.1, pp. 119-122
Prohorov et al. (eds)
1986 VNU Science Press

DETECTION AND DIAGNOSIS OF CHANGES IN THE A.R. PART OF AN A.R.M.A. MODEL WITH NONSTATIONARY UNKNOWN M.A. COEFFICIENTS

M. Basseville, A. Benveniste, G. Moustakides, A. Rougée
I.R.I.S.A. (CNRS & INRIA)
Campus de Beaulieu
35042 RENNES Cedex - France

Problem statement :

Let $$y_t = \sum_{i=1}^{p} a_i\, y_{t-i} + \sum_{j=0}^{q} b_j(t)\, e_{t-j} \tag{1}$$

be a scalar ARMA(p,q) model, where (e_t) is a gaussian white noise with constant variance σ_e^2, and where the MA coefficients $(b_j(t))_j$ are time-varying, possibly subject to jumps. The problems to be solved are :

i) given a single (long) record of observations $(y_t^0)_{0\leq t\leq s}$, estimate the AR parameters, without using the unknown time-varying MA parameters ; then, calling "reference model" the AR part characterized by these estimated parameters,

ii) given another record of $(y_t^1)_{0\leq t\leq s}$, test whether the reference AR model still fit these new observations or has changed, again without knowing the (highly) time-varying MA coefficients ("global test") ;

iii) decide which poles have changed, i.e. solve the diagnosis problem.

Notations :

For i = 0,1, let $R_k^i(s)$ be the empirical covariances :

$R_k^i(s) = \sum_{t=0}^{s-k} y_{t+k}^i\, y_t^i$ $(k \geq 0)$ and $\mathcal{H}_{p+1,N}^i(s)$ the (p+1)xN empirical Hankel matrices :

$$\mathcal{H}^{i}_{p+1,N} = \begin{pmatrix} R^{i}_{q-p+1}(s) & R^{i}_{q-p+2}(s) & \cdots & R^{i}_{q-p+N}(s) \\ R^{i}_{q-p+2}(s) & & & \vdots \\ \vdots & & & \vdots \\ R^{i}_{q+1}(s) & \cdots & \cdots & R^{i}_{q+N}(s) \end{pmatrix}$$

where $N \geq p$.

Theorem 1 |3| :

The least squares solution $\theta \triangleq (a^{0}_{i}(s))_{1 \leq i \leq p}$ of the equation :

$$(-a_p \; - a_{p-1} \; \cdots \; -a_1 \; 1) \; \mathcal{H}^{0}_{p+1,N}(s) = 0$$

is a *consistent* estimate of the true AR parameters, in the *nonstationary* case.

Precise conditions and proof (using the law of large numbers for martingales) can be found in |3|. In other words, the so-called instrumental variables method for identification is consistent even if the MA part is nonstationary, and thus solves problem i).

Let be : $U'_N(s) \triangleq (-\, a^{0}_{p} \; \cdots \; -\, a^{0}_{1} \; 1) \mathcal{H}^{1}_{p+1,N}(s)$

$$= \sum_{t=q+N}^{s} w_t \; Z_t$$

where : $w_t = y^1_t - a^0_1\, y^1_{t-1} - \cdots - a^0_p\, y^1_{t-p}$ is the "moving average" part of the record (y^1_t) and $Z_t = (y^1_{t-q-1} \cdots y^1_{t-q-N})'$. Let also be :

$$\hat{\Sigma}_N(s) = \sum_{t=q+N}^{s-q} \; \sum_{i=-q}^{q} w_t \, w_{t-i} \; Z_t \, Z'_{t-i} \; .$$

Theorem 2 |5| :

. *Nonstationary law of large numbers :* $\hat{\Sigma}_N$ is a *consistent* estimate of the true covariance matrix Σ_N of U_N, namely : $\Sigma_N^{-1}(s) \; \hat{\Sigma}_N(s) \xrightarrow{s\to\infty} I_N$, under both null hypothesis, i.e. the AR parameters are θ_0, and *local* alternative hypothesis, i.e. the parameters are $\theta_0 + \dfrac{\delta\theta}{\sqrt{s}}$, where $\delta\theta$ is fixed.

. *Central limit theorem :* $\Sigma_N(s)^{-1/2} \, . \, U_N(s) \xrightarrow{s\to\infty} \mathcal{N}(0, I_N)$ under $\mathbf{P}_{\theta_0}$

$$\Sigma_N(s)^{-1/2} \cdot (U_N(s) - \mathcal{H}^{1'}_{p,N}(s) \cdot \frac{\delta\theta}{\sqrt{s}}) \xrightarrow{s\to\infty} \mathcal{N}(0, I_N) \text{ under } \mathbb{P}_{\theta_0 + \frac{\delta\theta}{\sqrt{s}}} .$$

Precise conditions and proof (limit theorems for martingales) can be found in |5|. In other words, the use of the *local* approach for detecting changes (|4|, |6|, |2|) reduces problem ii) to the problem of detecting a change in the mean of a Gaussian process. A possible χ^2 test for solving ii) is thus :

$$t_0 = U_N' \; \hat{\Sigma}_N^{-1} \; \mathcal{H}'_{p,N} \, (\mathcal{H}_{p,N} \; \hat{\Sigma}_N^{-1} \mathcal{H}'_{p,N})^{-1} \mathcal{H}_{p,N} \; \hat{\Sigma}_N^{-1} \; U_N \qquad (2).$$

Let ψ be the set of the $m(\leq p)$ poles to be monitored, and J be the pxm Jacobian matrix $J_{p,m} = (\frac{\partial a_i}{\partial \psi_j})_{\{\psi_j\} = \psi_0}$. Then the corresponding *diagnosis test* (solving problem iii) is of the form (2) where $\mathcal{H}'_{p,N}$ is replaced by : $\mathcal{H}'_{p,N} \; J_{p,m}$ (|1|). This approach allows, for example, separate monitoring of poles or subsets of poles, without knowing in advance which poles will actually change.

Extension to the vector case-applications to offshore platforms :

All these results are also *valid* in the *vector case*. See |3| and |8| for theoretical results. In |7| can be found numerical results (on real data) concerning the identification problem i) for vibrating characteristics of offshore platforms ; the main conclusion is that high order poles and modes are correctly identified. Results of simulations concerning the test ii) and diagnosis iii) problems in the scalar case can be found in |1| ; the main conclusion is that it is possible to detect and diagnose *small changes* (1 %) in eigenfrequencies, provided the sample size is large enough. Experimentations for the vector case are described in |9| and lead to discrimination between physically different changes.

References :

1. Basseville, M., Benveniste, A., Moustakides, G. (1986). Detection and diagnosis of abrupt changes in modal characteristics of non stationary digital signals. To appear in IEEE Trans. on Information Theory.

2. Basseville, M., Benveniste, A., Editors. (1986).Detection of abrupt changes in signals and dynamical systems. LNCIS n°77, Springer-Verlag, Berlin.

3. Benveniste, A., Fuchs, J.J. (1985). Single sample modal identification of a nonstationary stochastic process. IEEE Trans. on Aut. Contr., Vol. AC-30, n°1, pp 66-74.

4. Ibragimov, I.A., Khas'minskii, R.Z. (1972). Asymptotic behavior of statistical estimators in the smooth case. 1. study of the likelihood ratio. Th. of Proba. and Appl., Vol.17, n°3, pp 445-462.

5. Moustakides, G., Benveniste, A. (1986). Detecting changes in the AR parameters of a nonstationary ARMA process. To appear in Stochastics.

6. Nikiforov, I.V. (1983). Sequential detection of abrupt changesin time series properties. Moscow : Naouka.

7. Prevosto, M., Benveniste, A., Barnouin, B. (1982). Identification of vibrating structures subject to nonstationary excitation : a nonstationary stochastic realization problem. INRIA Research Report n°130.

8. Rougée, A. (1985). Détection de changements dans les paramètres AR d'un processus ARMA vectoriel : application à la surveillance des vibrations. Thèse de 3ème Cycle, Univ. de Rennes I.

9. Basseville, M., Benveniste, A., Moustakides, G., Rougée, A. (1985) Detection and diagnosis of changes in the eigenstructure of nonstationary multivariable systems. Submitted for publication.

Prob. Theory and Math. Stat., Vol.1, pp. 123-143
Prohorov et al. (eds)
1986 VNU Science Press

NONPARAMETRIC ESTIMATION OF DISTRIBUTION FUNCTIONS BASED ON INCOMPLETE DATA

Yu.K.Belyayev
Moscow State University, Moscow, USSR

In [1], the so called multiplicative intensity model was suggested, which generalized previous models of statistical data such as the independent competing riscs model [2] and various models of life testing of elements without replacdment [3]. The properties of statistical procedures in the multiplicative intensity model were investigated by martingale methods [4, 5]. At the same time it was pointed out that many models important for applications cannot be studied within the multiplicative intensity model [6]. In [7-9] we outlined an approach connected with a notion of discrete factorization (see definition 2 below). The models of obtaining statistical data to be considered in the present paper are much more general then the multiplicative intensity model.

Definition 1. A portion of data $\underline{x}_j$ is a collection of (generally mutually dependent) random values $\underline{x}_j = \{s_{ij}, \gamma_{ij}, i = 1, \dots, i_j\}$, where j -is the number of the portion, s_{ij} are real numbers, γ_{ij} are labels, $\gamma_{ij} \in \{\pm 1, \dots, \pm r\}$. A file of data is obtained when mutially independent portions of data $\underline{x}_1, \dots, \underline{x}_m$ are united.

We shall consider the case where all $s_{ij} > 0$. Then the values s_{ij} can be regarded as observed life times. Labels γ_{ij} give us types of observations. Right censoring corresponds to negative labels. Portions $\underline{x}_j$

can have different distributions. The representation of the file of data $\underline{X}_m = \{\underline{x}_1, \cdots, \underline{x}_m\}$ as a union of mutially independent portions of data is typical for many problems in applied statistics.

Example. Let Y_s be a semimarkov process with a finite number of states $\mathcal{E} = \{1, \cdots, c\}$ determined by an irreducible matrix of transition probabilities $P = (p_{ab})$ and distribution functions (d.f.) $F_{ab}(u)$ of transition times from the state a to the state b. We number all pairs a, b with $p_{a,b} > 0$ by natural numbers from 1 to r, where r is the number of all such pairs. Let us write $(a, b) \sim \ell$ if the number of the pair a, b is ℓ. We write then $F_\ell(u)$ instead of $F_{ab}(u)$. Suppose that the initial state is a_{j_1}, i.e. $Y_0 = a_{j_1}$. The variable s_{1j} is the time of staying at the state a_{j_1}. If a_{j_2} is the next state then s_{2j} is the time of staying at the state a_{j_2} and so on. Just after each jump we know the next state. Then an observation of a realisation Y_s, $s \in [0, T_j]$ corresponds to a portion $\underline{x}_j$ in which s_{ij} is the complete transition time from the state a_{j_i} to the state $a_{j_{i+1}}$, if $\gamma_{ij} = \ell$ and $(a_{j_i}, a_{j_{i+1}}) \sim \ell$, s_{ij} is the incomplete time of observation of the transition from a_{j_i} to $a_{j_{i+1}}$ when $\gamma_{ij} = -\ell$. The interruption of observation arises because the transition from a_{j_i} to $a_{j_{i+1}}$ started before T_j was not finished by the time T_j. In such portion, only one label has negative sign.

The model of obtaining statistical data described in this example was investigated in [10].

Denote by $\mathcal{X}_j$ a set of all possible portions $\underline{x}_j$, by $\mathcal{L}_j$ the σ-algebra of subsets of $\mathcal{X}_j$, generated by subsets $\{s_{ij} \leq y_j, \gamma_{ij} = \ell_i, i \leq k_j = k\}$. Denote by $\mathcal{P}_j = \{P_{\underline{F}, \underline{\theta}, j}\}$ a family of distributions on $(\mathcal{X}_j, \mathcal{L}_j)$, each distribution $P_{\underline{F}, \underline{\theta}, j}$ being determined by a collection of d.f.'s $\underline{F}=\{F_1(s), \cdots, F_r(s)\}$ and a nuisance parameter $\underline{\theta}$, which also can be a collection of d.f.'s. Let $1(A)$ be the indicator of an event A. For any function $f(t)$ without discontinuities of the second kind we denote the jump at a point t by $\Delta f(t)$, $\Delta f(t) = f(t+) - f(t-)$.

Definition 2. The family $\mathcal{P}_j$ is called F_ℓ-descretely factorized, if (i) for every $\underline{x}_j \in \mathcal{X}_j$ one can find $\underline{F}, \underline{\theta}$ such that $P_{\underline{F},\underline{\theta},j}(\underline{x}_j) > 0$ and

$$P_{\underline{F},\underline{\theta},j}(\underline{x}_j) = \prod_{i=1}^{i_j} \left[\Delta F(s_{ij})\right]^{1(\gamma_{ij}=\ell)} \left[\Delta \bar{F}_\ell(s_{ij})\right]^{1(\gamma_{ij}\in A_\ell)} \cdot C_{\underline{F}',\underline{\theta}}(\underline{x}_j), \quad (1)$$

where $\bar{F}_\ell(s) = 1 - F_\ell(s)$, $A_\ell \subset \{\pm 1, \cdots, \pm r\} \setminus \ell$, and $C_{\underline{F}',\underline{\theta}}(\underline{x}_j)$ does not depend on the d.f. $F_\ell(s)$, $\underline{F}_\ell' = \{F_1(s), \cdots, F_{\ell-1}(s), F_{\ell+1}(s), \cdots, F_r(s)\}$, (ii) for every sizes of jumps $\Delta F_\ell(s_{ij}) > 0$, $\sum_{i=1}^{i_j} \Delta F_\ell(s_{ij}) \le 1$, there are $\underline{F}$, $\underline{\theta}$ which satisfy condition (i).

The family $\mathcal{P}_j$ is called discretely factorized if it is F_ℓ-discretely factorized for each $\ell = 1, 2, \cdots, r$.

The portion of data obtained by observation of a semimarkov process Y_s, $s \in [0, T_j]$ in the above example corresponds to a discretely factorized family $\mathcal{P}_j$ if the d.f.'s of transition times $F_{a,b}(s) = F_\ell(s)$, $(a,b) \sim \ell$, can be arbitrary, and $A_\ell = \{-\ell\}$. The structure of A_ℓ depends on the model of obtaining statistical data $\underline{x}_j$. In the model with r independent competing riscs, $\gamma_{ij} \in (1, \cdots, r)$, $A_\ell = \{1, \cdots, \ell-1, \ell+1, \cdots, r)$.

To a portion $\underline{x}_j$ and a value $\ell \in \{1, \cdots, r\}$, there corresponds a nonincreasing left continuous stepwize process

$$N_{\ell j}(u) = \sum_{1 \le i \le i_j} 1(s_{ij} \ge u, \gamma_{ij} \in A_\ell \cup \ell). \quad (2)$$

and a nondecreasing right continuous stepwise process

$$\mathcal{D}_{ej}(u)=\sum_{1\le i\le i_j} 1\,(s_{ij}\le u,\ \gamma_{ij}=e) \tag{3}$$

To a file of data $X_m=\{x_1,\dots,x_m\}$ there correspond stepwise processes

$$N_e(m,u)=\sum_{j=1}^{m} N_{ej}(u),\quad \mathcal{D}_e(m,u)=\sum_{j=1}^{m}\mathcal{D}_{ej}(u) \tag{4}$$

It follows from (2)-(4) that the set of values u which correspond to all jumps of $\mathcal{D}_e(m,u)$ is a subset of the set $\{z_{e_1}(m),\dots,z_{e_n}(m)\}$ of all jumps of the function $N_e(m,u)$. Let $\Delta\mathcal{D}_e(m,z_{ek}(m))=$ $=f_{ek}(m),\ |\Delta N_e(m,z_{ek}(m))|=c_{ek}(m)+f_{ek}(m)$. The statistical information on $F_e(s)$ contained in the file of data $\underline{X}_m$ can be represented as an ordered statistical table

$$\mathcal{D}_e(\underline{X}(m))=\begin{pmatrix} z_{e_1}(m), & \dots, & z_{e_n}(m) \\ f_{e_1}(m), & \dots, & f_{e_n}(m) \\ c_{e_1}(m), & \dots, & c_{e_n}(m) \end{pmatrix} \tag{5}$$

This representation of data is convenient for numerical calculations.

It is possible to use generalized principle of maximum likelihood for nondeminated families of probability distributions, see [8, 11], to obtain estimators for d.f.'s $F_e(s)$, or for tails of d.f.'s $\overline{F_e}(s)$, $e=1,\dots,r$.

<u>Theorem 1.</u> If each portion $\underline{x}_j$ of the file of data $\underline{X}_m$ has F_e -discretely factorized family $\mathcal{P}_j$,

then generalized maximum likelihood estimator for $\bar{F}_e(s)$ can be represented as

$$\hat{\bar{F}}_{em}(s) = \prod_{k=1}^{n(s)} \left(1 - \frac{f_{ek}(m)}{N_e(m,u)}\right) , \qquad (6)$$

where $n(s) = \max\{k : z_{ek}(m) \leq s\}$

Proof is based on the formula (1) and is similar to the proof of the formula (1.10) in [8].

Definition 3. The estimator $\hat{F}_{em}(s) = 1 - \hat{\bar{F}}_{em}(s)$, where $\hat{\bar{F}}_{em}(s)$ is defined by (6), is called a Product-Limit estimator (PL -estimator).

We denote the expectation of a random variable ξ by $\mathcal{E}\xi$, $\mathcal{E}N_{ej}(u) = k_{ej}(u)$, $k_e(m,u) = \mathcal{E}N_e(m,u)$, $\Lambda_e(u) = -\ln \bar{F}_e(u)$, $s_F = \sup\{s : \bar{F}(s) > 0\}$, $s_0 \leq s_F$.

Let us introduce now the following conditions.

$(\mathcal{Y}_0)$ D.f.'s $F_e(s)$ are continuous for $s < s_0$

$(\mathcal{Y}_1)$ For an $\varepsilon > 0$ and all j, $\mathcal{E}(N_{ej}(0))^{2+\varepsilon} \leq C_{2+\varepsilon} = C_{F,\theta}(\varepsilon) < \infty$

$(\mathcal{Y}_1')$ For all j, $\mathcal{E}N_{ej}(0)^2 \leq C_2 = C_{F,\theta} < \infty$.

$(\mathcal{Y}_2)$ There is a function $k_0(s) > 0$, $s < s_0$, such that $k_e(m,s) \geq m\,k_0(s)$ for all sufficiently large m.

$(\mathcal{Y}_3)$ For all j and $s < s_0$

$$\mathcal{E}D_{ej}(s) = \int_0^s \frac{k_{ej}(u)\,dF_e(u)}{\bar{F}_e(u)} \qquad (7)$$

$(\mathcal{Y}_4)$ The processes

$$M_{ej}(s) = D_{ej}(s) - \int_0^s N_{ej}(u)\,\frac{dF_e(u)}{\bar{F}_e(u)} \qquad (8)$$

have uncorrelated incremenets,

$(\mathcal{Y}_5)$ All processes $M_{ej}(s)$, $\ell=1,\dots,r$, $j=1,2,\dots$ are mutially uncorrelated .

Let

$$M_\ell(m,u)=\sum_{j=1}^{m} M_{\ell j}(u),\quad J_\ell(m,u)=1\!\!1\big(N_\ell(m,u)=0\big)$$

Now we investigate asymptotic properties of PL - estimators (6), as $m\to\infty$, which corresponds to the accumulation of statistical data.

Theorem 2. If conditions $(\mathcal{Y}_0)$, $(\mathcal{Y}_1')$, $(\mathcal{Y}_2)$, $(\mathcal{Y}_3)$ are satisfied then $\hat{F}_{\ell m}(s)$ for $s<s_0$ is a consistent estimator for $F_\ell(s)$ as $m\to\infty$.

For the sake of brevity we shall write P instead of $P_{\underline{F},\underline{\theta}}$ and $\mathcal{E}$ instead of $\mathcal{E}_{\underline{F},\underline{\theta}}$. The following lemmas will be used in the proof of Theorem 2.

Lemma 1. If conditions $(\mathcal{Y}_1')$ and $(\mathcal{Y}_2)$ are satisfied then

$$mP\{N_\ell(m,s)=0\}\to 0 \qquad \text{as } m\to\infty.$$

Proof. By the independence of portions $\underline{x}_j$ we have

$$P\{N_\ell(m,s)=0\}=\prod_{j=1}^{m}\{1-P\{N_{\ell j}(s)>0\}\}\le$$

$$\le \exp\Big\{-\sum_{j=1}^{m}P\{N_{\ell j}(s)>0\}\Big\} \tag{9}$$

Condition $(\mathcal{Y}_1')$ and the inequality $\mathcal{E}N_{\ell j}(s)\le \{\mathcal{E}N_{\ell j}^2\}^{1/2}\le\sqrt{C_2}$ imply $K_0(s)\le\sqrt{C_2}$. Let $A_{\ell m}=\{j;\ \mathcal{E}N_{\ell j}(s)\ge \frac{K_0(s)}{2}\}$ let $|A_{\ell m}|$ be the number of elements in the set A_{em} . We obtain from condition $(\mathcal{Y}_2)_m$ that

$$m k_0(s) \le \sum_{j=1}^{m} \mathcal{E} N_{e_j}(s) \le \frac{k_0(s)}{2}(m-|A_{em}|)+|A_{em}|\sqrt{C_2}$$

Consequently

$$|A_{em}| \ge m\left[k_0(s) / (2\sqrt{C_2} - k_0(s))\right] \tag{10}$$

The inequality

$$\left[\mathcal{E} N_{e_j}(s)\right]^2 = \left[\mathcal{E} N_{e_j}(s)\, 1(N_{e_j}(s)>0)\right]^2 \le$$

$$\le \mathcal{E} N_{e_j}^2(s)\, P\{N_{e_j}(s)>0\} \le C_2 P\{N_{e_j}(s)>0\}$$

and inequality (10) imply

$$\sum_{j=1}^{m} P\{N_{e_j}(s)>0\} \geqslant C_2^{-1} \sum_{j=1}^{m} \left[\mathcal{E} N_{e_j}(s)\right]^2 \geqslant$$

$$\geqslant C_2^{-1}\left(k_0(s)/2\right)^2 |A_{em}| \ge m C_2^{-1} k_0^3(s) / (8\sqrt{C_2} - 4k_0(s)) \tag{11}$$

Now lemma 1 follows from (9) and (11).

Lemma 2. If conditions $(\mathcal{Y}_1')$, $(\mathcal{Y}_2)$ are satisfied then for any $\varepsilon > 0$

$$P\left\{\left|\frac{N_e(m,s) - k_c(m,s)}{m}\right| > \varepsilon\right\} \longrightarrow 0 \quad ,$$

$$P\left\{ \frac{N_e(m,s)}{m} \geq K_0(s) - \varepsilon \right\} \longrightarrow 1,$$

as $m \to \infty$.

Proof. We obtain this statement by Tchebyshev's inequality and the inequality $K_e(m,s)/m \geq K_0(s) - \varepsilon$ which follows from condition $(\mathcal{Y}_2)$.

Lemma 3. If conditions $(\mathcal{Y}_1')$, $(\mathcal{Y}_2)$ are satisfied then the following inequality holds for $\delta \in (0, 1/16)$ and $C_3 = 54\, C_2^{3/2}$

$$P\left\{ \sup_{0 \leq u \leq s} \left| N_e(m,u) - K_e(m,u) \right| > m^{3/4-\delta} \right\} \leq C_3 m^{-\delta} \quad (12)$$

Proof. Let $K_e(m,u+) = \lim_{h \downarrow 0} K_e(m, u+h)$, $N_e(m,u+) = \lim_{h \downarrow 0} N_e(m,u+h)$. Condition $(\mathcal{Y}_1')$ implies the inequality $K_e(m,u) \leq \sum_{j=1}^{m} \mathcal{E} N_{ej}(0) \leq m\sqrt{C_2}$. In the strip $[0,\infty) \times [0, m\sqrt{C_2}]$, consider levels

$$\mathcal{U}_e = \{(u, \tfrac{1}{3} e m^{1/2+3\delta}) : u \geq 0\},$$

$e = 0, 1, \ldots, [3\sqrt{C_2}\, m^{1/2-3\delta}]$, where $[a]$ is the integer part of a. To every level $\mathcal{U}_e$ there corresponds the number

$$u_{me} = \inf\{u : K_e(m,u) \geq \tfrac{1}{3} e m^{1/2+3\delta} \geq K_e(m,u+)\},$$

$$u_{m,e+1} \leq u_{me}.$$

Consider intervals $\Delta_{me} = [u_{m,e+1}, u_{me}] \cap [0,s] \neq \emptyset$. The union of these intervals is the interval $[0, s^*]$ which contains not more then $[3\sqrt{C_2}, m^{1/2-3\delta}]$ points u_{me}. Define the event

$$A_m(\delta, s) = \left\{ \max_{u_{me} \in [0, s^*]} | N_e(m, u_{me}) - k_e(m, u_{me}) | > \right.$$

$$\left. > \tfrac{1}{3} m^{3/4 - \delta} \right\} \cup$$

$$\cup \left\{ \max_{u_{me} \in [0, s^*]} | N_e(m, u_{me}+) - k_e(m, u_{me}^+) | > \tfrac{1}{3} m^{3/4-\delta} \right\}$$

We, obtain using Tchebyshev's inequality and the inequality $\mathcal{E}[N_e(m,u) - k_e(m,u)]^2 \leq m C_2$, that

$$P\{A_m(\delta, s)\} \leq \sum_{u_{me} \in [0, s^*]} P\{ | N_e(m, u_{me}) - k_e(m, u_{me}) | >$$

$$> \tfrac{1}{3} m^{3/4-\delta} \} +$$

$$+ \sum_{u_{me} \in [0, s^*]} P\{ | N_e(m, u_{me+}) - k_e(m, u_{me}^+) | \} > \tfrac{1}{3} m^{3/4 - \delta} \leq$$

$$\leq 2 \cdot 3 \sqrt{C_2} \; m^{1/2 - 3\delta} \frac{m C_2}{(1/3)^2 m^{3/2 - 2\delta}} = C_3 m^{-\delta}$$

If the event $A_m^c(\delta, s)$ (the complement of $A_m(\delta, s)$) occurs then the graph of the nonincreasing function $N_e(m,u)$ on each interval $(u_{m,e+1}, u_{me}]$, $u_{m,e+1} > u_{me}$, $u_{me} \leq s$ lies in the rectangle

$$(u_{m,\ell+1}, u_{m\ell}] \times \left[\tfrac{1}{3}(\ell-1)m^{1/2+3\delta} - \tfrac{1}{3}m^{3/4-\delta}, \tfrac{1}{3}\ell m^{1/2+3\delta} + \tfrac{1}{3}m^{3/4-\delta}\right]$$

Consequently if the event $A_m^c(\delta, s)$ occurs and $\delta \in (0, 1/16)$ we have on each interval $(u_{m,\ell+1}, u_{m\ell}]$

$$\sup_{u_{m,\ell-1} \le u \le u_{m\ell}} |N_\ell(m,u) - k_\ell(m,u)| \le 2 \cdot \tfrac{1}{3} m^{3/4-\delta} + \tfrac{1}{3} m^{1/2+3\delta} \le m^{3/4-\delta}$$

Thus we have $A_m^c(s,\delta) \subseteq \left\{\sup_{0 \le u \le s^*} |N_\ell(m,u) - k_\ell(m,u)| \le m^{3/4-\delta}\right\}$

This implies

$$P\left\{\sup_{0 \le u \le s} |N_\ell(m,u) - k_\ell(m,u)| > m^{3/4-\delta}\right\} \le P\left\{A_m(\delta, s)\right\} \le C_3 m^{-\delta}$$

<u>Proof of Theorem 1</u>. It follows from condition (Y_0) that the process $\mathcal{D}_\ell(m,u)$ defined by (4) has jumps only of size 1, i.e. all $f_{\ell k}(m) = 1$. So we can write the PL -estimator for $\bar{F}_{\ell m}(s)$ in the following form

$$\bar{F}_{\ell m}(s) = \exp\left\{-I_{\ell m}(s)\right\}, \qquad (13)$$

where

$$I_{\ell m}(s) = -\sum_{k=1}^{n(s)} \ln\left[1 - \frac{1}{N_\ell(m, t_{\ell k}(m))}\right]$$

It is easy to check that the following identity is true

$$I_{em}(s) = \Lambda_e(s) + \int_0^s \frac{dM_e(m,u)}{K_e(m,u)} -$$

$$- \int_0^s \frac{N_e(m,u) - k_e(m,u)}{k_e^2(m,u)} \, dM_e(m,u) + \tag{14}$$

$$+ \int_0^s \frac{[N_e(m,u) - k_e(m,u)]^2}{k_e^2(m,u)\, N_e(m,u)} \, dM_e(m,u) - \int_0^s [1 - J_e(m,u)] \frac{dF_e(u)}{\bar{F}_e(u)} -$$

$$- \int_0^s \left\{ \ln\left(1 - \frac{1}{N_e(m,u)}\right) + \frac{1}{N_e(m,u)} \right\} dD_e(m,u)$$

Since $\bar{F}_e(s) = \exp\{-\Lambda_e(s)\}$, we see that the statement of Theorem 1 will follow from the convergence to zero in probability of all integrals in the right hand side of (14) as $m \to \infty$. We have from condition (Y_2) and Campbell's theorem [12] that the expectation of the first integral in (14) is equal to

$$\mathcal{E} \int_0^s \frac{dM_e(m,u)}{K_e(m,u)} = \sum_{j=1}^{m} \left\{ \mathcal{E} \int_0^s \frac{dD_{ej}(u)}{K_e(m,u)} - \int_0^s \frac{N_{ej}(u)\, dF(u)}{K_e(m,u)} \right\} = 0$$

We have from condition (Y_2) that

$$\int_0^s \frac{dM_{je}(u)}{K_e(m,u)} \leq \frac{N_{e_j}(0)[1 + \Lambda_e(s)]}{m k_0(s)}$$

for all sufficiently large m . Using this inequality we obtain the following inequality for the variance of the first integral in (14)

$$V_m = \mathcal{E}\left\{\int_0^s \frac{dM_e(m,u)}{K_e(m,u)}\right\}^2 = \sum_{j=1}^{m} \mathcal{E}\left\{\int_0^s \frac{dM_{ej}(u)}{K_e(m,u)}\right\}^2 \leq$$

$$\leq \frac{m\sqrt{C_2}\,[1+\Lambda_e(s)]}{m^2 K_0(s)} \longrightarrow 0 .$$

Consequently the first integral converges to zero in probability as $m \to \infty$. The second and the third integrals in (14) can be estimated similarly. Thus we consider only the third integral. Note that the variation of the process $M_e(m,u)$ on $[0,s]$ is not greater than $N_e(m,0)\times[1+\Lambda_e(s)]$. Both functions $N_e(m,u)$, $K_e(m,u)$ are non-increasing. Using condition (Y_2) we obtain

$$\int_0^s \frac{[N_e(m,u)-K_e(m,u)]^2}{K_e^2(m,u)N_e(m,u)}\, dM_e(m,u) \leq$$

$$\leq \sup_{0\leq u\leq s} |N_e(m,u)-K_e(m,u)|^2 \times$$

$$\times \frac{N_e(m,0)[1+\Lambda_e(s)]}{N_e(m,s)\, m^2 K_0^2(s)} =$$

$$= \frac{1}{m^{1/2+\delta}} \left\{ \frac{\sup_{0 \le u \le s} |N_e(m,u) - K_e(m,u)|}{m^{3/4-\delta/2}} \right\}^2 \frac{[N_e(m,0)/m]}{[N_e(m,s)/m]} \frac{1+\Lambda_e(s)}{K_0^2(s)} \quad (15)$$

It follows from Lemmas 2 and 3 that the right hand side of (15) converges to zero in probability as $m \to \infty$. The convergence to zero in probability of the fourth integral in (14) follows from Lemma 1. To estimate the last integral in (14), we use the inequality

$$2 \left| \ln(1-x) + x \right| \le x^2/(1-x) \quad , \quad 0 < x < 1$$

and obtain

$$\left| \int_0^s \left[\ln\left(1 - \frac{1}{N_e(m,u)}\right) + \frac{1}{N_e(m,u)} \right] d\mathcal{D}_e(m,u) \right| \le$$

$$\le \frac{1}{2m} \frac{[N_e(m,0)/m]}{[N_e(m,s)/m]\left[[N_e(m,s)-1]/m\right]} \quad (16)$$

It follows from lemma 2 that the right hand side of (16) converges to zero as $m \to \infty$. The proof of Theorem 1 is completed. The accuracy of PL -estimators can be described by corresponding asymptotic γ-confident intervals or γ-confident bands. The deviation of the PL -estimator from the true d.f. is

$$\hat{F}_e(s) - F_{em}(s) = \hat{\bar{F}}_{em}(s) \left\{ e^{I_{em}(s) - \Lambda_e(s)} - 1 \right\}, \quad (17)$$

thus we have to investigate the asymptotic behaviour of the difference $I_{em}(s) - \Lambda_m(s)$ as $m \to \infty$ where

$I_{em}(s)$ is defined by (14). It will be shown that the deviation (17) is determined by the integral

$$Q_e(m,s) = \sqrt{m} \int_0^s \frac{dM_e(m,u)}{K_e(m,u)} = \sum_{j=1}^{m} q_{ejm}(s), \tag{18}$$

where

$$q_{ejm}(s) = \sqrt{m} \int_0^s \frac{dM_{ej}(u)}{K_e(m,u)}, \quad j = 1,\dots,m$$

Lemma 4. If conditions $(Y_0) - (Y_4)$ are satisfied, then the variances of $q_{ejm}(s)$ and $Q_e(m,s)$ are equal respectively

$$\sigma^2_{ejm}(s) = m \int_0^s \frac{K_{ej}(u)\, dF_e(u)}{K_e^2(m,u)\bar{F}_e(u)}, \quad V_e(m,s) = m \int \frac{dF_e(u)}{K_e(m,u)\bar{F}_e(u)} \tag{19}$$

We omit the proof note only that it consists in the division of the interval $[0, s]$ into subintervals $\left(\frac{e-1}{n}s, \frac{e}{n}s\right)$ and investigation of the asymptotic behaviour of

$$\mathcal{E}\left[q_{ejm}\left(\frac{e}{n}s\right) - q_{ejm}\left(\frac{e-1}{n}s\right)\right]^2 \quad \text{as } n \to \infty$$

Lemma 5. If conditions $(Y_0) - (Y_4)$ are satisfied

$$\sigma^2_{ejm}(s) \le \frac{\Lambda_e(s)}{m k_0^2(s)}, \quad \frac{\Lambda_e(s)}{\sqrt{C_2}} \le V_e(m,s) \le \frac{\Lambda_e(s)}{k_0(s)}.$$

Proof. These inequalities are obtained from (19).the fact that the functions $k_{ej}(u)$, $k_e(m,u)$ are noninc- reasing the inequality in (Y_2) and the inequality

$k_e(m,u) \le m\sqrt{C_2}$ which follows from condition (Y_1').
Thus using lemma 5 we have

$$\frac{\sigma^2_{ejm}(s)}{V_e(m,s)} \le \frac{\sqrt{C_2}}{m k_o^2(s)}$$

Hence we can regard $\sqrt{m}\, Q_e(m,s)$ as a sum of m multyally independent summands with uniformly small variances.

Lemma 6. If conditions $(Y_0)-(Y_4)$ are satisfied then

$$\lim_{m\to\infty} P\left\{\frac{Q_e(m,s)}{\sqrt{V_e(m,s)}} \le y\right\} = \Phi(y)$$

where

$$\Phi(y) = \frac{1}{\sqrt{2\pi}}\int_{-\infty}^{y} e^{-z^2/2}\,dz$$

Proof. By the general theory for sums of independent random variables with uniformly small variances, we have to check the Lindeberg condition [13, 14], i.e. to check that for every $\varepsilon > 0$

$$V_e^{-1}(m,s)\sum\nolimits_1(l,m,\varepsilon,s) \to 0 \quad \text{as } m\to\infty \qquad (20)$$

where

$$\sum\nolimits_1(l,m,\varepsilon,s) = \sum_{j=1}^{m} \mathcal{E} q^2_{ejm}(s)\, 1\left(|q_{ejm}(s)| \ge \varepsilon V_e^{1/2}(m,s)\right)$$

From lemma 5 we have $V_e^{-1}(m,s) \le \sqrt{C_2}/\Lambda_e(s)$. Consequently if (20) does not hold then

$$\limsup_{m\to\infty} \sum\nolimits_1(l,m,\varepsilon,s) > 0$$

Then we can find a subsequence $\{m_n\}$, $m_n \to \infty$, $n \to \infty$ and a number $\delta(\varepsilon) > 0$ such that

$$\sum (\ell, m_n, \varepsilon, s') \geq \delta(\varepsilon)$$

This implies that for some index $j_n \leq m_n$ the following inequality holds

$$\frac{\delta(\varepsilon)}{m_n} \leq \mathcal{E} q^2_{e_{j_n} m_n}(s)\, 1\left(|q_{e_{j_n} m_n}(s)| > \varepsilon V_\ell^{1/2}(m_n, s)\right). \quad (21)$$

The inequalities

$$q_{e_j m}(s) \leq \frac{N_{e_j}(0)[1 + \Lambda_e(s)]}{\sqrt{m}\, k_0(s)} \text{ and } V_\ell(m,s) \geq \frac{\Lambda_e(s)}{\sqrt{C_2}},$$

and (21) for some $c' > 0$ imply

$$\mathcal{E} N^2_{e_{j_n}}(0)\, 1\left(N_{e_{j_n}}(0) > c'\sqrt{m_n}\right) \geq c' > 0 \quad (22)$$

From Hölder's and Tchebishev's inequalities we obtain

$$\mathcal{E} N^2_{e_{j_n}}(0)\, 1\left(N_{e_{j_n}}(0) > c'\sqrt{m_n}\right) \leq$$

$$\leq \left\{\mathcal{E} N^{2+\varepsilon}_{e_{j_n}}(0)\right\}^{2/(2+\varepsilon)} P\left\{N_{e_{j_n}}(0) > c'\sqrt{m_n}\right\}^{\varepsilon/(2+\varepsilon)} \leq$$

$$\leq \left\{\mathcal{E} N^{2+\varepsilon}_{e_{j_n}}(0)\right\}^2 \Big/ \left(c'^{\varepsilon} m_n^{3/2}\right) \to 0 \text{ as } n \to \infty$$

which contradicts (22). Thus condition (20) is fulfilled. The proof of Lemma 5 is completed.

Theorem 2. If conditions $(Y_0)-(Y_4)$ are satisfied then

$$\lim_{m\to\infty} P\left\{\sqrt{\frac{m}{V_e(m,s)}}\;\frac{\hat{F}_{em}(s)-F_e(s)}{\hat{\hat{F}}_{em}(s)}\leq y\right\}=\Phi(y)$$

Proof. The analysis of the behaviour of the integrals in (14) shows that

$$\alpha_e(m,s)=\sqrt{\frac{m}{V_e(m,s)}}\left[I_m(s)-\Lambda_e(s)\right]-\frac{Q_e(m,s)}{\sqrt{V_e(m,s)}}\xrightarrow{P}0$$

as $m\to\infty$ where $\xrightarrow{P}$ denotes the convergence in probability. From (17) we obtain the following identity

$$Z_e(m,s)=\sqrt{\frac{m}{V_e(m,s)}}\cdot\frac{\hat{F}_{em}(s)-F_e'(s)}{\hat{\hat{F}}_{em}(s)}=$$

$$=\left[\frac{Q_e(m,s)}{\sqrt{V_e(m,s)}}+\alpha_e(m,s)\right]\left[1+\beta_e(m,s)\right]$$

where

$$\beta_e(m,s)=\left[I_{em}(s)-\Lambda_e(s)\right]^{-1}\left\{e^{I_{em}-\Lambda_e(s)}-1-\left[I_{e_m}(s)-\Lambda_e(s)\right]\right\}.$$

In the proof of Theorem 1, we have shown that $I_{e_m}(s)\xrightarrow{P}\Lambda_e(s)$ as $m\to\infty$. Thus $\beta_e(m,s)\xrightarrow{P}0$ as $m\to\infty$. Consequently $Z_e(m,s)$ and $Q_e(m,s)/\sqrt{V_e(m,s)}$ have the same limit distribution as $m\to\infty$. The theorem follows now from Lemma 6.

Under the conditions of Theorem 1, a consistent estimator for $V_e(m,s)$ is

$$\hat{V}_e(m,s) = m \sum_{k=1}^{n(s)} \frac{1}{N_e^2(m, z_{ek}(m))}$$

Thus we have by Theorem 2 that

$$\left(\hat{\bar{F}}_{em}(s)\left[1-u_{\frac{1-\gamma}{2}}\sqrt{\hat{V}_e(m,s)/m}\,\right], \hat{\bar{F}}_{em}(s)\left[1+u_{\frac{1-\gamma}{2}}\sqrt{\hat{V}_e(m,s)/m}\,\right]\right)$$

is an asymptotically γ -confident interval for $\bar{F}_e(s)$ where $\Phi(-u_p)=p$, $p=(1-\gamma)/2$

Definition 4. A sequence of n -dimensional d.f.'s $\{G_m(\underline{x}), \underline{x}\in R^n\}$ weakly approaches the sequence of n -dimensional d.f.'s $\{H_m(\underline{x}), \underline{x}\in R^n\}$ if

$$\sup_{\underline{x}\in R^n} |G_m(\underline{x}) - H_m(\underline{x})| \to 0 \qquad \text{as } m\to\infty$$

Let $W_1(t), \dots, W_r(t)$ be mutually independent Wiener processes, $\mathcal{E}\, W_i(t)=0$, $\mathcal{E}\, W_i^2(t)=t$, $t\geq 0$

One can prove that it conditions $(\mathcal{Y}_0) - (\mathcal{Y}_5)$ are satisfied then the n -dimensional d.f.'s of the sequence of random processes $\{Y_1(m,s), \dots, Y_r(m,s)\}$ where

$$Y_e(m,s) = \sqrt{m}\;\frac{\bar{F}_e(s) - \hat{\bar{F}}_{em}(s)}{\hat{\bar{F}}_{em}(s)}$$

weakly approach the corresponding n -dimensional distributions of Wiener processes

$$W_1(V_1(m,s)), \dots, W_r(V_r(m,s))$$

In the above example of observations of a semimarkov

process all portions of data satisfy conditions (Y_1), $(Y_3)-(Y_5)$, [10]. The condition (Y_2) imposes restrictions on the uniformness of observations. For this example a theorem on weak convergence (in Skorokhod topology) of the normalized processes of deviations of PL -estimators of transition d.f.'s to the corresponding Wiener processes was proved in [10].

The proposed approach simplifies the construction of the fsoftware for computers [15]. The results obtained are a generalization of similar results for the multiplicative intensity model [1] to a more general class of models of accumulation of incomplete statistical data, including observations of renewal processes [16].

REFERENCES

1. Aalen O.O. Nonparametric inference for a family of counting processes. - Annals of statistics, 1978, v.6, p.701-726.
2. Kaplan E.L.,Meier P. Nonparametric estimation from incomplete observations. - Journal Amer.Statist. Assoc. 1958, v.53, p.451-481.
3. Barzilovich E.Yu., Belyayev Yu.K., Kashtanov V.A., Kovalenko I.N., Solovjev A.D., Ushakov I. A. Voprosi matematicheskoi teorii nadeznosti / pod red.Gnedenko B.V. - M.: Radio i cvyaz, 1983 (in Russian).
4. Gill R.D. Censoring and stochastic integrals. - Math. Centre Tracts, v.124; Mathematical centre Amsterdam, 1980.
5. Gill R.D. Large sample behaviour of the product-limit estimator on the whole line. - Annals of Statistics, 1983, v.11, p.49-58.
6. Anderson P.K., Borgan Ø. Counting process models for

life history data. - A rewiew 10-th Nordic conference on Math.Statistics, BolkesiØ, Norway, 17-21 june 1984.

7. Belyayev Yu.K. Neparametricheskie metodiv zadachah obrabotki dannih resultatov ispitaniy i ekspluatatcii. - M.Znanie, 1984 (in Russian).

8. Belyayev Yu.K. Mnozitelnii otcenki veroyatnosti bezotkaznoy roboti. - Izvestia AN SSSR, Technicheskaya kibernetika, 1985, N 4, s.45-59 (in Russian).

9. Belyayev Yu.K. Nonparametric estimation of distribution functions using censored statistical databases. - Abstracts of communications Fourth International Vilnius Conferense on Probability theory and Math.Statistics, v.IV (A-Z), Vilnius 1985, p.26-28.

10. Belyayev Yu.K., Zamyatin A.A. Mnozitelnii otcenki dlya funktcii raspredeleniya vremen perehodov polumarkovskogo protcessa. - Mezvuzavskiy sbornic "Verojatnostnie processi i ih prilozeniya. - M. MIEM, 1985, s.3-17 (in Russian).

11. Kiefer J., Wolfowitz J. Consistency of the maximum likelihood estimator in the presence of infinitely many incidental parameters. - Annals of Math.Statistics, 1957, v.27, p.887-907.

12. Matthes K., Kerstan J., Mecke J. Infinitely divisible Point Processes. - John Wiley, N.Y., 1978.

13. Gnedenko B.V., Kolmogorov A.N. Predelnii teoremi dlya summ nezavisimih sluchaynih velichin. - M. Fismatgis, 1949 (in Russian).

14. Loeve M. Probability theory, 2nd edition, D.van Nostrand N.Y., 1960.

15. Makarov A.P., Belyayev Yu.K., Belyayeva V.A., Dugina T.N. Matematicheskoe obespechenie obrabotki tcenzurirovannih dannih. - Tesisi III Vsesouznoi nauchnotehnicheskoy konferentcii "Primenenie mnogomernogo

statisticheskogo analiza", chast II, CEMI AN SSSR, Tartu, 1985, s.184-186.

16. Gill R.D. Testing with replacement and the product-limit estimator. - Annals of Statistics, 1981, v.9, p.853-860.

Prob. Theory and Math. Stat., Vol.1, pp. 145-169
Prohorov et al. (eds)
1986 VNU Science Press

DIFFUSION PROCESSES ON THE GROUP T^∞ AND ELLIPTIC EQUATIONS OF INFINITELY MANY VARIABLES

A.D.Bendikov, I.V.Pavlov
Institute of Construction Engineering, Rostov-on-Don, USSR

§1. INTRODUCTION

In the modern probability potential theory [1], [2] and in some problems of physics [3] became clear the necessity of investigation of diffusion processes and equations generated by them on more general phase spaces then the Euclidean space R^n or differentiable manifolds. As an example of such "nonclassic" phase space we take T^∞, the countable product of unit circles considered as a compact abelian group.

By the diffusion process on T^∞ we mean a strong Markov process with a.s. continuous sample paths, generated by the differential operator

$$\Delta: u(x) \longrightarrow \frac{1}{2}\sum_{i,j=1}^{\infty}(a_{ij}(x)u'_{x_i}(x))'_{x_j}, \quad x=(x_1,x_2,\dots)\in T^\infty.$$

In §2 we impose conditions on the matrix $A(x)=(a_{ij}(x))_{i,j=1}^{\infty}$ under which one can construct a process generated by the operator Δ. Besides, in this section we consider some properties of such process. Here we use the technique presented in [4].

In § 3, following [5], [6] and [13], we prove Littlewood-Paley inequalities to be used below for analysing weak solutions of the equations $\Delta u-\lambda u=f$, $\lambda\geq 0$.

Note that here we restrict ourselves to the case when the matrix $A(x)$ does not depend on $x \in T^\infty$.

Finally, in §4 we obtain under some conditions, imposed on a symmetric process $\mathcal{X}$ on T^∞, the infinite-dimensional analog of the second Beurling-Deny's formula on the structure of the Dirichlet form of the process $\mathcal{X}$ (in the case of R^n cf. [4, Th. 5.4.4. (ii)]).

The results of this paper were announced in [7].

§2. DIRICHLET FORMS AND SYMMETRIC HUNT PROCESSES ON T^∞

2.1. We present below the necessary definitions and facts from [4].

Let E be a locally compact separable Hausdorff space and let ν be a dense Random measure on E. Let $L_2 := L_2(E, \nu)$, $\|u\|_2 := \|u\|_{L_2}$, and let C_0 be the space of continuous functions on E with compact supports.

Definition 1. A closed symmetric bilinear form ($\mathcal{E}$, $D[\mathcal{E}]$) on L_2 is called a regular Dirichlet form, if

1) $D[\mathcal{E}] \cap C_0$ is dense in C_0 in the uniform metric and in $D[\mathcal{E}]$ the metric $\sqrt{\mathcal{E}_1(u)}$:

$$\mathcal{E}_1(u) = \mathcal{E}(u) + \|u\|_2^2 ; \qquad (1)$$

2) $\mathcal{E}$ is a Markov form (cf. [4, p.5]): $\forall \varepsilon > 0 \ \exists \varphi_\varepsilon(t): -\varepsilon \le \varphi_\varepsilon(t) \le 1 - \varepsilon$ for $t \in R^1$, $\varphi_\varepsilon(t) = t$ for $t \in [0,1]$, $0 \le \varphi_\varepsilon(t) - \varphi_\varepsilon(s) \le t - s$ for $s < t$, and $u \in D[\mathcal{E}] \Rightarrow \varphi_\varepsilon(u) \in D[\mathcal{E}]$, $\mathcal{E}(\varphi_\varepsilon(u)) \le \mathcal{E}(u)$.

Definition 2. Let $\mathcal{X}$ be a Markov process on E and (P_t) be its Markov semigroup. The process $\mathcal{X}$ is called symmetric, if

$$(P_t f, g) = (f, P_t g).$$

Theorem 1. 1) Given a regular Dirichlet form $\mathcal{E}$ on L_2, there exists a symmetric Hunt process $\mathcal{X}$ on E, whose Dirichlet form is the given on $\mathcal{E}$ (cf. [4, Th. 6.2.1]).

2) Let $\mathcal{X}_1$ and $\mathcal{X}_2$ be two symmetric Hunt processes on E, possessing a common regular Dirichlet space on L_2. Then $\mathcal{X}_1$ and $\mathcal{X}_2$ are equivalent (cf. [4, Th. 4.3.6]).

3) Process $\mathcal{X}$ has a.s. continuous sample paths, if the form $\mathcal{E}$ is local:

$$u, v \in D[\mathcal{E}] : u|_{\operatorname{supp} v} = 0 \Rightarrow \mathcal{E}(u,v) = 0$$

(cf. [4, Th. 6.2.2]).

2.2 Let $E = T^\infty$ be the countable product of the unit circles $T_k = T$ ($K = 1, 2, \dots$), which we consider as a compact abelian group. We shall often identify T with the interval $]-\pi, \pi]$. Then the group operation coincides with addition by mod (2π). Let dx_k be normalized Lebesque measure on $T_k =]-\pi, \pi]$, and let $dx = \bigotimes_{k=1}^{\infty} dx_k$. Then $E = T^\infty$ is a compact abelian group and dx is itssHaar measure.

Let Z be a set of integers, $Z^n = Z \times Z \times \dots \times Z$ be n-dimensional lattice in R^n, and $Z^{(\infty)} = \bigcup_{n>0} Z^n$ be discrete abelian group, which is naturally identified with the dual group $\widehat{T}^\infty$ of T^∞:

$$Z^{(\infty)} \ni \theta \longleftrightarrow e_\theta(x) = \exp(i\theta x) \in \widehat{T}^\infty$$

For a function $\varphi \in L_1$ and for a measure μ Fourier transforms $\widehat{\varphi}(\theta)$ and $\widehat{\mu}(\theta)$ are given by the formulas:

$$\hat{\varphi}(\theta)=\int_{T^{\infty}}\varphi(x)\,e_{\theta}(x)\,dx,\quad \hat{\mu}(\theta)=\int_{T^{\infty}}e_{\theta}(x)\,d\mu(x).$$

Below we denote by $\mathcal{L}$ the set of all cylindric infinitely-differentiable functions on T^{∞}.

2.2.1. Let $(a_{ij})_{i,j=1}^{\infty}$ be an infinite matrix of numbers such that for any $n>0$ $(a_{ij})_{i,j=1}^{n}\geq 0$ (i.e. the matrix $(a_{ij})_{i,j=1}^{n}$ is non-negative definite). Let

$$\mathcal{E}(u,v):=\frac{1}{2}\int_{T^{\infty}}\sum_{i,j=1}^{\infty}a_{ij}\,u'_{x_i}(x)\,v'_{x_j}(x)\,dx,\quad D[\mathcal{E}]=\mathcal{L};$$

$$\Psi(\theta):=\frac{1}{2}\sum_{i,j=1}^{\infty}a_{ij}\,\theta_i\,\theta_j,\quad \theta=(\theta_1,\theta_2,\dots)\in Z^{(\infty)}.$$

Proposition 1. The form $(\mathcal{E},\mathcal{L})$ is closable. If $\bar{\mathcal{E}}$ is its smallest closed extension, then

1) $\bar{\mathcal{E}}$ is a regular Dirichlet form;

2) $D[\bar{\mathcal{E}}]=\{u\in L_2:\ \bar{\mathcal{E}}(u)=\int_{Z^{(\infty)}}\Psi(\theta)|\hat{u}(\theta)|^2 d\theta<\infty\}.$

Proof. The form $\mathcal{E}$ is symmetric and non-negative definite. Using Parseval's equality and $\widehat{u'_{x_k}}(\theta)=i\theta_k\hat{u}(\theta)$, we get

$$\mathcal{E}(u)=\int_{Z^{(\infty)}}\Psi(\theta)|\hat{u}(\theta)|^2 d\theta,\quad u\in\mathcal{L}.$$

This formula proves a closability and item 2) of our proposition. According to [4, Th. 2.1.1], the proof of Markov property for $\bar{\mathcal{E}}$ is reduced to thepproof that $\mathcal{E}$ is Markovian. In view of [4, problem 1.2.1] the function $\varphi_{\varepsilon}(t)$ from the definition 1 can be chosen to be infinitely-differentiable. Then the Markov property is easily to verify: for $u\in\mathcal{L}$, $\varphi_{\varepsilon}(u)\in\mathcal{L}$ since $\varphi_{\varepsilon}(u)'_{x_i}=\varphi'_{\varepsilon}(u)u'_{x_i}$ then

$$\mathcal{E}(\varphi_\varepsilon(u)) = \frac{1}{2}\int_{T^\infty} |\varphi'_\varepsilon(u)|^2 \sum_{i,j=1}^{\infty} a_{ij}\, u'_{x_i} u'_{x_j}\, dx \leq \mathcal{E}(u)$$

2.2.2. Now let the matrix $(a_{ij})_{i,j=1}^{\infty}$ be diagonal: $(a_{ij})_{i,j=1}^{\infty} = diag(a_i)$. Let us denote by $W_2'\{a_k\}$ the Hilbert space of functions u from L_2 such that their derivatives (in the sense of distributioas) of the first order belong to L_2 and

$$\|u\|^2_{W_2'\{a_k\}} := \sum_{k=1}^{\infty} a_k \|u'_{x_k}\|_2^2 + \|u\|_2^2 < \infty .$$

It is not difficult to prove that $D[\bar{\mathcal{E}}] = W_2'\{a_k\}$ and

$$\bar{\mathcal{E}}(u) = \frac{1}{2}\int_{T^\infty} \sum_{k=1}^{\infty} a_k (u'_{x_k})^2 dx, \quad u \in D[\bar{\mathcal{E}}].$$

2.2.3. Let the matrix $(a_{ij}(x))_{i,j=1}^{\infty}$ consists of the bounded Borel functions and satisfies condition

$$\sum_{i,j=1}^{\infty} a_{ij}(x)\xi_i\xi_j \geq 0 \ (dx - a.e.)\ \forall \xi \in R^n, \forall n > 0.$$

Let us consider the bilinear form

$$\mathcal{E}(u,v) = \frac{1}{2}\int_{T^\infty} \sum_{i,j=1}^{\infty} a_{ij}(x)\, u'_{x_i}(x)\, v'_{x_j}(x)\, dx, \qquad D[\mathcal{E}] = \mathcal{L}.$$

It is easy to see that the form $\mathcal{E}$ is symmetric and non-negative difinite.

<u>Proposition 2</u>. Let us suppose that one of the following conditions is fulfilled:

a) there exists a numerical matric $(a^0_{ij})_{i,j=1}^{\infty}$ and $\delta > 0$, such that

$$\delta^{-2}\sum_{i,j=1}^{n} a^{0}_{ij}\xi_i\xi_j \le \sum_{i,j=1}^{n} a_{ij}(x)\xi_i\xi_j \le \delta^{2}\sum_{i,j=1}^{n} a^{0}_{ij}\xi_i\xi_j ,$$
$$\forall \xi\in R^n,\ \forall n>0;$$

b) the matrix $(a_{ij}(x))_{i,j=1}^{\infty}$ is quasi-diagonal and only the left part of the inequality from a) is true.

Then the form $\mathcal{E}$ is closable and its smallest closed extension $(\bar{\mathcal{E}}, D[\bar{\mathcal{E}}])$ is a regular Dirichlet form.

If condition a) is fulfilled with a diagonal matrix $(a^{0}_{ij})_{i,j=1}^{\infty} = diag(a_k)$, $a_k>0$ $(k=1,2,\dots)$, then $D[\bar{\mathcal{E}}]=W_2^1\{a_k\}$ and for each $u\in D[\bar{\mathcal{E}}]$

$$\bar{\mathcal{E}}(u) = \frac{1}{2}\int_{T^\infty}\sum_{i,j=1}^{\infty} a_{ij}(x)u'_{x_i}(x)u'_{x_j}(x)dx. \quad (2)$$

<u>Proof</u>. From a) it follows that $(\mathcal{E},\mathcal{L})$ is closable and $D[\bar{\mathcal{E}}]=D[\bar{\mathcal{E}}_0]$. If $(a^{0}_{ij})_{i,j=1}^{\infty} = diag(a_k), a_k>0$ $(k=1,2,\dots)$, then $D[\bar{\mathcal{E}}]=D[\bar{\mathcal{E}}^0]=W_2^1\{a_k\}$ (item 2.2.2).

Let us prove (2). Let $\{\varphi_k\}$ be δ-sequence from $\mathcal{L}$ and $u_k=u*\varphi_k$. Then for each $k>0$ $u_k\in\mathcal{L}$ and $u_k\mapsto u$ in metric $\bar{\mathcal{E}}^0_1$ (cf. (1)) and so in the metric $\bar{\mathcal{E}}_1$, i.e. $\bar{\mathcal{E}}(u)=\lim_{k\to\infty}\mathcal{E}(u_k)$. Let $\sum_{i,j=1}^{\infty} a_{ij}\xi_i\xi_j = (A\xi,\xi)$, $\nabla u := grad u$. We have:

$$|(A\nabla u_k,\nabla u_k)^{1/2}-(A\nabla u,\nabla u)^{1/2}|\le$$

$$\le (A\nabla(u-u_k),\nabla(u-u_k))^{1/2}\le \delta\Big(\sum_{n=1}^{\infty} a_n((u-u_k)'_{x_n})^2\Big)^{1/2}.$$

Then

$$\int_{T^\infty}|(A\nabla u_k,\nabla u_k)^{1/2}-(A\nabla u,\nabla u)^{1/2}|^2dx \le \delta^2\bar{\mathcal{E}}^0(u-u_k)\to 0$$

as $k\to\infty$, i.e. $(A\nabla u_k,\nabla u_k)\to(A\nabla u,\nabla u)$ in L_1 and

$$\bar{\mathcal{E}}(u)=\lim_{k\to\infty}\mathcal{E}(u_k)=\lim_{k\to\infty}\tfrac{1}{2}\|(A\nabla u_k,\nabla u_k)\|_{L_1}=$$

$$=\tfrac{1}{2}\|(A\nabla u,\nabla u)\|_{L_1}=\tfrac{1}{2}\int_{T^\infty}\sum_{i,j=1}^{\infty}a_{ij}(x)u'_{x_i}(x)u'_{x_j}(x)dx.$$

The formula (2) is proved.

Now let condition b) be satisfied. It is obvious that instead of $(\overset{\circ}{a}_{ij})_{i,j=1}^{\infty}$ one can chose there the matrix $diag(a_k)$. Let A_k be the k-block of the quasidiagonal matrix $(a_{ij}(x))_{i,j=1}^{\infty}$ and

$$\mathcal{E}_k(u)=\tfrac{1}{2}\int_{T^\infty}(A_k\nabla u,\nabla u)dx.$$

Then $\mathcal{E}=\sum_{k=1}^{\infty}\mathcal{E}_k$. Let a sequence of functions u_n from $\mathcal{L}$ be such that $u_n\to0$ in L_2 and $\mathcal{E}(u_n-u_m)\to 0$ as $m,n\to\infty$. From inequality $\delta^2\mathcal{E}(u)\geq\mathcal{E}^0(u)$ and from the closability of the form $\mathcal{E}^0$ it follows that there exists a subsequence n_ℓ such that u_{n_ℓ} (dx-a.e.) with their derivatives. Then

$$\mathcal{E}(u_m)=\sum_{k=1}^{\infty}\mathcal{E}_k(u_m)=\sum_{k=1}^{\infty}\mathcal{E}_k(\lim_{\ell\to\infty}(u_m-u_{n_\ell}))=$$

$$=\sum_{k=1}^{\infty}\lim_{\ell\to\infty}\mathcal{E}_k(u_m-u_{n_\ell}).$$

Using Fatou's lemma we have:

$$\mathcal{E}(u_m)\leq\lim_{\ell\to\infty}\sum_{k=1}^{\infty}\mathcal{E}_k(u_m-u_{n_\ell})=\lim_{\ell\to\infty}\mathcal{E}(u_m-u_{n_\ell}).$$

From the last inequality it follows that $\mathcal{E}(u_m)\to0$ when $m\to\infty$ (i.e. $\mathcal{E}$ is closable) and $D[\bar{\mathcal{E}}]\subset$

$dW_2'\{a_k\}$. As in point 2.2.1, it can be proved that $\overline{\mathcal{E}}$ is Markovian and so is a regular Dirichlet form.

2.3. This item presents a description of the Hunt process $\mathcal{X}$ on T^∞ which corresponds to the Dirichlet form from 2.2.3.

2.3.1. Suppose that one of conditions a) or b) is fulfilled and in addition:

c) for each $i,j=1,2,\ldots$ $a_{ij}(x)\in\mathcal{L}$;

d) for each $i,j=1,2,\ldots$ function $a_{ij}(x)$ depends only on the first $n=i\wedge j$ coordinates.

Using these assumptions and [8, Th.10.13 and point 10.29] one can prove that the process $\mathcal{X}$ is a projective limit (cf. [9]) of the sequence $\mathcal{X}_n$ of diffusion processes (the process $\mathcal{X}_n$ is generated by the Dirichlet form $\mathcal{E}_n:=\mathcal{E}|_{\mathcal{L}(T^n)}$ on the group T^n). Each of the processes $\mathcal{X}_n$ is a Feller process (and even a strong Feller one), hence the process $\mathcal{X}$ is also a Feller process. But $\mathcal{X}$ has not (in general case) strong Feller property (cf. 2.3.2). It is obvious that $\mathcal{L}$ is included in the domain of generator Δ of the process $\mathcal{X}$, $\Delta(\mathcal{L})\subset\mathcal{L}$ and for $u\in\mathcal{L}$

$$\Delta u(x)=\frac{1}{2}\sum_{i,j=1}^{\infty}\left(a_{ij}(x)u'_{x_i}(x)\right)'_{x_j}.$$

2.3.2. Now let us suppose that the matrix $(a_{ij}(x))_{i,j=1}^{\infty}$ does not depend on $x\in T^\infty$. In this case $\mathcal{X}$ is an infinite divisible symmetric process on the group T^∞ (cf. [10]). Its Markov semigroup (P_t) is generated by a weak continuous convolution semigroup (ν_t) of symmetric Gaussian measures on T^∞ (cf. [10],[11, Ch.5, §5.5]):

$$P_t f=f*\nu_t,\quad \hat{\nu}_t(\theta)=\exp(-t\,\Psi(\theta)),\ t>0,$$

where $\Psi(\theta) = \frac{1}{2}\sum_{i,j=1}^{\infty} a_{ij}\theta_i\theta_j$, $\theta \in Z^{(\infty)}$.

For each $t>0$ measure ν_t is a projective limit of the sequence $\nu_t^{(n)}$ of Gaussian measures on T^n. If for every $n>0$ matrix $A_n=(a_{ij})_{i,j=1}^{n}$ is non-degenerate, each measure $\nu_t^{(n)}$ is absolutely continuous with respect to Haar measure $dx^{(n)}$ on T^n and

$$\frac{d\nu_t^{(n)}}{dx^{(n)}} = \left(\frac{2\pi}{t}\right)^{n/2}(\det A_n)^{-1/2}\sum_{m\in Z^n} \exp\left\{-\frac{(A_n^{-1}(x-2mn),(x-2\pi m))}{2t}\right\}$$

(cf. [11, 5.5.9]).

In the diagonal case $(a_{ij})_{i,j=1}^{\infty}=diag(a_k)$, $a_k>0$, we have

$$\frac{d\nu_t^{(n)}}{dx^{(n)}} = \prod_{k=1}^{n} G_{a_k t}(x_k),\ x=(x_1,\dots)\in T^\infty$$

where

$$G_t(x)=\sqrt{\frac{2\pi}{t}}\sum_{k=-\infty}^{+\infty}\exp\left\{-\frac{(x-2\pi k)^2}{2t}\right\},\ x\in\,]-\pi,\pi].$$

The strong Feller property of the process $\mathcal{X}$ is equivalent to the absolute continuity of the measure ν_t with respect to the Haar measure dx (for every $t>0$). The last property is not always fulfilled. For example, in the diagonal case the absolute continuity of ν_t is equivalent to convergence of

$$\sum_{k=1}^{\infty}\exp(-2ta_k)$$

(cf. [10], [12], [11, 5.5.12 and Th. 5.5.13]).

2.4. Let us consider the process $\mathcal{X}=(x_t,\mathcal{M}_t,\mathcal{M},P_x)$, which is generated by the Dirichlet form $\mathcal{E}$ from 2.2.3. We suppose that the form $\mathcal{E}$ satisfies condition a) with the matrix $(a^0_{ij})^\infty_{i,j=1} = \mathrm{diag}(a_k)$, $a_k > 0$ $(k=1,2,\dots)$. We use a terminology from the book by Fukushima [4].

Let $u \in D[\bar{\mathcal{E}}]$. Then we have the Fukushima's decomposition (cf. [4, Th. 5.2.2]):

$$\tilde{u} \circ x_t - \tilde{u} \circ x_0 = M_t^{[u]} + N_t^{[u]},$$

in which $\tilde{u}$ is a quasi-continuous version of u (cf. [4, § 3.1, §4.2, Th. 4.3.2]), $N_t^{[u]}$ is a continuous additive functional of zero energy (cf. [4, 5.2 (iii)]) and $M_t^{[u]}$ is a martingale additive functional of finite energy (cf. [4, 5.2 (ii)]).

Proposition 3. 1) For each function $u \in D[\bar{\mathcal{E}}]$ and for each probability measure ν on T^∞, which does not charge exceptional sets (cf. [4, p.96]), the process $(M_t^{[u]}, \mathcal{M}_t, P_t)$ is a continuous square integrable martingale.

2) If a function u is bounded, then the quadratic variation $\langle M^{[u]}\rangle_t$ can be calculated by the formula

$$\langle M^{[u]}\rangle_t = 2\int_0^t \Big(\sum_{i,j=1}^\infty a_{ij}\, u'_{x_i}\, u'_{x_j}\Big)\circ x_s \, ds$$

Proof. 1) If $u \in \mathcal{L}$, then the continuity of $M_t^{[u]}$ follows from the continuity of x_t. Now let $u \in D[\bar{\mathcal{E}}]$ and a sequence $\varphi_k \in \mathcal{L}$ converges to u in the metric $\bar{\mathcal{E}}_1$. In view of [4, Cor.1, p.139] there exists a subsequence $n_k \to \infty$ such that P_ν -a.s. $M_t^{[\varphi_{n_k}]}$ converges to $M_t^{[u]}$ uniformly on any finite interval. Since $M_t^{[\varphi_{n_k}]}$ are continuous, $M_t^{[u]}$ is also continuous.

2) $\langle M^{[u]}\rangle$ is a positive continuous additive functional. According to [4, Th.5.1.3], it corresponds to the unique smooth measure $\mu_{\langle M^{[u]}\rangle}$ (cf. [4, p.72]) which can be calculated with the help of [4, Th.5.2.3]:

$$\mu_{\langle M^{[u]}\rangle}(dx) = \gamma(x)\,dx ,$$

$$\gamma(x) = 2\sum_{i,j=1}^{\infty} a_{ij}(x)\, u'_{x_i}(x)\, u_{x_j}(x).$$

In view of [4, Th.5.1.3] this measure corresponds to the unique positive additive functional $\int_0^t \gamma\cdot x_s\, ds$ Q.E.D.

Let us note that the results of this section hold true for the group $R^n \times T^\infty$ too. In this case instead of $\mathcal{L}$ it should be taken that set $\bar{\mathcal{L}}$ of cylindic infinitely-differentiable functions with compact supports.

§3. THE LITTLEWOOD-PALEY INEQUALITIES AND SOLUTIONS OF ELLIPTIC EQUATIONS ON T^∞.

3.1. Let Y be a Hunt process on T^∞ with a Direchlet form $\mathcal{E}_Y$:

$$\mathcal{E}_Y(u, v) = \frac{1}{2}\int_{T^\infty} \sum_{i,j=1}^{\infty} a_{ij}(x)\, u'_{x_i}(x)\, v'_{x_j}(x)\, dx,$$

$$\delta^{-2}\sum_{k=1}^{\infty} a_k \xi_k^2 \le \sum_{i,j=1}^{\infty} a_{ij}(x)\xi_i \xi_j \le \delta^{2}\sum_{k=1}^{\infty} a_k \xi_k^2 . \quad (3)$$

Let us consider the standard Wiener process Z on R^1 and the process $\mathcal{X} = (Y, Z)$ on $T^\infty \times R^1$. It is obvious that for functions $F, G \in \mathcal{L} = \mathcal{L}(T^\infty \times R^1)$ we have:

$$\mathcal{E}_{\mathcal{X}}(F,G)=\frac{1}{2}\int_{T^\infty\times R^1}\Big[\sum_{i,j=1}^{\infty}a_{ij}(x)F'_{x_i}(x,z)G_{x_j}(x,z)+F'_z(x,z)+G'_z(x,z)\Big]dx\,dz.$$

Let $T_+^\infty := T^\infty \times R_+^1$, $\partial T_+^\infty = T^\infty \times \{0\}$ and $\tau = \inf\{t>0: x_t \notin T_+^\infty\}$. Let $f \in \mathcal{L}$ and let $G \in \bar{\mathcal{L}}$ be such that $G|_{\partial T_+^\infty} = f$. We consider the function $F(x,z) = E_{(x,z)}(G \circ x_\tau)$. The function F is bounded, quasi-continuous and belongs to $D[\bar{\mathcal{E}}_x]$ (cf. [4, Th.4.4.1]) We have:

$$F\circ x_t - F\circ x_0 = M_t^{[F]} + N_t^{[F]}.$$

The function F is harmonic relative to $\mathcal{X}$ on T_+^∞. Then (cf. [4, Th.5.3.4]) $N_t^{[F]}=0$, for $t<\tau$ (P_ν - a.s.) for each probability measure ν, which charges no exceptional sets. From Proposition 3 it follows that $(M_t^{[F]}, \mathcal{M}_t, P_\nu)$ is a continuous square integrable matringale and

$$\langle M^{[F]}\rangle_t = 2\int_0^t\Big[\sum_{i,j=1}^{\infty}a_{ji}F'_{x_i}F'_{x_j}+(F'_z)^2\Big]\circ x_s\,ds.$$

Therefore $M_t := F\circ x_{t\wedge\tau} = F\circ x_0 + M_{t\wedge\tau}^{[F]}$ is continuous square integrable martingale and

$$\langle M\rangle_t = 2\int_0^{t\wedge\tau}\gamma_F\circ x_s\,ds,\quad \gamma_F := \sum_{i,j=1}^{\infty}a_{ij}F'_{x_i}F'_{x_j}+(F'_z)^2.$$

3.2. We give here some necessary facts to be used below.

1) Let $\nu_a := dx \otimes \varepsilon_a\ (a>0)$, where ε_a is the Diract measure in the point $a\in R^1$. It is easy to verify the inequality

$$\int u^2 d\nu_a \le c(\mathcal{E}_x(u)+\|u\|_2^2), \quad u\in\bar{\mathcal{L}},$$

where $c>0$ is a constant, not depending on u. From this inequality it follows that the measure ν_a possesses the finite energy integral (cf. [4, 3.2]) and therefore charges no exceptional sets (cf. [4, Lemma 3.2.2]).

2) The next equalities are true (cf. [5, p.130]):

$$F(x,z)=E_{(x,z)}(G\circ x_\tau)=E_{(x,z)}(f\circ Y_\tau)=$$

$$=\int_0^\infty E_x(f\circ Y_t)E_z(\tau\in dt)=\int_0^\infty (P_t f)(x)h_z(t)dt=Q_z f(x),$$

where (P_t) is Markov semigroup of the process Y, $h_z(t)=(4\pi)^{-\frac{1}{2}}t^{-\frac{3}{2}}\exp(-z^2/4t)$, (Q_z) is Cauchy semigroup.

3.3. Let $f\in\mathcal{L}$, $F(x,t)=Q_t f$, $\bar{f}=(f_1,f_2,\dots)$, $f_k\in\mathcal{L}$, $\bar{F}(x,t)=Q_t\bar{f}$. Let us consider the Littlewood-Paley functions:

$$K_f(x):=\left(\int_0^\infty t\,(Q_t(\sqrt{\gamma_F})(x))^2dt\right)^{1/2},$$

$$K_{\vec{f}}(x):=\left(\int_0^\infty t\,(Q_t(|\bar{F}_t'|)(x))^2dt\right)^{1/2}.$$

<u>Proposition 4.</u> The following inequality is true:

$$\|K_f\|_p\le C_p\|f\|_p \qquad (1<p<\infty).$$

<u>The scheme of the proof</u>. 1) Let $2<p<\infty$. Just in the same way as in [6, III, e), p.157] we obtain for $p=2r$:

$$C_\tau E_\nu[|f(Y_\tau)|^{2\tau}] \geq E_\nu[(E_\nu[2\int_0^\tau \gamma_F \cdot x_s ds | Y_\tau])^\tau],$$

where ν is a probability measure on T_+^∞, which charges no exceptional sets. Let $\nu=\nu_a$ (cf. 3.2). Then the left side of the inequality is equal to $C_\tau \|f\|_{2\tau}^{2\tau}$ and the right side - to $2^\tau \|\mathcal{I}\|_\tau^\tau$, where $\mathcal{I}(x)= = \int_0^\infty t \wedge a\, Q_t \gamma_F(x) dt$ (cf. [6, (3), p.156]). If $a \to \infty$, then we get:

$$C_\tau \|f\|_{2\tau}^{2\tau} \geq 2^\tau \int_{T^\infty} dx\, [\int_0^\infty t\, Q_t \gamma_F(x) dt]^\tau \geq$$

$$\geq 2^\tau \int_{T^\infty} dx [\int_0^\infty t (Q_t \sqrt{\gamma_F'}(x))^2 dt]^\tau = 2^\tau \|K_f\|_{2\tau}^{2\tau}.$$

2) Let $1 < p \leq 2$. In this case the proof is carried out in the same way as in [6, p.157-159], if we take into consideration item 1) of the Proposition 3.

Proposition 5. If a vector-function $\bar{f} \in \mathcal{L} \times \mathcal{L} \times \cdots$ is such that $\int_{T^\infty} \bar{f} dx = 0$, then

$$\|K_{\vec{f}}\|_p \geq C_p \||\bar{f}|\|_p \qquad (1 < p < \infty).$$

The proof follows from the corresponding inequality for $G^{\to}$ - function in [6, p.155] (cf. also 13, Th.3).

3.4. In the remaining items of § 3 we suppose that the matrix $(a_{ij}(x))_{i,j=1}^\infty$ does not depend on $x \in T^\infty$.

Let (P_t) be a Markov semigroup of the process Y, (Q_t) be Cauchy semigroup (cf. 3.2, 2)), Δ and B be L_2-generators of the semigroups (P_t) and (Q_t) correspodingly. It is easy to verify that the operators P_t, Q_t, Δ and B trahsform $\mathcal{L}$ into $\mathcal{L}$ and that for $f \in \mathcal{L}$ $\Delta f = -B^2 f$.

Definition 3. Let $\vec{L}: \mathcal{L} \to \mathcal{L} \times \mathcal{L} \times \ldots$ be a linear operator satisfying conditions:

1) $\vec{L} P_t = P_t \vec{L}, \forall t > 0$;

2) $\int \vec{L} f(x) dx = 0, \forall f \in \mathcal{L}$;

3) $|\vec{L} f|^2 \leq \sum_{i,j=1}^{\infty} a_{ij} f'_{x_i} f'_{x_j} + (Bf)^2, \forall f \in \mathcal{L}$.

The operator $\vec{L}$ is called admissible.

Proposition 6. If $\vec{L}$ is admissible, then

$$\| |\vec{L} f| \|_p \leq C_p \|Bf\|_p \quad (f \in \mathcal{L}, 1 < p < \infty).$$

Proof. Let $\bar{\tau} := \vec{L} f$, $u : Bf$, $\bar{R}_t := Q_t \bar{\tau}$ and $u_t := Q_t u$. Then

$$\bar{R}'_t = B Q_t \vec{L} f = \vec{L} Q_t Bf = \vec{L} u_t .$$

Therefore $|\bar{R}'_t|^2 \leq \gamma_{u_t}$ and $K_{\vec{\tau}} \leq K_u$. We have:

$$\| |\bar{\tau}| \|_p \leq C_p \| K_{\vec{\tau}} \|_p \leq C_p \| K_u \|_p \leq C_p \| u \|_p .$$

Q.E.D.

Example 1. Let (a_k) be the sequence of numbers from the inequality (3) and let

$$\nabla_a := \left(\sqrt{a_1} \frac{\partial}{\partial x_1}, \sqrt{a_2} \frac{\partial}{\partial x_2}, \ldots \right).$$

It is easy to verify the properties 2) and 3) from the definition 3. Since the matrix $(a_{ij})_{i,j=1}^{\infty}$ does not depend on $x \in T^{\infty}$, P_t is a convolution operator (cf. 2.3.2) i.e. $\nabla_a P_t = P_t \nabla_a \ (t > 0)$. Therefore ∇_a is admissible and using the Proposition 6 we have:

$$\| |\nabla_a f| \|_p \leq C_p \|Bf\|_p \quad (1 < p < \infty \qquad (4)$$

3.5. Let a vector-function $\bar{m}(\theta) = (m_1(\theta), m_2(\theta), \ldots)$

be defined on $Z^{(\infty)}$ and $|\overline{m}(\theta)|$ is bounded. Let us consider linear operator $\vec{R}=(R_1, R_2, \dots)$ on $\mathcal{L}$ such that

$$(\widehat{\vec{R}f})(\theta)=\overline{m}(\theta)\hat{f}(\theta), \quad \theta \in Z^{(\infty)}.$$

Definition 4. A vector-function $\overline{m}$ is called L_p-multiplier and the operator $\vec{R}$ -multiplier operator on L_p, if

$$\| |\vec{R}f| \|_p \le C_p \|f\|_p \qquad (f \in \mathcal{L}).$$

Let $\vec{L}$ be an admissible operator. For $\theta \in Z^{(\infty)}$ we define:

$$\overline{\lambda}(\theta) := \vec{L} e_\theta(0),$$

$$m_k(\theta) := \begin{cases} \dfrac{\lambda_k(\theta)}{\sqrt{\Psi(\theta)}}, & \theta \neq 0 \\ 0, & \theta = 0 \end{cases},$$

$$\overline{m}(\theta) = (m_1(\theta), m_2(\theta), \dots).$$

Proposition 7. The vector-function $\overline{m}(\theta)$ is L_p - multiplier for every p, $1 < p < \infty$.

The proof of this proposition is based on the Proposition 6 and is analogous to the proof of [13, Th.5].

Example 2. For admissible operator ∇_a from Example 1 the multiplier $\overline{m}(\theta) = (m_1(\theta), m_2(\theta), \dots)$ has a form:

$$m_k(\theta) = \begin{cases} \dfrac{i\sqrt{a_k}\,\theta_k}{\sqrt{\Psi(\theta)}}, & \theta \neq 0 \\ 0, & \theta = 0 \end{cases}$$

$(K=1,2,\dots)$. By the analogy with the classical case we shall call $\overline{m}(\theta)$ the Riesz's vector multiplier and corresponding multiplier operator $\vec{R}$ - the Riesz's vector operator.

3.6. Let W_p^k $(k=1,2,\dots)$ be a linear space of functions f of L_p whose derivatives (in the sense of distributions) up to k^{th} order inclusive belong to L_p $(1<p<\infty)$. The space W_p^k is a countably normed space.

Theorem 2. Let $u\in L_p$ be a solution of the equation

$$\Delta u-\lambda u=f,\ f\in W_p^k.$$

Then

1) $u\in W_p^{k+2}$;

2) $\Delta u=\sum_{i,j=1}^{\infty}a_{ij}u''_{x_ix_j}$ (the series converges in L_p).

Proof. 1) It is sufficient to consider the case $\lambda>0$. Then $u=R_\lambda f$, where (R_λ) is the resolvent of the semigroup (P_t) . Therefore we have to prove that $R_\lambda f\in W_p^{k+2}$. Obviously it is sufficient to consider the case $k=0$. In the following calculations we put:

$$\langle f,g\rangle:=\int f(x)g(x)dx.$$

For a function $\varphi\in\mathcal{L}$ we have:

$$\langle (R_\lambda f)'_{x_k},\varphi\rangle:=-\langle R_\lambda f,\varphi'_{x_k}\rangle=-\langle f,(R_\lambda\varphi)'_{x_k}\rangle,$$

where we have taken into account that $R_\lambda\varphi\in\mathcal{L}$ and $(R_\lambda\varphi)'_{x_k}=R_\lambda\varphi'_{x_k}$. Using the Hölder's inequality and inequality (4), we get $\left(\frac{1}{p}+\frac{1}{q}=1\right)$:

$$|\langle (R_\lambda f)'_{x_k},\varphi\rangle|\le\|f\|_p\|(R_\lambda\varphi)'_{x_k}\|_q\le C_q\|f\|_p\|BR_\lambda\varphi\|_q.$$

Since the operator BR_λ is bounded in L_p , then

$$|\langle (R_\lambda f)'_{x_k}, \varphi\rangle| \le C_p \|f\|_p \|\varphi\|_q. \tag{5}$$

From the inequality (5) it follows that the linear functional on $\mathcal{L}$

$$\varphi \to \langle (R_\lambda f)'_{x_k}, \varphi\rangle$$

can be extended to a continuous linear functional on L_q. Therefore there exists a function $g_k \in L_p$ such that

$$\langle (R_\lambda f)'_{x_k}, \varphi\rangle = \langle g_k, \varphi\rangle \quad (k=1,2,\dots),$$

i.e. $R_\lambda f \in W_p'$.

Let us consider the second order derivatives (for example $(R_\lambda f)''_{x_k x_k}$). Then

$$\langle (R_\lambda f)''_{x_k x_k}, \varphi\rangle = \langle R_\lambda f, \varphi''_{x_k x_k}\rangle = \langle f, (R_\lambda \varphi)''_{x_k x_k}\rangle.$$

It is easy to verify the following equality for $\varphi \in \mathcal{L}$:

$$(R_\lambda \varphi)''_{x_k x_k} = \frac{1}{a_k} R_k^2 \Delta R_\lambda \varphi .$$

Using the Hölder's inequality, Example 2 and the boundness of the operator ΔR_λ on L_p we obtain the estimate:

$$|\langle (R_\lambda f)''_{x_k x_k}, \varphi\rangle| \le C_p \|f\|_p \|\varphi\|_q ,$$

from which it follows the existence of a function $g_{kk} \in L_p$ such that

$$\langle (R_\lambda f)''_{x_k x_k}, \varphi\rangle = \langle g_{kk}, \varphi\rangle ,$$

i.e. $R_\lambda f \in W_p^2$.

2) For each function $g \in L_q$ and for $m, n = 1,2,\dots$ we have:

$$\mathcal{M}_n := \langle \sum_{i,j=1}^{n} a_{ij} u''_{x_i x_j}, g\rangle = \langle \sum_{i,j=1}^{n} \tilde{a}_{ij} R_i R_j \Delta u, g\rangle =$$

$$= \sum_{i,j=1}^{n} \tilde{a}_{ij} \langle R_i g, R_j^* \Delta u\rangle = -\sum_{i,j=1}^{n} \tilde{a}_{ij} \langle R_i g, R_j \Delta u\rangle =$$

$$= -\langle (\tilde{A}\vec{R}^{(n)} g, \vec{R}^{(n)} \Delta u), 1\rangle,$$

where $\tilde{a}_{ij} = a_{ij}/\sqrt{a_i a_j}$, $\tilde{A} = (\tilde{a}_{ij})_{i,j=1}^{\infty}$

$$\vec{R}^{(n)} = (R_1, R_2, \dots, R_n, 0, \dots).$$

Therefore

$$\mathcal{M}_{m,n} := \mathcal{M}_n - \mathcal{M}_m = \langle (\tilde{A}\vec{R}^{(n)} g, \vec{R}^{(n)} \Delta u), 1\rangle -$$

$$-\langle (\tilde{A}\vec{R}^{(m)} g, \vec{R}^{(m)} \Delta u), 1\rangle = \langle (\tilde{A}\vec{R}^{m,n} g, \vec{R}^{(n)} \Delta u),$$

$$, 1\rangle + \langle (\tilde{A}\vec{R}^{(m)} g, \vec{R}^{m,n} \Delta u), 1\rangle,$$

where $\vec{R}^{m,n} := (0, \dots, 0, R_{m+1}, \dots, R_n, 0, \dots)$.

From inequality (3) it follows that $\tilde{A}$ is the matrix of a bounded symmetric operator $\mathcal{A}$ in the natural base of Hilbert space l_2 and that $\|\mathcal{A}\| \le \delta^2$. Therefore

$$|\mathcal{M}_{m,n}| \le \delta^2 (\langle |\vec{R}^{m,n} g|, |\vec{R}^{(n)} \Delta u|\rangle + \langle |\vec{R}^{(m)} g|, |\vec{R}^{m,n} \Delta u|\rangle).$$

Using the Hölder's inequality and Example 2, we get the inequality:

$$|\mathfrak{M}_{m,n}| \le c_p \delta^{-2} (\| |\vec{R}^{m,n} g| \|_q \|\Delta u\|_p + \|g\|_q \| |\vec{R}^{m,n} \Delta u| \|_p),$$

from which it follows that $\mathfrak{M}_{m,n} \to 0$ as $m, n \to \infty$. Q.E.D.

Remark 1. The assertion, which are analogous to the Theorem 2, are true for the group $R^n \times T^\infty$ and also for domains $D \subset R^n \times T^\infty$.

Remark 2. In the contrast to the classical case, the belonging of the function φ to space W_p^k for large k (and even for $k = \infty$) does not imply its continuity. Let us give a counter-example.

Counter-example. Let $a_k = \frac{1}{2} \ln k$ $(k = 1, 2, \dots)$ and let (ν_k) be Gaussian measures semigroup, corresponding to the matrix $(a_{ij})_{i,j=1}^{\infty} = diag(a_k)$ (cf. 2.3.2). It is known (cf. [12]) that measure ν_t is singular (with respect to Haar measure), if $0 < t \le 1$, and that it is absolutely continuous, if $1 < t < \infty$, and that it has a continuous density, if $t > 2$. Let

$$g(x) := \frac{d\nu_2(x)}{dx}, \text{ then } g(x) = \prod_{k=1}^{\infty} G_{a_k}(x_k), \; x = (x_1, x_2, \dots).$$

Since the function $g(x)$ is lower semi-continuous and positive, it has a positive infimum. Therefore the function $\varphi(x) := \frac{1}{g(x)}$ is bounded, differentiable in each finite number of variables and all its derivatives are bounded too. For each $p > 1$ and $k = 1, 2, \dots$ we have: $\varphi \in W_p^k$. But the function $\varphi(x)$ is not continuous.

Let us note that if $a_k / k \to \infty$ and u is a solution of the equation $\Delta u - \lambda u = f$ in a domain D , then the number of continuous derivatives of u is equal to the number of continuous derivatives of f (cf. [15]).

§4. THE SECOND BEURLING-DENY'S FORMULA ON

Let us formulate the classical result of Beurling and Deny (cf. [4, Th.2.2.2 and Th.5.4.4]).

Theorem 3. Let E be a domain in R^n and let $\mathcal{E}$ be a regular Dirichlet form on $L_2(E)$ such that:

1) $\mathcal{E}$ is local, i.e.

$$\forall u,v \in \mathcal{D}[\mathcal{E}] \quad u=\text{const on supp } v \Rightarrow \mathcal{E}(u,v)=0;$$

2) $C_0^\infty(E) \subset \mathcal{D}[\mathcal{E}]$.

Then

$$\mathcal{E}(u,v)=\int_E \sum_{i,j=1}^n u'_{x_i} v'_{x_j}\, d\nu_{ij},\ u,v\in C_0^\infty(E),$$

where ν_{ij} are Radon measures on E such that foe each compact $K\subset E$ matrix $(\nu_{ij}(K))_{i,j=1}^n$ is non-negative definite. The form $\mathcal{E}$ defines measures ν_{ij} uniquely.

From the Theorem 3 it is easy to obtain the analogous result for T^n.

Proposition 8. Let $\mathcal{E}$ be a regular Dirichlet form on $L_2(T^n)$, such that:

1) $\mathcal{E}$ is local;

2) $\mathcal{L}(T^n) \subset \mathcal{D}[\mathcal{E}]$.

Then

$$\mathcal{E}(u,v)=\int_{T^n} \sum_{i,j=1}^n u'_{x_i} v'_{x_j}\, d\nu_{ij},\ u,v\in \mathcal{L}(T^n),$$

where ν_{ij} are Radon measures on T^n, such that for each compact $K\subset T^n$ the matrix $(\nu_{ij}(K))_{i,j=1}^n$ is non-negative definite. The form $\mathcal{E}$ defines measures ν_{ij} uniquely.

Now we prove the analogous result for T^∞.

Theorem 4. Let $\mathcal{E}$ be a regular Dirichlet form on $L_2 = L_2(T^\infty)$ such that:

11) $\mathcal{E}$ is local;

2) $\mathcal{L} \subset \mathcal{D}[\mathcal{E}]$.

Then

$$\mathcal{E}(u,v) = \int_{T^\infty} \sum_{i,j=1}^{\infty} u'_{x_i} v'_{x_j} d\nu_{ij}, \quad u,v \in \mathcal{L},$$

where ν_{ij} are a Radon measures on T^∞, such that for each compact $K \subset T^\infty$ the matrix $(\nu_{ij}(K))_{i,j=1}^{n}$ is non-negative definite for every $n = 1,2\dots$. The form $\mathcal{E}$ defines the measures ν_{ij} uniquely.

Proof. Let us denote by $\mathcal{L}_n$ a subset of functions from $\mathcal{L}$, depending on the first n coordinates. Such a function from $\mathcal{L}_n$ can be considered as an infinitely differentiable function on T^n. Let us define the forms $\mathcal{E}_n$ on $L_2(T^n)$:

$$\mathcal{D}[\mathcal{E}_n] = \mathcal{L}_n, \quad \mathcal{E}_n(u) = \mathcal{E}(u), \quad u \in \mathcal{L}_n.$$

It is obvious that $\mathcal{E}_n$ is symmetric non-negative definition, closable and posesses the Markov property. From the Proposition 8 we get:

$$\mathcal{E}_n(u) = \int_{T^n} \sum_{i,j=1}^{n} u'_{x_i} u'_{x_j} d\nu_{ij}^{(n)}, \quad u \in \mathcal{L}_n.$$

Using the analogous formula for $\mathcal{E}_{n+1}$ and taking into account that for $u \in \mathcal{L}_n \subset \mathcal{L}_{n+1}$ $\mathcal{E}_{n+1}(u) = \mathcal{E}_n(u)$ we obtain:

$$\nu_{ij}^{(n+1)} \Big|_{\mathcal{F}_n} = \nu_{ij}^{(n)}, \quad i,j \le n$$

where $\mathcal{F}$ is the σ-field of cylindric borel subsets of T^{n+1} with the bases in T^n. So for each fixed $i,j \ge 1$ the sequence of measures $\{\nu_{ij}^{(n)}\}_{n \ge i \vee j}$ is projective. We have for $i,j \le n$, $n > 0$

$$|\nu_{ij}^{(n)}(K)| \leq \sqrt{\nu_{ii}^{(n)}(K)}\sqrt{\nu_{jj}^{(n)}(K)} \leq \sqrt{\nu_{ii}^{(n)}(1)}\sqrt{\nu_{jj}^{(n)}(1)} =$$

$$= \sqrt{\mathcal{E}_n(e_{\theta_i})}\sqrt{\mathcal{E}_n(e_{\theta_j})} = \sqrt{\mathcal{E}(e_{\theta_i})}\sqrt{\mathcal{E}(e_{\theta_j})}.$$

From this inequality it follows that for fixed $i,j \geq 1$ the variations of measures $\{\nu_{ij}^{(n)}\}$ are uniformly bounded. Therefore there exists a measure ν_{ij} on T^∞ such that for each $n>0$ the restriction of measure ν_{ij} on the σ-field of cylindric Borel subsets of T^∞ with the bases in T^n coincides with bthe measure $\nu_{ij}^{(n)}$. If $u \in \mathcal{L}$, then $u \in \mathcal{L}_n$ for some $n>0$, and we have

$$\mathcal{E}(u) = \mathcal{E}_n(u) = \int_{T^n} \sum_{i,j=1}^{n} u'_{x_i} u'_{x_j} \, d\nu_{ij}^{(n)} =$$

$$= \int_{T^\infty} \sum_{i,j=1}^{\infty} u'_{x_i} u'_{x_j} \, d\nu_{ij}.$$

The fact that the matrix $(\nu_{ij}(K))_{i,j=1}^{n}$ is non-negative definite, followsffrom the projectivity of the sequence $\{\nu_{ij}^{(n)}\}_{n\geq \max\{i,j\}}$ and from the fact that matrix $(\nu_{ij}^{(n)}(K))_{i,j=1}^{n}$ is non-negative definite $(n=1, 2,\dots)$. Q.E.D.

Remark 3. If each measure ν_{ij} is absolutely continuous with respect to Haar measure dx (i.e. $d\nu_{ij}(x) = a_{ij}(x)dx$), then $\mathcal{E}$ has the same form as in 2.2.3.

REFERENCES

1. Stoica L. Local operators and Markov processes. -

Lect.Notes Math. 819, Springer-Verlag, 1980.
2. Berg C., Forst G. Potential theory on locally compact abelian groups. - Springer-Verlag, 1975.
3. Holley R., Stroock D. Diffusion on an infinite dimensional torus. - J.of Funct.Anal., 42, 1981, p. 29-63.
4. Fukushima M. Dirichlet forms and Markov processes. - North-Holland Math.Library, v.23, 1980.
5. Meyer P.A. Demonstration probabiliste de certaines inequalites de Littlewood-Paley. - Lect.Notes Math. 511, 1976, p.125-183.
6. Meyer P.A. Retour sur la theorie de Littlewood-Paley. - Lect.Notes Math. 850, 1981, p.151-166.
7. Bendikov A.D., Pavlov I.V. Diffusion processes on the group T^∞ and elliptic equations with infinite number of variables. - Fourth international Vilnius conference on prob.theory and math. statistics. Abstracts of communications, v.1, 1985, p.74-75 (in Russian).
8. Dynkin E.B. Markov processes. - Moscow: Fismatguis, 1963 (in Russian).
9. Bendikov A.D. On harmonic functions for some category of Markov processes and projective limits. - Uspechi Matem.Nauk, XXXI, 1976, 2 (188), p.209-210 (in Russian).
10. Bendikov A.D. On harmonic functions for some class of diffusion processes on a group. - Theor.Probab. Appl., X, 4, 1975, p.773-784 (in Russian).
11. Heyer H. Probability measures on locally compact groups. - Moscow: Mir, 1981 (in Russian).
12. Berg C. Potential theory on the infinite dimensional torus. - Invent.Math., 32, 1976, p.49-100.
13. Bendikov A.D., Pavlov I.V. Boundness in $L_p(T^\infty)$ of some class of vector multiplier operators. - Sibir.Math.J., N 1, 1986, p.3-10 (in Russian).

14. Stein E.M., Weiss G. Introduction to Fourier analysis on Euclidean spaces. - Moscow: Mir, 1974 (in Russian).
15. Bendikov A.D. On weak solutions of elliptic equations on the group. - The 8th school on operator's theory in functional spaces. Abstracts of communications, Riga, 1983, p.24.

Prob. Theory and Math. Stat., Vol.1, pp. 171-187
Prohorov et al. (eds)
1986 VNU Science Press

LOWER ESTIMATES OF THE CONVERGENCE RATE IN THE CLT IN BANACH SPACES

V. Bentkus

Institute of Mathematics and Cybernetics
of the Academy of Sciences of the Lithuanian SSR,
Vilnius State University
232021 Vilnius, Akademijos 4, USSR

Lower estimates of the convergence rate in the Central Limit Theorem (below briefly "estimates") in infinite-dimensional and finite-dimensional Banach space are obtained. For example, in the infinite-dimensional case the best possible upper estimate of Prohorov metric can not have the order better than $O(n^{-1/8})$. Our results emphasize the essential difference between infinite-dimensional and finite-dimensional cases. Proper upper estimates of Prohorov and some metrics in Banach spaces were obtained earlier by the author and A. Rachkauskas [1-3].

Suppose that B is a real separable Banach space with norm $|\cdot|_B$, B^* is the dual of B, $E \subset B$ is a Banach space such that the identical embedding $E \subset B$ is linear and continuous, $X, X_1, X_2, \ldots \in E$ are independent identically distributed random elements of the space E, $EX=0$, $Y \in E$ is a Gaussian random element, $EY=0$, the covariances of X and Y coincide, $cov X = cov Y$. Denote

$$S_n = S_n(X) = n^{-1/2}(X_1 + \ldots + X_n).$$

Let $\mathcal{A}$ be a class of subsets of the space B. Let us define the Prohorov-Lévy metric $\pi(X, Y; \mathcal{A}) =$ $= \inf\{\varepsilon > 0 : P\{X \in A\} \leq P\{Y \in A^{\varepsilon}\} + \varepsilon$ and

$P\{Y \in A\} \leq P\{X \in A^{\varepsilon}\} + \varepsilon$ simultaneously for all $A \in \mathcal{A}\}$, where $A^{\varepsilon} = \{x \in B: \exists y \in A$ such that $|x-y|_B < \varepsilon\}$. We write $\pi(\cdot,\cdot;\mathcal{A}) = \pi(\cdot,\cdot)$ if the class $\mathcal{A}$ coincides with the class of all Borel subsets.

Define the metric

$$\zeta(X,Y;\mathcal{F}) = sup_{f\in\mathcal{F}}\left|\int_B f(x)\,(L(X)-L(Y))\,(dx)\right|,$$

where $\mathcal{F}$ is a class of functions $f: B \to R^1$, $L(X)$ is the distribution of X. We write $\zeta(\cdot,\cdot;\mathcal{F}) = \zeta(\cdot,\cdot)$ if $\mathcal{F}$ coincides with the class of functions satisfying $|f(x)| \leq 1$, $|f(x)-f(y)| \leq |x-y|_B$. Recall that $\pi^2 \leq \zeta \leq 3\pi$ (see [4]).

In this paper we consider only lower estimates of π and ζ. Lower estimates of uniform metrics and lower estimates of the closeness of moments see in [5–8].

Denote $\eta(\cdot,\cdot) = \zeta(\cdot,\cdot;\mathcal{F}_0)$, where $\mathcal{F}_0$ is the class of functions $f: B \to R^1$ such that $f(x) = g(u(x))$ with some $u \in B^*$, $|u|_{B^*} = 1$ and $g: R \to R^1$, $|g(x)| \leq 1$, $|g(x)-g(y)| \leq |x-y|$. The metric η one can consider as a "minimal" among all "infinite-dimensional" metrics $\zeta(\cdot,\cdot;\mathcal{F})$. In particular $\eta \leq \zeta$.

Theorem 1. Suppose that the space E is infinite-dimensional, $\beta \geq 2$, a sequence $\theta_n \downarrow 0$. There exists a symmetric random element $X \in E$, $E|X|_E^{\beta} < \infty$ such that

$\eta(S_n, Y) \geq \theta_n n^{1/\beta - 1/2}$ for infinitely many n.

Proof of this and other theorems are given below.

Under some assumptions on spaces $E \subset B$ in [1–2] the estimate

$$\zeta(S_n, Y) = \mathcal{O}(n^{-1/6}(E|X|_E^3)^{1/3})$$

is obtained. Theorem 1 shows that this estimate has the unimprovable order in n.

Define the metric $L_0(X,Y) = sup\{L(z(X), z(Y))\}$, where $L(U,V)$ is the Lévy distance between real random variables $U, V \in R^1$ and the sup is taken over all $z \in B^*$, $|z|_{B^*} = 1$. Obviously, the metric L_0 coincides with the Prohorov metric on the class of all half-spaces of B. Thus one can consider the metric L_0 as the "minimal" among all "infinite-dimensional" Prohorov-Lévy metrics. In particular $L_0 \leqslant \pi$.

Theorem 2. Suppose that the space E is infinite-dimensional, $\beta \geqslant 2$, a sequence $\theta_n \downarrow 0$. There exists a symmetric random element $X \in E$, $E|X|_E^\beta < \infty$ such that

$$L_0(S_n, Y) \geqslant \theta_n n^{-(\beta-2)/(2\beta+2)} \quad \text{for infinitely many } n.$$

Under some conditions on the spaces $E \subset B$ in [1-2] the estimate

$$\pi(S_n, Y) = \mathcal{O}(n^{-1/8}(E|X|_E^3)^{1/4})$$

is obtained. Theorem 2 shows that this estimate has the unimprovable order in n.

When $\beta = 3$ and the space $E = B$ is a Hilbert one in [9], a result similar to Theorem 2 was obtained earlier.

Corollary 3. Suppose that H is a real separable infinite-dimensional Hilbert space, $\beta \geqslant 2$, $\theta_n \downarrow 0$, a sequence $\mu_s > 0$ is bounded. There exists a symmetric random element $X \in H$, $E|X|_H^\beta < \infty$ such that

$$\eta(S_n, Y) \geqslant \theta_n n^{1/\beta - 1/2}, \tag{1}$$

$$L_0(S_0, Y) \geqslant \theta_n n^{-(\beta-2)/(2\beta+2)} \tag{2}$$

for infinitely many n and the eigenvalues α_s of the covariance operator of X satisfy $\overline{\lim}_s \alpha_s/\mu_s < \infty$.

Under some restrictions in [10] upper estimates of Prohorov metric in Hilbert space were obtained. Corollary 3 gives lower bounds for the order in n in these estimates.

Let G and H be Hilbert spaces, $A: H \longrightarrow G$ be a (linear) operator. We say that A is of class S_p, $0 < p < \infty$ if there exists a commutative diagram

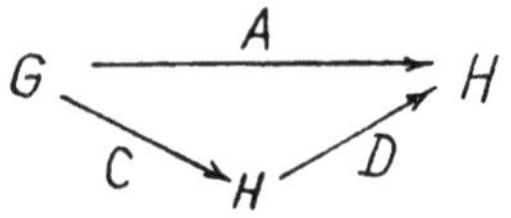

such that the operator C is continuous and the operator $(DD^*)^{p/2}: H \longrightarrow H$ is nuclear, $Tr\,(DD^*)^{p/2} < \infty$. According to the definition the class S_∞ coincide with the class of compact operators. Let $\ell_{2,\lambda}$ be the Hilbert space of sequences with the finite norm $|x|^2_{2,\lambda}$ = = $\sum_{i=1}^{\infty} x_i^2 / \lambda_i^2$. A simply verification shows that the identical embedding $\ell_{2,\lambda} \subset \ell_2$ is the operator of class S_p if the series $\sum_{i=1}^{\infty} |\lambda_i|^{p/2}$ is convergent (or $\lim_i \lambda_i = 0$ if $p = \infty$).

<u>Theorem 4</u>. Suppose that $0 < q < \infty$, a sequence $\theta_n \downarrow 0$. There exist an identical embedding $\ell_{2,\lambda} \subset \ell_2$ of class S_q and a symmetric bounded in the space $\ell_{2,\lambda}$ random element X , $P\{|X|_{2,\lambda} = 1\} = 1$ such that

$$\zeta(S_n, Y) \geqslant \theta_n n^{-1/q}, \tag{3}$$

$$\pi(S_n, Y;\ C) \geqslant \theta_n n^{-1/q} \tag{4}$$

for infinitely many n . Here $\pi(\cdot,\cdot;\ C)$ stands for Prohorov metric on the class C of all convex subsets of the space ℓ_2. The factor $n^{-1/q}$ in (3) and (4) should be omitted if $q = \infty$.

<u>Remark</u>. Suppose that $\theta_n \downarrow 0$. Theorem 4 yields the existence of a symmetric bounded random element $X \in \ell_2$ such that $\pi(S_n, Y;\ C) \geqslant \theta_n$ for infinitely many n . A similar result can be obtained in the case of metric

$\zeta(\cdot,\cdot;\mathcal{F})$, where the class $\mathcal{F}$ is the class of all convex functions $f\colon \ell_2 \longrightarrow R^1$ such that

$$\| f^{(k)}(x) - f^{(k)}(y) \| \leqslant |x-y|_{\ell_2}, \quad k = 0,1.$$

<u>Theorem 5</u>. Suppose that a Banach space $E \neq \{0\}$. Denote by R^1 a one-dimensional subspace of E .

i) Suppose that a class $\mathcal{F}$ of functions $f\colon B \longrightarrow R^1$ is such that the class $\mathcal{F}_1 = \{ f|_{R^1} \colon f \in \mathcal{F} \}$ of restrictions contains the class of all functions $g\colon R^1 \longrightarrow R^1$ such that $|g(s) - g(t)| \leqslant |s-t|$. Suppose that with some $\alpha, \gamma \geqslant 0,\ \beta > 2,\ c = c(E, B, \alpha, \beta, \gamma)$ the estimate

$$\zeta(S_n, Y; \mathcal{F}) \leqslant c n^{-\alpha} \nu_\beta^\gamma \tag{5}$$

is valid, where $\nu_\beta = \int_B |x|_E^\gamma \, |L(X) - L(Y)|(dx)$.
Then necessarily $\alpha \leqslant 1/2 - 1/\beta,\ \beta\gamma = 1$. In particular, $\alpha \leqslant 1/6,\quad \gamma = 1/3$ if $\beta = 3$.

ii) Suppose that a class $\mathcal{A}$ of convex subsets of B is such that the class $\mathcal{A}_1 = \{ A \cap R^1\colon A \in \mathcal{A} \}$ contains the class of all intervals $(-\infty, t)$. Suppose that the estimate

$$\pi(S_n, Y; \mathcal{A}) \leqslant c(E, B, \alpha, \beta, \gamma) n^{-\alpha} \nu_\beta^\gamma \tag{6}$$

is valid. Then necessarily $\gamma \leqslant 1/(\beta+1),\ \alpha \leqslant \gamma(\beta-2)/2$. In particular, $\gamma \leqslant 1/4,\ \alpha \leqslant \gamma/2$ if $\beta = 3$.

One can replace in (5) and (6) the pseudomoment ν_β by the moment $E\,|X|_E^\beta$.

Theorem 5 shows that upper estimates in [1,2] have a proper shape.

Let $\ell_q^k,\ q \geqslant 1$ be the k-dimensional Banach space of real sequences $x = (x_1, \ldots, x_k)$ with the norm $|x|_q^q = |x_1|^q + \ldots + |x_k|^q$.

Theorem 6. Suppose that $E = B = \ell_q^k$, $q \geqslant 1$ and that with some $\alpha, \theta \geqslant 0$, $\beta > 2$ the estimate

$$\zeta(S_n, Y) \leq c(\alpha, \theta, \beta, q) k^{\theta} n^{-\alpha} (E|X|_q^{\beta})^{1/\beta} \qquad (7)$$

is valid. Then necessarily $\alpha \leq 1/\beta - 1/2$, $\theta \geqslant \alpha - 1/2 + 1/q$. In particular, $\theta \geqslant 1/6$ if $\beta = 3$, $q = 2$, $\alpha = 1/6$. One can replace the moment $E|X|_2^{\beta}$ in (7) by the pseudomoment $\int_{\ell_2^k} |X|_2^{\beta} \, |L(X) - L(Y)|(dx)$ if $q = 2$.

Theorem 7. Suppose that $E = B = \ell_q^k$, $q \geqslant 1$ and that with some θ, $\alpha \geqslant 0$, $\beta > 2$ the estimate

$$\pi(S_n, Y; C) \leq c(\theta, \alpha, \beta, q) k^{\theta} n^{-\alpha} (E|X|_q^{\beta})^{1/(\beta+1)} \qquad (8)$$

is valid. Then necessarily $\alpha \leq (\beta - 2)/(2\beta + 2)$, $\theta \geqslant \alpha + \beta(1/q - 1/2)/(\beta + 1)$. In particular, $\theta \geqslant \alpha$ if $q = 2$. One can replace the moment in (8) by the corresponding pseudomoment if $q = 2$.

If $q = 2$ the estimate (8) in [11] is obtained with $\alpha = \theta = 1/8$, $\beta = 3$. Theorem 7 shows that this estimate depends on dimension in the proper way.

Let us define a real random variable

$$P\{X = -a\} = P\{X = a\} = p, \quad P\{X = 0\} = 1 - 2p, \qquad (9)$$

where $a > 0$, $0 < 2p \leq 1$. A simply verification shows that $EX = 0$, $\sigma^2 = EX^2 = 2a^2 p$, $E|X|^{\beta} = 2a^{\beta} p$. As usually by Y we denote a Gaussian random variable such that $EY = 0$, $\sigma^2 = EY^2$.

Lemma 8. Let $n \geqslant 1$ be an integer, the random variable X be defined in (9). Let us define the function $f: R^1 \longrightarrow R^1$

$$f(x) = \begin{cases} 0, & \text{if } x \leq 0 \text{ or } x \geqslant an^{-1/2}, \\ x, & \text{if } 0 < x \leq an^{-1/2}/2, \\ an^{-1/2} - x, & \text{if } an^{-1/2}/2 \leq x \leq an^{-1/2}. \end{cases}$$

Then

i) $|f(x)| \leq an^{-1/2}$,

ii) $Ef(S_n(X)) = 0$,

iii) $Ef(Y) \geq \pi^{-1/2} a p^{1/2}(1 - \exp\{-1/(16pn)\})$.

Proof. ii) Note that $P\{X = kan^{-1/2},\ k = 0, \pm 1, \ldots\} = 1$.

iii) Suppose that a random variable η is distributed normally with parameters 0 and 1. Then

$$Ef(Y) \geq EYI\{0 < Y < an^{-1/2}/2\} =$$

$$= \sqrt{2p}\, aE\eta I\{0 < \eta < (8pn)^{-1/2}\} = \sqrt{p/\pi}\, a(1 - \exp\{-1/(16pn)\}),$$

which yields iii). By $I\{A\}$ we denote the indicator function of the event A.

Lemma 9. Let a random variable X be defined in (9). Denote $\delta_n = P\{S_n(X) = 0\}$. If $pn \leq 1/2$, then $1 - 2pn \leq \delta_n \leq 1 - 2pn + 5(pn)^2$.

Proof. Obviously

$$\delta_n \geq P\{X_1 = \ldots = X_n = 0\} = (1-2p)^n \geq 1 - 2pn.$$

Let us prove the right-hand side inequality of Lemma. A little calculation involving characteristic functions shows that $\delta_n = (1 - 2p)^n + R$, where

$$R = \sum_{2 \leq 2k \leq n} \binom{n}{n-2k}\binom{2k}{k} p^{2k}(1-2p)^{n-2k} \leq (pn)^2/(1-(pn)^2).$$

Here we applied inequalities $1 - 2p \leq 1$, $\binom{n}{n-2k}\binom{2k}{k} \leq n^{2k}$ and summed a geometrical progression. Similarly

$$(1-2p)^n \leq 1 - 2pn + 2(pn)^2/(1-pn).$$

Collecting estimates and noting that $pn \leq 1/2$, we complete the proof.

Lemma 10. Let the random variable X be defined in (9). If $100pn \leq 1$, then

$$L(S_n(X), Y) \geq 2^{-1} \min\{an^{-1/2};\ pn\}.$$

Proof. According to the definition of Lévy metric it is sufficient to prove that for every $\varepsilon < 2^{-1}\min\{an^{-1/2};\ pn\}$ the inequality

$$P\{S_n \leq -an^{-1/2}\} > P\{Y \leq -an^{-1/2} + \varepsilon\} + \varepsilon \quad (10)$$

is valid (we consider distribution functions near the point $-an^{-1/2}$; in addition we can assume that these functions are left-continuous). We have $P\{Y \leq -an^{-1/2} + \varepsilon\} \leq P\{Y \leq an^{-1/2}/2\}$ since $2\varepsilon < an^{-1/2}$. Thus, the inequality

$$1 - P\{S_n = 0\} \geq P\{|Y| \geq an^{-1/2}/2\} + 2\varepsilon$$

implies (10). By Lemma 8 $1 - P\{S_n = 0\} \geq 2p - 5(pn)^2 \geq 3pn/2$ since $pn \leq 1/10$. Noting that $2\varepsilon \leq pn$ we see, that it is sufficient to prove that $pn \geq 2P\{|Y| \geq an^{-1/2}/2\}$.

Denote by η a random normally distributed with parameters 0 and 1 random variable. Put $x^2 = 1/(4pn)$. Evidently, now it is sufficient to show that $8x^2 P\{|\eta| \geq x\} \leq 1$. But this inequality is obvious since $x = (4pn)^{-1/2} \geq 5$ and $100pn \leq 1$.

Lemma 11. Suppose that the space $E \subset B$ is infinite-dimensional. There exist a sequence $(e_k)_{k \geq 1}$ of elements of E and a sequence $(e_k^*)_{k \geq 1}$ of elements of B^* such that $|e_k^*|_{B^*} = 1$, $e_k^*(e_k) = 1$, $e_k^*(e_j) = 0$ if $k \neq j$.

Proof of Lemma 11 is anlogous to the proof of similar Proposition I.f.3 from [12].

Proof of Theorem 1. Let $(e_k)_{k \geq 1}$ and $(e_k^*)_{k \geq 1}$

be sequences of elements of E and B from Lemma 11. Denote $\lambda_k = |e_k|_E$ and choose a sequence $n_s \uparrow \infty$ of integers such that

$$2\sum_{s=1}^{\infty} n_s^{-1} \leq 1, \quad \sum_{s=1}^{\infty} \lambda_s^{\beta}\, \theta_{n_s}^{\beta} < \infty, \quad n_s^{1/\beta - 1/2}\theta_{n_s} \leq 1.$$

Put $p_s = n_s^{-1}$, $a_s = n_s^{1/\beta}\theta_{n_s}$ and define the symmetric random element

$$P\{X = -a_s e_s\} = P\{X = a_s e_s\} = p_s, \quad P\{X = 0\} = 1 - 2\sum_{s=1}^{\infty} p_s.$$

We have

$$E|X|_E^{\beta} = 2\sum_{s=1}^{\infty} a_s^{\beta}\lambda_s^{\beta} p_s = 2\sum_{s=1}^{\infty} \lambda_s^{\beta}\theta_{n_s} < \infty.$$

Obviously $P\{e_s^*(X) = a_s\} = P\{e_s^*(X) = -a_s\} = p_s$ $P\{e_s^*(X) = 0\} = 1 - 2p_s$. Let $f = f_s$ be the function from Lemma 8 with parameters $a = a_s$, $n = n_s$. Consider the function $g_s(x) = f_s(e_s(x))$, $x \in B$. Using Lemma 8 we obtain

$$Eg_s(S_{n_s}(x)) = Ef_s(S_{n_s}(e_s^*(x))) = 0,$$

$$\eta(S_{n_s}(X), Y) \geq Eg_s(Y) = Ef_s(e_s^*(Y)) \geq$$

$$\geq \sqrt{p_s/\pi a_s}\,\left[1 - \exp\{-1/(16 p_s n_s)\}\right] \geq c\,\theta_{n_s} n_s^{1/\beta - 1/2},$$

where the constant $c > 0$ is absolute.

Proof of Theorem 2. Let $(e_k)_{k\geq 1}$ and $(e_k^*)_{k\geq 1}$ be sequences of elements of E and B^* from Lemma 11.

Denote $\lambda_k = |e_k|_E$ and choose a sequence $n_s \uparrow \infty$ of integers such that

$$\sum_{s=1}^{\infty} \lambda_s^{\beta}\, \theta_{n_s}^{\beta+1} < \infty, \quad 2\sum_{s=1}^{\infty} n_s^{-3\beta/(2\beta+2)}\theta_{n_s} \leq 1,$$

$$100\, n_s^{-(\beta-2)/(2\beta+2)}\,\theta_{n_s} \leq 1.$$

Put

$$\alpha_s = n_s^{3/(2\beta+2)}\theta_{n_s}, \qquad p_s = n_s^{-3\beta/(2\beta+2)}\theta_{n_s}.$$

Define the symmetric random element

$$P\{X = -\alpha_s e_s\} = P\{X = \alpha_s e_s\} = p_s, \quad P\{X = 0\} = 1 - 2\sum_{s=1}^{\infty} p_s.$$

We have

$$E\,|X|_E^{\beta} = 2\sum_{s=1}^{\infty} \lambda_s^{\beta}\alpha_s^{\beta} p_s = 2\sum_{s=1}^{\infty} \lambda_s^{\beta}\,\theta_{n_s}^{\beta+1} < \infty.$$

Obviously $P\{e_s^*(X) = \alpha_s\} = P\{e_s^*(X) = -\alpha_s\} = p_s$, $P\{e_s^*(X) = 0\} = 1 - 2p_s$ and $100\, p_s n_s \leq 1$. Using Lemma 10 we obtain

$$L_0(S_{n_s}(X), Y) \geq L\big(e_s^*(S_{n_s}(X)), e_s^*(Y)\big) = L\big(S_{n_s}(e_s^*(X)), e_s^*(Y)\big) \geq$$

$$\geq 2^{-1}\min\{\alpha_s n_s^{-1/2};\ p_s n_s\} = 2^{-1}\theta_{n_s} n_s^{-(\beta-2)/(2\beta+2)}.$$

Proof of Corollary 3. Obviously we can assume that $H = \ell_2$ and that the covariance operator of X is diagonal. Define the norm $|x|_1^2 = \sum_{s=1}^{\infty} x^2(s)/\mu_s$, where $x = (x(1), x(2), \dots) \in \ell_2$. According to the Theorems 1 and 2 there exists a symmetric random element $x \in \ell_2$,

$E|X|_1^\beta < \infty$ satisfying (1) and (2). We have

$$\alpha_s/\mu_s = EX^2(s)/\mu(s) \le (E|X|_1^\beta)^{\beta/2} < \infty,$$
$$E|X|_H^\beta \le \sup_s \mu_s^{\beta/2} E|X|_1^\beta < \infty,$$

which proves the corollary.

Let $(e_s)_{s=1}^k$ be the standard basis of R^k, $0 < p \le 1/(2k)$. Define a symmetric random element $X \in R^k$

$$P\{X=-ae_i\} = P\{X=ae_i\} = p,\ 1 \le i \le k,$$
$$P\{X=0\} = 1-2kp. \tag{11}$$

Define the function $g:\ R^1 \longrightarrow R^1$

$$g(t) = \begin{cases} 0, & \text{if } t \le 0 \text{ or } t \geqslant an^{-1/2}, \\ t, & \text{if } 0 < t < an^{-1/2}/2, \\ an^{-1/2} - t, & \text{if } an^{-1/2}/2 \le t \le an^{-1/2}. \end{cases}$$

Put

$$f(x) = k^{-1/2} \sum\nolimits_{s=1}^k g(x_s),\quad x = (x_1, \ldots, x_k) \in R^k. \tag{12}$$

__Lemma 12__. Let the random element X and the function be defined in (11) and (12). Then $Ef(S_n(X)) = 0$

$$Ef(Y) \geqslant \sqrt{kp/\pi}\ a\left(1 - \exp\{-1/(16pn)\}\right).$$

Here as usually the symmetric Gaussian random element has the covariance of X.

__Proof__. We have

$$Ef(Y) = k^{-1/2} \sum_{s=1}^k Eg(Y_s) = k^{1/2} Eg(Y_1).$$

Now Lemma 8 yields the estimate for $Ef(Y)$. Obviously $Ef(S_n) = 0$.

Lemma 13. Consider the Prohorov metric $\pi(\cdot,\cdot;C)$ on the class of convex subsets of ℓ_q^k, $q \geqslant 1$. The space ℓ_q^k is linearly isomorphic to R^k. Let the random element $X \in \ell_q^k$ be defined in (11). Then

$$\pi(S_n(X), Y; C) \geqslant \min\{1-2\pi/k;\ a k^{1/q}((p/\pi)^{1/2} - n^{1/2}/k)\}. \quad (13)$$

Proof. Consider the set $A = \{x \in \ell_q^k : |x|_1 \leqslant a n^{1/2}\}$, where $|\cdot|_q$ denotes the norm of ℓ_q^k. Let A^ε be the ε-neighbourhood of the set A in the space ℓ_q^k. We have

$$A^\varepsilon \subset \{|x|_1 < a n^{1/2} + k^{1-1/q}\}\varepsilon. \quad (14)$$

According to the definition of metric $\pi(\cdot,\cdot;C)$ it is sufficient to show that for every $\varepsilon > 0$ smaller than the right-hand side of (13) the inequality $P\{S_n \in A\} > \varepsilon + P\{Y \in A^\varepsilon\}$ is fulfilled. The inequality $|S_n|_1 \leqslant a\sqrt{n}$ and (14) show that it is sufficient to verify

$$1 > \varepsilon + P\{|Y|_1 < a n^{1/2} + k^{1-1/q}\varepsilon\}. \quad (15)$$

Suppose that independent random variables $\eta_1, \dots, \eta_k$ are distributed normally with parameters 0 and 1, $\xi_i = |\eta_i| - E|\eta_i|$. Then

$$|Y|_1 = a(2p)^{1/2}(|\eta_1| + \dots + |\eta_k|), \quad E|\eta_1| = (2/\pi)^{1/2}.$$

Thus,

$$P\{|Y|_1 < a n^{1/2} + k^{1-1/q}\varepsilon\} \leqslant P\{\ \xi_1 + \dots + \xi_k \quad < -k(2\pi)^{-1/2}\} \leqslant$$

$$\leqslant 2\pi k^{-2} E(\xi_1 + \dots + \xi_k)^2 \leqslant 2\pi k$$

since $\varepsilon < a k^{1/q}((p/\pi)^{1/2} - n^{1/2}/k)$.

Now the inequality $1 > \varepsilon + 2\pi k$ (which is fulfilled since ε is smaller than right-hand side of (13)) yields (15).

Proof of Theorem 4. Let $k_s \uparrow \infty$, $s = 1, 2, \ldots, k_0 = 0$ be a sequence of integers. Let I_s be blocks of lenght k_s of integers

$$I_s = \{ m : 0 < m - k_0 - \ldots \quad - k_{s-1} \leq k_s \}, \quad s = 1, 2, \ldots .$$

Let us suppose that $q < \infty$ and prove (13). Denote $n_s = 2^{s-1} k_s$ and select a sequence k_s such that

$$\sum_{s=1}^{\infty} 2^{q(s+1)/2 - s - 1} \theta_{n_s}^{q} < \infty .$$

Denote $p_s = 1/n_s$,

$$\alpha_s = \pi^{1/2} \left(1 - \exp\{-1/16\} \right)^{-1} 2^{(s+1)/2} n_s^{1/q} \theta_{n_s} .$$

Define a random element $X \in \ell_2$ by $P\{X = \alpha_s e_i\} = P\{X = -\alpha_s e_i\} = p_s$ if $i \in I_s$. The random element X is correctly defined since $2 \sum_{s=1}^{\infty} k_s p_s = 2 \sum_{s=1}^{\infty} k_s n_s^{-1} = 1$. Let us define the sequence of weights $\lambda_i = \alpha_s^2$ if $i \in I_s$. The inclusion $\ell_{2,\lambda} \subset \ell_2$ is the operator of class S_q since

$$\sum_{i=1}^{\infty} \lambda_i^{q/2} = \sum_{s=1}^{\infty} k_s \alpha_s^q < \infty$$

in view of choise of k_s. In addition $P\{ |X|_{2,\lambda} = 1\} = 1$.

Let

$$f(x) = f_s(x) = k_s^{-1/2} \sum_{i \in I_s} g(x_i), \quad x = (x_1, x_2, \ldots) \in \ell_2$$

be the function from Lemma 12 with $n = n_s$, $\alpha = \alpha_s$. We have $|f_s(x)| \leq k_s^{-1/2} \alpha_s (2n_s)^{-1/2} \leq 1$ for sufficiently large s. In addition

$$|f_s(x) - f_s(y)| \le k_s^{-1/2} \sum_{i \in I_s} |x_i - y_i| \le |x-y|_2 .$$

Therefore Lemma 12 yields $Ef_s(S_{n_s}(X)) = 0$,

$$\zeta(S_{n_s}(X), Y) \ge Ef_s(Y) \ge$$

$$\ge (k_s p_s/\pi)^{1/2} a_s (1 - exp\{-(16p_s n_s)^{-1}\}) = \theta_{n_s} n_s^{-1/q}.$$

The proof of (3) when $q < \infty$ is completed. The proof of (3) is similar if $q = \infty$.

Let us prove (4) when $q < \infty$. Choose sequences $n_s \uparrow \infty$ and k_s such that $n_s = 2^{-(s+5)} k_s$ and

$$1 - 2\pi/k_s \ge 1/2 , \quad n_s^{-1/q} \theta_{n_s} < 1/2 ,$$

$$\sum_{s=1}^{\infty} 2^{(1+q/2)(s+5)} \theta_{n_s}^q < \infty .$$

Denote

$$p_s = 2^{-s-1} k_s^{-1} , \qquad a_s = 2^{(s+5)/2} n_s^{-1/q} \theta_{n_s}$$

and define the random symmetric element $X \in \ell_2$ by $P\{X = -a_s e_i\} = P\{X = a_s e_i\} = p_s$ if $i \in I_s$.

The random element X is correctly defined since $2\sum_{s=1}^{\infty} k_s p_s = 1$.

Put $\lambda_i = a_s^2$ if $i \in I_s$. Then

$$\sum_{i=1}^{\infty} \lambda_i^{q/2} = \sum_{s=1}^{\infty} k_s a_s^q < \infty$$

and therefore the inclusion $\ell_{2,\lambda} \subset \ell_2$ is the operator of class S_q . Obviously $P\{|X|_{2,\lambda} = 1\} = 1$.

Define the convex set $A_s = \{x \in \ell_2 : \sum_{i \in I_s} |x_i| \leq a_s n_s^{1/2}\}$. Then Lemma 13 with $q = 2$ yields

$$\pi(S_{n_s}(X), Y; C) \geq$$
$$\geq \min\{1 - 2\pi/k_s;\ a_s k_s^{1/2}((p_s/4)^{1/2} - n_s^{1/2}/k_s)\} \geq$$
$$\geq \min\{1/2;\ n_s^{-1/q}\ \theta_{n_s}\} = \theta_{n_s} n_s^{-1/q}.$$

The proof of (4) is similar when $q = \infty$.

Proof of Theorem 5. During the proof of Theorem we can assume that $E = B = R^1$, $\zeta(\cdot,\cdot;\mathcal{F}) = \zeta$, $\pi(\cdot,\cdot;\mathcal{A}) = L$ since norms of E and B on the real line R^1 are equivalent. In addition on the real line R^1 $\nu_\beta \leq c(\beta) E|X|^\beta$ and therefore during the proof in (5) and (6) we can replace ν_β by $E|X|^\beta$.

Let us prove (5). Let the random variable X and the function f be the same as in Lemma 8. Lemma 8 and (5) imply $n^{1/2-\alpha-\gamma} a^{\beta\gamma-1} \geq c_1 > 0$. Putting $a \uparrow \infty$ we obtain $\beta\gamma \geq 1$. Putting $a \downarrow 0$ we ontain $\beta\gamma \leq 1$. Thus, $\beta\gamma = 1$. Therefore $n^{1/2-\alpha-\gamma} \geq c_1 > 0$ yields $\alpha \leq 1/2 - 1/\beta$.

Let us prove (6). Let the random variable X be the same as in Lemma 10. Put $a = n^{-s}$, $p = n^{-s-3/2}$, $s > -1/2$ in the estimate of Lemma 10. Then (6) yields $1/2 - \alpha - 3\gamma/2 + s(1 - \beta\gamma - \gamma) \geq 0$. Putting $s \uparrow \infty$ we obtain $\alpha \leq \gamma(\beta-2)/2$. Putting $s \downarrow -1/2$ we obtain $\alpha \leq \gamma(\beta-2)/2$.

Proof of Theorem 6. The inequality $\alpha \leq 1/2 - 1/\beta$ is valid in the view of Theorem 5.

Let us use Lemma 12. The function f from this Lemma satisfies

$$|f(x) - f(y)| \leq k^{-1/2} \sum_{s=1}^{k} |x_i - y_i| \leq k^{1/2-1/q} |x - y|_q .$$

Thus, (7) and Lemma 12 with $\alpha \leq 1$ imply $\theta \geq \alpha + 1/q - 1/2$. When $q = 2$ one can replace the moment in (7) by pseudo-moment since $E|Y|_2^{\beta} \leq c(\beta)(E|Y|_2^2)^{\beta/2} \leq c(\beta)E|X|_2^{\beta}$.

Proof of Theorem 7 is similar to that of Theorem 6, but Lemma 12 should be replaced by Lemma 13.

REFERENCES

1. Bentkus, V., Rachkauskas, A. An estimate of the convergence rate in the Central Limit Theorem in the Prohorov metric. - Dokl.Akad.Nauk USSR, 1982, vol. 264, No 3, pp. 521-524 (Russian).
2. Bentkus, V., Rachkauskas, A. Estimates of the distance between sums of independent random elements of Banach space. - Teor.Verojatn.i primen., 1984, vol. 29, No 1, pp. 49-64 (Russian).
3. Bentkus, V., Rachkauskas, A. On the convergence rate in the Central Limit Theorem in infinite-dimensional spaces. - Lit.matem.sb., 1981, vol. 21, No 4, pp. 9-18 (Russian).
4. Dudley, R.M. Distance of probability measures and random variables. - Ann.Math.Stat., 1968, vol. 39, No 5, pp. 1563-1572.
5. Bentkus, V. Lower estimates of the accuracy of normal approximation in the Central Limit Theorem in Banach spaces. - Lit.matem.sb., 1984, vol. 24, No 1, pp. 12-18 (Russian).
6. Bentkus, V. Lower estimates of the convergence rate in the Central Limit Theorem in Banach spaces. - Lit.matem.sb., 1985, vol. 25 (to appear, Russian).
7. Rhee, W.S. & Talagrand, M. Bad rates of convergence for the Central Limit Theorem in Hilbert space. - Ann.Probab., 1984, vol. 12, No 3, pp. 843-850.
8. Borisov, I.S. Lower estimates of the convergence rate in the Central Limit Theorem in Banach spaces. - Teor.Verojatn.i primen., 1984, vol. 29, No 3, pp. 604-605 (Russian).
9. Senatov, V.V. Lower estimates of the convergence rate in the Central Limit Theorem in Hilbert space. - Dokl.Akad.Nauk SSSR, 1981, vol. 256, No 6, pp. 1318-1321 (Russian).
10. Jurinskii, V.V. On the error of Gaussian approximation of convolutions. - Teor.Verojatn. i primen., 1977, vol. 22, No 2, pp. 242-253 (Russian).
11. Senatov, V.V. An estimate of the Lévy-Prohorov metric. - Teor.Verojatn. i primen., 1984, vol. 29, No 1, pp. 108-112 (Russian).

12. Lindenstrauss, J., Tzafriri, L. Classical Banach spaces. - Berlin-Heidelberg-New York, Springer-Verlag, 1977. - 188 p.

Prob. Theory and Math. Stat., Vol.1, pp. 189-198
Prohorov et al. (eds)
1986 VNU Science Press

WEAK LIMITS OF PROBABILITY MEASURES ON METRIC SCHAUDERSPACES

Harald Bergström
Mathematical Statistics
Chalmers University of Technology
and the University of Göteborg
S-412 96 Göteborg, Sweden

1. Schauderspaces

In [1] I have presented a reduction procedure for weak convergence of measures on normal spaces. For metric spaces method is submitted by Lemma III.6.1 in [1]. Here I shall consider this lemma for metric Schauderspaces which I am going to define. I then also give complements to the proofs of some theorems in Chapter V and Chapter VI of [1].

We say that a separable metric space S with metric ρ is a Schauderspace with respect to a given productspace $\tilde{S}$ if it has the following properties

1° S is mapped onto a finite dimensional product space $\tilde{S}^{(r)}$ by a measurable mapping π_r for $r = 1,2,\ldots$ where $\tilde{S}^{(r)}$ is a product space of separable metric spaces $\tilde{S}_i$, $i = 1,2,\ldots,k_r$, k_r positive integer, $k_r < k_s$ for $r < s$, the $\tilde{S}_i$ being factors in the infinite product space $\tilde{S}$. Under the usual product metric for product spaces S then is mapped by a continuous mapping (projection) onto $\tilde{S}^{(r)}$ and

$$\pi_r = \tilde{\pi}_{rs}\pi_s, \quad \tilde{\pi}_r = \tilde{\pi}_{rs}\tilde{\pi}_s \quad \text{for } r < s \tag{1}$$

where $\tilde{\pi}_{rs}$ is the projection of $\tilde{S}^{(r)}$ onto $\tilde{S}^{(s)}$ (Hence $\tilde{\pi}_{rs}$ is continuous).

2° $\tilde{S}^{(r)}$ is mapped by a continuous mapping V_r into S. Note that $V_r\pi_r$ is a measurable mapping of S into S and $V_r\tilde{\pi}_r$ is a continuous mapping of $\tilde{S}$ into S. Further the $\pi_r x$, $r = 1,2,\ldots$ determine an $\tilde{x} \in \tilde{S}$ such that

$$\tilde{\pi}_r\tilde{x} = \pi_r x, \quad r = 1,2\ldots \tag{2}$$

and consequently such that

$$V_r \pi_r x = V_r \tilde{\pi}_r \tilde{x} \qquad r=1,2,\ldots \tag{3}$$

We require that $V_r \pi_r x$ converges to x for every x as $r \to \infty$. Those $\tilde{x} \in \tilde{S}$ which satisfy [2] for some $x \in S$ (the same x for all r) form a subset of $\tilde{S}$ and the relations [2] define a mapping π of S into $\tilde{S}$. We call $V_r \pi_r x$ a Schauder approximation of x.

We say that μ is a measure on a topological or σ-topological space if μ is a measure, not necessarily σ-smooth on the algebra generated by the closed set. If μ is σ-smooth it can then be extended to a σ-smooth measure on the σ-algebra generated by the closed sets (compare [1] Section 1.13 and Section 1.14).

Our first problem for a Schauder space is the following one: When does a probability measure $\tilde{\mu}$ on $\tilde{S}$ determine a probability measure μ on S. We give an answer to this question:

Theorem 1: Let S be a metric Schauder space with respect to the infinite product space $\tilde{S}$. If $\{V_r \tilde{\pi}_r \tilde{x}\}$ converges a.s. $(\tilde{\mu})$ to elements in S for a probability measure $\tilde{\mu}$ on $\tilde{S}$, then $\{\tilde{\mu}\}$ determines a probability measure μ on S such that

(i) $$\lim_{r\to\infty} \int_{\tilde{S}} f(V_r \tilde{\pi}_r \tilde{x}) \tilde{\mu}(d\tilde{x}) = \int_S f(x) \mu(dx)$$

for any realvalued bounded continuous function from S.

Remark: The relation (i) means that the sequence $\{\tilde{\mu}(\tilde{\pi}_r^{-1} V_r^{-1} \cdot)\}$ of probability measures on S converges weakly to μ.

Corollary: Let S be given as in the theorem and suppose that Cauchy convergence of $\{V_r \tilde{\pi}_r \tilde{x}\}$ in the metric implies convergence to elements in S. If $\{V_r \tilde{\pi}_r \tilde{x}\}$ be Cauchy convergent in probability $(\tilde{\mu})$ in the metric, then $\tilde{\mu}$ determines a probability measure μ on S such that (i) in the theorem holds for a subsequence $\{r_k\}$ of the sequence $\{r\}$.

Remark: Note that the first condition is satisfied if S is complete in the metric.

Proof of the theorem: Let Ψ be the class of realvalued bounded continuous functions from S. Consider the functional

$$L^{(r)}(f) = \int_{\tilde{S}} f(V_r\tilde{\pi}_r\tilde{x})\tilde{\mu}(d\tilde{x}) \tag{4}$$

for $f \in \Psi$. By the made assumption $\{V_r\tilde{\pi}_r\tilde{x}\}$ converges a.s. $(\tilde{\mu})$ to elements in S. This means that there exists a measurable subset $\tilde{S}_0$ of S such that

$$\rho(V_r\pi_r x,\ x) \to 0 \qquad (r \to \infty) \tag{5}$$

for $\tilde{x} \in \tilde{S}_0$ with $x \in S$. Then by the continuity of f the sequence of functions

$$\tilde{f}_r(\tilde{x}) = f(V_r\tilde{\pi}_r x)$$

converges to a measurable function $\tilde{f}$ where

$$\tilde{f}(\tilde{x}) = f(x)$$

and by the dominated convergence theorem for integrals

$$L(f) = \lim_{r\to\infty} L^{(r)}(f) = \int_{\tilde{S}} \tilde{f}(\tilde{x})\tilde{\mu}(d\tilde{x}) \tag{6}$$

It follows by the properties of limits that L(f) so defined for $f \in \Psi$ is a bounded linear functional on Ψ. By Corollary 2 of Theorem II.4.1. in [1] (Alexandrov's first theorem) the functional L on Ψ determines a measure μ on S (not necessarily σ-smooth) such that

$$L(f) = \int_S f(x)\mu(dx) \tag{7}$$

For f identically equal to 1 on S we find by (4) - (6) that

$$\mu(S) = 1$$

Further, by the transformation

$$V_r\tilde{\pi}_r\tilde{x} = y$$

we may write

$$L^{(r)}(f) = \int_S f(y)\tilde{\mu}(\tilde{\pi}_r^{-1} V_r^{-1} dy) \tag{8}$$

Hence, by [6] and [8], the sequence of probability measures

$$\{\tilde{\mu}(\tilde{\pi}_r^{-1} V_r^{-1} \cdot)\}$$

on S converges weakly to μ and then μ is σ-smooth according to Alexandrov's third theorem (Theorem III.4.1 in [1]). □

Proof of the corollary: Let $\{V_r \tilde{\mu}_r \tilde{x}\}$ be Cauchy convergent a.s. $(\tilde{\mu})$ in the metric ρ. This means that

$$\tilde{\mu}\{x \in \tilde{S}: \ \rho(V_r \tilde{\pi}_r \tilde{x}, \ V_{r+\nu} \tilde{\pi}_{r+\nu} \tilde{x}) \geq \varepsilon\}$$

converges to 0, uniformly with respect to ν for $\nu = 1, 2, \ldots$ as $r \to \infty$ and this for any $\varepsilon > 0$. Proceeding as for realvalued random variables but considering the metric ρ and transforming the proof accordingly (for instance the proof in [4] p.116) we find that Cauchy convergence in probability $\tilde{\mu}$ of $\{V_r \tilde{\pi}_r \tilde{x}\}$ implies Cauchy convergence a.s. $\tilde{\mu}$ of a subsequence $\{V_{r_k} \tilde{\pi}_{r_k} \tilde{x}\}$ in the metric and by the made assumption this Cauchy convergence implies convergence a.s. $\tilde{\mu}$ to elements in S. □

2. Weak convergence of sequences of probability measures on Schauder spaces.

Theorem 2: Let $\{\mu_k\}$ be a sequence of probability measures on the metric Schauder space S with respect to the product space $\tilde{S}$ and suppose that $\{\mu_n(\pi_r^{-1} \cdot)\}$ converges weakly to a tight probability measure $\tilde{\mu}^{(r)}$ on $\tilde{S}^{(r)}$ for $r=1,2,\ldots$. Then the $\tilde{\mu}^{(r)}$ determine a probability measure $\tilde{\mu}$ on $\tilde{S}$ such that $\tilde{\mu}^{(r)} = \tilde{\mu}(\tilde{\pi}_r^{-1} \cdot)$ for $r=1,2,\ldots$. If furthermore

(i) $\{V_r \tilde{\pi}_r \tilde{x}\}$ converges a.s. $\tilde{\mu}$ to an element in S and

(ii) $\lim\limits_{r \to \infty} \limsup\limits_{n \to \infty} \mu_n\{x \in S: \rho(V_r \pi_r x, x) \geq \varepsilon\} = 0$ for any $\varepsilon > 0$

then $\{\mu_n\}$ converges weakly to a probability measure on S.

If Cauchy convergence of $\{V_r \tilde{\pi}_r \tilde{x}\}$ in the metric ρ implies convergence to an element in S, hence certainly if S is complete in ρ, then the theorem holds without the requirement (i).

Conversely, let the sequence $\{\mu_n\}$ of probability measures converge weakly to a probability measure μ and let the mappings π_r be continuous. Then $\{\mu_n(\pi_r^{-1} \cdot)\}$ converges weakly to a probability measure $\tilde{\mu}^{(r)}$ for all r and $\tilde{\mu}^{(r)} = \tilde{\mu}^{(s)}(\tilde{\pi}_{rs}^{-1}\cdot)$ for $r < s$. Further (ii) holds if

$$\lim_{r\to\infty} \rho(V_r \pi_r x, x) = 0$$

for $x \in S$.

Remark: If the sufficient conditions in the theorem are satisfied for a sequence $\{\mu_n\}$ of probability measures on S and thus this sequence converges weakly to a probability measure μ on S, then μ is uniquely determined by the $\tilde{\mu}^{(r)}$. This is an invariance principle. Hence if a sequence $\{\lambda_n\}$ of probability measures on S also satisfies the sufficient conditions and $\tilde{\mu}^{(r)} = \tilde{\lambda}^{(r)}$ for all r, then the two sequences $\{\mu_n\}$ and $\{\lambda_n\}$ converge to the same probability measure.

Proof: The proof is close to the proof of Lemma III. 6.1 in [1].

The weak convergence of $\{\mu_n(\pi_r^{-1}\cdot)\}$ to tight probability measures $\tilde{\mu}^{(r)}$ implies the weak convergence of $\mu_n(\pi_s^{-1}\tilde{\pi}_{rs}^{-1})$ to $\tilde{\mu}^{(s)}(\tilde{\pi}_{rs}^{-1}\cdot)$ since $\tilde{\mu}_{rs}$ is continuous but also to $\tilde{\mu}^{(r)}$ since $\tilde{\pi}_{rs}\pi_s = \pi_r$. Hence the $\tilde{\mu}^{(r)}$ form a consistent class of probability measures. By Kolmogorov's consistence theorem (Theorem 1.15.2 in [1]) the $\tilde{\mu}^{(r)}$ determine a probability measure $\tilde{\mu}$ on $\tilde{S}$ such that $\tilde{\mu}^{(r)} = \tilde{\mu}(\tilde{\pi}_r^{-1})$.

Applying Theorem 1 we find that $\tilde{\mu}$ under the made assumption determines a probability measure μ on S. It remains to proof that $\{\mu_n\}$ converges weakly to ν. To prove this it is sufficient to show that

$$\lim_{n\to\infty} \int_S f(x)\mu_n(dx) = \int_S f(x)\mu(dx)$$

for $f \in \Psi_0$ where Ψ_0 is the class of realvalued bounded uniformly continuous functions from S. This follows by the corollary of Theorem III. 1.1 in [1] (Alexandrov's second theorem). Compare now

$$L_n(f) = \int_S f(x)\mu_n(dx), \quad f \in \Psi_0$$

and

$$L_n^{(r)}(f) = \int_S f(V_r \pi_r x)\mu_n(dx), \quad f \in \Psi_0.$$

We get

$$|L_n(f) - L_n^{(r)}(f)| \leq \sup_{\rho(V_r\pi_r x,x)<\varepsilon} |f(x) - f(V_r\pi_r x)|$$

$$+ 2 \sup_{x\in S}|f(x)| \cdot \mu_n[x: \rho(V_r\pi_r x,x) \geq \varepsilon].$$

By (ii) the second term on the right hand side tends to 0 as in order $n\to\infty$, $r\to\infty$. The first term on this side is arbitrarily small for sufficiently small $\varepsilon > 0$. Hence

$$\lim_{r\to\infty} \limsup_{n\to\infty} |L_n(f) - L^{(r)}(f)| = 0, \quad f \in \Psi_0 \tag{1}$$

But

$$L^{(r)}(f) = \int_{\tilde{S}^{(r)}} f(V_r x^{(r)})\tilde{\mu}^{(r)}(dx^{(r)}) = \int_{\tilde{S}} f(V_r\tilde{\pi}_r\tilde{x})\tilde{\mu}(d\tilde{x})$$

and this relation holds for $f\in\Psi$. We have found in Section 1 that

$$L^{(r)}(f) \to L(f)$$

and that $L(f)$ determines a probability measure on S according to Theorem 1, provided that $\{V_r\tilde{\pi}_r\tilde{x}\}$ converges a.s. $(\tilde{\mu})$ to elements in S.

Consider now the case when we only require that Cauchy convergence of $\{V_r\tilde{\pi}_r x\}$ in the metric ρ implies convergence to elements in S. We shall show that the condition (ii) implies Cauchy convergence a.s. $(\tilde{\mu})$ in the metric ρ and so we may apply the corollary of Theorem 1 to show that $L^{(r)}(f)$ determines a probability measure on S and the reduction (1) follows as before.

The condition (ii) implies

$$\lim_{r\to\infty} \limsup_{n\to\infty} \mu_n\{\rho(V_r\pi_r x, V_s\pi_s x) \geq \varepsilon\} = 0 \tag{2}$$

uniformly for $s>r$. Now

$$\rho(V_r\pi_r x, V_s\pi_s x) = \rho(V_r\tilde{\pi}_{rs}x^{(s)}, V_s\tilde{x}^{(s)}) \tag{3}$$

where $V_r\tilde{\pi}_{rs}$ and V_s are continuous. Then the function

$$\tilde{x}^{(s)} \to \rho(V_r\tilde{\pi}_{rs}\tilde{x}^{(s)}, V_s\tilde{x}^{(s)})$$

is a continuous function from $\tilde{S}^{(r)}$ into R. This fact is easy to prove by Lemma I.8.1. in [1]. By the made assumption $\mu_n(\pi_s^{-1}\cdot)$ converges weakly to $\tilde{\mu}^{(s)}$. The function

$$h(\varepsilon) = \tilde{\mu}^{(s)} \{\tilde{x}^{(s)} \in \tilde{S}^{(s)} : \rho(V_r \tilde{\pi}_{rs} \tilde{x}^{(s)}, V_s \tilde{x}^{(s)} \geq \varepsilon\} \tag{4}$$

is non-increasing as ε increases and thus has only finitely many continuity points. For a continuity point ε of $h(\varepsilon)$ the set considered in (4) is a continuity set for $\tilde{\mu}^{(s)}$ (compare Theorem III. 1.1 in [1]). Hence

$$\mu_n\{x \in S: \rho(V_r \pi_r x, V_s \pi_s x) \geq \varepsilon$$

$$= \mu_n(\pi_s^{-1}\{\tilde{x}^{(s)} \in S^{(s)}: \rho(V_r \tilde{\pi}_{rs} \tilde{x}^{(s)}, V_s \tilde{x}^{(s)}) \geq \varepsilon\}$$

for such a point ε tends to

$$\tilde{\mu}^{(s)}\{\tilde{x}^{(s)} \in \tilde{S}^{(s)} : \rho(V_r \tilde{\pi}_{rs} \tilde{x}^{(s)}, V_s \tilde{x}^{(s)}) \geq \varepsilon\}$$

$$= \tilde{\mu}\{\tilde{x} \in \tilde{S} : \rho(V_r \tilde{\pi}_r \tilde{x}, V_s \tilde{\pi}_s \tilde{x}) \geq \varepsilon\}$$

and [2] implies

$$\lim_{r\to\infty} \tilde{\mu}\{\tilde{x} \in \tilde{S}: \rho(V_r \tilde{\pi}_r \tilde{x}, V_s \tilde{\pi}_s \tilde{x}) \geq \varepsilon\} = 0$$

for any $\varepsilon>0$, uniformly with respect to $s>r$, i.e. $\{V_r \tilde{\pi}_r \tilde{x}\}$ is Cauchy convergent in probability $\tilde{\mu}$ in the metric. □

3. Applications

It has been shown in [1], Section V.3, that the C-space and the D-space have the properties required for Schauder spaces and there $\tilde{S}$ is the product space $R^{(\infty)}$ of factors R. The C-space is complete in the uniform metric ρ_0. Hence the proof of Theorem V.4.1 in [1] follows directly by Theorem 2.1 above. The C-space is a Banach space. A sequence $\{x_j\}_{j=1}^{\infty}$ of elements x_j in a Banach space B over R is called a Schauder basis if for any $x \in B$ there exists a unique sequence $\{a_j\}_{j=1}^{\infty}$ of real numbers a_j so that

$$x = \sum_{j=1}^{\infty} a_j x_j \tag{1}$$

(For the properties of Banach spaces we refer to [3], Section 1 a). The given norm $\|x\|$ on a Banach space is equivalent to the norm

$$|||x||| = \sup_{r=1,2,..} \|\sum_{j=1}^{r} a_h x_j\|$$

in the meaning that the metrics $\|x-y\|$ and $|||x-y|||$ determine the same topology (σ-topology). The basis is said to be normalized if $\|x_j\| = 1$ for any j. In the following we only deal with a normalized basis. The mapping

$$x \to \{a_j\}_{j=1}^{r}$$

for x in (1) of B onto the product space $\tilde{S}^{(r)}$ of r factors R is continuous. This follows by Lemma 1.8.1 in[1] Clearly the mapping V_r defined by

$$(a_j)_{j=1}^{r} \to \sum_{j=1}^{r} a_j x_j$$

is also continuous and

$$P_r x = V_r \pi_r x = \sum_{j=1}^{r} a_j x_j$$

is then a continuous mapping of B into B. Further $V_r \pi_r x$ is a Schauder approximation of x and moreover

$$\|V_r \pi_r x - x\| \to 0 \qquad (r \to \infty).$$

Regarding the properties of a Banach space B over the infinite product space $\tilde{S} = R^{(\infty)}$ we find that B is a metric Schauder space. Applying Theorem 2.1 we then get, observing that B is complete:

Theorem 3: The Banach space B with normalized Schauder basis and with norm $\|x\|$ is a metric Schauder space. A sequence $\{\mu_n\}$ of probability measures on B converges weakly if and only if

(i) $\{\mu_n(\pi_r^{-1} \cdot)\}$ converges weakly for $r=1,2,\ldots$,

(ii) $\lim\limits_{r\to\infty} \limsup\limits_{n\to\infty} \mu_n\{x \in B: \|V_r \pi_r x - x\| \geq \varepsilon\} = 0\}$

for any $\varepsilon>0$.

Remark: In [2] I have considered weak convergence on Banach spaces and there also a condition $\tilde{\mu}(\tilde{S}_0) = 1$ was introduced but it was remarked that it could be removed. This is done here.

Consider now the D-space. In Section V.3 in [1] we have introduced Schauder approximations $\{V_r \pi_r x\}$ which are pure stepfunctions with steps only at points $t_j^{(r)} = j\, 2^{-r}$, $j=0,1,2,\ldots r$ $r=1,2,\ldots$. We denote the set of these points by Γ. The relation

$$\lim_{r\to\infty} \rho(V_r\pi_r x, x) = 0$$

holds in the Skorokhod metric ρ. The D-space is not complete in the weak form of the Skorokhod metric ρ considered in Section V of [1]. However we shall show that Cauchy convergence of $\{V_r\tilde{\pi}_r\tilde{x}\}$ in the metric ρ implies convergence to an element in S. Then Theorem 2 applies and Theorem V.4.2 in [1] follows.

Put $V_r\tilde{\pi}_r\tilde{x} = x_r$. Cauchy convergence of $\{x_r\}$ in the metric ρ means that there to r and s, $r<s$ exists a strictly increasing continuous function γ_{rs} from $[0,1]$ onto $[0,1]$ such that

$$\sup_{t\in[0,1]} |x_r(t)-x_s[\gamma_{rs}(t)]| < \varepsilon \tag{2}$$

$$\sup_{t\in[0,1]} |\gamma_{rs}(t)-t| < \varepsilon \tag{3}$$

for $r > r(\varepsilon)$ and any s. Here x_r is a pure right-continuous step-function with steps only at the points $t\in\Gamma_r$. By the definition of x_r we have $x_r(t) = \tilde{x}(t)$ for $t\in\Gamma_r$. Since Γ is dense in $[0,1]$ and $x_r(t) = \tilde{x}_r(\tau)$ with $\tau\in\Gamma_r$ for any $t\in[0,1]$, it follows from (2) and (3) that we may choose a sequence $\{r_k\}$ $k=1,2,\ldots$ such that

$$|x_{r_k}(t) - \tilde{x}(\tau)| < \frac{1}{k} \quad \text{for} \quad t \leq \tau < t + \frac{1}{k} \tag{4}$$

$\tau\in\Gamma$ and $t\in[0,1]$. We now define a function x on $[0,1]$ by putting $x(t) = x(t)$ for $t\in\Gamma$, $x(t) = \liminf_{\tau\to t^+,\tau\in\Gamma} x(\tau)$ for $t\notin\Gamma$. By (4) we then get

$$\sup_{t\in[0,1]} |x_{r_k}(t) - x(t)| \leq \frac{1}{k}$$

Hence x_{r_k} tends uniformly to x as $k\to\infty$. Since x_{r_k} belongs to D it follows from this inequality that also x belongs to D. The Cauchy convergence of $\{V_r\tilde{\pi}_r\tilde{x}\}$ then implies the convergence of $\{V_r\tilde{\pi}_r\tilde{x}\}$ to x. □

References:

Bergström, H. (1982). Weak convergence of measures, Academic Press New York.

Bergström, H. (1982). Weak convergence of convolution powers on certain Banach spaces; to be published in the proceedings of Colloquium on limit theorems in probability and statistics, June 21-26, Veszprem.

Lindenstrauss, J., Tzafriri, L. (1977). Classical Banach spaces 1, Ergebnisse der Mathematik und ihrer Grenzgebiete 92, Springer Verlag.

Loève, M. (1963). Probability theory, D. Van Nostrand comp.

Prob. Theory and Math. Stat., Vol.1, pp. 199-201
Prohorov et al. (eds)
1986 VNU Science Press

OPTIMAL CONSUMPTION AND INVESTMENT IN A STOCHASTIC MODEL

T. Björk
Department of Mathematics
Royal Institute of Technology
100 44 Stockholm
Sweden

1. INTRODUCTION

The method of dynamic programming has proved to be extremly successful in the theory of optimal control of completely observable diffusions. Both in its analytical (see e.g. Fleming-Rishel (1975)) and in its martingale (see e.g. Elliot (1982)) formulation it has provided quite general existence results. However, one major problem from a practical point of view is the well-known fact that generally you can not obtain an explicit solution of the Bellman equation.

In fact, only a handful of cases are known for which an analytical solution of the Bellman equation has been obtained. The best known example is of course the linear-quadratic Gaussian regulator (Fleming-Rishel (1975)) and its extensions, (Jacobsson (1977)). Another interesting example is Merton's optimal portfolio problem (Merton (1973)), (see also (Fleming-Rishel (1975)).

The object of this talk is to add one item to the (incomplete) list above, by presenting a model for which the optimal control law can be given explicitly. The model is roughly as follows. An economic agent (the consumer) has an initial wealth x(0), which heis allowed to spend in the time interval [0.T]. There are N consumer goods on which he can spend his money as he pleases, provided he does not get into debt. The prices of the consumer goods are stochastic processes, and the consumer's problem is to maximize his expected utility over the planning period [0,T].

Given a number of (rather restrictive) assumptions we are able to solve this problem explicitly, and it also turns out that Merton's model (Merton (1973)) can be adjoined to ours.

The talk is based on the paper (Björk et al.)

2. THE MODEL

To be more precise, let p be the price (vector-) process, let

$c(t,x,p)$ be your consumption rate at time t, (given wealth x and consumption c).

We assume that the components p_i $(i = 1,\dots,N)$ of the p process satisfies the following stochastic differential equations

(1) $$dp_i = \mu_i(p)dt + \sigma_i(p)dw_i. \qquad i = 1,\dots,N,$$

where $w_1,\dots,w_M$ are independent Wiener processes.

Given a fixed consumption policy, c, the wealth process x will satisfy

(2) $$dx(t) = [rx(t) - c(t,x(t),p(t))*p(t)]dt,$$

where $r = 1 + \sigma$ (σ = the rate of interest on capital) and * is the transpose operation.

Our problem is now to find the consumption policy c which maximizes

$$E[\int_0^{\tau\wedge T} U(t,c(t,x(t),p(t)))dt]$$

subject to the equations (1), (2) and the condition $c \geq 0$.

Here $T \in R_+$ is our planning horizon and τ is the stopping time $\inf\{t|x(t) = 0\}$.

3. THE MAIN RESULTS

Asume that U is of the form

(3) $$U(t,c) = e^{-\beta t} \prod_{i=1}^{N} c_i^{\alpha_i},$$

where $\alpha_i > 0$, $\sum_{i=1}^{N} \alpha_i = \alpha < 1$, $\beta > 0$.

Then the optimal consumption policy $\hat{c}$ is linear in x whereas the dependence on t and p is rather complicated.

Assume furthermore that the p_i process is of the form

(4) $$dp_i = \mu_i p_i dt + d_i p_i dw_i, \qquad i = 1,\dots,N,$$

where μ_i and d_i are constants $(i = 1,\dots,N)$.

Then the optimal control problem can be solved explicitly, and, in fact

$$\hat{c}_i(t,x,p) = \frac{x}{p_i} \cdot \frac{\alpha_i}{\alpha} \cdot g(t),$$

where g is an easily computable function.

REFERENCES

Björk, T., et al. Optimal Consumption with Stochastic Prices in Continuous Time. (To appear in the Journal of Applied Probability).

Elliott, R.J., (1982). Stochastic Calculus and Applications, Springer, New York.

Fleming, W.H., and Rishel, R.W.,(1975). Deterministic and Stochastic Optimal Control, Springer, New York

Jacobsson, D.H., (1977). Extensions of Linear Quadratic Control, Optimization and Matrix Theory. Academic Press, New York.

Merton, R.C., (1973). Optimum Consumption and Portfolio Rules in a Continuous Time Model, Journal of Economic Theory, Vol. 3, pp. 373-413.

Prob. Theory and Math. Stat., Vol.1, pp. 203-209
Prohorov et al. (eds)
1986 VNU Science Press

LARGE DEVIATIONS FROM CLASSICAL PATHS AND THE CLASSICAL LIMIT OF QUANTUM STOCHASTIC FLOWS.

Ph. Blanchard[1] Ph. Combe[2,3]

M. Sirugue[3] M. Sirugue-Collin[3,4]

1. Theoretische Physik, Universität Bielefeld - D 4800 Bielefeld - RFA
2. Dépt. de Physique, Université d'Aix-Marseille II - Marseille (France)
3. C.P.T. CNRS - Luminy, Case 907 - F 13288 - Marseille Cedex 9
4. U.E.R. de Physique, Université de Provence, Marseille (France)

INTRODUCTION:

In this paper we want to report on the following results :

1) It is possible to describe quantum mechanical time evolution as a stochastic flow on classical phase space. The underlying process being a stochastic jump process.

2) In the classical limit namely when the Planck constant tends to zero then the trajectories of the stochastic process approach the classical trajectories.

The main tool for this result is the Ventzel' theory of large deviations [Ventsel (1976)] which gives a strong estimate on the deviations from classical paths. Large deviation theory has already been used to study some problems connected with classical limit (see eg. Jona-Lasinio et al. (1981)) but in the context of diffusion processes. Introduction of jump processes for solving quantum problems has originally been done by Maslov and Chebotarev (1980) for the Schrödinger equation.

Quantum Mechanical Evolution as Stochastic Flow on Phase Space

If we consider a quantum system whose classical analogue has $\mathbb{R}^{2n}$ as phase space, a (pure) quantum state of the system is traditionally discribed by a normalized wave function $\psi \in L_2(\mathbb{R}^n, dx)$. However such an object is in general ill defined in the classical limit $\hbar \to 0$, where $\hbar$ is the Planck's constant. Another equivalent description of the state of the system is in term of its Wigner function $W(a)$, $a \in \mathbb{R}^{2n}$ (Wigner (1932)), such that any observable function f in $L_1(\mathbb{R}^{2n}, da)$ has an expectation in the state given by:

$$(2.1) \quad \langle f \rangle_{\text{quantum}} = \int_{\mathbb{R}^{2n}} f(a)\, W(a)\, da$$

In classical theory $w(a)\, da$ is a probability measure on phase space and if we define its Fourier transform as:

$$(2.2) \quad \widetilde{W}(a) = (2\pi)^{-n} \int_{\mathbb{R}^{2n}} W(b) \exp\{ i\sigma(a,b) \}$$

σ being the fundamental non degenerate symplectic form on $\mathbb{R}^{2n}$, $\widetilde{W}$ is completly characterized by the normalization $\widetilde{W}(o) = 1$ and the fact that it is of positive type. In quantum Mechanics it satisfies:

1- $\widetilde{W}(o) = 1$

$$2 - \sum_{i,j=1}^{N} \lambda_i \bar{\lambda}_j \exp\{-i\hbar/2\, \sigma(a_i, a_j)\}\, \widetilde{W}(a_i - a_j) \geq 0$$

$$\forall (\lambda_i)_{i=1\ldots N} \in \mathbb{C}, \ \forall \{a_i\}_{i=1..N} \in \mathbb{R}^{2n}$$

3 - $\lambda \in \mathbb{R} \to \widetilde{W}(a_1, \ldots, \lambda a_i, \ldots, a_{2n})$ is continuous $\forall i = 1,2 \ldots, 2n$

The standard Mackey-von Neumann theorem ensures that W is a convex combination of extremal elements of the form

(2.3) $$W(a) = (2\pi)^{-n} \int_{\mathbb{R}^n} d\xi\, \overline{\psi}(q+\hbar/2\,\xi)\, e^{ip.\xi}\, \psi(q-\hbar/2\xi)$$

for $a = (q,p)$, $q,p \in \mathbb{R}^n$ and some $\psi \in L_2(\mathbb{R}^n,dx)$, $\|\psi\| = 1$.

We consider those systems whose classical evolution if governed by an Hamiltonian function $H\ (q,p), q,p \in \mathbb{R}^n$ such that

(2.4 $$H\ (q,p) = p^2/2m + m\omega^2 q^2/2 + V(q,p)$$

for two positive constants m and ω and for a potential such that

(2.5) $$V(a) = \int_{\mathbb{R}^{2n}} d\mu(b) \cos(\sigma(a,b) + \varphi(b))$$

where μ is a bounded positive measure and φ is a smooth antisymmetric function. It corresponds to a (quantum) self adjoint operator defined on $C_0^\infty(\mathbb{R}^n)$

(2.6) $$(H\psi)(x) = -\frac{\hbar^2}{2m} \sum_{i=1}^{n} \frac{\partial^2}{\partial x_i^2} \psi(x) + \frac{m\omega^2}{2} \left(\sum_{i=1}^{n} x_i^2\right) \psi(x)$$

$$+\int_{\mathbb{R}^{2n}} d\mu(q',p')\ \exp\{i\varphi(a') + \frac{i\hbar}{2} q'.p' + ip'.x\}\, \psi(x-q')\ ,\ a' = (q',p') \in \mathbb{R}^{2n}$$

where q.p denotes the euclidean scalar product in $\mathbb{R}^n$ of q and p. With this definitionH extends to a self adjoint operator on $L_2(\mathbb{R}^n,dx)$. If ψ is a normalized element of $L_2(\mathbb{R}^n,dx)$, define $\psi_t = \exp\{-iH/\hbar\, t\}\psi$ and W_t the corresponding Wigner function (see formula (2.3) and also Moyal (1949).

Proposition : (2.1) if ψ is a normalized element of $L_2(\mathbb{R}^n,dx)$ in the domain of H, W_t the corresponding Wigner function, then it satisfies the following integrodifferentialequation

(2.7) $$\frac{\partial W_t}{\partial t}(q,p) = \sum_{i=1}^{n} \{ -p_i/m \frac{\partial}{\partial q_i} + m\omega^2 q_i \frac{\partial}{\partial p_i} \} W_t(q.p)$$

$$+ 2/\hbar \int_{\mathbb{R}^{2n}} d\mu(a') \sin(\sigma(a,a') + \varphi(a')) W_t(a+\hbar a'/2)$$

With an initialcondition $\lim_{t\to 0} W_t = W$

Proof is a matter of calculation.

Let us consider the set $C^b_k(\mathbb{R}^{p+1})$ of bounded complex functions on $\mathbb{R}^{2n+1}$ with k derivatives in the first p variables, continuous in the last. Then $F \to \hat{F}$

(2.8) $\hat{F}(a,s,t) = \exp\{4|\mu|t/\hbar + is/h\} F(a,t)$

where $|\mu|$ is the total mass of μ, is a mapping of $C^b_k(\mathbb{R}^{2n+1})$ into $C^b_k(\mathbb{R}^{2n+2})$

Lemma (2.2) with the same assumptions as in the previous proposition if W_t is a Wigner function then $\hat{W}_t$ satisfies the integro differential equation.

(2.9) $$\frac{\partial}{\partial t}\hat{W}_t(a,s) = \sum_{i=1}^{n} \{ -p_i/m \frac{\partial}{\partial q_i} + m\omega^2 q_i \frac{\partial}{\partial p_i} \} \hat{W}_t(a,s)$$

$$+ 1/\hbar \int_{\mathbb{R}^{2n+1}} d\nu_{a,s}(a',s') \{ \hat{W}_t(a+a',s+s') - \hat{W}_t(a,s) \}$$

where $d\nu_{a,s}$ is for each a,s a bounded positive measure on $\mathbb{R}^{2n+1}$

(2.10) $d\nu_{a,s}(a',s') = 2/\hbar\, d\mu(2a'/\hbar)\,(1-\sin(2/\hbar\,\sigma(a,a') + \varphi(2a'/\hbar))) + \mu(2a'/\hbar))\,\delta_{\hbar}(s') + 2/\hbar\, d\mu(2a'/\hbar)\,\delta_{\hbar/2}(s')$

Next we chose T>0 and we consider the extended Wigner function $\hat{W}$ for T-t, $t\in[0,T]$ the initial condition turns out to be replaced by a final condition and the corresponding equation is a backward Kolmogorov equation (Combe et al.(1984) (see eg. Gihman and Skorohod (1979)). Hence the proposition:

Proposition: (2.3) There exists a stochastic process $(Q_t(s), P_t(s), S_t(s))$ $t<s\leq T$ in $\mathbb{R}^{2n+1}$ such that

$$(Q_T(T), P_T(T), S_T(T)) = (q,p,s) \quad \text{a.s.}$$

whose generator has a modification $\partial_t\psi = A_t\psi$ for $\psi\in C_1^{\,b}(\mathbb{R}^{2n+1})$ with

$$(A_t\psi)(q,p,s) = \sum_{i=1}^{n} \{ q_i/m \frac{\partial}{\partial p_i} - m\omega^2 p_i \frac{\partial}{\partial q_i} \}\ \psi(q,p,s)$$

$$+ 1/\hbar \int_{\mathbb{R}^{2n+1}} d\nu_{qps}(q',p',s')\,\{\psi(q+q',p+p',s+s') - \psi(q,p,s)\}$$

such that the solution of eq. (2.9) is given by

(2.11) $W_T(a) = \exp\{4|\mu|\,T/\hbar\}\ \mathbb{E}\,[\exp\{ i S_0(T)/h\}\, W(Q_0(T), P_0(T))]$

To avoid the linear growth of the drift of the process we use the "interaction picture" namely, define $W_t^I(a)$ as $W_t(a_t)$, where a_t is the solution of the equation $da_t/dt = Ja_t$, with $Je_i = -1/m\, e_{n+i}$, $Je_{n+i} = m\omega^2 e_i$, i=1. n, in an orthonormal basis $\{e_i\}_{i=1..2n}$ of $\mathbb{R}^{2n}$. W_t^I has a representation in term of a stochastic process $(Q_t^I(s), P_t^I(s), S_t^I(s))$ very similar to the one given in the last proposition. However the generator of this last process has no drift. Its exponential generator (see e.g. Ventsel' (1976)) has the form :

(2.12) $G^{I}(a,s,z,t) = 2/\hbar \int_{\mathbb{R}^{2n}} d\mu(a'_t)\,(1-\sin(\sigma(a,a') + \varphi(a'_t))$

$\exp\{\hbar/2\, q'.z_1 + \hbar/2 p'.z_2\}\, ch(\hbar z_3) + 2/\hbar \int_{\mathbb{R}^{2n}} d\mu(a'_t)\, ch\,(\hbar/2\, z_3)$

$\exp\{\hbar/2 q'.z_1 + \hbar/2\, p'.z_2\} - 4|\mu|/\hbar$

for $z = (z_1,z_2,z_3) \in \mathbb{R}^n x \mathbb{R}^n x \mathbb{R}$. It is a positive continuous convex function of $z \in \mathbb{R}^{2n+1}$.

Lemma (2.4) : $G^{I*}(a,s,\xi,t)$ the Legendre transform of G^{I} with respect to z reaches its minimum for

(2.13) $\xi_1 = -(\partial V/\partial p)(a)$, $\xi_2 = (\partial V/\partial q)(a)$, $\xi_3 = 0$

independently of $\hbar$

Consequently it is possible to prove (see e.g. Ventsel' (1976)) that the probability for the trajectory of the process $(Q_t,(T),P_t(T),S_t(T))$ to make excursions outside a very small tublet around the classical path is very small and one has the proposition.

Proposition (2.5) [Blanchard, Sirugue (1985)] let Φ_t^h be the quantum flow defined by

$(\Phi_t^{\hbar} F)(q,p,s) = \mathbb{E}\,[F(Q_0(t), P_0(t),S_0(t))]$

with $F \in C_1^b(\mathbb{R}^{2n+2})$ then

$$\lim_{\hbar \to 0} (\Phi_t^{\hbar} F)(q,p,s) = F(q_t,p_t,s)$$

where $(q_t.p_t)$ are the solutions of the Hamilton equations

$q_t = - \partial H/\partial q \quad p_t = \partial H/\partial p$

References :

Blanchard Ph, Sirugue M, (1985)
Large deviation from Classical Paths. Hamiltonian Flows as classical limits of quantum Flows. Commun. Math. Phys. 101 173-185

Combe Ph, Guerra F, Rodriguez R, Sirugue M, Sirugue-Collin M, (1984)
Quantum dynamical time evolution as stochastic flows in phase space. Physica 124A 561-574.

Gihman I.I, Skorohod A.V. The theory of stochastic Processes. Springer Verlag (1979)

Jona-Lasinio G, Martinelli F, Scoppola E, (1981). New Approach to the semi classical limit of quantum mechanics. I multiple tunneling in one dimension. Commun. Math. Phys. 80 223 - 254

Maslov V.P. Chebotarev A.M. (1980) Jump type processes and their application to Quantum Mechanics. Journal of Soviet Math. 13 315-357.

Moyal J.E. (1949) Quantum mechanics as a statistical theory. Proc. Camb. Phil. Soc. 45 99-124.

Ventsel' A.D. (1976). Rough limit theorem on large deviations for Markov Stochastic processes .

I. Theory Prob. Appl. 21 227-242

II.Theory Prob. Appl. 21 499-512

III.Theory Prob. Appl. 24 675-692

IV.Theory Prob. Appl. 25 215-234

Wigner E, (1932) On the quantum correction for thermodynamic equilibrium. Phys. Rev. 40 749-759.

Prob. Theory and Math. Stat., Vol.1, pp. 211-229
Prohorov et al. (eds)
1986 VNU Science Press

LIMIT THEOREMS FOR MULTICOMPONENT HIERARCHICAL MODELS

P.M.Bleher
Keldysh Institute for Applied Mathematics,Moscow,USSR

Main results of this work were obtained by the author in collaboration with P.Major. In the paper limit theorems for sums $\sigma(1) + \dots + \sigma(N)$ of dependent multicomponent (multidimensional) random variables $\sigma(k) \in \mathbb{R}^m$ are studied. The random variables under consideration are distributed according to the Gibbs distributions of hierarchical models of statistical physics. For these models the existence of a critical point β_c is proved which the following properties. For $\beta < \beta_c$ the random variables $\sigma(k)$ are weakly dependent and obey the central limit theorem (CLT). For $\beta \geq \beta_c$ the random variables $\sigma(k)$ are strongly dependent and in that case we meet with new types of limit distributions which did not appear earlier in probability theory. The main aim of the work consists in the study of these new types of limit distributions and in the proof of corresponding limit theorems.

Let us recall the definition of the Gibbs distributions of hierarchical models. Let V_{kn} be a set consisting of 2^n subsequent natural numbers $\{(k-1)2^n+1, \dots, k2^n\}$, $k = 1, 2, \dots$; $n = 0, 1, 2, \dots$. The sets V_{kn} possess the following properties:

1) $V_{kn} \cap V_{k'n} = \emptyset$ when $k \neq k'$ 2) $\bigcup_{k=1}^{\infty} V_{kn} = \mathbb{N} = \{1, 2, 3, \dots\}$;

3) $V_{kn} = V_{2k-1, n-1} \cup V_{2k, n-1}$.

The collection of sets $\{V_{kn},\ k = 1, 2, \dots;\ n = 0, 1, \dots\}$ is called the hierarchical structure on the set of natural numbers $\mathbb{N}$. The hierarchical metrics on $\mathbb{N}$ is defined by the formula $d(i,j) = 2^{\min\{n \mid \exists k : \{i,j\} \subset V_{kn}\} - 1}$ when $i \neq j$, $d(i,i) = 0$. Denote for brevity $V_n = V_{1n} = \{1, \dots, 2^n\}$.

Let a sequence of m-dimensional $(m \geq 2)$ random variables $\sigma(1), \sigma(2), \dots, \sigma(2^n)$ have the distribution

$$\mu_n(d\sigma; \beta, \nu) = \Xi_n^{-1}(\beta, \nu) \exp\Big(\frac{\beta}{2} \sum_{\substack{i,j=1 \\ i \neq j}}^{2^n} \frac{(\sigma(i), \sigma(j))}{d^a(i,j)}\Big) \cdot \prod_{i=1}^{2^n} \nu(d\sigma(i)) \tag{1}$$

where
$$\Xi_n(\beta, \nu) = \int \exp\Big(\frac{\beta}{2} \sum_{\substack{i,j=1 \\ i \neq j}}^{2^n} \frac{(\sigma(i), \sigma(j))}{d^a(i,j)}\Big) \cdot \prod_{i=1}^{2^n} \nu(d\sigma(i)).$$

is the grand partition function; $\beta > 0$ is the "reciprocal temperature"; $(\cdot, \cdot)$ is the scalar product of vectors in $\mathbb{R}^m$; $a > 1$; $\nu(dx)$ is an arbitrary probability distribution in $\mathbb{R}^m$ possessing the following two properties:

1) $\nu(\mathcal{U}A) = \nu(A)$ for any Borel set $A \in \mathcal{B}(\mathbb{R}^m)$ and any orthogonal transformation $\mathcal{U} \in O(m)$ of the space $\mathbb{R}^m$, i.e. the distribution $\nu(dx)$ depends only on the radius $|x|$; 2) $E \exp(\alpha |x|^2) = \int \exp(\alpha |x|^2) \nu(dx) < \infty \quad \forall \alpha > 0.$

The distribution (1) is a finite Gibbs distribution of the hierarchical model.

Let us introduce a class of the probability distributions $\nu(dx)$ which we use in what follows. We shall assume that the distribution $\nu(dx)$ is absolutely continuous with respect to the Lebesque measure dx on $\mathbb{R}^m$. Let $u>0$. Consider a set S_u of probability distribution $\nu(dx)$ on $\mathbb{R}^m$ such that the density $\frac{\nu(dx)}{dx}$ exists and may be written in the form

$$\frac{\nu(dx)}{dx} = L^{-1} exp\left(-\frac{u}{4}|x|^4 - \frac{1}{2}|x|^2 - \tau(x)\right)$$ where

the function $\tau(x) \in C^4(\mathbb{R}^m)$ possesses the following properties: 1) $\tau(x)$ depends only on $|x|$, $\tau(x) = \tau_0(|x|)$;

2) $\|\tau(x)\|_{C^4(\mathbb{R}^m)} \leq u^{3/2}$. In what wollows we shall assume that $\nu \in S_u$ where u is sufficiently small (a condition of the smallness of u is included in the formulation of theorems).

The number $a>1$ and the probabilistic distribution $\nu(dx)$ are the parameters of hierarchical model and in the following they are assumed to be fixed. The reciprocal temperature $\beta = T^{-1}$ is a parameter of the Gibbs distribution and later we shall study limit theorems for sums of random variables $\sigma(k)$ distributed according to the Gibbs distributions (1) with various values of the parameter β. We prove a general statement which may be formulated as follows. Denote

$$G(x;\chi) = (2\pi\chi)^{-m/2} \cdot exp\left(-\frac{|x|^2}{2\chi}\right)$$

and $\|\cdot\| = \|\cdot\|_{C(\mathbb{R}^m)} + \|\cdot\|_{L^1(\mathbb{R}^m)}$.

Let $q_n^{(\alpha)}(x;\beta)_n$ be a density of the random variable $\sigma_n^{(\alpha)} = 2^{-n\alpha} \sum_{j=1}^{2^n} \sigma(j)$ where $\sigma(j)$ are μ_n- distributed. Note that the value of the exponent $\alpha = 1$ corresponds to the normalization of the large numbers law and $\alpha = 1/2$ does to the normalization of the CLT.

__Statement__ A. A (unique) critical point $\beta_c > 0$ exists such that

$$\lim_{n\to\infty} \| q_n^{(1/2)}(x;\beta) - G(x;\chi(\beta)) \| = 0, \text{ when } \beta < \beta_c,$$

$$\lim_{n\to\infty} \| q_n^{(a/2)}(x;\beta) - \beta^{m/2} f(x\sqrt{\beta}) \| = 0, \text{ when } \beta = \beta_c,$$

$$\lim_{n\to\infty} \| q_n^{(\alpha)}(x;\beta) - L^{-1} g(|x| - M_n^{1-\alpha}(\beta);\beta) \| = 0,$$

when $\beta > \beta_c$,

where
$$f(x) \in \mathcal{S}(\mathbb{R}^m),\ f(x) > 0,\ \int_{\mathbb{R}^m} dx\, f(x) = 1,$$
$$f(x) = f_0(|x|);\ g(x) \in \mathcal{S}(\mathbb{R}^1),\ g(x) > 0,\ L = L(\beta) =$$
$$= \int_{\mathbb{R}^m} dx\, g(|x| - M_n^{1-\alpha}(\beta);\beta);\ \lim_{n\to\infty} M_n(\beta) = M(\beta) > 0;$$
$$\lim_{\beta\to\beta_c} \chi(\beta) = \infty,\ \lim_{\beta\to\beta_c} M(\beta) = 0.$$

In terms of the μ_n-distributed random variables $\sigma(1), \ldots, \sigma(2^n)$ the statement A means that

$$\lim_{n\to\infty} \frac{\sigma(1) + \ldots + \sigma(2^n)}{(2^n)^{1/2}} = \xi_{G(x;\ \chi(\beta))},$$

when $\beta < \beta_c$,

$$\lim_{n\to\infty} \frac{\sigma(1) + \ldots + \sigma(2^n)}{(2^n)^{a/2}} = \frac{\xi_{f(x)}}{\sqrt{\beta}},$$

when $\beta = \beta_c$,

$$\lim_{n\to\infty} \frac{|\sigma(1) + \ldots + \sigma(2^n)| - 2^n M_n(\beta)}{(2^n)^{\alpha}} = \xi_{g(x;\beta)},$$

when $\beta > \beta_c$,

where $\xi_{\rho(x)}$ is a random variable with the distribution $\rho(x)dx$, and the convergence of random variables variables is understood in the sense of the convergence of densities of their distributions with respect to the norm $\|\cdot\|$

The form of the functions f, g and the value of the exponent α depens on the parameter a of the hierarchical model. The four cases should be differentiated: 1) $a \geq 2$, 2) $2 > a > 3/2$, 3) $a = 3/2$, 4) $3/2 > a > 1$. At first we consider the cases 1),4).

Theorem 1. For any $a \geq 2$ there exists $u_0 > 0$ such that for any $\nu \in S_u$, where $0 < u < u_0$, for any $\beta > 0$ the local CLT holds:

$$\lim_{n \to \infty} \| q_n^{(1/2)}(x;\beta) - G(x; \chi(\beta)) \| = 0 .$$

This theorem shown that for $a \geq 2$ the critical point β_c is absent.

Theorem 2. For any $3/2 > a > 1$ there exists $u_0 > 0$ such that for any $\nu \in S_u$, where $0 < u < u_0$, a critical point $\beta_c = \beta_c(\nu, a) > 0$ exists for which Statement A^{ν} holds, $f(x)$ being equal to $G(x; \chi_0)$, $\chi_0 = 1 - 2^{1-a}$, $\alpha = 1/2$, $g(x;\beta) = G(x; \chi(\beta))$.

Moreover $\chi(\beta) \sim C_{\pm} |\beta - \beta_c|^{-1}$ as $\beta \to \beta_c \pm 0$,

$M(\beta) \sim C(\beta - \beta_c)^{1/2}$ as $\beta \to \beta_c + 0$.

The notation $F(x) \sim G(x)$ when $x \to x_0$ being used here and later means that $\lim_{x \to x_0} F(x)/G(x) = 1$. By $C, C_+, C_-, C_1, C_2, \ldots$ we denote positive constants which may be different in different formulas. The notations $\beta \to \beta_c - 0$, $\beta \to \beta_c + 0$ mean $\beta \uparrow \beta_c$, $\beta \downarrow \beta_c$ respectively. Theorem 2 is proved in [1],[2].

By Theorem 2, when $1 < a < 3/2$, the exponent α takes the classical value $1/2$ and the limit densities $f(x)$, $g(x;\beta)$ are Gaussian. When $2 > a > 3/2$ the densities $f(x), g(x;\beta)$ are non-Gaussian. To construct them let us consider two auxiliary nonlinear integral equations

$$f(x) = Q(f)(x) = L^{-1}\int_{\mathbb{R}^m} du \exp\left(\frac{|x|^2}{2^{2-a}} - |u|^2\right) \cdot$$

$$\cdot f\left(\frac{x}{2^{1-a/2}} - u\right) \cdot f\left(\frac{x}{2^{1-a/2}} + u\right), \tag{2}$$

$$g(x) = T(g)(x) = L_0^{-1}\int_{-\infty}^{\infty} ds \int_{\mathbb{R}^{m-1}} dv \exp\left(-|v|^2\right) \cdot$$

$$\cdot g\left(\frac{x}{2^{2-a}} - s + \frac{|v|^2}{2}\right) g\left(\frac{x}{2^{2-a}} + s + \frac{|v|^2}{2}\right), \tag{3}$$

where the constants L, L_0 secure the condition of probability normalization $\int_{\mathbb{R}^m} dx\, Q(f)(x) = 1$, $\int_{-\infty}^{\infty} dx\, T(g)(x) = 1$. In subsequent formulas various constants L, L_n connected with the probability normalization appear also. These constants may be fifferent in different formulas. The equations (2),(3) are fixed point equations for the nonlinear operators Q and T. It may be checked directly that the Gaussian density $f^0(x) = G(x; \chi_0) \equiv$

$$\equiv (2\pi\chi_0)^{-m/2} \exp\left(-\frac{|x|^2}{2\chi_0}\right), \quad \chi_0 = 1 - 2^{1-a},$$

is a symmetric (i.e. dependent only on $|x|$) solution of the equation (2). Next theorem establishes the existence of non-Gaussian symmetric solutions of this for $a = 3/2 + \varepsilon$, where $0 < \varepsilon \ll 1$.

Theorem 3. There exists $\varepsilon_0 > 0$ such that for any $a = 3/2 + \varepsilon$, where $0 < \varepsilon < \varepsilon_0$, a non-Gaussian positive symmetric solution $f^1(x) \in \mathcal{S}(\mathbb{R}^m)$ of equation (2) exists satisfying the estimates $\|f^1(x) - f^0(x)\| \le C\varepsilon$,

$$0 < f^1(x) < Cf^0(x) \exp(-C_0 \varepsilon |x|^{\beta}),$$

where $\rho = \frac{2}{2-a} > 4$.

Proof of Theorem 3 for $m = 1$ is given in [3] (see also [4]). Its generalization to the case $m>1$ is given in [5].

Theorem 4. For any $1<a<2$, $a \neq \frac{3}{2}, \frac{4}{3}, \dots,$ a (unique) analitical positive solution $g(x) \in \mathcal{S}(\mathbb{R}^1)$ of equation (3) exists satisfying the estimates

$$0 < g(x) < C\exp x, \text{ for } x \le 0, \text{ and}$$
$$0 < g(x) < C\exp(-C_0(a)x^{\lambda}), \quad \lambda = \frac{1}{2-a},$$

for $x \geq 0$. For $3/2 < a < 2$ this solution is stable in the following sense: If $g_0(x) > 0$,

$$\int_{-\infty}^{\infty} dx\, g_0(x) = \int_{-\infty}^{\infty} dx\, g(x) = 1, \quad \int_{-\infty}^{\infty} dx\, x g_0(x) = \int_{-\infty}^{\infty} dx\, x g(x), \quad \int_{-\infty}^{\infty} x^4 g_0(x)\, dx < \infty,$$

then $\lim_{n\to\infty} T^n(g_0) = g$.

Consider besides the linear operator D_{f^1}:

$$h(x) \to 2L^{-1} \int_{\mathbb{R}^m} du \exp\left(\frac{|x|^2}{2^{2-a}} - |u|^2\right) \cdot f^1\left(\frac{x}{2^{1-a/2}} - u\right) h\left(\frac{x}{2^{1-a/2}} + u\right), \tag{4}$$

where

$$L = \int_{\mathbb{R}^m} dx \int_{\mathbb{R}^m} du \exp\left(\frac{|x|^2}{2^{2-a}} - |u|^2\right) \cdot f^1\left(\frac{x}{2^{1-a/2}} - u\right) \cdot f^1\left(\frac{x}{2^{1-a/2}} + u\right),$$

in the space of symmetric functions $h(x) = h_0(|x|)$. This operator is closely related to the linearization of the nonlinear operator Q at the fixed point $f^1(x)$. It is possible to prove that in the space $L^2_{sym}(\mathbb{R}^m;$

$\exp(-\gamma|x|^2)$ of symmetric functions which are square integrable with respect to the measure $\exp(-\gamma|x|^2)\,dx$ where $\gamma = \dfrac{2^a - 1}{1 - 2^{1-a}} - 2 > 0$, the operator D_{f^1} is a compact operator and its spectrum consists of the numbers $2, \lambda_1, \lambda_2, \dots$, where $\lambda_1 > 1$ and $|\lambda_i| < 1$, $i > 1$.

Theorem 5. There exists $\varepsilon_0 > 0$ such that for any $a = 3/2 + \varepsilon$, where $0 < \varepsilon < \varepsilon_0$, a number $u_0 > 0$ exists such that for any $\nu \in S_u$, where $0 < u < u_0$ a critical point $\beta_c = \beta_c(\nu, a) > 0$ exists for which Statement A holds, f being the non-Gaussian solution of equation (2), $\alpha = a - 1$ and $g(x;\beta) = g(x\sqrt{\beta M(\beta)})$, where $g(x)$ is the solution of equation (3). Moreover $\chi(\beta) \sim C_-|\beta - \beta_c|^{\gamma}$ as $\beta \to \beta_c - 0$, $M(\beta) \sim C|\beta - \beta_c|^{\omega}$ as $\beta \to \beta_c + 0$, where $\gamma = \dfrac{a-1}{\log_2 \lambda_1}$, $\omega = \dfrac{1-a/2}{\log_2 \lambda_1}$ and $\lambda_1 > 1$ is the eigenvalue of the operator D_{f^1}. For any $3/2 < a < 2$, $\beta_1 > \beta_0 > 0$ exist such that

$$\lim_{n\to\infty} \frac{\sigma(1) + \dots + \sigma(2^n)}{(2^n)^{1/2}} = \xi G(x; \chi(\beta)),$$

when

$$\beta < \beta_0,$$

$$\lim_{n\to\infty} \frac{|\sigma(1) + \dots + \sigma(2^n)| - 2^n M_n(\beta)}{2^{n(a-1)}} = \frac{\xi g(x)}{\sqrt{\beta M(\beta)}},$$

when $\beta > \beta_1$,

where $\lim\limits_{n\to\infty} M_n(\beta) = M(\beta) > 0$.

Let us now turn to the case $a = 3/2$.

Theorem 6. For $a = 3/2$ there exists $u_0 > 0$ such that for any $\nu \in S_u$, where $0 < u < u_0$ a critical point $\beta_c = \beta_c(\nu) > 0$ exists such that

$$\lim_{n\to\infty}\frac{\sigma(1)+\dots+\sigma(2^n)}{2^{n/2}}=\xi_{G(x;\,\chi(\beta))},\ \text{when}\ \beta<\beta_c,$$

$$\lim_{n\to\infty}\frac{\sigma(1)+\dots+\sigma(2^n)}{2^{na/2}}=\xi_{G(x;\,\chi_0)},\ \text{when}\ \beta=\beta_c,$$

$$\lim_{n\to\infty}\frac{|\sigma(1)+\dots+\sigma(2^n)|-2^n M_n(\beta)}{2^{n/2}n^{1/2}}=$$

$$=\xi_{G(x;\,\chi(\beta))},\quad \text{when}\ \beta>\beta_c,$$

where $\lim_{n\to\infty} M_n(\beta)=M(\beta)>0$ .Moreover $\chi(\beta)\sim$ $\sim C_\pm|\beta-\beta_c|^{-1}$ as $\beta\to\beta_c\pm 0$, $M(\beta)\sim C\sqrt{(\beta-\beta_c)|\ln(\beta-\beta_c)|}$ as $\beta\to\beta_c+0$.

Proofs of the above formulated theorems use the recurrent equations for functions $q_n^{(\alpha)}(x;\beta)$. From the form of the Gibbs distribution (1) and the fact that $d(i,j)=2^n$ as $1\le i\le 2^n$ and $2^n+1\le j\le 2^{n+1}$ it follows the recurrent equation

$$q_{n+1}^{(\alpha)}(x;\beta)=L_n^{-1}(\beta)\int_{\mathbb{R}^m}du\,\exp\Big(\beta 2^{(2\alpha-a)n}\cdot \tag{5}$$

$$\cdot\Big(\frac{|x|^2}{2^{2-2\alpha}}-|u|^2\Big)\Big)\cdot q_n^{(\alpha)}\Big(\frac{x}{2^{1-\alpha}}-u;\beta\Big)\cdot$$

$$\cdot\, q_n^{(\alpha)}\Big(\frac{x}{2^{1-\alpha}}+u;\beta\Big).$$

Write
$$q_n^{(\alpha)}(x;\beta) = \exp\left(-\beta 2^{(2\alpha-a)n}\frac{|x|^2}{2\chi_0}\right)\cdot$$
$$\cdot\, p_n^{(\alpha)}\left(x\sqrt{\beta a_1};\beta\right),$$
where $\chi_0 = 1-2^{1-a}$, $a_1 = \dfrac{\chi_0+1}{\chi_0}$. Then (5) implies

$$p_n^{(\alpha)}(x;\beta) = L_n^{-1}(\beta)\int_{\mathbb{R}^m} du\, \exp\left(-2^{(2\alpha-a)n}|u|^2\right)\cdot$$
$$\cdot\, p_n^{(\alpha)}\left(\frac{x}{2^{1-\alpha}}-u;\beta\right) p_n^{(\alpha)}\left(\frac{x}{2^{1-\alpha}}+u;\beta\right). \tag{6}$$

It is easy to see that $q_0^{(\alpha)}(x;\beta) = \dfrac{\nu(dx)}{dx}$. Assume for simplicity that $\dfrac{\nu(dx)}{dx} = L^{-1}\exp\left(-\frac{u}{4}|x|^4 - \frac{1}{2}|x|^2\right)$.

Then
$$p_0^{(\alpha)}(x;\beta) = L^{-1}\exp\left(\frac{|x|^2}{2(1+\chi_0)} - \frac{u}{4}\frac{|x|^4}{(\beta a_1)^2} - \frac{1}{2}\frac{|x|^2}{\beta a_1}\right) =$$
$$= L^{-1}\exp\left(\frac{|x|^2}{2(1+\chi_0)}\left(1-\frac{\chi_0}{\beta}\right) - \frac{u}{4}\cdot\frac{|x|^4}{(\beta a_1)^2}\right). \tag{7}$$

Now we may forget about the original sense of the functions $p_n^{(\alpha)}(x;\beta)$ and study recurrent equation (6) with initial condition (7). Equation (6) resembles usual convolution but with an extra exponential kernel $\exp\left(-2^{(2\alpha-a)n}\cdot|u|^2\right)$. This kernel does not depend on β and all the dependence on β is contained in the initial condition (7). The functions $p_0^{(\alpha)}(x;\beta)$ are (spherically) symmetric. For small β the function $p_0^{(\alpha)}(x;\beta)$ is delta-shaped, i.e. it is localized near the origin, while for large β the function $p_0^{(\alpha)}(x;\beta)$

is localized near the sphere of the radius $M_0=\beta a_1\left(\frac{1-\chi_0/\beta}{u(1+\chi_0)}\right)^{1/2}$.

For $d=a/2$ the equation (6) has the form $p_{n+1}^{(a/2)}=R(p_n^{(a/2)})$, where

$$R(p)(x)=L^{-1}\int_{\mathbb{R}^m} du\, \exp(-|u|^2)\, p\left(\frac{x}{2^{1-a/2}}-u\right)\cdot \tag{8}$$
$$\cdot\, p\left(\frac{x}{2^{1-a/2}}+u\right).$$

The function $p^0(x)\equiv 1$ is a fixed point of the operator R. The normalizing constant L is equal to $\pi^{m/2}$ in this case. Let

$$R_0(p)(x)=\pi^{-m/2}\int_{\mathbb{R}^m} du\, \exp(-|u|^2)\, p\left(\frac{x}{2^{1-a/2}}-u\right)\cdot$$
$$\cdot\, p\left(\frac{x}{2^{1-a/2}}+u\right).$$

The linearization of the operator R_0 at the fixed point $p^0(x)\equiv 1$ has the form

$$D_0 h(x)=2\pi^{-m/2}\int_{\mathbb{R}^m} du\, \exp(-|u|^2)\, h\left(\frac{x}{2^{1-a/2}}-u\right). \tag{9}$$

The operator D_0 is the Gause operator (see [6]). It is compact and self-adjoint in the Hilbert space $L^2(\mathbb{R}^m; \exp(-\gamma|x|^2))$, $\gamma=1-2^{a-2}$, defined by the scalar product $(h(x),g(x))=\int_{\mathbb{R}^m} dx\, h(x)\overline{g(x)}\exp(-\gamma|x|^2)$. Its spectrum in the space of symmetric functions $h(x)\in$ $\in L^2_{sym}(\mathbb{R}^m; \exp(-\gamma|x|^2))$ consists of the numbers $\lambda_k=2^{1+k(a-2)}$, $k=0,1,2,\dots$. Corresponding eigen-functions are the Hermite polynomials $h_k(x)=$ $=:|x|^{2k}:$ with respect to the Gaussian measure $\exp(-\gamma|x|^2)\,dx$. The linearization of the operator R at $p^{\circ}(x)\equiv 1$ has the form $\mathcal{D}h=\mathcal{D}_{o}h-\ell(h)\cdot 1$, where $\ell(h)$ is a linear functional. The spectrum of the operator $\mathcal{D}$ consists of the numbers $\lambda_{o}=0$, $\lambda_{k}=2^{1+k(a-2)}$, $k=1,2,\ldots$.

Properties of the convergence of the iteration $p_{n+1}^{(a/2)}=R(p_{n}^{(a/2)})$ to p° is connected with the presence of stable, $|\lambda_{k}|<1$, and unstable, $|\lambda_{k}|>1$, eigenvalues of the operator $\mathcal{D}$. The eigenvalue $\lambda_{1}=2^{a-1}>1$ is unstable so the fixed point $p^{\circ}(x)\equiv 1$ is hyperbolic. For $a<3/2$ the eigenvalue $\lambda_{2}=2^{2a-3}<1$ and all the subsequent eigenvalues λ_{k} are stable. For small values of $u>0$ as is seen from the formula (7), the initial family $p_{0}^{(a/2)}(x;\beta)$ goes (for $\beta=\chi_{o}$) near the fixed point $p^{\circ}(x)\equiv 1$ and intersects transversally the stable hyperplane of the operator $\mathcal{D}$, $\{p(x)=p^{\circ}(x)+\sum_{k=2}^{\infty}a_{k}e_{k}(x)\}$, where $e_{k}(x)$ are the eigenfunctions of the operator $\mathcal{D}$. So one can assume for $1<a<3/2$ a (unique) point $\beta_{c}>0$ exists such that $\lim_{n\to\infty}R^{n}(p_{0}^{(a/2)})(x;\beta_{c})=p^{\circ}(x)$. The existence of the critical point β_{c} is proved rigorously in Theorem 2. For $\beta\neq\beta_{c}$ the iterations $p_{n}^{(a/2)}(x,\beta)=R^{n}(p_{0}^{(a/2)})(x;\beta)$ "go out" from the fixed point $p^{\circ}(x)$. Their behaviour as $n\to\infty$ will be considered below.

The value $a=3/2$ is a bifurcation one. For this value $\lambda_{2}=1$. In accord with the general theory of bifurcations for $a=3/2+\varepsilon$ a branch of fixed point

$$p^{1}(x;\varepsilon)=p^{\circ}(x)+\sum_{k=1}^{\infty}\varepsilon^{k}b_{k}(x)$$ of the operator R

should exist. A necessary condition for the existence of such a branch is a non-degeneracy condition. For the operator R the non-degeneracy condition has the form

$\int_{\mathbb{R}^{m}}dx\, R(h_{4})(x)h_{4}(x)\exp(-\gamma|x|^{2})\neq 0$. A direct calculation shown that this non-degeneracy condition of the operator R is fulfilled. A rigorous proof of the existence of the branch of fixed points $p^1(x;\varepsilon)$ for $0<\varepsilon<\varepsilon_0$ (for some $\varepsilon_0>0$) is given in Theorem 3.

For $a=3/2+\varepsilon$, where $0<\varepsilon<\varepsilon_0$, the spectrum of the linearization of the operator R at the fixed point $p^1(x;\varepsilon)$ contains a single unstable eigenvalue $\lambda_1>1$. Therefore as above one may assume that a (unique) critical point $\beta_c>0$ exists such that $\lim_{n\to\infty} R^n(p_0^{(a/2)})(x;\beta_c)=p^1(x;\varepsilon)$. The existence of β_c for $a=3/2+\varepsilon$, where $0<\varepsilon<\varepsilon_0$, is proved in Theorem 5.

For $a=3/2$ the eigenvalue $\lambda_2=1$ is neutral. The convergence of the iterations $\lim_{n\to\infty} R^n(p_0^{(a/2)})(x;\beta)=p^0(x)$ at the critical point occurs owing to the quadratic part of the operator R at the fixed point $p^0(x)$. An essential role plays here the non-degeneracy condition of the operator R. It is noteworthy that for $a=3/2$ the iterations $R^n(p_0^{(a/2)})(x;\beta_c)$ converge to the limit $p^0(x)$ as $n\to\infty$ with a powerlike (and not with exponential) rate in n: $R^n(p_0^{(a/2)})(x;\beta_c)=p^0(x)+\frac{c}{n}h_2(x)+O\left(\frac{1}{n^2}\right)$. A proof of the existence of the critical point β_c for $a=3/2$ is given in Theorem 6.

At $\beta\neq\beta_c$ as was mentioned above the iteration $R^n(p_0^{(a/2)})(x;\beta)$ "go out" from the fixed points $p^0(x)$ for $1<a\le 3/2$ and $p^1(x;\varepsilon)$ for $3/2<a<3/2+\varepsilon_0$.

Now we give a sketch of the proof of the fact that at $\beta<\beta_c$ the CLT holds. For $\alpha=1/2$ the equation (6) has the form

$$p_{n+1}^{(1/2)}(x;\beta)=L_n^{-1}(\beta)\int_{\mathbb{R}^m} du\,\exp\left(-2^{(1-a)n}|u|^2\right).$$

$$\cdot\, p_n^{(1/2)}\left(\frac{x}{\sqrt{2}}-u;\beta\right) p_n^{(1/2)}\left(\frac{x}{\sqrt{2}}+u;\beta\right) \equiv S_n\left(p_n^{(1/2)}\right)(x;\beta) \quad (10)$$

For $a>1$ the kernel $\exp\left(-2^{(1-a)n}|u|^2\right)\to 1$ as $n\to\infty$. If $n\to\infty$ then the operator S_n tends to the operator

$$S(p)(x)=2^{m/2}\int_{\mathbb{R}^m} du\, p\left(\frac{x}{\sqrt{2}}-u\right)p\left(\frac{x}{\sqrt{2}}+u\right)=$$

$$=2^{m/2}\, p * p(\sqrt{2}x).$$

Gaussian densities $G(x;\chi)=(2\pi\chi)^{-m/2}\exp\left(-\frac{|x|^2}{2\chi}\right)$ are the symmetric fixed points of the operator S. Let us study the stability of these fixed points in the linear approximation. Let $p(x)=G(x;\chi)(1+\varepsilon h(x))$. Then $S(p)(x)=G(x;\chi)(1+\varepsilon D^{\circ}h(x)+O(\varepsilon^2))$, where

$$D^{\circ}h(x)=2(\pi\chi)^{-m/2}\int_{\mathbb{R}^m} du\,\exp\left(-\frac{|u|^2}{\chi}\right) h\left(\frac{x}{\sqrt{2}}-u\right).$$

The operator D° is compact and self-adjoint in the space $L^2\left(\mathbb{R}^m;\exp\left(-\frac{|u|^2}{2\chi}\right)\right)$. Its spectrum in the space of symmetric functions $L^2_{sym}\left(\mathbb{R}^m;\exp\left(-\frac{|u|^2}{2\chi}\right)\right)$ consists of the numbers $\lambda_k=2^{1-k}$, $k=0, 2, \ldots$ Corresponding eigen-functions are the Hermite polynomials $h_k(x)=:|x|^{2k}:$. Note that the perturbation $\varepsilon h(x)$ has no projection on $h_0(x)\equiv \text{const}$ because of the normalization condition $\int_{\mathbb{R}^m} p(x)dx=1$. Moreover by changing the parameter χ one can achieve that $\varepsilon h(x)$ would have no projection on $h_2(x)=$ $=|x|^2-\text{const}$. In the subspace orthogonal to

$\{h_d(x),\ h_2(x)\}$ the spectrum of the operator $\mathcal{D}^\circ$ has the form $\lambda_k = 2^{1-k} < 1$, $k = 2, 3, \dots$, i.e. the operator S is stable in the linear approximation. It is verifies that for any $\beta < \beta_c$, $n = n(\beta)$ exists such that the density $p_n^{(1/2)}(x;\beta)$ is close to a Gaussian density $G(x; \gamma_n(\beta))$. After that using the stability of the operator S one establishes the convergence $p_n^{(1/2)}(x;\beta)$ to $G(x; \gamma(\beta))$, $\gamma(\beta) \approx \gamma_n(\beta)$ as $n \to \infty$.

The case $\beta > \beta_c$ is more complicated. Let us consider for $\beta > \beta_c$ the behaviour of the finctions $p_n^{(1)}(x;\beta)$ as $n \to \infty$. Assume that for some n the function $p_n^{(1)}(x;\beta)$ is localized near a sphere of a radius $M > 0$. More precisely let $p_n^{(1)}(x;\beta) = g_n(c^n(|x| - M))$, where $c = 2^{2-a}$ and $g_n(t) \in \mathcal{S}(\mathbb{R}^1)$, $g_n(t) > 0$, $\int_{-\infty}^{\infty} t g_n(t)\,dt = 0$, $g_n(t) < C g_n(0) \exp(-C_0|t|)$. Let us check if the function $p_{n+1}^{(1)}(x;\beta)$ possesses similar properties. By the formula (6),

$$p_{n+1}^{(1)}(x;\beta) = L_n^{-1}(\beta)\int_{\mathbb{R}^m} du \exp(-2^{(2-a)n}|u|^2)\cdot \tag{11}$$
$$\cdot\, g_n(c^n(|x-u| - M))\, g_n(c^n(|x+u| - M)).$$

Let $x = (M + tc^{-n-1}, 0) \in \mathbb{R}^m$, $u = (c^{-n}s, c^{-n/2}v)$,

where $t, s \in \mathbb{R}^1$, $v \in \mathbb{R}^{m-1}$. Then

$$c^n(|x \pm u| - M) = c^n\left(\sqrt{(M + t\cdot c^{-n-1} \pm c^{-n}s)^2 + c^{-n}\cdot|v|^2}\ -\right.$$
$$\left. -\ M\right) = \frac{t}{c} \pm s + \frac{|v|^2}{2M} + \delta_n(t, \pm s, v);$$

$$\delta_n(t, \pm s, v) = O\left(\frac{t^2 + s^2 + |v|^4/M^2}{c^n M}\right).$$

Substituting these expressions in to the equation (11) we get

$$g_{n+1}(t) = p_{n+1}^{(1)}\left(M + t\cdot c^{-n-1}, 0; \beta\right) =$$

$$= L_n^{-1}(\beta) \int_{-\infty}^{\infty} ds \int_{\mathbb{R}^{m-1}} dv \exp\left(-c^{-n} s^2 - |v|^2\right) g_n\left(\frac{t}{c} - \right.$$

$$\left. - s + \frac{|v|^2}{2M} + \delta_n(t, -s, v)\right) \cdot g_n\left(\frac{t}{c} + s + \frac{|v|^2}{2M} + \delta_n(t, s, v)\right).$$

Assume that the number n is large enough and so we may neglect the terms $\delta_n(t, \pm s, v)$ and $c^{-n} s^2$ in the last formula. Then $g_{n+1}(t) \approx T(g_n)(t)$ where

$$T(g)(t) = L^{-1} \int_{-\infty}^{\infty} ds \int_{\mathbb{R}^{m-1}} dv \exp\left(-|v|^2\right) g\left(\frac{t}{c} - s + \frac{|v|^2}{2M}\right) \cdot \qquad (12)$$

$$\cdot\, g\left(\frac{t}{c} + s + \frac{|v|^2}{2M}\right).$$

One can give a probabilistic interpretation of this relation. Consider the probabilistic density

$$\rho_M(t) = \begin{cases} \dfrac{M}{\Gamma\left(\frac{m-1}{2}\right)} (Mt)^{(m-3)/2} \exp(Mt), & t < 0, \\ 0, & t \geq 0 \end{cases} \qquad (13)$$

The density $\rho_M(-t)$ is nothing else that the density of the χ^2-distribution with $(m-1)$ degrees of freedom (see [7]). Then one can rewrite the formula (12) as

$$T(g)(t) = \frac{2}{c} g * g * \rho_M\left(\frac{2}{c} t\right) = 2^{a-1} g * g * \rho_M(2^{a-1} \cdot t).$$

This relation shows that if random variables $\xi_n^{(1)}$, $\xi_n^{(2)}$, η are independent, $\xi_n^{(1)}$, $\xi_n^{(2)}$ have the same distribution $g_n(t)\,dt$ and η has the distribution $\rho_M(t)\,dt$ then the random variable

$$\xi_{n+1} = \frac{\xi_n^{(1)} + \xi_n^{(2)} + \eta}{2^{a-1}} \tag{14}$$

has the distribution $T(g_n)(t)\,dt$.

Equation (14) implies, in particular, that

$$\int_{-\infty}^{\infty} t\,T(g_n)(t)\,dt = E\,\xi_{n+1} = E\,\eta/2^{a-1} = -2^{-a}(m-1)/M.$$

To fulfil the condition $\int_{-\infty}^{\infty} t\,g_{n+1}(t)\,dt = 0$ we change the parameter M and define $g_{n+1}(t) =$

$$= \rho_{n+1}^{(1)}(M' + t\cdot c^{-n-1}, 0; \beta) = \rho_{n+1}^{(1)}(M + (t + (M'-M)c^{n+1})\cdot c^{-n-1}, 0; \beta) \approx T(g_n)(t + (M'-M)c^{n+1}).$$

The value of M' is determined by the condition

$$0 = \int_{-\infty}^{\infty} t\,g_{n+1}(t)\,dt \approx \int_{-\infty}^{\infty} t\,T(g_n)(t + (M'-M)c^{n+1})\,dt =$$

$$= (M - M')c^{n+1} - 2^{-a}(m-1)/M,$$

hence $M' \approx M - \frac{1}{4}c^{-n}(m-1)/M$.

These calculations show that if $\rho_n^{(1)}(x;\beta) = g_n(c^n(|x| - M_n))$, then $\rho_{n+1}^{(1)}(x;\beta) = g_{n+1}(c^{n+1}(|x| - M_{n+1}))$, where

$$M_{n+1} \approx M_n - c^{-n}(m-1)/(4M_n), \tag{15}$$

$$g_{n+1}(t) \approx 2^{a-1} g_n * g_n * \hat{\rho}_{M_n}(2^{a-1}t),$$

$$\hat{\rho}_M(t) = \rho_M(t - 2^{-a}(m-1)/M), \int_{-\infty}^{\infty} t\hat{\rho}_M(t)dt = 0. \quad (16)$$

An analysis of recurrent equation (16) is based on the following lemma.

Lemma 7. Let ξ_0 be a real random variable, $E\xi_0 = 0$, $E\xi_0^4 < \infty$. Let random variables $\xi_1, \xi_2, \ldots$ be constructed be the recurrent equation

$$\xi_{n+1} = \frac{\xi_n^{(1)} + \xi_n^{(2)} + (\eta - E\eta)}{2^{a-1}}$$

where the random variables $\xi_n^{(1)}, \xi_n^{(2)}, \eta$ are independent and $\xi_n^{(1)}, \xi_n^{(2)}$ have the same distribution as ξ_n and η has the distribution $\rho_M(t)dt$.

Then:

1) If $2 > a > 3/2$ then $\lim_{n\to\infty} \xi_n = \xi_* = \sum_{k=1}^{\infty} \frac{\eta_k - E\eta_k}{2^{k(a-1)}}$, where η_k are independent and have the distribution $\rho_M(t)dt$

2) If $a = 3/2$ then $\lim_{n\to\infty} \frac{\xi_n}{\sqrt{n}} = \xi^0$, where ξ^0 is normally distributed;

3) If $3/2 > a > 1$ then $\lim_{n\to\infty} \frac{\xi_n}{2^{(3/2-a)n}} = \xi^0$, where ξ^0 is normally distributed.

The relations 1), 2), 3) determine the limit distribution of normalized sums of μ_n- distributed random variables at $\beta > \beta_c$ in the cases $a > 3/2$, $a = 3/2$ and $a < 3/2$.

References

1. Bleher P.M. Large deviations theorem in vicinity of the critical point of the φ^4 -hierarchical model - - Theor,Probab.Appl., 1985,30,N3, 499-510(Russian).

2. Bleher P.M.,Major P. Renormalization of Dyson's hierarchical vector-valued φ^4-model at low temperatures - Commun.Math.Phys., 1984,95,N4,487-532.

3. Bleher P.M., Sinai Ya.G. Critical indices for Dyson's asymptotically hierarchical models- Commun. Math.Phys.,1975, 45,N3,247-278.

4. Collet P., Eckmann J.-P. A Renormalization group analysis of the hierarchical model in statistics mechanics - Lecture Notes in Physics, N74,Berlin, Heidelberg, New York: Springer, 1978, 1-199.

5. Gawedzki K., Kupiainen A. Non-Gaussian fixed points of the block spin transformation. Hierarchical model approximation- Commun.Meth.Phys., 1983,84, N2, 191-220.

6. Bateman G., Erdelyi A. Higher transcendental functions, Vol.2. New York, Toronto, London: McGraw-Hill, 1955.

7. Feller W. An introduction to probability theorem and its applications, Vol.2. New York e.a. John Wiley and Sons Inc.1971.

Prob. Theory and Math. Stat., Vol.1, pp. 231-250
Prohorov et al. (eds)
1986 VNU Science Press

ON LIMIT THEOREMS FOR RANDOM VECTORS CONTROLLED BY A MARKOV CHAIN

I.V.Bokuchava, Z.A.Kvatadze, T.L.Shervashidze
Tbilisi Mathematical Institute, Tbilisi, USSR

Many authors studied sums of random variables with joint distributions determined by a controlling sequence of random elements [1-11]. This is a natural generalization of problems of summation of independent random variables. The choice of an ergodic controlling sequence and the conditionally m -dependent controlled sequence of random vectors has led the authors to the theorems on asymptotic normality of sums of such vectors presented below. The convergence of the distributions of sums is considered in detail in the case when a finite ergodic Markov chain is the controlling sequence and the controlled random vectors are conditionally independent.

1. ERGODIC CONTROLLING SEQUENCE AND THE CENTRAL LIMIT THEOREM FOR CONDITIONALLY m -DEPENDENT VECTORS

On a probability space $(\Omega, \mathcal{F}, \mathbb{P})$ consider a stationary two-component sequence

$$(\xi_j, X_j)_{j\geq 1} \qquad (1)$$

where ξ_j belong to a set Ξ and X_j take values in R^k . Let $\xi = (\xi_1, \xi_2, \ldots)$

be an ergodic controlling sequence. The ergodicity is understood as the convergence almost surely of arithmetic means of the values of measurable functions on $\xi_1, \dots, \xi_n$ to the mathematical expectations of these functions as $n \to \infty$.

Denote $\xi_{1n} = (\xi_1, \dots, \xi_n)$; $P_y, P_{y|\xi_{1n}}, p_y(x)$ and $p_{y|\xi_{1n}}(x)$ denote, respectively, the absolute and conditional (for a given ξ_{1n}) distribution and the density of a random vector $y \in R^{kn}, n \geq 1$. Let Φ_B and $\varphi_B(x)$ denote, respectively, the normal distribution in R^k with parameters $(0, B)$ and the corresponding density.

The sequence of the second componenets of (1)

$$X = (X_1, X_2, \dots)$$

is called the conditional m -depend random vectors (controlled by the sequence ξ) if for any n and given ξ_{1n} the random vectors $X_1, \dots, X_n$ become m -dependent and for any $j_1, \dots, j_r$, $1 \leq j_1 < \dots < j_r \leq n$, we have

$$P_{(X_{j_1}, \dots, X_{j_r})|\xi_{1n}} = P_{(X_{j_1}, \dots, X_{j_r})|(\xi_{j_1}, \dots, \xi_{j+l})}$$

in particular, for $l > m$, $j + l \leq n$

$$P_{(X_j, X_{j+l})|\xi_{1n}} = P_{X_j|\xi_j} * P_{X_{j+l}|\xi_{j+l}}$$

If $m = 0$, then $X_1, X_2, \dots$ are called conditionally independent random vectors (controlled by the sequence ξ).

Denote $\mu(\xi_j) = E(X_j|\xi_{1n}), R(\xi_j, \xi_l) = E\{[X_j - \mu(\xi_j)][X_l - \mu(\xi_l)]' | \xi_{1n}\}, R(\xi_j) = R(\xi_j, \xi_j), \mu = EX_j = E\mu(\xi_j)$,

$$R_m = E[R(\xi_1) + \sum_{e=1}^{m} R(\xi_1, \xi_{e+1}) + R(\xi_{e+1}, \xi_1))]$$

We consider the sums $S_n = n^{-1/2} \sum_{j=1}^{n} (X_j - \mu)$, $S_{n1} = n^{-1/2} \sum_{j=1}^{n} (X_j - \mu(\xi_j))$, $S_{n2} = n^{-1/2} \sum_{j=1}^{n} (\mu(\xi_j) - \mu)$.

The main idea of our approach is to use the representation

$$S_n = S_{n1} + S_{n2} \tag{2}$$

It can be easily seen that S_{n1} and S_{n2} are uncorrelated and, as shall see later, asymptotically independent.

Let us state the main theorem. The sign $\xrightarrow{w}$ denotes the weak convergence as $n \to \infty$ (passing to the usual ($\mathbb{P}$ - a.s.) limits is meant below also as $n \to \infty$).

<u>Theorem 1.</u> If $sp(R_m) < \infty$, then

a) $P_{S_{n1}|\xi_{1n}} \xrightarrow{w} \Phi_{R_m}$ $\mathbb{P}$ - a.s.;

b) $P_{S_{n1}} \xrightarrow{w} \Phi_{R_m}$

c) if $P_{S_{n2}} \xrightarrow{w} Q$, then $P_{S_n} \xrightarrow{w} \Phi_{R_m} * Q$

Note that R_m can be degenerate. For $m = 0$ and $k = 1$ this theorem was first published in [8] (see also [13]) and proved in [12] by a method which is demonstrated below in the proof of theorems on the density c convergence and the convergence in variation. This method assumes that the limit normal distribution is nondegenerate. For random vectors in the case $m = 0$ the Theorem was proved in [14] by the method of characteristic functions, which enables one to reject the requirement $\det(R_0) > 0$. The case $m > 0$ is treated then by means of Bernstein's partitioning technique. The use of ergodicity enables one to avoid the requirement of moments of order $2+\delta$.

2. CENTRAL LIMIT THEOREM WITH A MARKOV CHAIN AS A CONTROLLING SEQUENCE

As a controlling sequence we take a finite homogeneous stationary Markov chain with a single ergodic class (possibly with periodic subclasses), and with the state space $\{1,\dots,s\}$. Denote the matrix of transition probabilities by P , the stationary probabilities by $\pi_1,\dots,\pi_s$ and the fundamental matrix by $Z=(z_{\alpha\beta})_{\alpha,\beta=1,\dots,s}$. In this case it is possible to write an explicit form of the limit distribution Q of the sums S_{nz} of vectors $\mu(\xi_j)$ forming a Markov chain (it should be noted that when there are several periodic subclasses the convergence of the transition probabilities $p_{\alpha\beta}^{(n)} \to \pi_\beta$ and the relation $Z = I + \sum_{j=1}^{n}(P^j - \Pi)$, where I is the identity matrix and Π is the matrix with the rows $(\pi_1,\dots,\pi_s)$, is understood in the Cesaro sense [15]).

Theorem 2. If $(f_1,\dots,f_k)=f:\{1,\dots,s\}\to R^k$, then for the normed sums of functions of the states of the Mrkov chain

$$\mathcal{U}_n = n^{-1/2}\sum_{j=1}^{n}[f(\xi_j)-Ef(\xi_j)]$$

we have

$$P_{\mathcal{U}_n}\xrightarrow{w}\Phi_{T_f}$$

where T_f is the matrix with the elements

$$t_{f,ij} = \sum_{\alpha,\beta=1}^{s} \pi_\alpha (2 z_{\alpha\beta} - \pi_\alpha - \delta_{\alpha\beta}) f_i(\alpha) f_j(\beta), \; i,j = 1,\dots,k \tag{3}$$

and $\delta_{\alpha\beta}$ is Kronecker's delta.

This theorem for $k=1$ appeared in [16] and [17] (in our case all moment restrictions are satisfied); in [12, 13] for $k=1$ the limit variance is gven in an explicit form in terms of the fundamental matrix. Cramer-Wold's approach leads then to the multidimensional case.

If in Theorem 2 we put

$$f(\alpha) = \mu(\alpha) = (\mu_1(\alpha), \dots, \mu_k(\alpha)) = E(X_j \mid \xi_j = \alpha), \tag{4}$$

(i.e. $\mu = \sum_{\alpha=1}^{s} \pi_\alpha \mu(\alpha)$) then we obtain a limit theorem for the conditional mathematical expectations:

$$P_{S_{n2}} \xrightarrow{w} \Phi_{T_\mu}$$

Moreover, denoting

$$R(\alpha) = E\{[X_1 - \mu(\alpha)][X_1 - \mu(\alpha)]' \mid \xi_1 = \alpha\}, \; \alpha = 1,\dots,s,$$

we have

$$R_0 = \sum_{\alpha=1}^{s} \pi_\alpha R(\alpha) \tag{5}$$

These arguments sherpen Theorem 1 in the following way.

Theorem 3. If a conditionally m-dependent sequence $X = (X_1, X_2, \dots)$ is constrolled by a finite homogeneous Markov chain and if $sp(R_m) < \infty$, then in addition to assertions a) and b) of Theorem 1, we have

c) $P_{S_{n2}} \xrightarrow{w} \Phi_{T_\mu}$;

d) $P_{S_n} \xrightarrow{w} \Phi_{R_m + T_\mu}$

e) if $m=0$, then $P_{S_n} \longrightarrow \Phi_{R_0 + T_\mu}$

where R_0 is given by (5) and

$$R_0 + T_\mu = \lim_{n\to\infty} E(S_n S_n') \tag{6}$$

The validity of the last assertion can be easily shown. The necessary computations can be found in [12] for The proof of e) above for $k=1$ in the case of a chain without periodic subclasses (when ξ and X are strongly mixing) is given in [5]. In the proof Ibragimov's results [18] are used. Among the papers devoted to the central limit problem in the case of a regular controlling chain we can mention a paper by Aleshkevichus [2], where the class of possible limit laws is described, but it is assumed that $\mu(\xi_j) \equiv 0$ and, actually, the proof of assertion b) of Theorem 1 for conditionally independent variables is given, although a more general controlling chain is considered. In a number of papers the conditional distributions are assumed to depend on several previous states. As it is indicated in [5] and [19], this case can be reduced to the case of single state dependence by passing to a complex chain.

Our method allows to investigate the density convergence and the convergence in variation and, besides, to obtain estimates of the rate of weak convergence, which are uniform in convex sets, as well as estimates of the density convergence rate and of the rate of convergence in variation. The point is that if one does not casre for exact constants in the estimates, the estimates can be constructed on the basis of the known estimates of normal approximation of sums of independent vectors [20] and the estimates for sums of random vectors forming a regular Markov chain (see 21). Such estimates for distribution functions are given in [3] and [6] (non-uniform estimates).

Below we consider a conditionally independent sequence of random vectors controlled by a finite ergodic Markov chain.

3. CONVERGENCE OF DENSITIES AND CONVERGENCE IN VARIATION IN THE CASE OF CONTROL BY AN ERGODIC CHAIN

The mathematical expectation $\mu(\alpha)$ and the covariance matrix $R(\alpha)$, $\alpha = 1, \dots, S$ are well defined and finite under the assumption $sp(R_0) < \infty$ of Theorem 3.

Theorem 4. If at least one of the conditional distributions P_α, $\alpha = 1, \dots, S$, has the characteristic function belonging to $L^r(R^k)$ for some positive integer r, then

a) for almost all $\omega \in \Omega$ there exists a number $N = N(\omega)$ such that for $n \geq N$ the distribution $P_{S_{n1}|\xi_{1n}}$ has a density and

$$\sup_{x \in R^k} |p_{S_{n1}|\xi_{1n}}(x) - \varphi_{R_0}(x)| \to 0$$

b) there is a representation $P_{S_n} = P_{S_n}^{(1)} + P_{S_n}^{(2)}$ which, in generally differs from the decomposition of P_{S_n} into an absolutely continuous $(P_{S_n}^{as})$ and a singular $(P_{S_n}^{s})$ components with respect to the Lebesgue measure such that

$$P_{S_n}^{s}(R^k) \leq P_{S_n}^{(2)}(R^k) \leq c/n$$

where the constant c depends on r and the chain parameters;

c) the density $p_{S_n}^{(1)}(x)$ of the measure $P_{S_n}^{(1)}$ satisfies the relation

$$\sup_{x \in R^k} |p_{S_n}^{(1)}(x) - \varphi_{R_0 + T_\mu}(x)| \to 0$$

Theorem 5. If all conditional distributions P_α, $\alpha = 1, \ldots, s$, have characteristic functions which belong to $L^r(R^k)$ for some positive integer r, then

a) for $n \geq r$ there exists the conditional density $p_{S_{n1}|\xi_{1n}}(x)$ which uniformly converges $\mathbb{P}$ -a.s. to $\varphi_{R_0}(x)$;

b) for $n \geq r$ $p_{S_n}(x)$ exists and

$$\sup_{x \in R^k} |p_{S_n}(x) - \varphi_{R_0 + T_\mu}(x)| \to 0$$

The distance in variation between the probability measures Q_1 and Q_2 is denoted by

$$v(Q_1, Q_2) = \sup_{A \in \mathcal{B}} |Q_1(A) - Q_2(A)|$$

where $\mathcal{B}$ is the class of Borel subsets of R^k.

Theorem 6. If for a positive integer r the r-fold convolution of at least one of the conditional distributions has a non-zero absolutely continuous component, then

a) $v(P_{S_{n1}|\xi_{1n}}, \Phi_{R_0}) \to 0$ $\mathbb{P}$ - a.s.

and for almost all $\omega \in \Omega$ at some random moment $N = N(\omega)$ $P_{S_{n1}/\xi_{1n}}$ has a non-zero absolutely continuous component $P^{ac}_{S_{n1}|\xi_{1n}}$ and for its density we have

$$\int |p^{ac}_{S_{n1}|\xi_{1n}}(x) - \varphi_{R_0}(x)|\, dx \to 0 \quad \mathbb{P}\text{- a.s.,}$$

b) $$v(P_{S_n}, \Phi_{R_0 + T_\mu}) \to 0.$$

4. CONVERGENCE RATE ESTIMATES

Since our estimates are derived from the estimates

of normal approximation for random vectors forming a regular Markov chain 21 in this section we assume that the controlling finite chain has a single ergodic class without periodic subclasses. Assume also that

$$\beta(\alpha) = E[\,|X_1 - \mu(\alpha)|^3 \,|\, \xi_1 = \alpha\,] < \infty, \quad \alpha = 1, \dots, s.$$

Denote by $\mathcal{H}$ the class of convex Borel subsets of R^k

Theorem 7. If the matrices $R(\alpha), \alpha = 1, \dots, s$, and the matrix T_μ are non-degenerate then

$$\sup_{A \in \mathcal{H}} |P_{S_n}(A) - \Phi_{R_0 + T_\mu}(A)| < c/\sqrt{n}$$

where the constant c depends on $k, \mu(\alpha), R(\alpha), \beta(\alpha)$ and characteristics of the chain ξ.

Theorem 8. If all conditional distributions P_α, $\alpha = 1, \dots, s$ have bounded densities, or their characteristic functions belong to $L^r(R^k)$ for some positive integer r and $\det(T_\mu) > 0$ then for $n \geq r$

$$\sup_{x \in R^k} |p_{S_n}(x) - \varphi_{R_0 + T_\mu}(x)| < \frac{c}{\sqrt{n}}$$

where c depends on k, conditional distributions and characteristics of the chain ξ.

Theorem 9. If for a positive integer r r-fold convolutions of all conditional distributions $P_\alpha, \alpha = 1, \dots, s$ have absolutely continuous components and $\det(T_\mu) > 0$, then

$$v(P_{S_n}, \Phi_{R_0 + T_\mu}) < C(\ln n)^{k/2}/\sqrt{n}$$

where C depends on k, conditional distributions and characteristics of the chain ξ.

5. PROOF OF THEOREMS 4-9

Constants independent of n which vary throughout the text are denoted by the same letter C.

Assertion a) of Theorem 4 follows from the multidimensional analogue [22] of Gnedenko's theorem on necessary and sufficient conditions for the uniform convergence of the densities of normalized sum of identically distributed random variables and Theorem 3. Note that here and in the subsequent arguments the representation

$$P_{S_{n1}|\xi_{1n}} = \mathop{\boxed{*}}_{\alpha=1}^{s} P_{S^{\alpha}_{n1}|\xi_{1n}} \tag{7}$$

where S^{α}_{n1} is the sum of summands from S_{n1} corresponding to the state α of the chain ξ at moments $1,\dots,n$ is of primary importance.

Let $|\int \exp[i(t,y)]P_1(dy)| \in L^r(R^k)$. Denote by $\nu_n(\alpha)$ the frequency of the state α of the chain ξ during n steps $\nu_n(1)+\dots+\nu_n(s)=n$. Denote also

$$P^{(1)}_{S_n} = E\,P_{S_n|\xi_{1n}}\, I(\nu_n(1)\geq r),$$
$$P^{(2)}_{S_n} = E\,P_{S_n|\xi_{1n}}\, I(\nu_n(1)< r), \tag{8}$$

where $I(M)$ is the indicator of the event M.

According to [15]

$$D\,\frac{\nu_n(\alpha)}{n} = E\left(\frac{\nu_n(\alpha)}{n} - \pi_\alpha\right)^2 < \frac{c}{n},\quad \alpha=1,\dots,s, \tag{9}$$

where c depends on the chain parameters. By Chebyshev's inequality we have $P^{(2)}_{S_n}(R^k) = \mathbb{P}(\nu_n(1)<r) < c/n$

and assertion b) of Theorem 4 follows from the fact that $P_{S_n}^{(1)}$ has the density

$$p_{S_n}^{(1)}(x) = E\, p_{S_n|\xi_{1n}}(x)\, I(\nu_n(1) \geq r) \tag{10}$$

and the absolutely continuous measure $P_{S_n}^{ac}$ can also be formed on other trajectories, which are not consider in the sum (8).

If $Y_1, Y_2, \ldots$ are independent random vectors with the same distribution P_1, then the density of the sum $\nu^{-1/2}(Y_1 + \ldots + Y_\nu - \nu\mu(1))$ as $\nu \to \infty$ converges to $\varphi_{R(1)}(x)$ and, hence, for $\nu \geq r$ this density is bounded. Thus,

$$p_{S_{n1}|\xi_{1n}}(x)\, I(\nu_n(1) \geq r) \leq C \sqrt{\frac{n}{\nu_n(1)}}\, I(\nu_n(1) \geq r) \tag{11}$$

By assertion a) of Theorem 4 we have

$$\zeta_n = \sup_x \left| p_{S_{n1}|\xi_{1n}}(x - S_{n2}) - \varphi_{R_0}(x - S_{n2}) \right| \cdot$$
$$\cdot\, I(\nu_n(1) \geq r) \to 0 \quad \mathbb{P}\text{-a.s.}$$

Applying (11), (9) and the fact that for a fixed $\varepsilon > 0$ and $|\nu_n(1)/n - \pi_1| \leq \varepsilon$

$$\zeta_n \leq c(\pi_1 - \varepsilon)^{-1/2} + (2\pi)^{-k/2}[\det(R_0)]^{-1/2} = c,$$

we obtain that $E\zeta_n \leq c\sqrt{\frac{n}{r}}\, \mathbb{P}(\nu_n(1) \geq r,$

$$\left|\frac{\nu_n(1)}{n} - \pi_1\right| > \varepsilon) + E\zeta_n\, I(\nu_n(1) \geq r,$$
$$\left|\frac{\nu_n(1)}{r} - \pi_1\right| \leq \varepsilon) \to 0 \tag{12}$$

Since
$$E\zeta_n \geq \sup_x \left| E p_{S_n|\xi_{1n}}(x)\, I(\nu_n(1) \geq r) - \right.$$
$$\left. - E\varphi_{R_0}(x - S_{n2})\, I(\nu_n(1) \geq r) \right|,$$

assertion c) or Theorem 4 follows from (10), (12) and the weak convergence $P_{S_{n2}} \xrightarrow{w} \Phi_{T_\mu}$ stated in Theorem 3. Theorem 4 is proved. The proof of Theorem 5 is now evident.

Now to prove Theorem 8 we denote $\hat{R}_0 = \sum_{\alpha=1}^{s} \frac{\nu_n(\alpha)}{n} R(\alpha)$ and write

$$\sup_x |P_{S_n}(x) - \varphi_{R_0+T_\mu}(x)| \leq$$

$$\leq \sup_x |E\{E[(P_{S_{n1}|\xi_{1n}}(x - S_{n2}) - \varphi_{\hat{R}_0}(x - S_{n2}))|\xi_{1n}]\}| +$$

$$+ \sup_x |E\{E[\varphi_{\hat{R}_0}(x - S_{n2}) - \varphi_{R_0}(x - S_{n2}))|\xi_{1n}]\}| +$$

$$+ \sup_x |E\varphi_{R_0}(x - S_{n2}) - \varphi_{R_0+T_\mu}(x)| = \sigma_1 + \sigma_2 + \sigma_3$$

We assume first that the distributions P_α, $\alpha = 1, \ldots, s$ have densities bounded by constants A_α. This enables us to apply the uniform convergence rate estimate for densities from [23]. After some elementary computations we obtain the estimate $\sigma_1 < c/\sqrt{n}$. If the characteristic functions of all conditional distributions P_α belong to $L^r(R^k)$ for some positive integer r one would have to refer to paper [24] which would lead to estimates of the same order starting with some n.

To estimate σ_2 an estimate of the density difference $\varphi_{\alpha_1 B}(x) - \varphi_{\alpha_2 B}(x)$ is to be applied, where $\det(B) > 0$, $\min(\alpha_1, \alpha_2) = \alpha > 0$, which is obtained by means of Lagrange's formula

$$\sup_{x \in R^k} |\varphi_{\alpha_1 B}(x) - \varphi_{\alpha_2 B}(x)| \leq \frac{k}{2}(2\pi)^{-\frac{k}{2}}[\det(B)]^{-1/2} \cdot \alpha^{-\frac{k+2}{2}} |\alpha_1 - \alpha_2| \qquad (13)$$

Since the determinant of a convex combination of the positive definite matrices R_α is not less than $\min_\alpha \det(R_\alpha)$ the expression under the sign of mathematical expectation in σ_2 does not exceed

$$2(2\pi)^{-k/2}[\min_\alpha \det(R_\alpha)]^{-1/2}$$

Define

$$M = \bigcap_{\alpha=1}^{s} \left\{ \left| \frac{\nu_n(\alpha)}{n} - \pi_\alpha \right| \leq \frac{\pi_\alpha}{2} \right\} \tag{14}$$

By (9) $\mathbb{P}(\bar{M}) < c/n$ and applying (13) we have

$$\sigma_2 \leq \frac{c}{n} + \sum_{\alpha=1}^{s} E \sup_x \left| \varphi_{\frac{\nu_n(\alpha)}{n} R(\alpha)}(x) - \right.$$

$$\left. - \varphi_{\pi_\alpha R(\alpha)}(x) \right| I(M) \leq \frac{c}{n} + C \sum_{\alpha=1}^{n} E \left| \frac{\nu_n(\alpha)}{n} - \pi_\alpha \right| < C/\sqrt{n}$$

Further,

$$\sigma_3 = \sup_x \left| \int \varphi_{R_0}(x-y)(P_{S_{n2}} - \Phi_{T_\mu})(dy) \right|$$

is estimated by means of a weak convergence rate estimate, i.e. an estimate of the integral

$$\int g(y)(P_{S_{n2}} - \Phi_{T_\mu})(dy)$$

which is uniform with respect to a class of bounded Lipschitz functions g (see [21], Ch.IV). It turns out that $\sigma_3 < C/\sqrt{n}$. This completes the proof of theorem88.

Now we shall prove Theorem 6.

Assertion a) follows from the multidimensional analogue of Prokhorov's theorem on necessary and sufficient conditions for convergence in variation [25]. Let P_1^{h*} have a non-zero absolutely continuous component. Then

$$v(P_{S_n}, \Phi_{R_0+T_\mu}) \leq v(P_{S_n}, E\,\Phi_{R_0}(\cdot - S_{n2}))$$

$$+ v(E\,\Phi_{R_0}(\cdot - S_{n2}), \Phi_{R_0+T_\mu}) = \tau_1 + \tau_2 , \tag{15}$$

where $Q(\cdot - a)$ denotes the shift by $a \in R^k$ of the measure Q .

Denoting $S_{n1}^{-} = S_{n1} - S_{n1}^{1}$, $R_0^{-} = R_0 - \pi_1 R(1)$ we obtain

$$\tau_1 \leq E\, v\,(P_{S_{n1}|\xi_{1n}}, \Phi_{R_0}) \leq$$

$$\leq E\, v\,(P_{S^1_{n1}|\xi_{1n}} * P_{S^-_{n1}|\xi_{1n}}, \Phi_{\Pi_1 R(1)} * P_{S^-_{n1}|\xi_{1n}}) +$$

$$+ E\, v\,(\Phi_{\Pi_1 R(1)} * P_{S^-_{n1}|\xi_{1n}}, \Phi_{\Pi_1 R(1)} * \Phi_{R_0^-}) \leq$$

$$\leq E\, v\,(P_{S^1_{n1}|\xi_{1n}}, \Phi_{\Pi_1 R(1)}) + \frac{1}{2} E \left(\int \left| \int \varphi_{\Pi_1 R(1)}((x-y)(P_{S^-_{n1}|\xi_{1n}} - \Phi_{R_0^-})(dy) \right| dx \right) = \tau_{11} + \tau_{12}$$

The convergence of τ_{11} to zero as $n \to \infty$ is evident. By Theorem 3 $P_{S^-_{n1}|\xi_{1n}} \xrightarrow{w} \Phi_{R_0^-}$ which implies the convergence to zero of the inner integral in τ_{12}. Thi integral is a density difference. Hence, by Sheffé's theorem, the outer integral also converges to zero and since it is bounded, $\tau_{12} \to 0$. The converges to zero of is proved by the same arguments, but now of major importance is the weak convergence of $P_{S_{n2}} \xrightarrow{w} \Phi_{T_\mu}$.

Theorem 9, which estimates the convergence rate in Theorem 6, is proved by means of the following lemma (cf. [26, 27]).

Lemma. If $\det(B_i) > 0$, $i = 1, 2$ then

$$v\,(\Phi_{B_1}, \Phi_{B_2}) \leq \{1 - 2^k / \det[(B_1 B_2)^{1/2}(B_1^{-1} + B_2^{-1})]\}^{1/2}.$$

In particular, if $B_i = \alpha_i B$, $\min(\alpha_1, \alpha_2) = \alpha > 0$, then

$$v\,(\Phi_{\alpha_1 B}, \Phi_{\alpha_2 B}) \leq C(k)\, \alpha^{-1} |\alpha_1 - \alpha_2|,$$

where

$$C(k) = \sup_{0 < x \leq 1} \left\{ \left[\sum_{j=0}^{k-1} (1+x^2)^{k-1-j} (2x)^j \right] (1+x^2)^{-k} \cdot (1+x)^{-2} \right\}^{1/2}$$

The lemma follows from the relation

$$1-[v(\Phi_{B_1},\Phi_{B_2})]^2 \leq \left[\int(\varphi_{B_1}(x)\varphi_{B_2}(x))^{1/2}dx\right]^2$$

Let us estimate now τ_1 from (15) (the matrix $\hat{R}_0$ was introduced earlier and M was defined in (14)).

$$\tau_1 \leq E\,v(P_{S_{n1}|\xi_{1n}},\Phi_{R_0}) \leq E\,v(P_{S_{n1}|\xi_{1n}},\Phi_{\hat{R}_0})+$$
$$+E\,v(\Phi_{\hat{R}_0},\Phi_{R_0}) \leq 2\,\mathbb{P}(M) +$$
$$+\sum_{\alpha=1}^{s} E\,v\left(P_{S_{n1}^{\alpha}|\xi_{1n}},\Phi_{\frac{\nu_n(\alpha)}{n}R(\alpha)}\right)I(M) +$$
$$+\sum_{\alpha=1}^{s} E\,v\left(\Phi_{\frac{\nu_n(\alpha)}{n}R(\alpha)},\Phi_{\pi_\alpha R(\alpha)}\right)I(M)=\tau_{11}^{\circ}+\tau_{12}^{\circ}+\tau_{13}.$$

We know that $\tau_{11}^{\circ}<c/n$. By the estimate of the rate of convergence in variation for sums of independent random vectors [28] we have

$$\tau_{12}^{\circ} \leq \frac{c}{\sqrt{n}}\sum_{\alpha=1}^{s} E\left(\frac{\nu_n(\alpha)}{n}\right)^{-1/2} I(M) < \frac{c}{\sqrt{n}},$$

where c depends on the conditional distributions and on k.

Applying the Lemma together with (9) we obtain

$$\tau_{13} \leq c\sum_{\alpha=1}^{s} E\left|\frac{\nu_n(\alpha)}{n}-\pi_\alpha\right| < \frac{c}{\sqrt{n}}$$

In the expression

$$\tau_2=\frac{1}{2}\int\left|\int\varphi_{R_0}(x-y)(P_{S_{n2}}-\Phi_{T_m})(dy)\right|dx$$

the modulus of the inner integral, according to [21], uniformly does not exceed $c/\sqrt{n}$.

Let $u(x)$ be the density of some measure $\mathcal{U}$ and $\sup_{x\in R^k}|u(x)-\varphi_I(x)|=\varepsilon$. Then it can be seen easily that

$$v(\mathcal{U},\Phi_I)=\frac{1}{2}\int|u(x)-\varphi_I(x)|dx \leq V_k H^k\varepsilon+\Phi_I\{x:|x|>H\} \quad (16)$$

where $H>0$ and V_k is the volume of k -dimensional sphere with radius 1. By the exponential Chebyshev inequality we have

$$\Phi_I\{x:|x|>H\} \leq \left(\frac{e}{k}\right)^{k/2} H^k e^{-H^2/2}$$

If now we substitute this estimate and $H=(2\ln\frac{1}{\varepsilon})^{1/2}$ in (16), we shall obtain

$$v(\mathcal{U},\Phi_I) \leq \left(V_k+\left(\frac{e}{k}\right)^{k/2}\right)\varepsilon\left(\ln\frac{1}{\varepsilon^2}\right)^{k/2} \tag{17}$$

which leads to the inequality

$$\tau_2 < c(\ln n)^{k/2}/\sqrt{n}$$

where c depends on k, the chain parameters and the matrices R_0 and T_μ. This completes the proof of Theorem 9.

In order to prove Theorem 7 Sazonov's theorem on the convergence rate estimate uniform in convex sets for independent random vectors in R^k (see [20]) and a similar theorem from Sirazdinov and Formanov's book [21] are applied together with the Lemma and the technique demonstrated above.

6. PRELIMINARY VERSIONS OF THEOREMS 4, 7 AND 8 WERE ANOUNCED IN [29]

In conclusion we note that Theorem 3 was applied to problems of statistical estimation when both components of a two-component sequence $(\xi_j, X_j)_{j\geq 1}$, $k=1$ are observable. Here $\xi=(\xi_1,\xi_2,\dots)$ is a Markov chain and $X=(X_1,X_2,\dots)$ are conditionally independent random variables. As a result an analogue of the empirical pro cess was constructed and its convergence to a multidimensional Brownian bridge was proved. It allowed to suggest non-parametric criteria to test the hypothesis on conditional distributions of X_1 (see [14]). Theorem 3 also enables us to write down the limit normal distributions of the random variables $\sqrt{n}(\hat{\sigma}^2-\sigma^2)$, $\sqrt{n}(\hat{\rho}_m-\rho_m)$ and the random vector $\sqrt{n}[(\hat{\sigma}^2,\hat{\rho}_m)-(\sigma^2,\rho_m)]$ where

$$\sigma^2 = E(X_1-\mu)^2,\ \rho_m = E(X_1-\mu)(X_{m+1}-\mu),\ \mu = EX_1$$

and

$$\hat{\sigma}^2 = \frac{1}{n}\sum_{j=1}^{n}(X_j-\mu)^2,\ \hat{\rho}_m = \frac{1}{n-m}\sum_{j=1}^{n-m}(X_1-\mu)(X_{m+1}-\mu)$$

are unbiased estimators of σ^2 and ρ_m .

REFERENCES

1. Hasminsky R.Z. On limit distributions for sums of conditionally independent random variables (in Russian). - Teoria Verojatn. i primen., 1961, v.6, N 1, p.119-125.
2. Aleshkevichus G. On the central limit problem for sums of random variables defined on a Markov chain (in Russian). - Liet.mat.rink., 1966, v.V, N 1, p.15-22.
3. Aleshkevichus G. Some limit theorems for sums of random variables defined on the homogeneous Markov chain (in Russian). - Liet.math.rink., 1966, v.VI, N 3, p.297-311.
4. Bezaeva Z.I. Limit theorems for conditional Markov chains (in Russian). - Teoria verojatn. i primen., 1971, v.XVI, N 3, p.437-445.
5. O'Brein G.L. Limit theorems forsums of chain-depen-

dent processes. - J.Appl.Probab., 1974, 11, p.582-587.
6. Skvortsov A.E. On the non-uniform estimate of the convergence rate in the central limit theorem. (in Russian). Deponent manuscript, N 1981-74 Dep.VINITI 1974.
7. Grigorescu S., Oprisan G. Limit theorems for processes with a general state space. - Z.Wahrscheinlichkeitstheorie verw. Geb., 1976, Bd.35, h.1, p. 65-73.
8. Bokuchava I.V. Central limit theorem for conditionally independent random sequences. (in Russian) - Soobsh.AN Gruz.SSR, 1978, v.91, N 2, p. 305-308.
9. Kaplan E.I., Silvestrov D.S. General limit theorems for sums of controlled random variables with arbitrary space of controlling sequence (in Russian). - Liet.mat.rink., 1980, v.10, N 4, p.61-72.
10. Silvestrov D.S. Limit theorems for processes of step sums of conditionally independent random variables (in Russian). - Teoria verojatn. i primen., 1982, v.27, N 3, p.611-613.
11. Anisimov V.V. Limit theorems for nonhomogeneous weakly dependent schemes of summation. (in Russian). - Teorya verojatn. i matem.statistika, 1982, issue 27, p.10-22.
12. Bokuchava I.V. On the central limit theorem for conditionally independent sequences (in Russian). - In: Investigations on Probability Theory and Mathematical Statistics. Tbilisi: Metsniereba, 1982, p.3-15.
13. Bokuchava I.V. Limit theorems for conditionally independent sequences (in Russian). - Teorya verojatn. i primen., 1984, v. XXIX, N 1, p.192-193.
14. Bokuchava I.V., Chitashvili R.J., Shervashidze T.L. On a limit theorem for conditionally independent random vectors and its application in statistics

(in Russian). - In: Limit theorems and stochastic equations. Tbilisi: Metsniereba, 1984, p.54-70.
15. Kemeny J.G., Snell J.L. Finite Markov chains. Princeton: University press, 1959.
16. Doob J. Stochastic processes. New York - London: Wiley - Chapmen, 1953.
17. Sirazdinov S.H. Limit theorems for homogeneous Markov chains (in Rusina). Tashkent: Publ.House of the Acad.of Sci. of the Uzb.SSR, 1955.
18. Ibragimov I.A., Linnik Y.V. Independent and stationary dsequences of random variables. Groningen: Wolters-Noordhoff, 1971.
19. Korolyuk V.S., Turbin A.F. Semi-Markov processes and their applications (in Russian). Kiev: Naukova Dumka, 1976.
20. Sazonov V.V. Normal approximation. Some recent advances. - Lecture Notes in Math., v.879, Berlin etc. Springer, 1981.
21. Sirazdinov S.H., Formanov Sh.K. Limit theorems for sums of random vectors connected in a Markov chain. (in Russian). Tashkent: Fan, 1979.
22. Heckendorf H. Multidimensional local theorem for densities. (in Russian). - Ukrain.Mat.Journ., 1964, v.XVI, N 3, p.365-373.
23. Shervashidze T.L. On the uniform estimate of the rate of convergence in a multidimensional local limit theorem (in Russian). - Teorya verojatn. i primen. 1971, v.XVI, N 4, p.765-767.
24. Shervashidze T.L. On the multidimensional local limit theorems for densities (in Russian). - In: Limit theorems and stochastic equations. Tbilisi: Metsnierega, 1984, p.15-53.
25. Bikelis A. On the central limit theorem in (in Russian). - Liet.matem.rink., 1971, v.XI, N 1, p.27-59.

26. Halikov M.K. Local theorem for sums of independent random vectors (in Russian). 9 Izv.AN Uz.SSR, ser. Fiz.Mat., 1958, 2, p.95-105.
27. Barsov S.S. On the accuracy of normal approximation of the distributions of random sums of random vectors (in Russian). - Teorya Veroyatn.i primen., 1985, v.XXX, N 2, p.351-354.
28. Bikelis A. On the central limit theorem in (In Russian). Liet.matem.rink., 1972, v.XII, N 3, p.19-35.
29. Kvatadze Z.A., Omiadze K.T., Shervashidze T.L. On the asymptotics of distribution of sums of random variables controlled by a Markov Chain (in Russian). - XVII colloquium on probability theory and mathematical statistics. Abstracts of communications. Tbilisi: Tbilisi State University, 1984, p.21.

Prob. Theory and Math. Stat., Vol.1, pp. 251-268
Prohorov et al. (eds)
1986 VNU Science Press

UPPER AND LOWER ESTIMATES OF THE CONVERGENCE RATE IN THE INVARIANCE PRINCIPLE FOR EMPIRICAL MEASURES

I.S.Borisov
Institute of Mathematics, Novosibirsk, USSR

Let $\{X_k; k \geq 1\}$ be a sequence of independent identicaly distributed random variables (r.v.) taking values in an arbitrary measurable space $(\mathcal{X}, \mathcal{A})$ and with distribution P on $\mathcal{A}$. Denote by P_m the empirical distribution based on the sample $X_1, X_2, \dots, X_m$. The paper is dedicated to the investigation of proximity of the distributions of the following two-parametric random fields (r.f.'s)

$$S_n(t,A) = n^{-1/2}[nt]\left(P_{[nt]}(A) - P(A)\right),$$

$$W_n(t,A) = n^{-1/2} \sum_{i \leq nt} G_i(A) \tag{1}$$

where $t \in [0,1]$, $A \in \mathcal{A}$, $\{G_i(\cdot); i \geq 1\}$ are independent Gaussian r.v.'s with mean zero and convariance

$$E\, G_i(A) G_i(B) = P(A \cap B) - P(A)P(B).$$

Let $\mathcal{B}$ be an arbitrary subclass of $\mathcal{A}$. Suppose that for every $\varepsilon > 0$ there are sets $A_i \in \mathcal{A}$, $i \leq N(\varepsilon)$, with the following property: for any $B \in \mathcal{B}$ there are

$\mathcal{A}_{i_1}, \mathcal{A}_{i_2}$ such that

$$A_{i_1} \subseteq B \subseteq A_{i_2}, \quad P(A_{i_2} \setminus A_{i_1}) \le \varepsilon. \tag{2}$$

The logarithm of the minimal number of $\{A_i\}$ satisfying (2) is called the metric entropy with inclusion (see [5]) and is denoted by $H_1(\varepsilon, \mathcal{B}, P)$.

Let a family of sets $\mathcal{A}_N = \{A_i \in \mathcal{A}; i \le N\}$ satisfy (2). Denote by $\mathcal{C}(\mathcal{A}_N)$ some finite class $\{B_j\}$ of measurable sets generating the family $\mathcal{A}_N$, that is

$$A_i = \bigcup_{k=1}^{M_i} \bigcup_{j=j_k^{(i)}}^{j_{k+1}^{(i)}-1} B_j, \quad i \le N, \tag{3}$$

where $B_j \cap B_i = \emptyset, i \ne j, \{j_k^{(i)}\}$ is an increasing in k subsequence of the sequence of natural numbers.

Let $M(\mathcal{A}_N, \mathcal{C}(\mathcal{A}_N)) = \max_{i \le N} M_i$, where M_i are defined in (3). Obviously $M(\cdot)$ depends on $\mathcal{A}_N, \mathcal{C}(\mathcal{A}_N)$ and on the enumeration of elements from $\mathcal{C}(\mathcal{A}_N)$. Denote

$$\gamma(\varepsilon, \mathcal{B}, P) = \inf M(\mathcal{A}_N, \mathcal{C}(\mathcal{A}_N)) \tag{4}$$

where the infimum in (4) is taken over all families $\mathcal{A}_N$ satisfying (2) with $N \ge \exp\{H_1(\varepsilon, \mathcal{B}, P)\}$ and over all classes $\mathcal{C}(\mathcal{A}_N)$ and all enumerations of subsets $B_j \in \mathcal{C}(\mathcal{A}_N)$. The sequence $B_k, B_{k+1}, \dots, B_{k+m}$ will be called a chain of subsets. Therefore $\gamma(\varepsilon, \mathcal{B}, P)$ is the minimal upper estimate of the number of chains of subsets $\{B_j\}$ which are contained in every set $A_k \in \mathcal{A}_N$ under an appropriate enumeration of elements from $\mathcal{C}(\mathcal{A}_N)$. Some examples of estimation of $\gamma(\cdot)$ are contained in [1]. As a rule, the value $\gamma(\cdot)$ in real problems is much smaller than 2^N (this rough upper estimate follows from (4)). But, generally speaking, the estimate 2^N cannot be improved essentially. The following proposition corrects Lemma 3 in [1].

Lemma 1. There exists a family of sets $\mathcal{A}_N=\{A_1,\dots,A_N\}$ such that

$$M(\mathcal{A}_N,\mathcal{C}(\mathcal{A}_N))\geq(2^{N-1}-1)N^{-1}.$$

Proof. Consider the class of all nonzero N-dimensional vectors $\mathcal{C}(\mathcal{A}_N)=\{a_k;\ k\leq 2^N-1\}$ with coordinates "0" or "1". Let A_k be the set of a_j's with k-th coordinate equal to 1. It is easy to see that $\mathcal{C}(\mathcal{A}_N)$ is the minimal class generating $\mathcal{A}_N$.

Let $\{a_k\}$ be numbered arbitrarily. Denote by $a_k^{(i)}$ the i-th coordinate of the vector a_k, $\|a_k\|=\sum_{i\leq N}|a_k^{(i)}|$. Then the following inequalities hold

$$2^N-2\leq\sum_{k\leq 2^N-2}\|a_{k+1}-a_k\|=\sum_{i\leq N}\sum_{k\leq 2^N-2}|a_{k+1}^{(i)}-a_k^{(i)}|\leq$$

$$\leq 2NM(\mathcal{A}_N,\mathcal{C}(\mathcal{A}_N)).$$

The lemma is proved.

In what follows C, C_k denote some absolute positive constants. Sometimes the parameters on which the constants depend are indicated in parentheses.

The following statements are corollaries of the author's results in [1]-[4]. The results in [1], [2] one imply.

Theorem 1. Let for some $\mathcal{B}$ and P Dubley's condition be fulfilled:

$$\int_0^1(H_1(\varepsilon^2,\mathcal{B},P)^{1/2}d\varepsilon<\infty. \tag{5}$$

Then one can construct the r.v.'s $X_1,\dots,X_n$ and r.f.'s $G_1(\cdot),\dots,G_n(\cdot)$ on a common probability space such that

$$P_r\{\Delta_n^*(\mathcal{B})>C\min_{m:\,H_1(2^{-m},\mathcal{B},P)\le n2^{-m}-1}\alpha_n(m)\}\le \le C_1 n^{-2} \tag{6}$$

where

$$\alpha_n(m)=n^{-1/2}(\log n)^2\gamma(2^{-m},\mathcal{B},P)+ + \left(1+\frac{\log n}{H_1(2^{-m},\mathcal{B},P)+1}\right)\sum_{k>m}\left(H_1(2^{-k},\mathcal{B},P)2^{-k}\right)^{1/2} \tag{7}$$

and $\Delta_n^*(\mathcal{B})$ is a measurable upper estimate of

$$\Delta_n(\mathcal{B})=\sup_{t\in[0,1]}\sup_{A\in\mathcal{B}}|S_n(t,A)-W_n(t,A)|.$$

Remark 1. Inequality (6) will hold if the function $H_1(\cdot)$ in (6) and (7) is replaced by some upper estimate for which. Dudley's condition (5) is fulfilled.

As an example, consider the case when $\mathcal{X}=[0,1]\times\times[0,1]$, $\mathcal{A}$ is the class of Borel subsets of $\mathcal{X}$, P is an absolute continuous distribution on $\mathcal{X}$ with a bounded density,

$$\mathcal{B}=\{(x,y)\in\mathcal{X}: y\le f(x);\ f\in C_L^1[0,1]\}$$

where $C_L^1[0,1]$ is the family of all absolute continuous on $[0,1]$ functions such that $\sup_x|f'(x)|+ + \int_0^1|df'(x)|\le L$

Lemma 2. For $\mathcal{X},\mathcal{B}$ and P above

$$H_1(\varepsilon,\mathcal{B},P)\le C(L,P)\varepsilon^{-1/2}, \tag{8}$$

$$\gamma(\varepsilon, \mathcal{B}, P) \le C_1(L, P)\varepsilon^{-3/2}. \tag{9}$$

Proof. Relation (8) follows directly from [6] since each function from $C_L^1[0,1]$ can be represented as the difference of two-non-negative nondecreasing convex functions. Note however that (8) will be proved authomatically when we prove (9).

Therefore it is sufficient to construct a family of sets $\mathcal{A}_N$ for the subclass $\mathcal{B}_1 \subset \mathcal{B}$ consisting of such subsets of $\mathcal{X}$ which are generated by non-negative nondecreasing convex functions. Let $f(x)$ be a function generating some set from $\mathcal{B}_1$. Denote by $f_m^+(x)$, $f_m^-(x)$ continuous broken lines with vertices at the points $(k/m, f(k/m))$, $k=0,1,\ldots,m$, $(k/m, f(\frac{k-1}{m}) + f'(\frac{k-1}{m})\frac{1}{m})$, $k=1,\ldots,m$, respectively. Define also $f_m^-(0) = f(0)$. It is easy to verify that $f_m^{\pm}$ are convex functions and

$$f_m^-(x) \le f(x) \le f_m^+(x)$$

for all $x \in [0,1]$ and natural m.

Define by $B(f)$ the set from $\mathcal{B}_1$ which is generated by f. Then

$$\Lambda(B(f_m^+) \setminus B(f)) = \sum_{k<m}\left\{ \frac{f(\frac{k}{m}) + f(\frac{k+1}{m})}{2m} - \int_0^{1/m} f\left(\frac{k}{m} + x\right)dx \right\} \equiv R \tag{10}$$

where $\Lambda(\cdot)$ is the Lbesgue measure on $\mathcal{X}$. Using the formula

$$f\left(\frac{k}{m}+z\right)=f\left(\frac{k}{m}\right)+f'\left(\frac{k}{m}\right)z+z\int_0^1(1-\theta)d_\theta f'\left(\frac{k}{m}+\theta z\right) \qquad (11)$$

which is true whenever $f'(x)$ has a bounded variation, we see that (10) implies $R\leq Lm^{-2}$.

$\Lambda(B(f)\setminus B(f_m^-))$ is estimated in a similar way. Therefore under the conditions of the Theorem we have $B(f_m^-)\subseteq B(f)\subseteq B(f_m^+)$ and

$$P(B(f_m^+)\setminus B(f_m^-))\leq C(P)Lm^{-2} \qquad (12)$$

where $C(P)=2\sup\frac{dP}{d\Lambda}$

Let $\varphi_m^\pm$ be the functions with the following properties: $\varphi_m^-\leq f_m^-$, $f_m^+\leq\varphi_m^+$ and $\varphi_m^\pm$ is the closest to $f_m^\pm$ continuous broken line with vertices at the points $(k/m, S/m^2)$, $k\leq m, S\leq m^2(L+1)$. Consider the class of all $\varphi_m^\pm$ constructed for each $f\in\mathcal{B}_1$. The broken lines $\varphi_m^\pm$ may not be convex functions, but for any $x\neq k/m$, $k=0,\dots,m-1$,

$$\frac{d}{dx}\left\{\varphi_m^\pm\left(x+\frac{1}{m}\right)-\varphi_m^\pm(x)\right\}\geq -1/m. \qquad (13)$$

Moreover from the construction of $\varphi_m^\pm$ and (12) it follows that

$$P(B(\varphi_m^+)\setminus B(\varphi_m^-))\leq C_1(P)Lm^{-2}.$$

Consider the class of derivatives $\left\{\frac{d}{dx}\varphi_m^\pm\right\}$. The construction of $\varphi_m^\pm$ imply that $\frac{d}{dx}\varphi_m^\pm(x)$ is equal to S/m with $S\in\{0,1,\dots,[(L+1)m]\}$ on the intervals $\left(\frac{k}{m},\frac{k+1}{m}\right)$. Moreover it follows from (13) that $\frac{d}{dx}\varphi_m^\pm$ "almost" increase. Thus one can assign to each trajectory $\frac{d}{dx}\varphi_m^\pm(x)$ some vector of dimension not exceeding $[(L+1)m]+2m$ with coordinates -1, 0,

1. Hence, the number of different broken lines $\varphi_m^{\pm}$ intersecting the unit square $\mathfrak{X}$ does not exceed $m^2 3^{[(L+1)m]+2m}$. The number of various functions which can be represented as the difference of two functions from $\{\varphi_m^{\pm}\}$ does not exceed $m^4 9^{[(L+1)m]+2m}$.

Thus, we have constructed a finite family of subsets $\mathcal{A}_N = \{B(\varphi_m)\}$ satisfying (2) with $\varepsilon = C(P)Lm^{-2}$ and $N = \log \operatorname{Card} \mathcal{A}_N \leq C(L,P)\varepsilon^{-1/2}$. It is very important for the rest of the proof that the "upper" boundary of an arbitrary set from $\mathcal{A}_N$ has the form $\max\{0, \min\{1, \varphi_m(x)\}\}$ where $\varphi_m(x)$ is a continuous broken line with vertices at the points $(k/m, S/m^2)$, $k=0,1,\dots,m$, $S \in \{0, \pm 1, \dots, [(L+1)m]\}$ and $\sup_x |\varphi_m'(x)| \leq L+1$.

To construct the corresponding family $\mathcal{C}(\mathcal{A}_N)$ we have to estimate from above the numberof different projections onto the x-axis of the intersection points of the boundaries of various sets from $\mathcal{A}_N$ (if there exists a common part of the broken lines φ_m and $\tilde{\varphi}_m$, then we fix two intersection points: the beginning and the end of the common part). It is clear that the set of all projections on the x-axis of the intersection points of the boundaries of sets from $\mathcal{A}_N$ is contained in the set of solutions of the following linear equations:

$$\frac{\ell_1}{m}\left(x - \frac{k_1}{m}\right) + \frac{S_1}{m^2} = \frac{\ell_2}{m}\left(x - \frac{k_2}{m}\right) + \frac{S_2}{m^2} \qquad (14)$$

where $\ell_i = 0, \pm 1, \dots, \pm [(L+1)m]$; $k_i = 0, \dots, m$; $S = 0, \dots, [(L+1)m^2]$. It follows that the set of solutions of (14) has the form $x = k(m\ell)^{-1}$ where $\ell = 1, \dots, [(\ell+1)m]$, $k = 0, 1, \dots, [(L+1)m^2]$. It is easy to show that

$$\operatorname{Card}\{k(m\ell)^{-1} \in [0,1]; \ell \leq [(L+1)m], k \leq [(L+1)m^2]\} \leq \qquad (15)$$
$$\leq C(L)m^3,$$

the order on m of this estimation being sharp.

Denote $B_k = \{(x,y)\in \mathcal{X} : z_{k-1}\le x \le z_k\},\ k\ge 1,$ where $\{z_k\}$ is the ordered family of all projections on the x-axis of the intersection points of the boundaries of sets from $\mathcal{A}_N$. Let $\mathcal{C}(\mathcal{A}_N)$ be the minimal class of subsets generating the family $\mathcal{A}_N \cup \{B_k\}$. Now let us number (order) the elements of $\mathcal{C}(\mathcal{A}_N)$ contained in the set B_1 in the positive direction of the y-axis. Then we similarly continue the enumeration of the elements from $\mathcal{C}(\mathcal{A}_N)$ in the set B_2, and so on. From this and (15) we obtain relation (9). The Lemma is proved.

Theorem 2. Under the conditions of Lemma 2 one can construct the r.f.'s $S_n(\cdot)$ and $W_n(\cdot)$ on a common probability space such that

$$P_r\{\Delta_n(\mathcal{B}) > C(L,P)(\log n)^{2/7} n^{-1/14}\} \le Cn^{-2}. \qquad (16)$$

Proof. Denote by $\widehat{H}_1(\cdot)$ and $\widehat{\gamma}(\cdot)$ the right-hand sides of (8) and (9) respectively and by m^* the solution of the equation

$$\sum_{k>m} (\widehat{H}_1(2^{-k}, \mathcal{B}, P)2^{-k})^{1/2} = \gamma(2^{-m}, \mathcal{B}, P) n^{-1/2}(\log n)^2.$$

In other words, m^* satisfies the equation

$$C_1(L,P)2^{-m/4} = 2^{3m/2} n^{-1/2}(\log n)^2,$$

hence $m^* = C_2(L,P) + \frac{4}{7}\log_2(n^{1/2}(\log n)^{-2})$, and, moreover

$$\min_{m:\ \widehat{H}_1(2^{-m},\cdot)\le n2^{-m}} \alpha_n(m) \le \alpha_n(m^*),$$

$$\alpha_n(m^*) \le C(L,P)(\log n)^{2/7} n^{-1/14}.$$

The Theorem is proved.

Remark 2. By the above method one can obtain some estimates of order $O(n^{-\alpha})$ (which are weaker than $O(n^{-1/14})$) of the convergence rate in the case when $\mathcal{B}$ is the class of all closed convex subsets of $\mathcal{X}$ or $\mathcal{B}=\{(x,y)\in\mathcal{X}: f_1(y)\le x\le f_2(y)\}, g_1(x)\le y\le g_2(x); f_i, g_i \in C_L^1[0,1]\}$ which imporve the corresponding results of [7]-[9]. One can also obtain analogous estimates in the case when $\mathcal{B}$ is the class of subsets of the K-dimensional unit cube which are defined by a family of functions on the $k-1$-dimensional unit cube.

Remark 3. Using the results of [2] and [3] it is easy to show that relation (16) is still true if $P(\cdot)=\nu P^{(1)}(\cdot)+(1-\nu)P^{(2)}(\cdot)$, $\nu\in[0,1]$ where $P^{(1)}$ satisfies the conditions of Lemma 2 and $P^{(2)}\equiv\{p_k\}$ is an arbitrary discrete distribution on $\mathcal{X}$ such that $p_k = O(k^{-2(1+\alpha)})$ as $k\to\infty$ and $\alpha > 1/11$. Moreover, the constant $C(L,P)$ does not depend on ν.

Analogous remark can be made concerning the results of [8] and [9].

It is known that Dudley's condition (5) (which is equivalent to the convergence of the series $\sum_{k\ge 1}(H_1(2^{-k},\mathcal{B},P)2^{-k})^{-1/2})$ is sufficient for the central limit theorem (and for the invariance principle) for empirical measures (see [5]). In general this condition cannot be weakened (see [3]). If the metric entropy with inclusion $H_1(\cdot)$ is very large (but satisfies Dudley's condition), then the estimates of $\alpha_n(m)$ in (7) will be rather bad. The following statements show that the estimates of $\alpha_n(m)$ cannot be improved in this case.

Theorem 3. For every $\gamma \in (0,1)$ there exist $(\mathcal{X}, \mathcal{A})$, $P \equiv P_\gamma$ and a class $\mathcal{B} \subseteq \mathcal{A}$ for which $H_1(\varepsilon, \cdot) \leq C_0^\gamma(\gamma)\varepsilon^{-\gamma}$. Moreover

1) the r.f. $S_n(\cdot)$ and $W_n(\cdot)$ can be defined on a common probability space such in a way that

$$\Pr\left\{ \Delta_n(\mathcal{B}) > C(\gamma) n^{\frac{\gamma-1}{2(\gamma+1)}} (\log n)^{\frac{2(1-\gamma)}{(\gamma+1)}} \right\} \leq C n^{-2};$$

2) for any method of construction of $S_n(\cdot)$ and $W_n(\cdot)$ on a common probability space

$$\lim_{n \to \infty} \Pr\left\{ \Delta_n(\mathcal{B}) > C_1(\gamma) n^{\frac{\gamma-1}{2(\gamma+1)}} \right\} = 1.$$

Proof. Let $\mathcal{X} = \mathcal{N}$ be the set of natural numbers, $\mathcal{A}$ be the family of all subsets of $\mathcal{N}$ and $P \equiv \{p_k\}$ be a discrete distribution in $\mathcal{N}$. Put $p_k = C(\alpha) \cdot k^{-2(1+\alpha)}$, $\alpha > 0$. It is shown in [1] that in this case $H_1(\sum_{k>m} p_k, \mathcal{A}, P) \leq m$ or, $H_1(C_1(\alpha) m^{-(1+2\alpha)}, \mathcal{A}, P) \leq m$. It follows that for any $\varepsilon > 0$

$$H_1(\varepsilon, \mathcal{A}, P) \leq C_2(\alpha) \varepsilon^{-\frac{1}{1+2\alpha}}. \tag{17}$$

Besides it follows from [1] and [2] that

$$\Pr\left\{ \Delta_n(\mathcal{B}) > C \min_{m : m < 4n \sum_{k>m} p_k} \beta_n(m) \right\} \leq C n^{-2} \tag{18}$$

where $\beta_n(m) = (1 + m^{-1} \log n) \sum_{k>m} \sqrt{p_k} + m n^{-1/2} (\log n)^2$. It is easy to see that

$$\min_{m:m<4n\sum_{k>m}p_k} \beta_n(m) \le$$

$$\le \beta_n(m(n)) \equiv \beta_n\left(\left[n^{\frac{1}{2(1+\alpha)}}(\log n)^{-\frac{2}{1+\alpha}}\right]\right),$$

$$\sum_{k>m(n)} \sqrt{p_k} \sim C(\alpha)m(n)^{-\alpha} \sim C_1(\alpha)n^{-\frac{\alpha}{2(1+\alpha)}}(\log n)^{\frac{2\alpha}{1+\alpha}}, \tag{19}$$

$$m(n)n^{-1/2}(\log n)^2 \sim C_2(\alpha)n^{-\frac{\alpha}{2(1+\alpha)}}(\log n)^{\frac{2\alpha}{1+\alpha}}. \tag{20}$$

Taking $\gamma = \frac{1}{1+2\alpha}$ we obtain from (17)-(20) the first two statements of the theorem.

Let us prove the last statement. Suppose that r.f. $S_n(\cdot)$ and $W_n(\cdot)$ are defined on a common probability space by a certain method. Then

$$\Delta_n(\mathcal{A}) \ge \sup_{A\in\mathcal{A}} |S_n(1,A) - W_n(1,A)| \equiv \tilde{\Delta}_n(\mathcal{A}).$$

It is shown in [1] that for any method of construction of r.f. $S_n(\cdot)$ and $W_n(\cdot)$ on a common probability space with a probability exceeding $1 - 5\exp\{-n \times \sum_{k>m} p_k/8\}$ the following inequality holds:

$$\tilde{\Delta}_n(\mathcal{A}) \ge \frac{1}{\sqrt{2\pi}} \sum_{k>m} \sqrt{p_k} - 2\sqrt{n} \sum_{k>m} p_k \sim$$

$$\sim C_1(\alpha) m^{-\alpha} - C_2(\alpha)\sqrt{n}\, m^{-1-2\alpha} \equiv \delta_n(m).$$

Put $m = m_1(n) \equiv C_0 [n^{\frac{1}{2(1+\alpha)}}]$. Then

$$\delta_n(m_1(n)) = (C_1(\alpha) C_0^{-\alpha} - C_2 C_0^{-1-2\alpha}) n^{-\frac{\alpha}{2(1+\alpha)}}$$

and $\delta_n(m_1(n)) > 0$ for sufficiently large C_0. Note that $n \sum_{k > m_1(n)} p_k \to \infty$ as $n \to \infty$. The Theorem is proved.

In [1], [4] some cases are considered when $H_1(\varepsilon, \cdot) \sim \sim \Psi(\varepsilon)\varepsilon^{-1}$ where $\lim_{\varepsilon \to 0} \Psi(\varepsilon)\varepsilon^{-\alpha} = \infty$ for any $\alpha > 0$, $\sum_{k \geq 1} \sqrt{\Psi(2^{-k})} < \infty$ (Dudley's condition). In these cases the convergence rate in the central limit theorem and invariance principle can be arbitrarily bad. Denote

$$L_1(x) = \log \max\{x, 2\},$$

$$L_r(x) = L_1(L_{r-1}(x)), \quad r = 2, 3, \ldots$$

The following theorems make up an essential supplement to Theorem 3.

Theorem 4. For every $r \in \mathcal{N}$ there exist $(\mathcal{X}, \mathcal{A}), P \equiv P_r$ and $\mathcal{B} \subseteq \mathcal{A}$ such that $H_1(\varepsilon, \mathcal{B}, P) \sim \Psi_r(\varepsilon)\varepsilon^{-1}$ as $\varepsilon \to 0$ where $\Psi_r(\varepsilon)$ satisfies the above conditions and, moreover

1) r.f. $S_n(\cdot)$ and $W_n(\cdot)$ can be defined on a common probability space in such a way that

$$Pr\{\Delta_n(\mathcal{B}) > C_1(r) L_r^{-1}(n)\} \leq C n^{-2};$$

2) for any method of construction of r.f.'s $S_n(\cdot)$ and $W_n(\cdot)$ on a common probability space

$$\lim_{n\to\infty} Pr\{\Delta_n(\mathcal{B}) > C_2(r)L_r^{-1}(n)\} = 1$$

Proof. Put in the proof of Theorem 3

$$p_k = C(r)\left(k L_r^2(k)\prod_{j=1}^{r-1} L_j(k)\right)^{-2},$$

$$m(n) = \sqrt{n}(\log n)^{-2} L_r^{-1}(n),$$

where $\prod_{j=1}^{0} = 1$. It is easy to verify that

$$\sum_{k>m(n)} \sqrt{p_k} \sim C_3(r) L_r^{-1}(n)$$

and $m(n) < 4n \sum_{k>m(n)} p_k$ for sufficiently large n. From here the first statement of the Theorem follows (see also the proof of Theorem 3).

To obtain a lower estimate we put in the proof of Theorem 3 $m_1(n) = \sqrt{n}\left(\prod_{j\le r} L_j(n)\right)^{-2}$. Theorem 4 is proved.

Theorem 5. For an arbitrary sequence $b_n \to 0$ there exist $(\mathcal{X}, \mathcal{A})$, $P \equiv P\{b_n\}$ and a class $\mathcal{B} \subseteq \mathcal{A}$ such that Dudley's condition (5) is fulfilled and for any method of construction of r.f.'s $S_n(\cdot)$ and $W_n(\cdot)$ on a common probability space

$$\inf\{\varepsilon : Pr\{\Delta_n(\mathcal{B}) > \varepsilon\} < \varepsilon\} > b_n$$

for all sufficiently large n.

The statement follows from [4] and the inequality

$$Pr\{\Delta_n(\mathcal{B}) > \varepsilon\} \ge$$

$$\geq \sup_x \left| Pr\{\sup_{A\in\mathcal{B}} |S_n(1,A)| < x\} - Pr\{\sup_{A\in\mathcal{B}} |W_n(1,A)| < x\}\right| -$$

$$- \sup_x Pr\{x \leq \sup_{A\in\mathcal{B}} |W_n(1,A)| \leq x+\varepsilon\}$$

where, of course, it is supposed that the appropriate measurability conditions are fulfilled.

Now we will describe a general approach to obtain lower estimates of the convergence rate in the invariance principle (and the central limit theorem) for empirical measures in the case of arbitrary $\mathcal{X}$, $\mathcal{B}$ and P.

<u>Theorem 6</u>. Let $\{A_k; k\geq 1\}$ be a sequence of pairwise disjoint non-empty measurable subsets of an arbitrary measurable space $(\mathcal{X}, \mathcal{A}, P)$. Suppose that $p_k = P(A_k)$ decrease as $k\to\infty$ and that $\bigcup_i A_{k_i} \in \mathcal{B}$ for any subsequence $\{k_i\}$ of natural numbers. Then for any method of construction of r.f.' $S_n(\cdot)$ and $W_n(\cdot)$ on a common probability space

$$\Delta_n(\mathcal{B}) \geq \max_{m\geq 0}\left(\frac{1}{\sqrt{2\pi}}\sum_{k>m}\sqrt{p_k} - 2\sqrt{n}\sum_{k>m} p_k\right) \tag{21}$$

with a probability exceeding $1-5\exp\{-n\sum_{k>m_0(n)} p_k/8\}$ where $m_0(n)$ is a value of m at which the maximum in (21) is attained.

<u>Proof</u>. Denote $A_0 = \mathcal{X}\setminus\bigcup_k A_k$. Let $A_0 \neq \emptyset$. Choose in each A_k a point a_k and denote by $\bar{P}$ the discrete distribution for which $\bar{P}(a_k) = p_k$, $k=0,1,\ldots$, $p_0 = 1-\sum_{k\geq 1} p_k$. Then it is easy to construct (see [1]) independent r.v. $\{\bar{X}_k\}$ with the distribution $\bar{P}$ for which $\bar{S}_n(t,A_j) = S_n(t,A_j)$, $\bar{W}_n(t,A_j) \overset{d}{=} W_n(t,A_j)$, $j=0,1,\ldots$, where $\bar{S}_n(\cdot)$ and $\bar{W}_n(\cdot)$ are defined in (1) for the sequence $\{\bar{X}_k\}$. Thus, for

any method of construction of r.f.'s $S_n(\cdot)$ and $W_n(\cdot)$ on a common probability space

$$\Delta_n(\mathcal{B}) \geq \sup |\bar{S}_n(1, \bigcup_i A_{k_i}) - \bar{W}_n(1, \bigcup_i A_{k_i})| \geq$$

$$\geq \frac{1}{\sqrt{2\pi}} \sum_{k>m} \sqrt{p_k} - 2\sqrt{n} \sum_{k>m} p_k \equiv R_n(m)$$

with a probability exceeding $1 - 5\exp\{-n \sum_{k>m} p_k / 8\}$ (see [1]) where m is an arbitrary natural number. Note that $R_n(0) < 0$, $\lim_{m \to \infty} R_n(m) = 0$ and $R_n(m_*) > 0$ for $m_* = \min\{m : p_m n < 1/36\}$. Thus, $m_0(n) < \infty$. The Theorem is proved.

Corollary. If under the conditions of Theorem 2 P is the Lebesgue measure in $\mathfrak{X}$, then for any method of construction of r.f.'s $S_n(\cdot)$ and $W_n(\cdot)$ on a common probability space

$$\Delta_n(\mathcal{B}) \geq C n^{-1/6} \tag{22}$$

with a probability tending to 1.

Proof. Let $\{z_k; k \geq 0\}$ be a strictly increasing sequence of numbers belorging to the unit interval, $z_0 = 0$, $\lim_{k \to \infty} z_k = 1$ and the values $\Delta_k = z_k - z_{k-1}$ decrease monotonically as $k \to \infty$. Denote

$$\varphi_k(x) = \begin{cases} \Delta_k (x - z_{k-1}) & \text{if } z_{k-1} \leq x \leq \frac{z_{k-1} + z_k}{2}, \\ -\Delta_k (x - z_k) & \text{if } \frac{z_{k-1} + z_k}{2} < x < z_k, \\ 0 & \text{if } x \notin [z_{k-1}, z_k), \end{cases}$$

$$\Psi_e(x) = \sum_{k \geq 1} e_k \varphi_k(x),$$

where $e = (e_1, e_2, \dots)$ is an infinite-dimensional vector with the coordinates 1 or 0. It is easy to see that for any e we have

$$\int_0^1 |d \Psi_e'(x)| \leq 1 .$$

Note also that

$$\{ B(\Psi_e); e \in \ell^\infty \} = \{ \bigcup_i B(\varphi_{k_i}); \{k_i\} \subseteq \mathcal{N} \}$$

where sets of the form $B(f)$ are defined in Leem 2, and

$$p_k = P(B(\varphi_k)) = \Delta_k^3 / 4 .$$

Since $\sum_{k \geq 1} \Delta_k = 1$ and $\{\Delta_k\}$ decrease monotinically, the following representation holds $\Delta_k = \varepsilon(k) k^{-1}$, where $\lim_{k \to \infty} \varepsilon(k) = 0$. Thus

$$\sum_{k > m} \sqrt{p_k} = \varepsilon_1(m) m^{-1/2},$$

$$\sqrt{n} \sum_{k > m} p_k = \varepsilon_2(m) \sqrt{n} m^{-2} \tag{23}$$

where $\varepsilon_i(m) \to 0$ as $m \to 0$, and $\varepsilon_i(m)$ may tend to zero arbitrarily slowly. Moreover one can assume that $\varepsilon_2(m)/\varepsilon_1(m) \to 0$. Taking in (23) $m = n^{1/3}$ we obtain (22). The Theorem is proved.

Remark.4. For the classes $\mathcal{B}$ considered in [8], [9] one can obtain by the above method the lower estimates of the convergence rate which may be arbitrarily had in its dependence on the behaviour of the function $H_1(\varepsilon,\cdot)$ in the neighbrourhood of zero. The lower estimate $O(n^{-1/6})$ is true in the case when $\mathcal{B}$ is the class of all closed convex subsets of $\mathcal{X}=[0,1]\times[0,1]$ and P is the Lebesgue measure on $\mathcal{X}$.

REFERENCES

1. Borisov I.S. Rate of convergence in central limit theorem for empirical measures. - In the book: Limit Theorems for Sums of Random Variables, Novosibirsk, "Nauka", 1984, p.125-143 (in Russian).
2. Borisov I.S. A new approach to the approximation problem of sum distributions of independent random variables in linear spaces. - In the book; Limit Theorems of Probability Theory, Novosibirsk, "Nauka", 1985 (in Russian).
3. Borisov I.S. To the problem of approximation accuracy in the central limit theorem for empirical measures. - Siberian Math. J., 1983, t.24, No 6, p.14-25 (in Russian).
4. Borisov I.S. A remark about convergence rate in the central limit theorem in Banach spaces. - Siberian Math.J., 1985, v.26, N 2, p.29-35 (in Russian).
5. Dudley R.M. Central limit theorems for empirical measures. - Ann.Probab., 1978, v.6, N 6, p.899-929.
6. Dudley R.M. Metric entropy of some classes of sets with differential boundaries. - J.Approx.Theory, 1974, v.10, N 3, p.227-236.
7. Dudley R.M., Philipp W. Invariance principles for

sums of Banach space valued random elements and empirical processes. - Z.Wahrscheinlichkeitstheorie verw. Gebiete, 1983, B.62, H.4, S.509-552.
8. Revesz P. On strong approximation of the multidimensional empirical processes. - Ann.Probab., 1976, v.4, N 5, p.729-743.
9. Revesz R. Three theorems of multivariate empirical processes. - Lecture Notes in Math., 1976, v.566, p.106-126.

Prob. Theory and Math. Stat., Vol.1, pp. 269-289
Prohorov et al. (eds)
1986 VNU Science Press

THE CONTRACTION PRINCIPLE FOR c_0-SUMMING OPERATORS AND SLLN FOR WEIGHED SUMS

V.V.Buldygin, S.A.Solntsev
Institute of Mathematics, Kiev, U.S.S.R.

1°. The contraction principle for c_0-summing matrices.

Let $\xi = (\xi_i, i \geq 1)$ be a sequence of independent symmetric (i.e. ξ_i and $-\xi_i$ have the same distribution, $i \geq 1$) random variables; $A = (a_{ni})_{n,i=1}^{\infty}$ an infinite real matrix.

Definition 1.1. A matrix A is said to be c_0-summing for the sequence ξ, if

1) $\forall n \geq 1$ the series $\sum_{i=1}^{\infty} a_{ni}\xi_i$ converges almost surely (a.s.);

2) $\lim_{n\to\infty} \zeta_n = 0$ a.s.

where

$$\zeta_n = \sum_{i=1}^{\infty} a_{ni}\xi_i \ , \quad n \geq 1$$

We denote the class of c_0-summing matrices by $\mathcal{A}_0 = \mathcal{A}_0(\xi)$.

The problem of complete description of the class $\mathcal{A}_0$ is quite difficult. In terms of the theory of summation of independent random variables we deal with the strong law of large numbers (SLLN) for weighed sums. If ξ is a standard Gaussian sequence, the description of the class $\mathcal{A}_0$ is equivalent to establishment of necessary and sufficient conditions for a.s. convergence of sequence of centered jointly Gaussian random variables to zero.

Transformations with respect to which class $\mathcal{A}_0$ is closed play an essential role in studying the class

$\mathcal{A}_0$. The simple examples of such transformations are: addition of matrices, multiplication of matrix by real number, rearrangement of rows etc.

Consider the matrix $\widetilde{A} = (\tilde{a}_{ni})_{n,i=1}^{\infty}$ where

$$\tilde{a}_{ni} = \beta_{ni}\, a_{ni}, \qquad n, i \geq 1,$$

and $(\beta_{ni})_{n,i=1}^{\infty}$ is an infinite matrix of uniformly bounded real numbers

$$\sup_{n,i\geq 1} |\beta_{ni}| < \infty. \tag{1.1}$$

Let $A \in \mathcal{A}_0(\xi)$. The following examples show that the condition (1.1) is not sufficient for $\widetilde{A}$ to belong to $\mathcal{A}_0(\xi)$ even if $\beta_{ni} \in \{-1, 1\}$, $n,i \geq 1$ (or $\beta_{ni} \in \{0,1\}$, $n,i \geq 1$) and ξ is a standard Gaussian sequence.

<u>Example 1.1</u>. We pick up in the set $\mathcal{N}^2$ the square arrays I_K, $K \geq 1$, where

$$I_K = \mathcal{N}_K \times \mathcal{N}_K, \qquad \mathcal{N}_K = \{2^{K-1}, \dots, 2^K - 1\}.$$

Let

$$\delta_{n,i}^{(K)} = \begin{cases} 1, & (n,i) \in I_K, \\ 0, & (n,i) \notin I_K; \end{cases}$$

$$a_{ni} = \frac{\delta_{n,i}^{(K)}}{\sqrt{K \cdot 2^K}}, \qquad (n,i) \in \mathcal{N}_K \times \mathcal{N}.$$

Consider the matrix $A = (a_{ni})_{n,i=1}^{\infty}$ and show that $A \in \mathcal{A}_0(\gamma)$ where $\gamma = (\gamma_i, i \geq 1)$ is a sequence of independent $\mathcal{N}(0,1)$-distributed random variables. At first we notice that

$$\zeta_n = \sum_{i=1}^{\infty} a_{ni}\gamma_i = \frac{1}{\sqrt{K \cdot 2^K}} \sum_{i=2^{K-1}}^{2^K - 1} \gamma_i,$$

provided $n \in \mathcal{N}_K$, $K \geq 1$. Hence

$$\lim_{n\to\infty} \zeta_n = 0 \quad \text{a.s.} \tag{1.2}$$

iff

$$\lim_{k \to \infty} \zeta_{2^{k-1}} = 0 \quad \text{a.s.} \tag{1.3}$$

As $(\zeta_{2^{k-1}}, \ k \geq 1)$ is a sequence of independent Gaussian random variables and

$$\forall \varepsilon > 0 \quad \sum_k \exp(-\varepsilon / D\zeta_{2^{k-1}}) = \sum_k \exp(-2\varepsilon k) < \infty,$$

then ([1]) relation (1.3), and so (1.2) hold true. Thus $A \in \mathcal{A}_0(\gamma)$.

Next, consider $2^{k-1} \times 2^{k-1}$- matrices

$$V_k = (\delta_{ni}^{(k)})_{(n,i) \in I_k}, \quad k \geq 1.$$

Together with matrices V_k, $k \geq 1$ we introduce matrices

$$U_k = (u_{ni}^{(k)})_{(n,i) \in I_k}, \quad k \geq 1,$$

constructed by the following recurrent procedure:

$$U_1 = (1),$$

$$U_{k+1} = \begin{pmatrix} U_k & U_k \\ U_k & -U_k \end{pmatrix}, \quad k \geq 1.$$

Elements of the matrix U_k take values ± 1 and thus only their signs may differ from those of the corresponding elements of matrix V_k. By induction, it is easy to show that rows of matrix U_k, $k \geq 2$ are pairwise orthogonal. Set

$$\beta_{ni}^{(k)} = \begin{cases} u_{ni}^{(k)}, & (n,i) \in \mathcal{N}_k \times \mathcal{N}_k \\ 0, & n \in \mathcal{N}_k, \ i \notin \mathcal{N}_k \end{cases}$$

$$\tilde{a}_{ni} = \beta_{ni}^{(k)} a_{ni}, \qquad (n,i) \in \mathcal{N}_k \times \mathcal{N}$$

It is obvious that

$$|\tilde{a}_{ni}| = |a_{ni}| \qquad ((n,i) \in \mathcal{N}^2).$$

Let us show that $\tilde{A} = (\tilde{a}_{ni})_{n,i=1}^{\infty} \in \mathcal{A}_0(\gamma)$.
We remark that if $n \in N_K$

$$\tilde{\zeta}_n = \sum_{i=1}^{\infty} \tilde{a}_{ni}\gamma_i = \sum_{i=2^{K-1}}^{2^K-1} u_{ni}^{(K)} a_{ni}^{(K)} \gamma_i .$$

Hence

$$E\tilde{\zeta}_n \tilde{\zeta}_m = 0, \quad n \in N_K, \, m \in N_p, \, K \neq p.$$

Furthermore, if $n, m \in N_K$ then

$$E\tilde{\zeta}_n\tilde{\zeta}_m = \frac{1}{K \cdot 2^K} \sum_{i=2^{K-1}}^{2^K-1} u_{ni}^{(K)} u_{mi}^{(K)} = \begin{cases} \frac{1}{K \cdot 2^K}, & n=m \\ 0, & n \neq m \end{cases}$$

Thus, $(\tilde{\zeta}_n, \, n \geq 1)$ is a sequence of independent Gaussian random variables. If we choose $\varepsilon \in (0, \frac{\ln 2}{4})$, then

$$\sum_{n=1}^{\infty} \exp(-\varepsilon / D\tilde{\zeta}_n) = \sum_{K=1}^{\infty} \sum_{n \in N_K} \exp(-\varepsilon / D\tilde{\zeta}_n) =$$

$$= \sum_{K=1}^{\infty} 2^{K-1} \exp(-2\varepsilon K) = \infty .$$

Therefore

$$\tilde{\zeta}_n \underset{n \to \infty}{\not\longrightarrow} 0 \quad \text{a.s.}$$

and $\tilde{A} \notin \mathcal{A}_0(\gamma)$.

Example 1.2. Let A and $\tilde{A}$ be matrices constructed in example 1.1. Set

$$C = \frac{1}{2}(A + \tilde{A}).$$

According to construction of A and $\tilde{A}$ one can see that the matrix C differs from A in the following sense: some elements of A are replaced by 0. At the same time $A \in \mathcal{A}_0(\gamma)$, $C \notin \mathcal{A}_0(\gamma)$, otherwise $\tilde{A} \in \mathcal{A}_0(\gamma)$ and this contradicts example 1.1.

It turns out that the restrictions under which $\tilde{A} \in \mathfrak{A}_0(\xi)$ are closely related not only to condition (1.1) but also to the behaviour of the variation of matrix (β_{ni}) ([2]).

Definition 1.2. Let $(b_{ni}) = (b_{ni})_{n,i=1}^{\infty}$ be an infinite real matrix. By variation of matrix (b_{ni}) we mean the quantity

$$Var(b_{ni}) = \sup_{n\geq 1} \sup_{m\geq 1} \left(\sum_{i=1}^{m-1} |b_{ni} - b_{n,i+1}| + |b_{nm}| \right).$$

(Set $\sum_{i=1}^{0} = 0$).

Theorem 1.1. Let $(\beta_{ni})_{n,i=1}^{\infty}$ be an infinite real matrix such that

a) $\beta_{ni} = b_{ni} c_i , \quad n, i \geq 1$;

b) $\sup_{i\geq 1} |c_i| < \infty$;

c) $Var(b_{ni}) < \infty$.

Then $A \in \mathfrak{A}_0(\xi)$ implies $\tilde{A} \in \mathfrak{A}_0(\xi)$.

Theorem 1.1 is the particular case of the following general contraction principle.

2°. The contraction principle in the scheme of series.

Let $(\mathfrak{X}, \|\cdot\|)$ be a separable Banach space; $B(\mathfrak{X})$ be the Borel σ-field of the space $\mathfrak{X}$; $\mathfrak{X}^{\infty}$ - the space of all sequences $(x_n , n \geq 1) \subset \mathfrak{X}$ equipped with the product topology and corresponding Borel σ-field $B(\mathfrak{X}^{\infty})$; $c_0(\mathfrak{X}) = \{ (x_n) \in \mathfrak{X}^{\infty} : \lim_{n\to\infty} \| x_n \| \}$; $(\Omega, \mathcal{F}, P)$ - the main probability space; as usually, the $(\mathcal{F}, B(\mathfrak{X}))$ - measurable transformation $X : \Omega \longrightarrow \mathfrak{X}$ is called the $\mathfrak{X}$-valued random element ($\mathfrak{X}$- v.r.e.).

Let $(Y_{ni})_{n,i=1}^{\infty}$ be an infinite matrix of $\mathfrak{X}$-v.r.e. which satisfies the following conditions:

1) $\forall\, n \geq 1$ the series $\sum_{i=1}^{\infty} Y_{ni}$ converges a.s. in the $\mathfrak{X}$-norm;

2) sequences $(Y_{ni},\ n \geq 1)$, $i \geq 1$ are independent and symmetric as $\mathfrak{X}^{\infty}$- v.r.e.

Let $(\beta_{ni})_{n,i=1}^{\infty}$ be an infinite real matrix. Set

$$Z_n = \sum_{i=1}^{\infty} Y_{ni}\ , \qquad \tilde{Z}_n = \sum_{i=1}^{\infty} \beta_{ni} Y_{ni}\ , \qquad n \geq 1.$$

Kahane's contraction principle ([3]).

Let $(X_i\ ,\ i \geq 1)$ be a sequence of independent symmetric $\mathfrak{X}$-v.r.e; $(c_i\ ,\ i \geq 1)$ - a real sequence such that

$$\sup_{i \geq 1} |c_i| < \infty .$$

If the series converges a.s. in $\mathfrak{X}$-norm, then the series $\sum_{i=1}^{\infty} c_i X_i$ converges a.s. in $\mathfrak{X}$-norm.

If the matrix (β_{ni}) satisfies the condition (1.1), then according to the Kahane's contraction principle the random elements $\tilde{Z}_n$, $n \geq 1$ are defined correctly.

Lemma 2.1. Let matrix (Y_{ni}) satisfy the conditions 1), 2) and matrix (β_{ni}) satisfy (1.1). Then

$$\| Z_n \| \xrightarrow[n \to \infty]{P} 0$$

implies

$$\| \tilde{Z}_n \| \xrightarrow[n \to \infty]{P} 0 .$$

The above examples 1.1, 1.2 show that one cannot change the convergence in probability to the a.s. convergence in lemma 2.1.

Theorem 2.1. Let matrix (Y_{ni}) satisfy conditions 1), 2) and matrix (β_{ni}) satisfy a)-c) of theorem 1.1. Then

$$\lim_{n\to\infty} \| Z_n \| = 0 \quad \text{a.s.}$$

implies

$$\lim_{n\to\infty} \| \widetilde{Z}_n \| = 0 \quad \text{a.s.}$$

Complete demonstration is given in [4]. That is why we only outline the scheme of the proof.

I. According to one variant of K.Ito and M.Nisio theorem ([5]) the following statement holds true.

Lemma 2.2 Let matrix (Y_{ni}) satisfy conditions 1), 2). Then the following conditions are equivalent:

i) $\lim_{n\to\infty} \| Z_n \| = 0$ a.s.;

ii) $\forall i \geq 1 \quad W_i = (Y_{ni}, n \geq 1) \in c_0(\mathfrak{X})$ a.s.

and the series $\sum_{i=1}^{\infty} W_i$ converges a.s. in $c_0(\mathfrak{X})$-norm;

iii) $\forall i \geq 1 \quad W_i \in c_0(\mathfrak{X})$ a.s. and

$$\lim_{K\to\infty} \sup_{m\geq K} \sup_{n\geq 1} \Big\| \sum_{i=K}^{m} Y_{ni} \Big\| = 0 \quad \text{a.s.}$$

II. Further, lemma 2.2 implies the assertion of theorem 2.2 if one uses Kahane's contraction principle and Abel's inequality according to which

$$\sup_{m\geq K} \sup_{n\geq 1} \Big\| \sum_{i=K}^{m} b_{ni} Y_{ni} \Big\| \leq$$

$$\leq \left(\operatorname{Var}(b_{ni})\right)\Big(\sup_{m\geq K} \sup_{n\geq 1} \Big\| \sum_{i=K}^{m} Y_{ni} \Big\|\Big).$$

Theorem 2.2. Let matrices (Y_{ni}), (β_{ni}) be the same as in the theorem 2.1. Then

$$\sup_{n\geq 1} \|Z_n\| < \infty \quad \text{a.s.}$$

implies

$$\sup_{n\geq 1} \|\tilde{Z}_n\| < \infty \quad \text{a.s.}$$

Theorem 2.2 follows from theorem 2.1 by applying the following auxiliary lemma.

Lemma 2.3 ([6]). Let $(\eta_n, n\geq 1)$ be a sequence of random variables. Then

$$\sup_{n\geq 1} |\eta_n| < \infty \quad \text{a.s.}$$

iff for every real sequence $(b_n, n\geq 1)$ monotonously tending to zero

$$\lim_{n\to\infty} b_n \eta_n = 0 \quad \text{a.s.}$$

Theorem 2.1. is closely related to Kahane's contraction principle for series of independent symmetric random elements in Banach spaces. More precisely, this principle implies the existence of multipliers c_i, $i\geq 1$ in a), b). At the same time condition c) reflects specific features of weighed sum scheme and explains effects which arise in examples 1.1, 1.2.

3°. SLLN for weighed sums.

Let us consider some simple corollaries of the theorem 2.1. Let $(X_i, i\geq 1)$ be a sequence of independent symmetric $\mathfrak{X}$-v.r.e.

Corollary 3.1. Let (a_{ni}) be a c_0-summing matrix for sequence (X_i), i.e. for every $n\geq 1$ the series $\sum_{i=1}^{\infty} a_{ni}X_i$ converges a.s. in $\mathfrak{X}$-norm and

$$\lim_{n\to\infty} \| \sum_{i=1}^{\infty} a_{ni} X_i \| = 0 \quad \text{a.s.}$$

If matrix (β_{ni}) satisfies conditions of the theorem 2.1, then

$$\lim_{n\to\infty} \| \sum_{i=1}^{\infty} \beta_{ni} a_{ni} X_i \| = 0 \quad \text{a.s.}$$

This proposition obviously follows from theorem 2.1 if we set

$$Y_{ni} = a_{ni} X_i , \qquad n, i \geq 1 .$$

Corollary 3.1 implies theorem 1.1 if $\mathfrak{X} = R$.

<u>Corollary 3.2.</u> Let $(\alpha_{ni} \ ; \ i=1,\dots,n, \ n\geq 1) \subset R$; $V_n = \operatorname{Var}(\alpha_{ni}, \ i\geq 1)$. Then

$$\lim_{n\to\infty} \| V_n \sum_{i=1}^{n} X_i \| = 0 \quad \text{a.s.}$$

implies

$$\lim_{n\to\infty} \| \sum_{i=1}^{n} \alpha_{ni} X_i \| = 0 \quad \text{a.s.}$$

<u>Corollary 3.3.</u> Let (α_{ni}), (V_n) be the same as in the corollary 3.2 and

$$\sup_{n\geq 1} n V_n < \infty .$$

Then

$$\lim_{n\to\infty} \| \frac{1}{n} \sum_{i=1}^{n} X_i \| = 0 \quad \text{a.s.}$$

implies

$$\lim_{n\to\infty} \| \sum_{i=1}^{n} \alpha_{ni} X_i \| = 0 . \quad \text{a.s.}$$

<u>Corollary 3.4.</u> Let matrix (α_{ni}) be such that

every its row can be divided into finite number of intervals of monotonicity;

$$\alpha_n^+ = \max_{i=1,\dots,n} |\alpha_{ni}| \qquad \alpha_n^- = \min_{i=1,\dots,n} |\alpha_{ni}| .$$

In order that

$$\lim_{n\to\infty} \| \sum_{i=1}^{n} \alpha_{ni} X_i \| = 0 \quad \text{a.s.}$$

it is sufficient that

$$\lim_{n\to\infty} \| \alpha_n^+ \sum_{i=1}^{n} X_i \| = 0 \quad \text{a.s.}$$

and it is necessary that

$$\lim_{n\to\infty} \| \alpha_n^- \sum_{i=1}^{n} X_i \| = 0 \quad \text{a.s.}$$

Corollary 3.5. Let

$$\lim_{n\to\infty} \| \sum_{i=1}^{\infty} \alpha_{ni} X_i \| = 0 \quad \text{a.s.}$$

Then for all $K_1(n) \leqslant K_2(n)$, $n \geqslant 1$

$$\lim_{n\to\infty} \| \sum_{i=K_1(n)}^{K_2(n)} \alpha_{ni} X_i \| = 0 \quad \text{a.s.}$$

Various applications and improvements of theorem 2.1 are given in ([4], [7]).

4°. SLLN in the Prohorov-Loève form for sums of independent random vectors normed by operators.

Let $(X_i, \ i \geqslant 1)$ be a sequence of independent symmetric random vectors in R^m $(m \geqslant 1)$; $S_n = X_1 + X_2 + \dots + X_n$, $n \geqslant 1$. In the study of limit theorems there are two possible types of normalizations: scalar and operator. Limit theorems with scalar normalizations are investigated quite well and form the basis of many classical theorems in probability theory. At the same time the operator normalizations (especially in strong limit theorems) have become the object of study not so long ago [8 - 12], though they reflect more adequately the asymptotical behaviour of sums of random vectors.

Let $(A_n, n \geq 1)$ be a sequence of linear operators from R^m to R^m (possibly degenerate).

We shall say 'SLLN in the Prohorov-Loève form' to mean the statements which establish the conditions under which the following statements are equivalent:

I) $\lim\limits_{n \to \infty} \| A_n S_n \| = 0$ a.s.

II) $\forall (n_k, k \geq 1) \in \mathfrak{N}$

$$\lim_{k \to \infty} \| A_{n_{k+1}} (S_{n_{k+1}} - S_{n_k}) \| = 0 \quad \text{a.s.}$$

where $\mathfrak{N}$ is some subclass of the class $\mathfrak{N}_\infty$ of all nondecreasing and unbounded sequences of natural numbers; $\|\cdot\|$ -the euclidean norm in R^m.

In case of real sums $(S_n, n \geq 1)$ and scalar normalizations the above form of SLLN dates back to the papers of Yu.V.Prohorov [13] and M.Loève [14] and was then generalized in [15 - 17].

<u>SLLN in Prohorov-Loève form with scalar normalizations ([17]).</u> Let $(X_i, i \geq 1)$ be a sequence of independent symmetric $\mathfrak{X}$-v.r.e. where $\mathfrak{X}$ is a separable Banach space; $(a_n, n \geq 1) \subset R$. Then

$$\lim_{n \to \infty} \| a_n S_n \| = 0 \quad \text{a.s.}$$

iff

i) $\lim\limits_{n \to \infty} a_n = 0$ or $\forall i \geq 1 \quad P\{X_i = 0\} = 1$,

ii) $\lim\limits_{k \to \infty} \| a_{n_{k+1}} (S_{n_{k+1}} - S_{n_k}) \| = 0$ a.s.

where

$$n_k = \max\{n : |a_n| \geq \lambda^{-k}\}$$

and $\lambda > 1$ is an arbitrary fixed number.

This proposition is proved in [17] only for $\mathfrak{X} = R$. The general case can be easily obtained in a similar manner.

Situation becomes much more complicated if we pass from scalar normalizations to operator ones. In this case, as a rule, the class $\mathfrak{N}$ consists of more than one sequence and convergence to zero of the sequence of operators (A_n) or degeneracy of summands (X_i) are not necessary for condition I) to be true.

Consider the necessary conditions for SLLN in Prohorov-Loève form with operator normalizations.

<u>Theorem 4.1.</u> Let $(X_i,\ i \geq 1)$ be a sequence of independent $\mathfrak{X}$- v.r.e., where $\mathfrak{X}$ is a separable Banach space; $(A_n,\ n \geq 1)$ is the sequence of bounded linear operators from $\mathfrak{X}$ into some separable Banach space $\mathcal{Y}$. Then

$$\lim_{n\to\infty} \|A_n S_n\|_{\mathcal{Y}} = 0 \quad \text{a.s.}$$

implies

i) $\forall i \geq 1 \qquad \lim_{n\to\infty} \|A_n X_i\| = 0 \quad$ a.s.

ii) $\forall (n_k) \in \mathfrak{N}_\infty$

$$\lim_{k\to\infty} \|A_{n_{k+1}}(S_{n_{k+1}} - S_{n_k})\|_{\mathcal{Y}} = 0 \quad \text{a.s.}$$

<u>Proof.</u> Let $(n_k) \in \mathfrak{N}_\infty$. Set

$$Y_{ni} = \begin{cases} A_n X_i, & i \leq n, \\ 0, & i > n, \quad n \geq 1; \end{cases}$$

$$\beta_{ni} = \begin{cases} 1, & \text{if} \quad n = n_{k+1},\ n_k < i \leq n_{k+1}, \\ 0, & \text{otherwise.} \end{cases}$$

Let us use theorem 2.1. Then $Z_n = A_n S_n$, $\tilde{Z}_{n_{k+1}} = A_{n_{k+1}}(S_{n_{k+1}} - S_{n_k})$ and

$$Var(\beta_{ni}) = 2.$$

This yields ii). Condition $\mathfrak{D}$- can be proved by the same argument.

There are examples ([4]) which show that even in case $\mathfrak{X} = \mathcal{Y} = H$, where H -separable Hilbert space the conditions i), ii) are not sufficient. However, if $\mathfrak{X} = \mathcal{Y} = R^m$ a complete solution to the problem can be obtained.

<u>Theorem 4.2.</u> Let

i) $\forall i \geq 1 \qquad \lim_{n\to\infty} \|A_n X_i\| = 0$ a.s.

Then there exists a finite collection of sequences $\mathfrak{N} \subset \mathfrak{N}_\infty$ (depending only on normalizing operators A_n) such that

ii) $\forall (n_k) \in \mathfrak{N}$

$$\lim_{k\to\infty} \|A_{n_{k+1}}(S_{n_{k+1}} - S_{n_k})\| = 0 \quad \text{a.s.}$$

implies

$$\lim_{n\to\infty} \|A_n S_n\| = 0 \quad \text{a.s.}$$

The proof of the theorem is cumbersome; the details can be found in [4].

The class of sequences $\mathfrak{N}$ is constructed in the proof of theorem 4.2. However its constructive discription when m is large enough is difficult. That is why it is often convenient to pass from the class $\mathfrak{N}$ to $\mathfrak{N}_\infty$.

<u>Theorem 4.3.</u>

$$\lim_{n\to\infty} \|A_n S_n\| = 0 \quad \text{a.s.}$$

iff the conditions i), ii) of theorem 4.2 are satisfied for $\mathfrak{N} = \mathfrak{N}_\infty$. SLLN in Kolmogorov's form follows from theorem 4.2.

Corollary 4.1. Let

i) $\forall i \geq 1$ $$\lim_{n\to\infty} \|A_n X_i\| = 0 \quad \text{a.s.},$$

ii) $$\sum_{i=1}^{\infty} \sup_{n \geq i} E\|A_n X_i\|^2 < \infty .$$

Then

$$\lim_{n\to\infty} \|A_n S_n\| = 0 \qquad \text{a.s.}$$

To prove this it is sufficient to establish that condition ii) of the corollary 4.1 implies condition ii) of theorem 4.1 for $\mathfrak{N} = \mathfrak{N}_\infty$. In fact let $(n_k) \in \mathfrak{N}_\infty$. Then

$$P(\|A_{n_{k+1}}(S_{n_{k+1}} - S_{n_k})\| > \varepsilon) \leq \frac{1}{\varepsilon^2} E\|A_{n_{k+1}}(S_{n_{k+1}} - S_{n_k})\|^2 =$$

$$= \frac{1}{\varepsilon^2} \sum_{i=n_k+1}^{n_{k+1}} E\|A_{n_{k+1}} X_i\|^2 \leq \frac{1}{\varepsilon^2} \sum_{i=n_k+1}^{n_{k+1}} \sup_{n \geq i} E\|A_n X_i\|^2.$$

Therefore for every $\varepsilon > 0$

$$\sum_{k=1}^{\infty} P(\|A_{n_{k+1}}(S_{n_{k+1}} - S_{n_k})\| > \varepsilon) \leq \frac{1}{\varepsilon^2} \sum_{i=1}^{\infty} \sup_{n \geq i} E\|A_n X_i\|^2 < \infty.$$

Application of Borel-Cantelli lemma concludes the proof.

Remark. In case $m=1$ condition i) of theorem 4.2 is equivalent to i) of SLLN in Prohorov-Loéve form with scalar normalization. Simple examples show that if $m \geq 2$ the condition

$$\lim_{n\to\infty} \|A_n\| = 0 \quad \text{or} \quad \forall i \geq 1 \quad P(X_i = 0) = 1.$$

($\|A_n\|$ is operator norm) does not follow in general from i) of theorem 4.2. Moreover in this case the condition i) in theorem 4.2 cannot be weakened.

Under additional conditions on the distribution of vectors $(X_i, i \geq 1)$ class $\mathcal{N}$ contains only one sequence.

Definition 4.1. Random vector in R^m is called spherically symmetrical if its distribution is invariant with respect to any transformation from the rotation group of the space R^m.

Theorem 4.4. Let $(X_i, i \geq 1)$ be a sequence of independent spherically symmetrical vectors in R^m $(m > 1)$. The following propositions are equivalent:

I) $\lim_{n \to \infty} \|A_n S_n\| = 0$ a.s.

II) i) $\lim_{n \to \infty} \|A_n\| = 0$ or $\forall i \geq 1$ $X_i = 0$ a.s.

ii) $\lim_{k \to \infty} \|A_{n_{k+1}}(S_{n_{k+1}} - S_{n_k})\| = 0$ a.s.

where

$$n_k = \max\{n : \|A_n\| > \lambda^{-k}\}, \quad k \geq 1,$$

$\lambda > 1$ is an arbitrary fixed number.

Theorem 4.4 improves the corresponding result from [12].

Corollary 4.2. Let $(X_i, i \geq 1)$ be a sequence of independent Gaussian vectors in R^m $(m \geq 1)$; $EX_i = 0$, $Cov(X_i) = B_i$, $i \geq 1$; $(A_n, n \geq 1)$ is the sequence of norming operators, A_n^* - the operator conjugate to A_n; $Tr\, C$ the trace of matrix C. Then

$$\lim_{n \to \infty} \|A_n S_n\| = 0 \quad \text{a.s.}$$

iff the following conditions hold:

i) $\forall i \geq 1$ $\lim_{n \to \infty} Tr(A_n B_i A_n^*) = 0$

ii) $\forall (n_k) \in \mathfrak{N}_\infty, \quad \forall \varepsilon > 0$

$$\sum_{k=1}^{\infty} \exp\Big(-\varepsilon \Big/ \sum_{i=n_k+1}^{n_{k+1}} \operatorname{Tr}\big(A_{n_{k+1}} B_i A^*_{n_{k+1}}\big)\Big) < \infty.$$

According to theorem 4.2 the sufficiency part of corollary 4.2 can be weakened if we replace the class of sequences $\mathfrak{N}_\infty$ in the condition ii) by some finite class $\mathfrak{N} \subset \mathfrak{N}_\infty$.

5°. Gaussian m-Markovian sequences

Definition 5.1. The sequence $(\xi_n, n \geq 1)$ of centered jointly Gaussian random variables is called the Gaussian m-Markovian if it satisfies the system of stochastic difference equations

$$\xi_n = a_{n1}\xi_{n-1} + a_{n2}\xi_{n-2} + \dots + a_{nm}\xi_{n-m} + b_n\gamma_n \tag{5.1}$$

(here $\xi_0 = \xi_{-1} = \dots = \xi_{1-m} = 0$), where $(\gamma_n, n \geq 1)$ is a standard Gaussian sequence and $(a_{nk};\ n \geq 1,\ k = 1, \dots, m)$ and $(b_n, n \geq 1)$ are some given arrays of real numbers. Moreover, for simplicity we impose the following condition of nondegeneracy:

$$a_{nm} \neq 0, \quad n \geq 1 \tag{5.2}$$

In case $m = 1$, i.e. for Gaussian Markov sequence, the following criterion is valid.

Criterion for a.s. convergence of Gaussiam Markov sequence to zero. [16]

Let $(\xi_n, n \geq 1)$ be a Gaussian Markov sequence. Then

$$\lim_{n\to\infty} \xi_n = 0 \quad \text{a.s.}$$

iff the following conditions hold:

i) $\lim\limits_{n\to\infty} (E\xi_n^2) = 0$

ii) $\forall (n_k) \in \mathfrak{N}_\infty$, $\forall \varepsilon > 0$

$$\sum_{k=1}^{\infty} \exp(-\varepsilon / D(\xi_{n_k}/\xi_{n_{k-1}})) < \infty$$

where

$$D(\xi_{n_k}/\xi_{n_{k-1}}) = E\left(\xi_{n_k} - E(\xi_{n_k}/\xi_{n_{k-1}})\right)^2, \quad k \geq 1,$$

while $\mathfrak{N}_\infty$ (as in section 4°) is the class of all nondecreasing and unbounded real sequences. Furthermore, in the sufficiency part of the criterion the class $\mathfrak{N}_\infty$ may be replaced by some special sequence $(n_k) \in \mathfrak{N}_\infty$.

The proof of the above criterion makes use of the SLLN in Prohorov-Loève form with scalar normalizations. In the case of Gaussian m-Markovian sequence the question of convergence to zero can be reduced to SLLN with operator normalizations. Note that such approach was discussed in [18].

Let $(\xi_n, n \geq 1)$ be a Gaussian m-Markovian sequence, $m > 1$. Set

$$Y_n = \begin{pmatrix} \xi_{n-m+1} \\ \vdots \\ \xi_n \end{pmatrix}, \quad Z_n = \begin{pmatrix} 0 \\ 0 \\ \vdots \\ b_n \gamma_n \end{pmatrix}, \quad n \geq 1.$$

Choosing matrices C_n, $n \geq 1$ (which according to (5.2) are nondegenerate) one can write the equivalent vector form of system (5.1) as follows:

$$Y_1 = Z_1, \qquad Y_n = C_n Y_{n-1} + Z_n, \quad n \geq 2.$$

Then

$$Y_n = C_n C_{n-1} \cdots C_2 Z_1 + C_n C_{n-1} \cdots C_3 Z_2 + \cdots + Z_n , \quad n \geq 2 \quad (5.3)$$

Let

$$A_1 = I , \quad A_n = C_n C_{n-1} \cdots C_2 , \quad n \geq 2,$$

$$X_1 = Z_1 , \quad X_k = (C_k \cdots C_2)^{-1} Z_k , \quad k \geq 2,$$

where I is the unit matrix. In new notation (5.3) can be rewritten as

$$Y_n = A_n S_n , \quad n \geq 1,$$

where $S_n = X_1 + \cdots + X_n$. Since

$$\lim_{n \to \infty} \xi_n = 0 \quad \text{a.s.}$$

iff

$$\lim_{n \to \infty} \| Y_n \| = 0 \quad \text{a.s.}$$

where $\| \cdot \|$ is the euclidean norm in R^m, corollary 4.2 implies the following assertion.

<u>Theorem 5.1.</u> Let $(\xi_n, \ n \geq 1)$ be a Gaussian m-Markovian sequence.
Then

$$\lim_{n \to \infty} \xi_n = 0 \quad \text{a.s.}$$

iff

i) $$\lim_{n \to \infty} (E \xi_n^2) = 0 ;$$

ii) $$\forall (n_k) \in \mathfrak{N}_\infty , \ \forall \varepsilon > 0$$

$$\sum_{k=1}^{\infty} \exp(-\varepsilon / d_k^2) < \infty,$$

where $d_k^2 = \mathbb{D}(\xi_{n_{k+1}} / \xi_i,\ i = n_k - m + 1, \dots, n_k) =$

$$= E(\xi_{n_{k+1}} - E(\xi_{n_{k+1}} / \xi_i,\ i = n_k - m + 1, \dots, n_k))^2$$

Furthermore, in the sufficiency part the class $\mathfrak{N}_\infty$ can be replaced by subclass $\mathfrak{N}$ which contains only finite numbers of sequences and depends on covariant characteristics of the sequence (ξ_n).

Remark. Theorem 5.1 holds true if we replace

$\lim_{n \to \infty} \xi_n = 0$ by $\sup_{n \geq 1} |\xi_n| < \infty$,

condition i) by $\sup_{n \geq 1} (E\xi_n^2) < \infty$,

$\forall \varepsilon$ in ii) by $\exists \varepsilon$

References

1. N.N.Vahanija Probability distributions in linear space, "Metsniereba", Tbilisi, 1971 (Russian).
2. V.V.Buldygin, S.A.Solntsev The comparison theorems and strong laws of large numbers for weighed sums, Proc. XVIII School: Probability Theory and Math. Statist. (Bakuriani, 1984), Tbilisi, 1984, p.7 (Russian).
3. J.-P. Kahane Some random series of functions, D.C. Heath, Lexington, Mass. U.S.A., 1968.
4. V.V.Buldygin, S.A.Solntsev The contraction principle for c_0-summing matrices and strong laws of large numbers with operator normalizations, Preprint No 85.68, Inst. Math. Acad. Sci. Ukrain.SSR, Kiev, 1985, 60 pp. (Russian).

5. V.V.Buldygin On some properties of the random series in Banach spaces, Teor. Verojatnost i Mat. Statist., 1973, N9, 37-46 (Russian).
6. V.L.Egorov On the law of the iterated logarithm, Teor. Verojatnost. i Primenen., 1969, 14, N4, 722-729 (Russian).
7. S.A.Solntsev On the strong limit theorems for weighted sums of independent random variables, Preprint of Kiev State Univ. N246-Uk-85 Dep., 20 pp. (Russian).
8. M.Hahn The generalized domain of attraction of a Gaussian law on Hilbert space. - Lect.Note Math., 1979, 709, 125-144.
9. M.Hahn, M.Klass The multidimensional central limit theorem for arrays normed by affine transformations. - Ann. Probab., 1981, 9, N4, 611-623.
10. S.V.Semovskiĭ Bernstein-Feller's theorem in $\mathbb{R}^p$, Teor. Verojatnost. i Primenen., 1984, 29, N3, 566-570 (Russian).
11. O.M. Jadrenko Strong law of large numbers with matrix normalizations and condition for convergence of some class of Gaussian sequences to zero. In the book: "Some questions on the theory of stochastic processes", Inst. Math. Acad. Sci. Ukrain. SSR, Kiev, 1982, 117-133 (Russian).
12. V.V.Buldygin, S.A.Solntsev, O.M.Jadrenko Strong laws of large numbers with operator normalization, Preprint No 84.34, Inst. Math. Acad. Sci. Ukrain. SSR, Kiev, 1984, 43 pp. (Russian).
13. Yu.V.Prohorov On the strong law of large numbers, Izv. Acad. Sci. USSR, 1950, 14, N6, 523-536 (Russian).
14. M.Loève Probability theory 2nd ed., Van Nostrand, Princeton, N.J., 1960.
15. A.I.Martikaĭnen, V.V.Petrov On necessary and suf-

ficient conditions for the law of the iterated logarithm, Teor. Verojatnost. i Preminen, 1977, 22, N1, 18-26 (Russian).

16. V.V.Buldygin The strong law of large numbers and convergence to zero of Gaussian sequences. Teor. Verojatnost. i Mat. Statist., 1978, 19, 33-41 (Russian).

17. A.I.Martikaĭnen On necessary and sufficient conditions for strong law of large numbers. Teor. Verojatnost. i Primenen, 1979, 24, N4, 814-820 (Russian).

18. O.M.Jadrenko On the convergence of some class of Gaussian sequences, Ukrain. Math. J., 1984, 36, N3, 358-364 (Russian).

Prob. Theory and Math. Stat., Vol.1, pp. 291-306
Prohorov et al. (eds)
1986 VNU Science Press

LIMIT THEOREMS FOR STOCHASTIC INVENTORY MODELS

E.V.Bulinskaya
Moscow State University, 117234 Moscow, USSR

The asymptotic behaviour of a stochastic process representing the water level in a controlled multipurpose finite dam, is investigated. Functional limit theorems for the inventory (dams, queueing, etc.) models described by random walks with two boundaries are also proved.

INTRODUCTION

Each inventory model (as well as dams, queueing, insurance risk and some other models) may be considered as a system S with input X(t) and output Y(t). Here X(t) is the amount supplied and Y(t) is the quantity demanded during the time interval (0, t), if S represents an inventory model; X(t) is inflow and Y(t) is outflow for a dam; in a queueing model X(t) and Y(t) describe the arrival and the departure of customers, etc.

There exists a certain rule giving for a pair (X(t), Y(t)) of (in general) stochastic processes a new process Z(t), an inventory level at time t (resp. a water level, waiting time, etc.).

One of the most interesting (as well as significant for applications) problems is the investigation of the asymptotic (as $t \to \infty$) behaviour of such systems. In particular, it is important to establish the existence of the steady state, to obtain the limit distribution of the process Z(t) and the rate of convergence to this distribution.

The results described above provide answers to most of the important questions concerning the system. Namely, they are useful for the study of systems' stability and for the optimization. The optimization is the choice of the system parameters and/or one of the processes X(t) or Y(t) in order to obtain the prescribed quality of the system. The quality of the system can be evaluated by its reliability or by some other objective function, for example, by the costs incurred by the operation of the system.

This paper is organized as follows. Section 1 is devoted to the analysis of multipurpose finite dams with the two-stage output procedure. The explicit form of the limit distribution of Z(t) as $t \to \infty$ is obtained. The optimization problems are also mentioned. Section 2 deals with a finite capacity store (or queueing models with restrictions on the waiting time). We assume that the arrival and demand processes are sequences of random variables, i.e., the basic process is a random walk with two impenetrable boundaries.Control parameters are mean supply and mean demand per unit time. Two functional limit theorems (for the controlled and uncontrolled processes) are proved. The limit processes are similar to that treated in section 1.

1. THE CONTROL OF A MULTIPURPOSE FINITE DAM

The storage theory (or theory of dams) is intensively developing during the last 30 years (see, e.g. [1]). However, there are few works dealing with controlled dam models or potimization problems and they are not included in [1]. One of the first controlled dam models was introduced by Faddy [2] in 1974. He considered a dam of capacity α designed to supply water for generating electricity. The release of water gives a reward A for each unit of output. It is assumed that water can be released at one of the two rate 0 or M units per unit time. The cost of increasing the output rate from 0 to M (at any time) is

KM. The problem is to control the output in such a way as to minimize the long-run average cost per unit time of operating the system. The cumulative input X(t) during the time interval (0, t) was supposed to be a Wiener process with drift μ ($\mu=1$) and variance parameter σ^2 ($\sigma^2=2$). The content Z(t) at time t was allowed to be negative, i.e., the author assumed a reflecting boundary for the process at level a and no boundary at level 0. The optimal policy was chosen in the class of monotone policies P_b^M of the following type. Assume that the dam is initially empty. The release rate is kept at 0 until the water level Z(t) in the dam increases to b ($0 < b \leq a$). As soon as this occures, the water is released at rate M. The output rate is kept at M as long as the dam is non-empty and it must be decreased to 0 if the water level becomes 0. Then the above procedure is repeated once more. A similar model was treated in [3] and [4]. In particular, it was shown in [4] that the monotone policies P_b^M are optimal in a more wide class of piece-wise continuous policies.

In [5] an objective function is obtained in explicit form under the P_b^M policy in the case of two reflecting boundaries (at 0 as well as at a).

In [6] under the same assumptions (two reflecting boundaries, arbitrary μ and σ^2) the asymptotic behaviour of the system is investigated. Namely, the limit distributions of the processes Z(t) and $\zeta(t)$ ($\zeta(t)$ being the output rate at time t) are obtained in the explicit form. The rate of convergence to a limit distribution and its dependence on control parameters are also considered.

In this paper we treat a multipurpose dam, i.e., designed not only for supplying water to a city or for generating electricity, or for irrigation but also used for navigation or for fish-farming, etc. It is obvious that the content Z(t) of such a dam would not be allowed to be 0. In order to achieve this, one can suppose that in addition to the above mentioned costs, there is a penalty cost $h(\cdot)$ per unit time depending on the water level $Z(t)$. The most simple case is $h(x)=h(d-x)^{+}$ where $d<a$ and $(y)^{+}=\max(0,y)$. Even if we assume the non-zero cost $\widetilde{K}M$ for decreasing the output rate from M to 0, it is more reasonable to consider the generalization of the P_b^M policy, namely, the policy $P_{b,c}^M$ depending on two parameters b and c ($b>c\geqslant 0$). It is the same as the policy P_b^M expect that whenever the water is released at rate M, the output rate is kept at M only as long as the dam content is more than c and it must be decreased to 0 if the water level is equal to c. When $Z(t)$ reaches b the water is once more released at rate M, etc.

Under the policy $P_{b,c}^M$ the process $Z(t)$ is delayed regenerative (i.e., the distribution of the first regeneration period is different from that of the following periods). This enables us to obtain the explicit form of the limit distribution. Hence, it is possible to calculate the long-run average cost per unit time. We shall also establish the rate of convergence to the limit distribution and dependence of the latter on the parameters b and c of the policy $P_{b,c}^M$.

Theorem 1. The process $Z(t)$ has a limit distribution $P(B)=\lim\limits_{t\to\infty} P_t(B)$ with a density $p(x)$. Here $P_t(B)=$ $=P\{Z(t)\in B\}$ and B is a Borel set.

Furthermore,

$$p(x)=\begin{cases} \dfrac{1}{g\mu}\left(e^{2\mu(x-c)\sigma^{-2}}-e^{2\mu(x-b)\sigma^{-2}}\right) & \text{for } 0<x\leq c \\ \dfrac{1}{g(M-\mu)}\left(1-e^{-2(M-\mu)(x-c)\sigma^{-2}}\right)+ & \\ \quad +\dfrac{1}{g\mu}\left(1-e^{2\mu(x-b)\sigma^{-2}}\right) & \text{for } c\leq x\leq b \\ \dfrac{1}{g(M-\mu)}\left(e^{-2(M-\mu)(x-b)\sigma^{-2}}-e^{-2(M-\mu)(x-c)\sigma^{-2}}\right) & \text{for } b\leq x<a, \end{cases} \tag{1}$$

where

$$g = E(\theta_2 - \tau_2) + E(\tau_2 - \theta_1)$$

and for $\ell \geqslant 2$

$$E(\theta_\ell - \tau_\ell) = \frac{b-c}{M-\mu} - \frac{\sigma^2}{2(M-\mu)^2}\left(e^{-2(M-\mu)(a-b)\sigma^{-2}} - e^{-2(M-\mu)(a-c)\sigma^{-2}}\right), \quad (2)$$

$$E(\tau_\ell - \theta_{\ell-1}) = \frac{b-c}{\mu} - \frac{\sigma^2}{2\mu^2}\left(e^{-2\mu c\sigma^{-2}} - e^{-2\mu b\sigma^{-2}}\right), \quad (3)$$

random variables θ_ℓ, τ_ℓ being given by

$$\theta_0 = 0, \quad \tau_\ell = \inf\{t > \theta_{\ell-1}, Z(t) = b\}, \quad \theta_\ell = \inf\{t > \tau_\ell, Z(t) = c\}. \quad (4)$$

(Formulae (1)-(3) are written for $\mu \neq 0$, $M-\mu \neq 0$, with obvious changes needed in other cases.)

The proof of this theorem proceeds almost along the same lines as that of Theorem 1 in [6], but the calculations are more involved and tedious, therefore some details will be omitted.

Since the process Z(t) is delayed regenerative and θ_i ($i \geqslant 1$) are its regeneration points, in order to obtain the limit distribution we use the following theorem due to Smith (see, e.g. [7]).

Let Z(t) be an aperiodic (delayed) regenerative process and $P\{\theta_1 < \infty\} = 1$. If the function $\psi_B(t) = P\{Z(\theta_{\ell-1}+t) \in B, \theta_\ell - \theta_{\ell-1} > t\}$ $(\ell \geqslant 2)$ is directly Riemann integrable, then there exists

$$\lim_{t\to\infty} P\{Z(t) \in B\} = \left(E(\theta_\ell - \theta_{\ell-1})\right)^{-1} \int_0^\infty \psi_B(t)\,dt. \quad (5)$$

As proved in [7], if $E(\theta_\ell - \theta_{\ell-1}) < \infty$ and $\psi_B(t)$ is integrable over each finite interval, then it is directly Riemann integrable.

Therefore, it suffices to calculate the right part of (5) and verify the validity of Smith's theorem conditions.

As mentioned above, all the regeneration periods (except the first one) are identically distributed. They consist of two parts with lengths $\tau_\ell - \theta_{\ell-1}$ and $\theta_\ell - \tau_\ell$ respectively. In the first part the behaviour of $Z(\theta_{l-1}+t)$ is that of a Wiener process $Z_1(t)$ with parameters μ and σ^2, initial state c, reflecting boundary at 0 and an absorbing one at b. In the second part the behaviour of $Z(\tau_\ell+t)$ is that of a Wiener process $Z_2(t)$ with parameters μ-M and σ^2, initial state b, reflecting boundary at a and an absorbing one at c.

The processes $Z_i(t)$ (i=1,2) are Markovian. Denote

$$\psi_{iB}(t) = P(x_i, t, B),$$

where $P_i(x,t;B)$ is the transition probability function of $Z_i(t)$ and $x_1=c$, $x_2=b$.

It is easy to verify (see, [6]) that

$$\int_0^\infty \psi_B(t)dt = \int_0^\infty \psi_{1B}(t)dt + \int_0^\infty \psi_{2B}(t)dt. \qquad (6)$$

Further, if f(y) is an arbitrary bounded continuous function on (r_{i1}, r_{i2}) and

$$v_i(t,x) = \int_{r_{i1}}^{r_{i2}} P(x,t,dy)f(y) \qquad (7)$$

has two bounded continuous derivatives $\partial v_i / \partial x$ and $\partial^2 v_i / \partial x^2$ then there exists $\partial v_i / \partial t$ and v_i satisfies the backward Kolmogorov equation

$$\frac{\partial v_i}{\partial t} = \frac{\sigma^2}{2}\frac{\partial^2 v_i}{\partial x^2} + \mu_i \frac{\partial v_i}{\partial x}$$

with the initial condition

$$\lim_{t\to 0} v_i(t,x) = f(x)$$

and boundary conditions

$$\frac{\partial v_1(t, r_{11})}{\partial x} = 0, \quad v_1(t, r_{12}) = 0, \quad (9_1)$$

$$v_2(t, r_{21}) = 0\,, \qquad \frac{\partial v_2(t, r_{22})}{\partial x} = 0\,, \qquad (9_2)$$

here $r_{11}=0$, $r_{12}=b$, $r_{21}=c$, $r_{22}=a$ and $\mu_1 = \mu$, $\mu_2 = \mu - M$.

Taking in (8) a Laplace transform with respect to t

$$v_{i\lambda}(x) = \int_0^\infty e^{-\lambda t} v_i(t, x)\, dt \qquad (10)$$

gives

$$\frac{\sigma^2}{2} \frac{d^2 v_{i\lambda}}{dx^2} + \mu_i \frac{d v_{i\lambda}}{dx} - \lambda v_{i\lambda} = -f\,. \qquad (11)$$

Instead of (9) we have, respectively

$$\frac{d v_{1\lambda}(0)}{dx} = 0\,, \qquad v_{1\lambda}(b) = 0\,, \qquad (12_1)$$

$$v_{2\lambda}(c) = 0\,, \qquad \frac{d v_{2\lambda}(a)}{dx} = 0\,. \qquad (12_2)$$

As it is well-known (see, e.g. [8]), equation (11) can be written in the form

$$(\lambda - D_{m_i} D_{s_i}^{+})\, v_{i\lambda} = f\,,$$

where $s_i(x)$ is a canonical scale and $dm_i(x)$ a canonical measure. This enables us to write for $\lambda > 0$ the solution of (11) satisfying the conditions (12^i) in the form

$$v_{i\lambda}(x) = \int_{r_{i1}}^{r_{i2}} K_{i\lambda}(x,y) f(y)\, dm_i(y)\,. \qquad (13)$$

The function $K_{i\lambda}(x,y)$ is given by

$$K_{i\lambda}(x,y) = \begin{cases} \dfrac{u_{i1}(x)\, u_{i2}(y)}{W(u_{i1}, u_{i2})} & \text{if } x \le y\,, \\[2ex] \dfrac{u_{i1}(y)\, u_{i2}(x)}{W(u_{i1}, u_{i2})} & \text{if } x \ge y\,, \end{cases}$$

where $W(u_{i1}, u_{i2}) = (D_{s_i}^{+} u_{i1}) u_{i2} - (D_{s_i}^{+} u_{i2}) u_{i1}$ is a constant, $u_{i1}(x)$ is a solution of the equation

$$(\lambda - D_{m_i} D_{s_i}^{+})\, v_{i\lambda} = 0$$

satisfying the first of the conditions (12_i) at the left

-hand end r_{i1} of the considered interval, and $u_{i2}(x)$ is a solution satisfying the second condition at the righ-hand end r_{i2}.

Now write (13) in the following more convenient form

$$v_{i\lambda}(x) = \int_{r_{i1}}^{r_{i2}} K_{i\lambda}(x,y)f(y)\,dy. \qquad (14)$$

It follows from (7), (10) and (14) that

$$\int_0^\infty e^{-\lambda t}\psi_{iB}(t)\,dt = \int_B K_{i\lambda}(x_i,y)\,dy. \qquad (15)$$

(Note that if there exists a transition density $q_i(x,t;y)$ then its Laplace transform is the Green function $K_{i\lambda}(x,y)$)

Furthermore, since

$$E(\tau_\ell - \theta_{\ell-1}) = \int_0^\infty P\{\tau_\ell - \theta_{\ell-1} > t\}\,dt$$

and for $\ell \geq 2$

$$P\{\tau_\ell - \theta_{\ell-1} > t\} = \int_0 P_1(c,t;dy),$$

we have

$$E(\tau_\ell - \theta_{\ell-1}) = \int_0^\infty \psi_{1,(0,b)}(t)\,dt. \qquad (16)$$

Similarly

$$E(\theta_\ell - \tau_\ell) = \int_0^\infty \psi_{2,(c,a)}(t)\,dt. \qquad (17)$$

As it is easily seen from (5), (6), (15), (16) and (17), in order to complete the proof, it remains to obtain

$$K_{i0}(x_i,y) = \lim_{\lambda\to 0} K_{i\lambda}(x_i,y).$$

After some tedious calculations we find that

$$K_{10}(c,y) = \begin{cases} \frac{1}{\mu}\left(e^{2\mu(y-c)\sigma^{-2}} - e^{2\mu(y-b)\sigma^{-2}}\right) & \text{for } 0 < y \leq c, \\ \frac{1}{\mu}\left(1 - e^{2\mu(y-b)\sigma^{-2}}\right) & \text{for } c \leq y < b. \end{cases}$$

$$K_{20}(b,y)=\begin{cases}\frac{1}{M-\mu}\left(1-e^{-2(M-\mu)(y-c)\sigma^{-2}}\right) & \text{for } c<y\leq b,\\ \frac{1}{M-\mu}\left(e^{-2(M-\mu)(y-b)\sigma^{-2}}-e^{-2(M-\mu)(y-c)\sigma^{-2}}\right) & \text{for } b\leq y<a.\end{cases}$$

These results easily yield that $E(\theta_\ell-\tau_\ell)$ and $E(\tau_\ell-\theta_{\ell-1})$ ($\ell\geqslant 2$) are given by (2) and (3), respectively, and the limit distribution density is given by (1). (Note that $E\tau_1$ is obtained by putting in (3) c=0.) Thus, the proof if completed.

Corollary 1. For each $n\geqslant 1$ there exists a constant c_n such that

$$V(P_t,P)\leq c_n t^{-n},$$

where V is the variational distance between two distributions P_t and P.

The proof follows immediately from the results [9].

Corollary 2. The process $\zeta(t)$ has a limit distribution

$$p_k=\lim_{t\to\infty}P\{\zeta(t)=k\},\qquad (k=\overline{0,M}).$$

Furthermore,

$$p_0=1-p_M=\frac{E(\tau_\ell-\theta_{\ell-1})}{E(\theta_\ell-\theta_{\ell-1})}. \tag{18}$$

where θ_ℓ,τ_ℓ are defined by (4).

The proof is obtained using the properties of alternating renewal processes.

It is clear from (1)-(3) and (18) that the limit distributions of Z(t) and $\zeta(t)$ depend continuously on system parameters.

Corollary 3. The long-run average cost per unit time is given by

$$\mathcal{L}(b,c)=\frac{(K+\tilde{K})M-AE(\theta_\ell-\tau_\ell)}{E(\theta_\ell-\theta_{\ell-1})}+\int_0^a h(x)p(x)dx. \tag{19}$$

The proof is based on the properties of regenerative processes.

An analogous model was treated independently in [10]. But in that paper only the long-run average cost per unit time (in a slightly different form than (19)) is determined by the method used previously in [5].

2. RANDOM WALKS WITH TWO BOUNDARIES

As mentioned in the introduction, many problems of inventory, dams, queueing theory, etc. are described by random walks with boundaries (more details can be found in [11]). A lot of authors studied such processes from different view-points. For example, the asymptotic behaviour of limit distribution of the random walk with two impenetrable boundaries at 0 and at β (as $\beta \to \infty$) was investigated in [12], [13] (as well as some optimization problems). It was assumed there that a distribution inducing the random walk does not depend on β.

In this paper we consider random walks induced by the series of random variables (the distribution of random variables depends on the upper boundary, in particular, its mean tends to zero as the boundary tends to the infinity) and prove functional limit theorems. For the sake of brevity we treat here only two simple cases, more involved ones will appear in the subsequent papers.

Let $\{\xi_{nk}\}_{k\geq 1}$ for each n=1,2,... be a sequence of independent identically distributed random variables with $E\xi_{nk} = \frac{\mu}{\sqrt{n}}$ and $D\xi_{nk} = \sigma^2$, let also the Lindeberg condition be satisfied. The random variable ξ_{nk} may be considered as net input (i.e., the difference between the supply and the demand) during the k-th period to the n-th store with capacity $a\sqrt{n}$. Then the random walk

$$Z_{nk} = \min\left(a\sqrt{n}, (Z_{n,k-1} + \xi_{nk})^+\right), \quad Z_{n0} = 0, \qquad (20)$$

represents the inventory level at the end of the k-th period if the store was initially empty.

Below we prove the weak convergence on $[0,\infty)$ of the normalized random walks to a Wiener process $\widetilde{W}(t)$ with parameters μ and σ^2 having reflecting boundaries at 0 and at a.

We formulate the desired result more precisely as follows.

Theorem 2. Let $\{\xi_{nk}\}$ be as described above and $\{Z_{nk}\}$ be given by (20). Consider the step function

$$Z_n(t) = Z_{nk} \qquad \text{for} \quad k \leq t < k+1, \quad k = 0,1,\dots$$

Then

$$\widetilde{Z}_n(t) = \frac{Z_n(nt)}{\sqrt{n}} \xrightarrow{\mathcal{D}} \widetilde{W}(t), \qquad n \to \infty,$$

where $\xrightarrow{\mathcal{D}}$ denotes the weak convergence in $D[0, \infty)$.

Proof. It is clear that the processes $\widetilde{Z}_n(t)$ belong to $D[0, \infty)$, their sample functions being right continuous on $[0, \infty)$ with limits from the left. The space $D[0, \infty)$ is assumed to be endowed with the extension of Skorohod's J_1 topology from $D[0, 1)$ to $D[0, \infty)$ (see, e.g. [14]). As follows from the results in [14], in order to establish the weak convergence of $\widetilde{Z}_n$ to $\widetilde{W}$ in $D[0, \infty)$, it is sufficient to show it in $D[0,T]$ for arbitrary T. This can be done using [15].

Remark 1. It is obvious that $\widetilde{W}(t)$ is a particular case of Z(t) treated in section 1 (obtained by putting M=0, c=0, b=a). Therefore, it follows from Theorem 1 that there exists a limit distribution of $\widetilde{W}(t)$ and its density $\widetilde{p}(x)$ is given for $0 < x < a$ by

$$\widetilde{p}(x) = \begin{cases} \dfrac{\mu}{\sigma^2}\left(sh\,\dfrac{\mu a}{\sigma^2}\right)^{-1} e^{\mu(2x-a)\sigma^{-2}} & \text{if} \quad \mu \neq 0, \\ a^{-1} & \text{if} \quad \mu = 0. \end{cases}$$

Thus, we see that although $\frac{\mu}{\sqrt{n}} \to 0$ as $n \to \infty$ for each μ, the limit distributions are different for $\mu > 0$, $\mu = 0$ and $\mu < 0$.

Remark 2. It is possible to consider random walks with the upper boundary $\beta \neq a\sqrt{n}$. We obtain similar results changing appropriately means and normalization.

Remark 3. In such a simple case as treated in Theorem 2 it is easy to verify the weak convergence of $\tilde{Z}_n$ to $\tilde{W}$ in $D[0,T]$ directly. But using [15], we shall be able to extend the results on dependent variables and to obtain not only Wiener limit processes.

Up to now we considered an uncontrolled random walk, below we shall deal with a controlled one. Namely, it will be possible to control the supply and/or the demand per period. We use a two-level policy analogous to the $P^M_{b,c}$ policy from section 1 (with obvious modifications appropriate to a discrete case). Assume for simplicity $c=0$ and $b=a\sqrt{n}$.

Let $\{\xi^{(1)}_{nk}\}_{k\geq 1}$ and $\{\xi^{(2)}_{nk}\}_{k\geq 1}$ be for each $n=1,2,\ldots$ two independent sequences of independent identically distributed variables with $E\,\xi^{(1)}_{nk} = \frac{\mu}{\sqrt{n}}$, $E\,\xi^{(2)}_{nk} = \frac{\mu - M}{\sqrt{n}}$ and $D\xi^{(i)}_{nk} = \sigma^2$ $(i=1,2)$, let also the Lindeberg condition be satisfied. We construct a controlled random walk as follows

$$\tilde{Z}_{n0} = 0 \quad \text{and} \quad \tilde{Z}_{nk} = \min\left(a\sqrt{n}, (\tilde{Z}_{n,k-1} + \tilde{\xi}_{nk})^+\right), \tag{21}$$

where

$$\tilde{\xi}_{nk} = \begin{cases} \xi^{(1)}_{nk} & \text{for } \tilde{\theta}_{n,i-1} \leq k < \tilde{\tau}_{ni}, \\ \xi^{(2)}_{nk} & \text{for } \tilde{\tau}_{ni} \leq k < \tilde{\theta}_{ni}, \quad i=1,2,\ldots \end{cases}$$

and

$$\tilde{\theta}_{n0} = 0, \quad \tilde{\tau}_{ni} = \inf\{k > \tilde{\theta}_{n,i-1},\ \tilde{Z}_{nk} = a\sqrt{n}\},$$

$$\tilde{\theta}_{ni} = \inf\{k > \tilde{\tau}_{ni},\ \tilde{Z}_{nk} = 0\}.$$

The random variable $\tilde{Z}_{nk}$ may be considered as the inventory level at the end of the k-th period in a store of capacity $a\sqrt{n}$ under the following control policy. Let the store be initially empty. Then the net input per period is a random variable with mean $\frac{\mu}{\sqrt{n}}$ as long as the inventory level is less than $a\sqrt{n}$. As soon as the store is completely filled, we change the net input per period in order to have the mean equal to $\frac{\mu - M}{\sqrt{n}}$. Such an input is used until the store is empty. Then the procedure is repeated once more.

It is interesting to mention that $\{\tilde{Z}_{nk}\}_{k\geq 0}$ defined by (21) will not be a Markov chain (unlike $\{Z_{nk}\}_{k\geq 0}$ given by (20)). But it is a regenerative process with the regeneration points $\tilde{\theta}_{ni}$ (i=1,2,...).

We have the following result.

Theorem 3. Let $\{\xi_{nk}^{(i)}\}$ be as described above and $\{\tilde{Z}_{nk}\}$ be given by (21). Consider the step function

$$Z_n'(t) = \tilde{Z}_{nk} \quad \text{for} \quad k \leq t < k+1, \quad k=0,1,\dots$$

Then
$$\hat{Z}_n(t) = \frac{Z_n'(nt)}{\sqrt{n}} \xrightarrow{\mathcal{D}} \hat{Z}(t), \quad n \to \infty,$$

where $\hat{Z}(t)$ is the process $Z(t)$ investigated in section 1 (under the policy $P^M_{0,a}$).

Proof. Since $\hat{Z}_n(t)$ are realized in $D[0, \infty)$, it is sufficient (as in the proof of Theorem 2) to show the weak convergence of these processes in $D[0,T]$ for arbitrary T. Taking into account the results [15]-[17] we obtain the tightness of $\{\hat{Z}_n\}$ in $D[0,T]$.

Now (see, e.g. [16]) it remains to prove the convergence of the finite dimensional distributions. Let us define a sequence ζ_{nk} as follows

$$\zeta_{nk} = \begin{cases} 0 & \text{for } \tilde{\theta}_{n,i-1} \leq k < \tilde{\tau}_{ni}, \\ M & \text{for } \tilde{\tau}_{ni} \leq k < \tilde{\theta}_{ni} \end{cases}$$

and two processes

$$\zeta_n(t) = \zeta_{nk} \text{ for } k \leq t < k+1, \quad k=0,1,\ldots$$

and $$\hat{\zeta}_n(t) = \zeta_n(nt).$$

First we consider

$$P\{\hat{Z}_n(t)\in B, \hat{\zeta}_n(t)=0\} = \sum_{k=0}^{\infty} P\{\hat{Z}_n(t)\in B, \theta_{nk} \leq t < \tau_{n,k+1}\}, \quad (22)$$

where $\theta_{nk} = \dfrac{\tilde{\theta}_{nk}}{n}$ and $\tau_{nk} = \dfrac{\tilde{\tau}_{nk}}{n}$.

The right side of (22) we rewrite as

$$\sum_{k=0}^{\infty} \int_0^t P\{\hat{Z}_n(t)\in B, \theta_{nk} \leq t < \tau_{n,k+1} \mid \theta_{nk}=s\}\, dP\{\theta_{nk} \leq s\}.$$

Since $\hat{Z}_n(t)$ is a regenerative process and θ_{nk} are its regeneration points, we have

$$P\{\hat{Z}_n(t)\in B, \theta_{nk} \leq t < \tau_{n,k+1} \mid \theta_{nk}=s\} = \psi^1_{nB}(t-s),$$

where

$$\psi^1_{nB}(t) = P\{\hat{Z}_n(t)\in B, t<\tau_{n1}\} = P\{\hat{Z}^1_n(t)\in B, t<\tau_{n1}\}$$

and $\hat{Z}^1_n(t)$ is the normalized walk with the impenetrable boundary at 0 induced by the sequence $\{\xi^{(1)}_{nk}\}_{k\geq 1}$. Namely,

$$\hat{Z}^1_n(t) = \frac{Z^1_n(nt)}{\sqrt{n}},$$

$$Z^1_n(t) = Z^1_{nk} \text{ for } k \leq t < k+1 \text{ and}$$

$$Z^1_{n0}=0, \quad Z^1_{nk} = (Z^1_{n,k-1} + \xi^{(1)}_{nk})^+.$$

Thus, (22) can be written as

$$P\{\hat{Z}_n(t)\in B, \hat{\zeta}_n(t)=0\} = \int_0^t \psi^1_{nB}(t-s)\, dH_{n\theta}(s), \quad (23)$$

where $H_{n\theta}(t) = \sum_{k=0}^{\infty} F^{k*}_{n\theta}(t)$ is the renewal function corresponding to $F_{n\theta}(t) = P\{\theta_{n1} \leq t\}$.

Let $\hat{\tau}_\ell$, $\hat{\theta}_\ell$ be defined by (4) with c=0, b=a. Since $\hat{Z}^1_n(t)$ weakly converges on [0,T] for arbitrary T to

$\hat{Z}_1(t)$ (a Wiener process with parameters μ and σ^2 having a reflecting boundary at 0), it follows immediately that $F_{n\tau}(t)=P\{\tau_{n1}\leqslant t\}$ tends uniformly on $[0, \infty)$ to $F_\tau(t)=P\{\hat{\tau}_1 \leqslant t\}$. Similarly we prove the uniform convergence of $P\{\theta_{n1}-\tau_{n1}\leqslant t\}$ to $P\{\hat{\theta}_1-\hat{\tau}_1\leqslant t\}$, hence that of $F_{n\theta}(t)$ to $F_\theta(t)=P\{\hat{\theta}_1 \leqslant t\}$ as $n\to\infty$.

Furthermore, we have

$$H_{n\theta}(t) \longrightarrow H_\theta(t) \tag{24}$$

uniformly on each fixed interval $[0,T]$. It follows from the inequality

$$\sum_{k=L+1}^{\infty} F_{n\theta}^{k*}(t) \leqslant \frac{(1-\alpha)^{L+1}}{\alpha},$$

α being equal to $1/2\ P\{\hat{\tau}_1 > T\} > 0$.

It is also easily seen that

$$\psi_{nB}^1(t) \longrightarrow \psi_B^1(t) = P\{\hat{Z}_1(t)\in B,\ t<\hat{\tau}_1\}.$$

From this fact and (23), (24) it readily follows that

$$P\{\hat{Z}_n(t)\in B,\ \hat{\xi}_n(t)=0\}\to P\{\hat{Z}(t)\in B,\ \hat{\xi}(t)=0\}$$

(the process $\hat{\xi}(t)$ being $\zeta(t)$ from section 1 under the policy $P^M_{0,a}$). The proof of the relation

$$P\{\hat{Z}_n(t)\in B,\ \hat{\xi}_n(t)=M\}\to P\{\hat{Z}(t)\in B,\ \hat{\xi}(t)=M\}$$

is similar. Hence, the convergence of one-dimensional distributions is proved. Using the induction argument, we obtain the convergence of all ℓ-dimensional distributions ($\ell \geqslant 2$) along the same lines.

Remark. As seen from the proof of Theorem 3, its result is obviously valid if we put

$$\xi_{nk}^{(1)} = \xi_k + \frac{\mu}{\sqrt{n}}, \quad \xi_{nk}^{(2)} = \xi_k + \frac{\mu - M}{\sqrt{n}}$$

$\{\xi_k\}_{k\geqslant 1}$ being a sequence of independent identically distributed random variables with $E\xi_k=0$, $D\xi_k=\sigma^2$.

REFERENCES

1. Jankiewicz M. Elementy teorii tam (in Polish). - Rocz. Pol.tow.mat.,ser.3, 1982, 18, p.5-57.
2. Faddy M.J. Optimal control of finite dams: discrete (2-stage) output procedure. - J.Appl.Prob., 1974, 11, p.111-121.
3. Zuckerman D. Two-stage output procedure of a finite dam. - J.Appl.Prob., 1977, 14, p.421-425.
4. Zuckerman D. Characterization of the optimal class of output policies in a control model of a finite dam. - J.Appl.Prob., 1981, 18, p.913-923.
5. Attia F.A., Brockwell P.J. The control of a finite dam. - J.Appl.Prob., 1982, 19, p.815-825.
6. Bulinskaya E.V. On one model of the theory of dams (in Russian). - In: Stability problems for stochastic models, Moscow, Institute for Systems Studies, 1984, p.34-39.
7. Afanas'eva L.G., Bulinskaya E.V. Stochastic processes in queueing and inventory theory (in Russian). - Moscow Univ.press, 1980.
8. Ito K. Stochastic processes, II, Moscow, 1963.
9. Anichkin S.A. Coupling of renewal processes and its applications (in Russian). - In: Stability problems for stochastic models, Moscow, Institute for Systems Studies, 1982, p.4-24.
10. Yeh L. Optimal control of a finite dam: average cost case. - J.Appl.Prob., 1985, 22, p.480-484.
11. Afanas'eva L.G., Bulinskaya E.V. Random walks with boundary conditions and their applications (in Russian). - In: Prahu N. Stochastic storage processes, Mir, 1984, p.155-184.
12. Afanas'eva L.G., Bulinskaya E.V. Some asymptotic results for random walks in a band (in Russian). - Teor.veroyat. i primen., 1984, 29, 4, p.654-668.
13. Afanas'eva L.G., Bulinskaya E.V. Optimization of store capacity (in Russian). - Techn.Cybern., 1981, 5, p.58-68.
14. Lindvall T. Weak convergence of probability measures and random functions in the function space $D[0,\infty)$. - J.Appl.Prob., 1973 , 10, p.109-121.
15. Borovkov A.A. Asymptotic methods in queueing theory (in Russian). - Nauka, 1980.
16. Billingsley P. Convergence of probability measures (in Russian), Moscow, Nauka, 1977.
17. Gihman I.I., Skorohod A.V. Theory of stochastic processes III (in Russian). - Nauka, 1975.

Prob. Theory and Math. Stat., Vol.1, pp. 307-326
Prohorov et al. (eds)
1986 VNU Science Press

LIMIT THEOREMS UNDER WEAK DEPENDENCE CONDITIONS

A.V. Bulinskii
Moscow State University
117234 Moscow, USSR

An approach permitting a unified treatment and comparison of number of dependence coefficients for random fields is outlined in this paper. Some specific features of this approach are demonstrated in the applications to the problem of asymptotic normality of a series of additive random functions.

Let $X(v)$ be a random field, i.e. a family of functions defined on the probability space $(\Omega, \mathcal{F}, P)$ and on some system $\mathcal{R}$ of sets $v \subset \mathbb{R}^m$. An important role in the study of such functions is played by mixing coefficients expressing weak (in certain sense) dependence of $X(\cdot)$ for sufficiently separated sets. To any set $V \subset \mathbb{R}^m$ there corresponds the minimal σ-algebra $\mathfrak{A}_X(V) = \sigma\{X(v), v \subset V, v \in \mathcal{R}\}$ generated by the field $X(\cdot)$ on V while $\sigma\{\emptyset\} = \emptyset$. The dependence (or mixing) coefficients are often used to describe the connections between σ-algebras $\mathfrak{A}_X(V_1), \dots, \mathfrak{A}_X(V_k)$, $k \in \mathbb{N}$, $V_i \subset \mathbb{R}^m$, $i = 1, \dots, k$. In specifying such a coefficient there are two essential points: 1) the choice of dependence measure for a k-tuple of σ-algebras $\mathfrak{A}_X(V_i)$ and 2) taking into account the "spatial location" of these σ-algebras, i.e. the location of the corresponding sets V_i.

It is convenient to operate not only with the systems of events depending on the subsets of $\mathbb{R}^m$, but also with the systems of random variables $\mathcal{X}(v)$, $v \in \mathcal{R}$ and to consider the dependence of the latter systems.

The set of indicator random variables $\mathcal{X}=\{\mathbb{1}_A, A\in\mathfrak{A}\}$ can be associated with any system of events $\mathfrak{A}$ (including the σ-algebra case). On the other hand, dealing with a system $\mathcal{X}(v)$, consisting of a single element $X(v)$, provides an opportunity of "individual" description of the dependence for a given field, i.e. allows us to consider the covariance for $X(v')$ and $X(v'')$ or joint characteristic functions of $X(v_1),\dots,X(v_k)$.

Now we induce a nonnegative function $\varkappa(Y)$ defined on a set $\mathcal{Y}$ of random vectors $Y=(Y_1,\dots,Y_k)$, e.g. $\varkappa$ being a function of mixed moments, cumulants, etc. For the systems $\mathcal{Y}_i$, $i=1,\dots,k$, of random variables define

$$\varkappa[\mathcal{Y}_1,\dots,\mathcal{Y}_k] = \sup\{\varkappa(Y):\ Y\in\prod_{i=1}^{k}\mathcal{Y}_i\cap\mathcal{Y}\}. \quad (1)$$

Taking indicators of certain events as the systems of random variables, we obtain the coefficients of connection for the systems $\mathfrak{A}$ of events defined by

$\varkappa\{\mathfrak{A}_1,\dots,\mathfrak{A}_k\} = \varkappa[\mathcal{Y}_1,\dots,\mathcal{Y}_k]$, where $\mathcal{Y}_i=\{\mathbb{1}_A, A\in\mathfrak{A}_i\}$.

We get various coefficients of dependence by choosing functions $\varkappa(\cdot)$ and proper systems of random variables. In some cases one can compare such coefficients. More specifically, leaning upon the properties of functions $\varkappa'(\cdot)$ and $\varkappa''(\cdot)$ we estimate $\varkappa'[\cdot]$ in terms of $\varkappa''[\cdot]$ or $\varkappa''\{\cdot\}$. For example, let us consider for σ-algebras $\mathfrak{A}$, $\mathfrak{B}$ in $(\Omega,\mathcal{F},P)$ the coefficients of dependence

$$\varrho_{\gamma,\delta}\{\mathfrak{A},\mathfrak{B}\} = \sup\left\{\frac{|P(AB)-P(A)P(B)|}{(P(A))^{\gamma}\,(P(B))^{\delta}}:\ A\in\mathfrak{A}, B\in\mathfrak{B}\right\},\ \gamma,\delta\in[0,1] \quad (2)$$

and the maximal correlation coefficient

$$\varrho\{\mathfrak{A},\mathfrak{B}\} = \sup\{|\mathrm{corr}(f,g)|:\ f\in L_2(\Omega,\mathfrak{A},P),\ g\in L_2(\Omega,\mathfrak{B},P)\},$$

where $\mathrm{corr}(f,g)$ denotes the correlation coefficient for f and g.

For $\gamma=\delta=0$ formula (2) coincides with the definition of Rosenblatt's α-mixing coefficient; with $\gamma=0$, $\delta=1$ (2) gives us Ibragimov's φ-mixing coefficient, while the $\gamma=\delta=1$ case corresponds to the so-called ψ-mixing coefficient. The coefficients (2) were introduced in [4] and some properties thereof were studied in [5]. The same coefficients were proposed independently and examined in [21].

The following theorem sharpening the results of [20] and [3] can be found in [4]. The same statement is contained among the results of [21].

Theorem 1. For any σ-algebras $\mathfrak{A}$, $\mathfrak{B}$ in the probability space $(\Omega, \mathcal{F}, P)$ the following inequalities

$$\rho_{1/2,1/2}\{\mathfrak{A},\mathfrak{B}\} \le \rho\{\mathfrak{A},\mathfrak{B}\} \le c\,\rho_{1/2,1/2}\{\mathfrak{A},\mathfrak{B}\}\left(\left|\ln \rho_{1/2,1/2}\{\mathfrak{A},\mathfrak{B}\}\right|+1\right) \quad (3)$$

hold with some absolute constant c.

Another example of the comparison of different dependence coefficients is provided by the well-known Kolmogorov-Rozanov theorem [11] on the connections between α and ρ coefficients for the σ-algebras generated by a stationary Gaussian process. We also recall that a variety of inequalities was obtained for the estimation of covariances of random variables ξ and η taking values in $L_p(\Omega,\mathfrak{A},P)$ and $L_q(\Omega,\mathfrak{B},P)$, respectively, in terms of their moments and dependence coefficients for $\mathfrak{A}$, $\mathfrak{B}$ (see, e.g. [9,10,21]).

Note that in some cases the employed coefficients describe the dependence of the systems of events without the σ-algebra property, e.g. for the events connected with crossing the levels by a stochastic process (see [22]). We do not tackle here the measures of dependence based on conditional expectations (see, e.g.[26,29]).

Passing to the measures of dependence for a random field, we fix a k-tuple $\mathfrak{M}=(\mathfrak{M}_1,\dots,\mathfrak{M}_k)$ of systems $\mathfrak{M}_i$ of subsets $V\subset\mathbb{R}^m$, $i=1,\dots,k$. The mixing coefficient for the field X and set $\mathcal{U}\subset\mathbb{R}^m$ is defined by

$$\varkappa_X(\mathcal{U},\mathfrak{M},\tau) = \sup \varkappa\{\mathfrak{A}_X(V_1\cap\mathcal{U}),\dots,\mathfrak{A}_X(V_k\cap\mathcal{U})\}, \qquad (4)$$

where supremum is taken over all $V_i \in \mathfrak{M}_i, i=1,\dots,k$ satisfying some additional condition τ. Of course, as it was pointed out above, one can use in (4) functions $\varkappa$ defined on the systems of random variables indexed by the subsets of $\mathbb{R}^m$. Moreover, when $\mathfrak{M}_i$, $i=1,\dots k$ contain, respectively, a single V_i, we get a nonuniform description of the dependence on the sets $\mathcal{U}\cap V_i$, $i=1,\dots,k$. Condition τ is ususally specified by means of several parameters appearing then as arguments of the mixing coefficient, while the "nonparametric" part of condition τ, whenever it exists, is labelled as condition Γ.

Most frequently "pair dependence" coefficients are used, i.e. $k=2$ in (4). Let $d(V_1,V_2)$ be a distance between the sets with respect to metric d in $\mathbb{R}^m$, $d(V_1,V_2) = \inf\{d(x,y): x\in V_1, y\in V_2\}$. If the sets $V_i \in \mathfrak{M}_i$, $i=1,2$ satisfy a certain condition Γ and $d(V_1,V_2)\geqslant r$, we denote the resulting mixing coefficient by $\varkappa_X(r,\mathcal{U},d,\Gamma,\mathfrak{M}_1,\mathfrak{M}_2)$. The fixed arguments of $\varkappa_X$ will be omitted along with Γ when the additional condition is not imposed. The notion of a mixing coefficient comprises some coefficients introduced earlier (with traditionally distinct symbols).

In an one-dimensional case the system $\mathfrak{M}_1$ ($\mathfrak{M}_2$, respectively) usually represents the "past" (the "future", respectively) of the process, i.e. $\mathfrak{M}_1=\{(-\infty,a], a\in\mathbb{R}\}$, $\mathfrak{M}_2=\{[b,\infty), b\in\mathbb{R}\}$. However, $\mathbb{R}^m$ for $m>1$ allows only a partial ordering, therefore one often deals with the systems $\mathfrak{M}_1=\mathfrak{M}_2$ formed by arbitrary Borel sets, or by arbitrary halfspaces, or by all the halfspaces cut out by hyperplanes parallel to the coordinate planes, and so on. Additional condition Γ may appear in the form of "mutual adjustment" of $\mathfrak{M}_1$ and $\mathfrak{M}_2$, for example, choose V_1 (V_2, respectively) in (4) to be the interior (resp. exterior) of two concentric cubes with parallel faces. Restrictions on diameters of the sets V_1, V_2, on their volumes (or the number of elements for the $\mathbb{Z}^m$ case) also appear in the literature. All such conditions lead to a variety of mixing coefficients (see, [1,6,7,8,10,12,19,25]).

We shall use below the α-mixing coefficients for random fields X. In other words, we take in (4) Rosenblatt's coefficient of connection among σ-algebras $\mathcal{A}, \mathcal{B}$ as the measure of dependence $\varkappa\{\cdot\}$, since this coefficient is the least restrictive among $\rho_{r,\delta}$ defined by formula (2). (Obviously, we have $\alpha\{\mathcal{A},\mathcal{B}\}=\rho_{0,0}\{\mathcal{A},\mathcal{B}\}\leq \rho_{r,\delta}\{\mathcal{A},\mathcal{B}\}$ for all $r,\delta\in[0,1]$). Thus, we consider the coefficients $\alpha_X(r,\mathcal{U},d,\Gamma,\mathfrak{M}_1,\mathfrak{M}_2)$ where d is the Euclidean distance. Let us define $\hat{\alpha}_X(r,\mathcal{U})$ considering that $\mathfrak{M}_1=\mathfrak{M}_2$ consists of halfspaces, corresponding to hyperplanes parallel to coordinate planes (without condition Γ). Including into $\mathfrak{M}_1=\mathfrak{M}_2$ all the halfspaces of $\mathbb{R}^m$ we arrive at the coefficient $\alpha_X(r,\mathcal{U})$, and $\hat{\alpha}(r,\mathcal{U})\leq\alpha(r,\mathcal{U})$ trivially. We also define $\tilde{\alpha}_X(r,\mathcal{U})$ choosing the system $\mathfrak{M}_1$ of all the cubes $K=\prod_{i=1}^m[a_i,a_i+\ell]$ and $\mathfrak{M}_2$ consisting of the closed complements to these cubes. Introducing in definition of $\tilde{\alpha}_X(r,\mathcal{U})$ the additional requirement (condition Γ) to include in $\mathfrak{M}_1$ only the cubes with the edge $\ell\leq q$, we get the coefficient $\tilde{\alpha}(r,q,\mathcal{U})$. Thus, we have $\tilde{\alpha}_X(r,\mathcal{U})=\tilde{\alpha}_X(r,\infty,\mathcal{U})$. Note that for $m=1$ we can use $\hat{\alpha}_X(r,\mathcal{U})=\alpha_X(r,\mathcal{U})$ instead of $\tilde{\alpha}_X(r,\mathcal{U})$.

Next we turn to the study of asymptotic normality of additive random functions $A_n(v)$, $n\in\mathbb{N}$ defined on probability spaces $(\Omega_n,\mathcal{F}_n,P_n)$ and subsets $v\in\mathbb{R}^m$ forming systems $\mathcal{R}_n$. Using here the α-coefficients introduced in the preceding paragraph, we write α_n instead of α_{A_n}. The additivity property of A_n means that

$A_n(v'\cup v'')=A_n(v')+A_n(v'')$, (P_n a.e.),

for all $v', v'' \in \mathcal{R}_n$ such that $v' \cap v'' = \emptyset$ and $v' \cup v'' \in \mathcal{R}_n$
Take, for example, the fields $X_t^{(n)}$, $t \in V_n \subset \mathbb{R}^m$, $n \in \mathbb{N}$ and σ -finite measures ν_n on $(V_n, \mathcal{B}_n)$. For any n we consider the system $\mathcal{R}_n$ of the sets v such that there exists the integral

$$A_n(v) = \int_{v \cap V_n} X_t^{(n)} \, d\nu_n(t) \tag{5}$$

A special case of (5) is the summation scheme

$$A_n(v) = \sum_{t \in v \cap \mathcal{K}_n} X_t^{(n)},$$

where $\mathcal{K}_n$ are denumerable sets in $\mathbb{R}^m$.

The investigation of asymptotic behaviour of additive functions of the intervals in $\mathbb{R}^1$ was initiated by Yu.A. Rozanov [14]. A vast literature is devoted to the problem of asymptotic normality of processes and fields possessing mixing properties (see, e.g.[2,6-10,14-19,23-31]). The rest of this paper is devoted to the examination of asymptotic normality of the centered and normalized variables $A_n(v_n)$ as $v_n \to \infty$ (i.e., when the sets v_n increase infinitely in a certain manner). We obtain wide sufficient conditions for the convergence to the normal law in the uniform metric and also estimate the rate of convergence. These conditions are unimprovable in some aspects. The proofs exploit the method of "partition and separation" traced up to S.N. Bernstein (see, [10]) as well as the method of "local separation" introduced by Ch. Stein in [30] and developed by A.N. Tichomirov [17,18] and other auothors. We do not apply here the technique of moments and cumulants. Concerning this method refer, for example, to [2,29].

It was explained in [6] how under natural assumptions the investigation of asymptotic behaviour of $A_n(v_n)$ for increasing $v_n \subset \mathbb{R}^m$ is reduced to the study of the limit distribution (as $n \to \infty$) of sums of random variables calculated over finite subsets of the lattice $\mathbb{Z}^m$. Therefore further we shall consider the sums

$$A_n(\mathcal{U}) = \sum_{j \in \mathcal{U}} X_j^{(n)} \tag{6}$$

where $X_j^{(n)}$ are arrays of random fields defined on $(\Omega_n, \mathcal{F}_n, P_n)$ for j belonging to finite sets $\mathcal{U}_n \subset \mathbb{Z}^m$ and $\mathcal{U} \subset \mathcal{U}_n$, $n \in \mathbb{N}$. (Of course, we can assume $X_j^{(n)}$ as defined everywhere in $\mathbb{Z}^m$ with $X_j^{(n)} = 0$ for $j \notin \mathcal{U}_n$).

Now suppose that for all $j \in \mathcal{U}_n$ and some $s > 2$

$$E_n X_j^{(n)} = 0 \ , \quad \| X_j^{(n)} \|_{s,n} < \infty , \tag{7}$$

where $E_n X = \int_{\Omega_n} X \, d P_n$ and $\|X\|_{p,n}$ denotes the norm in $L_p(\Omega, \mathcal{F}, P)$, $1 \le p \le \infty$.

We choose $s > 2$ in condition (7), since no rate of strong mixing is sufficient, in general, for the validity of the central limit theorem for the sums of random variables having only second finite moments. This no-go result is true even for stationary in a strict sense sequences of random variables with asymptotically linear groowth of variance of partial sums, see, [23].

For $\mathcal{U} \subset \mathbf{Z}^m$ set

$$B_n^2(\mathcal{U}) = E_n A_n^2(\mathcal{U}) \ , \qquad C_{s,n}(\mathcal{U}) = \max_{j \in \mathcal{U}} \| X_j^{(n)} \|_{s,n} . \tag{8}$$

Examining the asymptotic behaviour of $A_n(\mathcal{U}_n)$ as $n \to \infty$, we shall consider various families of finite sets $\mathcal{U}_n \subset \mathbb{Z}^m$. First we assume that

$$\underline{\lim}_n \ |\mathcal{U}_n|^{-1} B_n^2(\mathcal{U}_n) > 0 , \tag{9}$$

where $|.|$ is the number of elements. Note that some of the results below do not rely on requirement (9).

Let all the parallelepipeds of the form $\Pi = \prod_{i=1}^m (a_i, b_i]$ have $a_i, b_i \in \mathbf{Z}$, $i = 1, \dots, m$. Then $\mathcal{U}_n = \prod_{i=1}^m (a_i^{(n)}, b_i^{(n)}] \cap \mathbb{Z}^m \to \infty$ in our notation is equivalent to $\min_{1 \le i \le m} (b_i^{(n)} - a_i^{(n)}) \to \infty$, $n \to \infty$. Moreover, assume

$$\overline{\lim}_n \ \sup_{I_n} |\Pi|^{-1} B_n^2(\Pi) < \infty , \tag{10}$$

where $I_n = \{ \Pi = \prod_{i=1}^m (a_i, b_i] \cap \mathbb{Z}^m \subset \mathcal{U}_n :$

$b_i - a_i \geq \min\{r_n, b_i^{(n)} - a_i^{(n)}\}, i=1,\dots,m\}$ and $r_n \geq 1$, $n \in \mathbb{N}$ is a numerical sequence.

If the field $X_j^{(n)} = X_j$, $j \in \mathbb{Z}^m$, $n \in \mathbb{N}$ is stationary, then a standard hypothesis (see, [10,19])

$$\sum_j |\operatorname{cov}(X_j, X_0)| < \infty, \quad \sum_j \operatorname{cov}(X_j, X_0) \neq 0$$

implies (9) and (10) for any sequence $r_n \to \infty$.

Defining $\gamma = 2/s$ $(2/\infty = 0)$ and

$$\hat{T}_n^2 = \sum_{i \geq r_n} i^{m-1} \hat{\alpha}_n^{1-\gamma}(i, U_n) + r_n^{(2-\gamma)m} |U_n|^{\gamma-1} \tag{11}$$

we can formulate the following theorem.

Theorem 2. Assume conditions (7), (9) and (10) be valid for some sequence $r_n \geq 1$, $n \in \mathbb{N}$ simultaneously with

$$\max\{\hat{T}_n, C_{s,n}(U_n)\hat{T}_n\} \to 0 \quad \text{as } n \to \infty. \tag{12}$$

It follows that

$$B_n^{-1}(U_n) A_n(U_n) \xrightarrow{\mathcal{D}} N(0,1) \quad \text{as } n \to \infty \tag{13}$$

Note that without loss of generality we can regard $r_n \to \infty$ as $n \to \infty$. In the case $X_j^{(n)} = X_j$ and $U_n \subset U_{n+1}$, $n \in \mathbb{N}$ condition (12) obviously turns into the requirement $C_{s,n}(U_n)\hat{T}_n \to 0$ as $n \to \infty$.

Now suppose that U_n are, in general, distinct from "integer" parallelepipeds considered above. We shall say that $U_n \to \infty$ in a regular fashion if a) $|U_n| \to \infty$ as $n \to \infty$ and b) for all sufficiently large n and some $a > 0$ the following inequality

$$|U_n| \geq a(\operatorname{diam} U_n)^m, \tag{14}$$

holds, where $\operatorname{diam} U_n$ denotes the diameter of U_n. Note that the definition of regular (growth) of sets $U_n \subset \mathbb{Z}^m$ in [25], involving the boundaries U_n is more restrictive. For $U_n \to \infty$ in the sense of Fisher condition (14) is valid.

Let $M_n = \{\Pi = \prod_{i=1}^m (a_i, b_i] \cap \mathbb{Z}^m : \min(b_i - a_i) \geq r_n\}$.

Now we require that the fields should satisfy the condition

$$\overline{\lim_n} \sup_{M_n} |\Pi|^{-1} B_n^2(\mathcal{U}_n \cap \Pi) < \infty , \tag{15}$$

for some sequence $r_n \geq 1$, $n \in \mathbb{N}$ and the corresponding M_n. A wide class of random fields fulfills (15) due to the estimate $B_n^2(U) \leq c|U|$, $c = const$.

Theorem 3. Let $\mathcal{U}_n \to \infty$ in a regular fashion. Assume the validity of (7), (9) and, moreover, assume that conditions (12) and (15) are fulfilled simultaneously for some sequence $r_n \geq 1$, $n \in \mathbb{N}$. It follows that for such random fields and sets $\mathcal{U}_n$ property (13) holds.

Note that the results of [25] can be deduced from this theorem. Next we consider arbitrary finite subsets of $\mathbb{Z}^m$ as $\mathcal{U}_n$, denoting by T_n^2 the right-hand side of (11) after substituting α_n for $\hat{\alpha}_n$. For some sequence $r_n \geq 1, n \in \mathbb{N}$ we define $G_n = \{U \subset \mathcal{U}_n : |U| \geq r_n^m\}$ and introduce a new requirement

$$\overline{\lim_n} \sup_{G_n} |U|^{-1} B_n^2(U) < \infty . \tag{16}$$

Theorem 4. Assume that (7) and (9) are satisfied for $\mathcal{U}_n$ such that $|\mathcal{U}_n| \to \infty$ as $n \to \infty$. Assume also that for some sequence $r_n \geq 1, n \in \mathbb{N}$ condition (16) is fulfilled simultaneously with

$$\max\{T_n, C_{s,n}(\mathcal{U}_n)T_n\} \to 0 \quad \text{as } n \to \infty. \tag{17}$$

It follows that for such fields and sets $\mathcal{U}_n$ property (13) holds.

In [6] Theorems 2-4 appear in the formulation oriented to $\mathcal{U}_n \subset \mathbb{R}^m$. The proofs of these theorems rely on the method of "partition and separation", see,[2,8,10, 24]. Namely, starting from the conditions of the theorems we construct $\mathbb{N}$-valued sequences p_n and q_n such that as

$$p_n^{-1} q_n \to 0, \ p_n^m |\mathcal{U}_n|^{-1} \to 0, \ C_{s,n}^2(\mathcal{U}_n) \sum_{i \geq q_n} i^{m-1} \hat{\alpha}_n^{1-\gamma}(i, \mathcal{U}_n) \to 0,$$

$$p_n^{(2-\gamma)m} |\mathcal{U}_n|^{\gamma-1} C_{s,n}^2(\mathcal{U}_n) \to 0 \ , \ p_n^{-m} |\mathcal{U}_n| \hat{\alpha}_n(q_n, \mathcal{U}_n) \to 0. \quad (18)$$

($\hat{\alpha}_n$ can be replaced by α_n). Further on, every set $\mathcal{U}_n$ is represented as a union of "solid" parts, each containing the number of points of the order p_n^m , and properly separated from each other by the distances of the order q_n . Applying the inequalities to the estimation of covariances of dependent variables and taking into account the conditions, imposed on the behaviour of variances of the functions $A_n(\cdot)$, we show that the sum of $X_j^{(n)}$ occuring in the separating "channels" is negligible, whereas the sums taken over "solid" parts are asymptotically independent and satisfy Lindeberg's condition.

Theorems 2-4 have a number of corollaries but we mention only one of them to illustrate the accuracy of the obtained results. The field $X_j^{(n)}$ will be called $\hat{m}_n$ -dependent on the set $\mathcal{U}_n$ if $\hat{\alpha}_n(\hat{m}_n, \mathcal{U}_n) = 0$.

<u>Corollary 1</u>. Let $\mathcal{U}_n \to \infty$ in $\mathbb{Z}^m$ in a regular fashion. Assume the fields $X_j^{(n)}$ to be $\hat{m}_n$-dependent on $\mathcal{U}_n, n \in \mathbb{N}$ and satisfy conditions (7), (9) and (15) for $r_n = \hat{m}_n$. Assume also that $\varliminf_n C_{s,n}(\mathcal{U}_n) > 0$. It follows that (13) holds provided that

$$C_{s,n}(\mathcal{U}_n) = o\left(|\mathcal{U}_n|^{\frac{1-\gamma}{2}} \hat{m}_n^{m(\frac{\gamma}{2}-1)} \right), \quad (19)$$

as $n \to \infty$.

Modification to arbitrary $m \in \mathbb{N}$ of the example constructed in [24] for $m=1$, shows that $o(\cdot)$ cannot be replaced, in general, by $O(\cdot)$ in condition (19). Indeed, there exist fields and sets $\mathcal{U}_n$ such that the conditions of Corollary 1 are fulfilled with $O(\cdot)$ in (19) but (13) is violated.

For $\mathcal{U} \subset \mathbb{Z}^m$ we set

$$\Delta_n(\mathcal{U}) = \sup_x |P_n(A_n(\mathcal{U}) \leq x B_n(\mathcal{U})) - \Phi(x)|, \quad (20)$$

where $\Phi(x)$ is the distribution function of the standard normal law.

Though other families of sets are feasible, in the next statement we shoose "integer" parallelepipeds $\mathcal{U}_n = \prod_{i=1}^{m} (a_i^{(n)}, b_i^{(n)}] \cap \mathbb{Z}^m$ such that $\min_{1 \le i \le m} (b_i^{(n)} - a_i^{(n)}) \to \infty$ as $n \to \infty$. Let $X_j^{(n)}$ be the fields satisfying (7), (9) and the following conditions with $s \in (2,3]$

$$\overline{\lim_n}\, C_{s,n}(\mathcal{U}_n) < \infty, \quad \overline{\lim_n} \sup_{\mathcal{U} \subset \mathcal{U}_n} |\mathcal{U}|^{-1} B_n^2(\mathcal{U}) < \infty,$$

$$\overline{\lim_n} \sup_{\mathcal{U} \subset \mathcal{U}_n} \| B_n^{-1}(\mathcal{U}) A_n(\mathcal{U}) \|_{s,n} < \infty \tag{21}$$

(sufficient conditions for the validity of (21) are given, e.g. in [8,10]). Assume also that $\hat{\alpha}_n(r, \mathcal{U}_n) \le c_k r^{-k} (1 + \operatorname{diam} \mathcal{U}_n)^\ell$ for all k and some ℓ.

Theorem 5. For fields $X_j^{(n)}$ and sets $\mathcal{U}_n$ described above, the estimate

$$\Delta_n(\mathcal{U}_n) \le C_\lambda |\mathcal{U}_n|^{-\lambda} \tag{22}$$

is true with any λ less that $\frac{s-2}{2}\left(1 + sm\frac{s-2}{s+1}\right)^{-1}$ and some $C_\lambda = \mathrm{const}$.

Thus, for s near to 2 the estimate of the rate of convergence is close to the unimprovable estimate (with the exponent $-(s-2)/2$) established for the systems of independent random variables. The proof of this theorem is analogous to that of the preceding theorems, however, it is necessary here to estimate more exactly the contribution of summands corresponding to separating "channels" accompanied by the application of Esseen's inequality.

An estimate sharper that (22) for the rate of convergence to the normal law can be obtained with the help of the "local separations" method making use of the coefficient $\tilde{\alpha}_n$.

We begin with the study of a single field X_j on $(\Omega, \mathcal{F}, P)$ defined for j in finite $\mathcal{U} \subset \mathbb{Z}^m$, therefore, omitting the index n in all the conditions above. Assume (7) is valid for some $s \in (2,3]$. Then there exist $\vartheta > 0$ and $\mu = \mu(\vartheta, m)$ such that for all $V \subset \mathbb{Z}^m$ we have

$$\| \sum_{j \in V \cap U} X_j \|_s \leq \mu \, C_s(U) \, |V \cap U|^{\vartheta} . \tag{23}$$

To be sure, μ and ϑ depend on the properties of the field X_j though inequality (23) evidently holds with $\mu = \vartheta = 1$. If X_j are independent (or weakly dependent in a proper way) one can put $\vartheta = 1/2$ in (23), see, [13], p. 79.

We exclude henceforth the trivial case $B(U) = 0$ and define $R_s(U) = C_s(U) / B(U)$.

<u>Theorem 6</u>. For the field X_j described above, there exists a constant $c = c(m)$ such that the following estimate is fulfilled for any $k \geq 2$ and every $h \geq 1$ satisfying $(2h+1)^m < |U|$

$$\begin{aligned} \Delta(U) \leq c \{ & |U| R^s L^{s-1} + R L k^{\vartheta(m-1)} + |U| R^2 H + |U| R^2 L k^{\vartheta(m-1)} k 4^k + \\ & + |U|^{1/2} R^2 L k^{-\vartheta(m-1)} \tilde{H} + R (|U| + |U|^{1/2} H^{1/2}) \alpha^{\frac{s-2}{s}} + \\ & + R (|U| k 2^k + |U|^{1/2} \tilde{H} k^{m/2+1} 2^k) \alpha^{\frac{s-1}{s}} \}, \end{aligned} \tag{24}$$

where $R = R_s(U)$, $\alpha = \tilde{\alpha}(h, 2kh, U)$, $H = \sum_{l > h} l^{m-1} \alpha^{\frac{s-2}{s}}(l, 2kh, U)$, $\tilde{H} = h^{m/2} + H^{1/2}$, $L = \mu h^{\vartheta m}$.

We emphasize the generality of estimate (24). In fact, the fields involved are, in general, nonstationary, posessing finite absolute moment of the order $s \in (2,3]$ and finite sets $U \subset \mathbb{Z}^m$ of arbitrary configurations. No assumptions on the relationship between $B^2(U)$ and $|U|$ are made.

For the proof of this theorem formulated in [7], we first of all develop notation. Define

$$Y_j = X_j / B(\mathcal{U}),\ j \in \mathcal{U}, \quad Z = \sum_{j\in\mathcal{U}} Y_j,\ f(t) = E\exp\{itZ\},\ t \in \mathbb{R}.$$

Let $S_j^{(0)} = Z$ for any $j \in \mathcal{U}$ and for $\ell \in \mathbb{N}$, $h > 1$ let

$$S_j^{(\ell)} = \sum_{p \in \mathcal{U}_j^{(\ell)}} Y_p, \quad Z_j^{\ell} = \sum_{p \in \mathcal{U} \setminus \mathcal{U}_j^{(\ell)}} Y_p,$$

here $\mathcal{U}_j^{(\ell)} = \{p \in \mathcal{U}: d_o(p,j) > \ell h\}$, $d_o(p,j) = \max_{1 \le q \le m} |p_q - j_q|$ and $p = (p_1, \dots, p_m)$, $j = (j_1, \dots, j_m)$.

Denote for $r \in \mathbb{N}$ and $t \in \mathbb{R}$

$$\eta_j^{(r)} = \exp\{-itZ_j^{(r)}\}, \quad \xi_j^{(r)} = \exp\{it(S_j^{(r-1)} - S_j^{(r)})\}, \quad \zeta_j^{(r)} = Y_j \prod_{\ell=1}^{r} \xi_j^{(\ell)}.$$

We have $f'(t) = \sum_{j\in\mathcal{U}} EY_j \exp\{itS_j^{(0)}\}$ for all $t \in \mathbb{R}$. Hence the right-hand side of this relation can be transformed identically to obtain

$$f'(t) = A(t) f(t) + B(t) \tag{25}$$

where

$$A(t) = i \sum_{j\in\mathcal{U}} \sum_{r=1}^{k-1} E\zeta_j^{(r)} E\eta_j^{(r+1)},$$

$$B(t) = i\Big[\sum_{j\in\mathcal{U}} EY_j \exp\{itS_j^{(1)}\} + \sum_{j\in\mathcal{U}} E\zeta_j^{(k)} \exp\{itS_j^{(k)}\} +$$

$$+ \sum_{j\in\mathcal{U}} \sum_{r=1}^{k-1} \operatorname{cov}(\zeta_j^{(r)}, \exp\{-itS_j^{(r+1)}\}) + \sum_{j\in\mathcal{U}} \sum_{r=1}^{k-1} E\zeta_j^{(r)} E(\eta_j^{(r+1)} - E\eta_j^{(r+1)}) \exp\{itS_j^{(0)}\}\Big], \tag{26}$$

here $\operatorname{cov}(\xi, \eta) = E\xi\bar{\eta} - E\xi E\bar{\eta}$.

Equation (25) with coefficients (26) extends to an arbitrary $m \in \mathbb{N}$ equation from [15] deduced for $m = 1$.

Denote $V_j^{(\ell)} = \mathcal{U}_j^{(\ell-1)} \setminus \mathcal{U}_j^{(\ell)}$ (assuming $\mathcal{U}_j^{(0)} = \mathcal{U}$ for all $j \in \mathcal{U}$) and for $r \in \mathbb{N}$, $h \ge 1$ define

$$M_r(h) = \max_{j\in\mathcal{U},\, \ell=1,\dots,r} \Big\| \sum_{p \in V_j^{(\ell)}} Y_p \Big\|_s \tag{27}$$

According to (23) for r and h under consideration we have

$$M_r(h) \le c_o \mu R(r^{m-1} h^m)^{\vartheta}, \quad c_o = c_o(m) \le 4^m. \tag{28}$$

Let $\theta(\cdot)$, $\theta_i(\cdot)$, $i=1,2,\dots$ stand for complexvalued functions of the real variable bounded by 1 in absolutevalue, and let c (unindexed) denote some positive constant having, in general, different meanings in different formulas.

<u>Lemma 1.</u> Assume

$$|t| < T = (4 c_0 \mu R (k^{m-1} h^m)^{\vartheta})^{-1}. \tag{29}$$

It follows that $A(t) = -t + c\,\theta(t)(a_0 + a_1|t| + a_2|t|^{s-1} + a_3|t|^2)$ where

$$a_0 = |U| R (\alpha^{\frac{s-2}{s}} + k 2^k \alpha^{\frac{s-1}{s}}), \quad a_1 = |U| R^2 H,$$

$$a_2 = |U| R^s h^{\vartheta m(s-1)} \mu^{s-1}, \quad a_3 = |U| R^3 h^{2\vartheta m} \mu^2.$$

The proof is based on the estimation of the terms in representation (26) for $A(t)$. We have $e^{ix} = 1 + ix + c(s)|x|^{s-1}\theta_1(x)$, $x \in \mathbb{R}$, $c(s) \le 2$, hence

$$i \sum_{j\in U} E\zeta_j^{(1)} = -t - t\sum_{j\in U} EY_j S_j^{(1)} + \theta_2(t)|t|^{s-1} \sum_{j\in U} E|Y_j||S_j^{(0)} - S_j^{(1)}|^{s-1}. \tag{30}$$

Now we note that

$$\Big|\sum_{j\in U} EY_j S_j^{(1)}\Big| = B_n^{-2}(U)\Big|\sum_{j\in U}\sum_{p\in U_j^{(1)}} \operatorname{cov}(X_j, X_p)\Big| \le c|U|R^2 H. \tag{31}$$

We have exploited the estimate of covariances of two random variables (see, [9, 10]) and the fact that $\tilde\alpha(r, q_1, U) \le \tilde\alpha(r, q_2, U)$ for $q_1 \le q_2$ in order to arrive at (31).

Using Hölder's inequality, we obtain

$$E|Y_j||S_j^{(0)} - S_j^{(1)}|^{s-1} \le R\|Z_j^{(1)}\|_s^{s-1} \le R M_1^{s-1}(h). \tag{32}$$

In an analogous way taking into account Liapunov's inequality, we get

$$E|\zeta_j^{(1)}| \le |t| R M_1(h). \tag{33}$$

For $r = 2, 3, \dots$

$$E|\zeta_j^{(r)}| \le \|\Pi'\xi_j^{(\ell)}\|_s \cdot \|Y_j \Pi''\xi_j^{(\ell)}\|_{\frac{s-1}{s}}$$

where Π' (resp. Π'')designate the product in odd (resp. even)indices $\ell \in \{1,\dots,r\}$). Denote by $q = q(r)$ (resp. $Q = Q(r)$) the number of even (resp. odd) numbers among $\{1,\dots,r\}$. Thus, we conclude

$$|E|\Pi'\xi_j^{(\ell)}|^s - \Pi' E|\xi_j^{(\ell)}|^s| \le 16q\,2^{sq}\tilde{\alpha}(h, 2rh, U).$$

With the help of the estimate $\|\xi_j^{(\ell)}\|_s \le |t| M_r(h)$ for $1 \le \ell \le r$ we have

$$\|\Pi'\xi_j^{(\ell)}\|_s \le c[q^{1/s}2^q\alpha^{1/s} + (|t|M_r(h))^q].$$

Similar arguments lead to the estimate

$$\|Y_j \Pi''\xi_j^{(\ell)}\|_{\frac{s}{s-1}} \le cR[Q^{\frac{s-1}{s}}2^Q\alpha^{\frac{s-2}{s}} + (|t|M_r(h))^Q].$$

Taking into account (28), for t satisfying condition (29) and $r = 2,\dots,k$ we have

$$E|\zeta_j^{(r)}| \le cR[(|t|M_r(h))^2 4^{-r} + r2^{-r/2}\alpha^{\frac{s-2}{s}} + r2^r\alpha^{\frac{s-1}{s}}]. \qquad (34)$$

Further on for $r = 1,2,\dots$ we get

$$\|\eta_j^{(r)} - 1\|_s \le |t|\,\|Z_j^{(r)}\|_s \le c|t|\mu R(rh)^{\vartheta m}. \qquad (35)$$

Hence $E|\eta_j^{(2)} - 1| \le c|t|\mu R h^{\vartheta m}$. In view of $E|\eta_j^{(r)}| = 1$ and estimates (28),(29),(33),(34) we deduce

$$\Big|\sum_{j\in U} E\zeta_j^{(1)} E(\eta_j^{(2)} - 1) + \sum_{j\in U}\sum_{r=2}^{k-1} E\zeta_j^{(r)} E\eta_j^{(r+1)}\Big| \le$$

$$\le c|U|R(R^2h^{2\vartheta m}\mu^2t^2 + \alpha^{\frac{s-2}{s}} + k2^k\alpha^{\frac{s-1}{s}}). \qquad (36)$$

Now the statement of Lemma 1 follows from (30)-(32) and (36).

<u>Lemma 2</u>. For t satisfying condition (29) one has $B(t) = c\,\theta(t)(b_0 + b_1t^2)$ where

$$b_0 = R\alpha^{\frac{s-2}{s}}(|\mathcal{U}|k2^{-k/2}+|\mathcal{U}|^{1/2}\tilde{H}) + R\alpha^{\frac{s-1}{s}}k2^k(|\mathcal{U}|+|\mathcal{U}|^{1/2}k^{m/2}\tilde{H}),$$

$$b_1 = R^3\mu^2 h^{2\vartheta m}(|\mathcal{U}|k^{2\vartheta(m-1)}4^{-k}+|\mathcal{U}|^{1/2}\tilde{H}).$$

The proof again exploits representation (26). At first we have

$$\sum_{j\in\mathcal{U}}|EY_j\exp\{itS_j^{(1)}\}| \le c|\mathcal{U}|R\alpha^{\frac{s-1}{s}} \tag{37}$$

From (34) it follows that

$$|\sum_{j\in\mathcal{U}}E\zeta_j^{(k)}\exp\{itS_j^{(k)}\}| \le \sum_{j\in\mathcal{U}}E|\zeta_j^{(k)}| \le c|\mathcal{U}|R[(t\mu R(k^{m-1}h^m)^{\vartheta})^2 4^{-k}+$$

$$+k2^{-k/2}\alpha^{\frac{s-2}{s}}+k2^k\alpha^{\frac{s-1}{s}}]. \tag{38}$$

Next we get

$$\sum_{j\in\mathcal{U}}\sum_{r=1}^{k-1}|\operatorname{cov}(\zeta_j^{(r)},\exp\{-itS_j^{(r+1)}\})| \le$$

$$\le c\sum_{j\in\mathcal{U}}\sum_{r=1}^{k-1}\alpha^{\frac{s-1}{s}}\|\zeta_j^{(r)}\|_s \le c|\mathcal{U}|R2^k\alpha^{\frac{s-1}{s}}. \tag{39}$$

It is easy to see that

$$|\sum_{j\in\mathcal{U}}E\zeta_j^{(r)}E(\eta_j^{(r+1)}-E\eta_j^{(r+1)})\exp\{itS_j^{(0)}\}|^2 \le$$

$$\le\sum_{j,p\in\mathcal{U}}E\zeta_j^{(r)}E\bar{\zeta}_p^{(r)}\operatorname{cov}(\eta_j^{(r+1)},\eta_p^{(r+1)}). \tag{40}$$

In (40) the sum over the set $\{j,p\in\mathcal{U}\}$ is represented as sums over $\{j\in\mathcal{U}, p\in\mathcal{U}_j^{(3rh)}\}$ and $\{j\in\mathcal{U}, p\in\mathcal{U}\setminus\mathcal{U}_j^{(3rh)}\}$. Estimating these sums separately, we get for $r=1,2,\ldots$

$$[\sum_{j,p\in\mathcal{U}}E\zeta_j^{(r)}E\bar{\zeta}_p^{(r)}\operatorname{cov}(\eta_j^{(r+1)},\eta_p^{(r+1)})]^{1/2} \le$$

$$\le c|\mathcal{U}|^{1/2}\max_{j\in\mathcal{U}}|E\zeta_j^{(r)}|\cdot\max_{j\in\mathcal{U}}\|\eta_j^{(r+1)}-1\|_s[(rh)^{m/2}+H^{1/2}].$$

With the help of (34), (35) and taking into account that $|\eta_j^{(r)}|=1$ (for $r=2,3,\ldots$) we attain the following inequality

$$\left|\sum_{j\in\mathcal{U}}\sum_{r=1}^{k-1} E\zeta_j^{(r)} E(\eta_j^{(r+1)} - E\eta_j^{(r+1)}) \exp\{itS_j^{(0)}\}\right| \leq$$

$$\leq c|\mathcal{U}|^{1/2} R(t^2\mu^2 R^2 h^{2-\vartheta m} + \alpha^{\frac{s-2}{s}} + k^{m/2+1} 2^k \alpha^{\frac{s-1}{s}}). \quad (41)$$

Thus, Lemma 2 follows from (37-(39) and (41).

The solution of the linear differential equation (25) has the form

$$f(t) = \exp\left\{\int_0^t A(x)dx\right\} + \int_0^t B(y)\exp\left\{\int_y^t A(x)dx\right\}dy.$$

Therefore with the aid of Lemmas 1, 2 we can estimate for t satisfying condition (29) the proximity of $f(t)$ to the characteristic function $\exp\{-t^2/2\}$ of the standard normal law. Next we apply Esseen's inequality (see, [15], [17]). Consequently, we have $\Delta(\mathcal{U}) \leq c(\sum_{i=0}^{3} a_i + b_0 + b_1 T_1 + T_1^{-1})$, $T_1 = \min\{T, (6a_3)^{-1}, a_0^{-1}, (6a_2)^{-1/(s-2)}\}$ and hence

$$\Delta(\mathcal{U}) \leq c\left(\sum_{i=0}^{3} a_i + b_0 + b_1 T + T^{-1}\right). \quad (42)$$

Inserting the expressions for coefficients a_i, b_j finally we arrive at the desired result.

Making various hypotheses on the behaviour of the coefficient $\tilde{\alpha}$, we deduce the corollaries of the theorem (see, [7]).

Corollary 2. Let the field X_j be ℓ-dependent on the set $\mathcal{U}$ (i.e. $\tilde{\alpha}(\ell, \mathcal{U}) = 0$) and assume $|\mathcal{U}| > (2\ell+1)^m$. Then

$$\Delta(\mathcal{U}) \leq c\,\ell^t \{R^s |\mathcal{U}| + R(\ln|\mathcal{U}|^{(m-1)/2}\},$$

where $t = t(m,s) \leq \max\{2m, (s-1/2)m\}$.

Theorem 6 and Corollary 2 are obviously applicable to the scheme of series, i.e. to the fields $X_j^{(n)}$, $n \in \mathbb{N}$ on spaces $(\Omega_n, \mathcal{F}_n, P_n)$ defined for $j \in \mathcal{U}_n \subset \mathbb{Z}^m$. In order to agjust the formulations it sufficies to label every object involved by the index n, denoting the number of the series.

Corollary 3. Let the fields $X_j^{(n)}$ be defined on the sets $U_n \subset \mathbb{Z}$ and assume that $|U_n| \to \infty$ as $n \to \infty$. Assume also that

$$\tilde{\alpha}_n(r, U_n) \leq \delta\, |U_n|^{\beta} e^{-\lambda r},$$

where $\delta, \beta \geq 0, \lambda > 0$, and assume, in addition, that $s_n = s \in (2, 3], n \in \mathbb{N}$. It follows that for all sufficiently large n

$$\Delta_n(U) \leq c\,\{ \mu_n^{s-1} |U_n| R_n^s (\ln |U_n|)^{m(s-1)\vartheta_n} + \mu_n R_n (\ln |U_n|)^{(2m-1)\vartheta_n} + \mu_n |U_n|^{1/2} R_n^2 (\ln |U_n|)^{\frac{m}{2}+\vartheta_n}\}.$$

where μ_n and ϑ_n are the same as in relation (23) for the fields $X_j^{(n)}$, $\mu_n \geq 1$ being chosen in (23).

In conclusion we remark that the results close to the statements of Corollaries 2, 3 have been independently achieved by J. Sunklodas in [16].

REFERENCES

1. Булинский А.В., Журбенко И.Г. Центральная предельная теорема для аддитивных случайных функций. - Теория вероятн.и ее примен., 1976, XXI, № 4, с. 707-717.
2. Булинский А.В. Предельные теоремы для случайных процессов и полей. - М.:Изд-во МГУ, 1981. - 64 с.
3. Булинский А.В. О мерах зависимости, близких к максимальному коэффициенту корреляции. - ДАН СССР,1984, т. 277, № 6, с. 1296-1298.
4. Булинский А.В. Об условиях перемешивания случайных полей. - Теория вероятн.и ее примен., 1984, XXIX, № 4, с. 512-513.
5. Булинский А.В., Тарасевич-Николаев К.К. Описание зависимости систем случайных величин новыми коэффициентами перемешивания. - В сб.Теория вероятн., теория случайн.процессов и функц.анализ. Изд-во МГУ, 1985, с. 9-12.
6. Булинский А.В. Асимптотическая нормальность случайных полей. - ДАН СССР, 1985, т. 285, с. 1044-1048.
7. Булинский А.В. Оценка скорости сходимости в центральной предельной теореме для случайных полей. - ДАН СССР, 1986.

8. Городецкий В.В. Центральная предельная теорема и принцип инвариантности для слабозависимых случайных полей. - ДАН СССР, 1984, т. 276, № 3, с. 528-531.

9. Давыдов Ю.А. О сходимости распределений, порожденных стационарными случайными процессами. - Теория вероятн.и ее примен., 1968, XIII, № 4, с. 730-737.

10. Ибрагимов И.А., Линник Ю.В. Независимые и стационарно связанные величины. - М.:Наука, 1965. - 524 с.

11. Колмогоров А.Н., Розанов Ю.А. Об условиях сильного перемешивания гауссовского стационарного процесса. - Теория вероятн.и ее примен., 1960, У, № 2, с. 222-227.

12. Леоненко Н.Н., Ядренко М.И. О принципе инвариантности для однородных случайных полей. - Теория вероятн.и ее примен., 1979, XXIУ, № 1, с. 175-181.

13. Петров В.В. Суммы независимых случайных величин. М.:Наука, 1972. - 414 с.

14. Розанов Ю.А. О центральной предельной теореме для аддитивных случайных функций. - Теория вероятн. и ее примен., 1960, У, № 2, с. 243-246.

15. Сунклодас Й. О скорости сходимости в центральной предельной теореме для случайных величин с сильным перемешиванием. - Лит.матем.сб., 1984, т. 24, № 2, с. 174-185.

16. Сунклодас Й. Об оценке скорости сходимости в центральной предельной теореме для слабо зависимых случайных полей. - IУ Междунар.Вильнюсск.конф.по теории вероятн.и матем.статистике. Тез.докладов, 1985, т.Ш, с. 164-166.

17. Тихомиров А.Н. О скорости сходимости в центральной предельной теореме для слабо зависимых величин. - Теория вероятн.и ее примен., 1980, XXУ, № 4, с. 800-818.

18. Тихомиров А.Н. О нормальной аппроксимации сумм векторных случайных полей с перемешиванием. - ДАН СССР, 1983, т. 272, № 2, с. 312-314.

19. Bolthausen E. On the central limit theorem for stationary mixing random fields. - Ann.Probab., 1982, v. 10, No 4, p. 1047-1050.

20. Bradley R.C. Equivalent measures of dependence. - J.Multiv.Anal., 1983, v. 13. p. 167-176.

21. Bradley R.C., Bryc W. Multilinear forms and measures of dependence between random variables. - J. Multiv.Anal., 1985, v. 16, No 3, p. 335-367.

22. Davis R.A. Maxima and minima of stationary sequences.

- Ann.Probab., 1979, v. 7, No 3, p. 453-460.

23. Herrndorf N. An example on the central limit theorem for mixing sequences. - Ann.Probab., 1983, v. 11, No 3, p. 809-813.

24. Herrndorf N. A functional central limit theorem for weakly dependent sequences of random variables. - Ann.Probab., 1984, v. 12, No 1, p. 141-153.

25. Neaderhouser C.C. Limit theorems for multiply indexed mixing random variables, with applications to Gibbs random fields. - Ann.Probab., 1978, v. 6, No 2, p. 207-215.

26. Philip W., Stout W. Almost sure invariance principles for partial sums of weakly dependent random variables. - Mem.Amer.Soc., 1975, No 161.

27. Rosenblatt M. A central limit theorem and a strong mixing condition. - Proc.Nat.Sci.USA, 1956, v. 42, p. 43-47.

28. Schneider E. On the speed of convergence in the random central theorem for -mixing processes. - Z. Wahrscheinlichkeitstheor.verw.Geb., 1981, B. 58, S. 125-138.

29. Statulevičius V. Limit theorems for dependent random variables under various regularity conditions. - Proc.Int.Congr.Math., Vancouver, 1972, v. 2, p. 173-181.

30. Stein Ch. A bound for the error in the normal approximation of a sum of dependent random variables. - Proc. 6th Berkeley Symp. Math.Stat.and Probab., 1972, v. 2, p. 583-602.

31. Takahata H. On the rates in the central limit theorem for weakly dependent random fields. - Z. Wahrscheinlichkeitstheor.verw.Geb., 1983, B. 64, H. 4, S. 445-456.

Prob. Theory and Math. Stat., Vol.1, pp. 327-330
Prohorov et al. (eds)
1986 VNU Science Press

NON COMMUTATIVE INTEGRATION AND PROBABILITY ON VON NEUMANN ALGEBRAS

Carlo Cecchini
Istituto di Matematica dell'Università di Genova - Via L.B.Alberti 4
Genova Italy

The aim of this note is to give an outline of a theory of non commutative L^p spaces with respect to a von Neumann algebra $\mathcal{M}$ and a normal faithful state ω defined on $\mathcal{M}$, and a result on non commutative martingale theory obtained in this framework. For the proofs references is made to the original papers.

Let $\mathcal{M}$ be a von Neumann algebra on which a normal faithful state ω is defined, with commutant $\mathcal{M}'$ and acting on a Hilbert space $\mathcal{H}$.

Let $D(\mathcal{H}, \omega) = \{\xi \in \mathcal{H} : \|a\xi\|^2 < \alpha\, \omega(a^+ a)$ for some $\alpha > 0$ and all $a \in \mathcal{M}\}$

For each $\xi \in D(\mathcal{H}, \omega)$ the operator $R_\omega(\xi)\, a\,\Omega = a\xi$ is a bounded operator from the Hilbert space $\mathcal{H}_\omega$ of the standard representation of $\mathcal{M}$ with respect to ω with cyclic and separating vector ω to $\mathcal{H}$ (see [5]) and $|R_\omega(\xi)|^2$ is in the commutant of the standard representation itself. Using the standard notation, we can associate to each $\xi \in D(\mathcal{H}, \omega)$ an operator $\Theta_\omega(\xi)$ in $\mathcal{M}_+$ by setting

$$\Theta_\omega(\xi) = \pi_\omega^{-1}(J_\omega |R_\omega(\xi)|^2 J_\omega).$$

So, for any $\varphi \in \mathcal{M}_*$ we can define a complex form q_φ on $D(\mathcal{H}, \omega)$ by setting $q_\varphi(\xi) = \varphi(\Theta_\omega(\xi))$.

On the other hand, as the linear span of the set $\{\Theta_\omega(\xi) : \xi \in D(\mathcal{H}, \omega)\}$ is weak operator dense in $\mathcal{M}$, the mapping $\varphi \longrightarrow q_\varphi$ is injective from $\mathcal{M}_*$ to $L(1; \mathcal{M}, \omega) = \{q_\varphi : \varphi \in \mathcal{M}_*\}$. By setting on $L(1; \mathcal{M}, \omega)$ the norm $\|q_\varphi\|_{L(1;\mathcal{M},\omega)} = \|\varphi\|_{\mathcal{M}_*}$ it becomes a Banach space.

We also set, for $a \in \mathcal{M}$, $\xi \in D(\mathcal{H}, \omega)$: $q_a(\xi) = \langle \xi, a\xi \rangle$; define

$L(\infty;\mathcal{M},\omega) = \{q_a : a \in \mathcal{M}\}$, and set $\|q_a\|_{L(\infty;\mathcal{M},\omega)} = \|a\|_{\mathcal{M}}$. It turns out that $L(\infty;\mathcal{M},\omega)$ is a dense subset of $L(1;\mathcal{M},\omega)$ and the inclusion mapping implements the same embedding of $\mathcal{M}$ into $\mathcal{M}_*$ as described by Terp in [8]. We can therefore define the spaces $L(p;\mathcal{M},\omega)$ as the complex interpolation spaces at 1/p relative to the couple $L(\infty;\mathcal{M},\omega)$ and $L(1;\mathcal{M},\omega)$. Those spaces coincide with those defined by Zoletarev [9]. Setting for a closed, densely defined operator T on $\mathcal{H}$ with polar decomposition $T = U|T|$ $\xi \in \mathcal{D}(|T|^{\frac{1}{2}})$, $\eta \in \mathcal{D}(|T^+|^{\frac{1}{2}})$: $v_T(\xi,\eta) = \langle U^+|T|^{\frac{1}{2}}\xi, |T^+|^{\frac{1}{2}}\eta\rangle$, we get the following

Theorem 1. Let ω' be a normal, faithful, semifinite weight on $\mathcal{M}$, $p \in [1,\infty]$ and $d = \frac{d\omega}{d\omega'}$ as defined by Connes in [5]. Then $L(p;\mathcal{M},\omega)$ coincides with the set of quadratic forms $q \in L(1;\mathcal{M},\omega)$ whose restriction to $D(\mathcal{H},\omega) \cap \mathcal{D}(d^{-\frac{1}{2p}})$ is of the type

$$(*) \quad q(\xi) = v_T(d^{-\frac{1}{2p}}\xi, d^{\frac{1}{2p}}\xi)$$

for some $T \in L^p(\mathcal{M},\omega')$ defined by Hilsum in [6]. Moreover the correspondence between forms in $L(p;\mathcal{M},\omega)$ and operators in $L^p(\mathcal{M},\omega')$ given by (*) defines a linear isometry in $L^p(\mathcal{M},\omega')$ onto $L(p;\mathcal{M},\omega)$.

Using the above described correspondence and [6] we can define products between elements of $L(p_1;\mathcal{M},\omega)$ and $L(p_2;\mathcal{M},\omega)$ for $1/p_1+1/p_2=1/r$, $1<p_1,p_2,r<\infty$ which have the usual continuity properties and enable us to implement the usual duality between $L(p_1;\mathcal{M},\omega)$ and $L(p_2;\mathcal{M},\omega)$ when $1/p_1+1/p_2=1$. Unlike the classical cases, however various possible products we can consider between $q_1,q_2 \in L(\infty;\mathcal{M},\omega) = \bigcap_{1<p<\infty} L(p;\mathcal{M},\omega)$ depend on the particular spaces $L(p_1;\mathcal{M},\omega)$ and $L(p_2;\mathcal{M},\omega)$ of which they are seen as elements. This dependence has some regularity properties which are studied in [3]. It is also possible to give explicit Radon-Nikodym type theorems connecting the spaces $L(p;\mathcal{M},\omega_1)$ and $L(p;\mathcal{M},\omega_2)$, when ω_1, ω_2 are two different normal faithful states on $\mathcal{M}$. Those results will be contained in a forthcoming paper by the author.

The ω-conditional expectation studied in [1] can be lifted to a linear contraction at the $L(p;\mathcal{M},\omega)$ level (see [2] and [3]); using this fact and working in the above sketched framework, in [9] D.Petz and the author have proved analogues of classical L^p martingale convergence theorems for martingales in $L(p;\mathcal{M},\omega)$ spaces formed with ω-conditional expectations.

Namely, let $\mathcal{M}$ be a von Neumann algebra on which a normal faithful state ω is defined, and $\mathcal{M}_n$ an increasing sequence of von Neumann subalgebras such that $\mathcal{M}$ is generated by $\bigcup_{n=1}^{\infty}\mathcal{M}_n$. Let $\omega_n = \omega|\mathcal{M}_n$. $\mathcal{E}_n^{\omega}$ the lifting of the ω-conditional expectation from $\mathcal{M}$ to $\mathcal{M}_n$ to a mapping from $L(1;\mathcal{M},\omega)$ to $L(1;\mathcal{M}_n,\omega_n)$ and for $q \in L(1;\mathcal{M}_n,\omega_n)$, $K_n q = q|_{D(\mathcal{H},\omega)}$. Then K_n is a linear contraction from $L(p;\mathcal{M}_n,\omega_n)$ to $L(p;\mathcal{M},\omega)$ for $1 \leq q \leq \infty$. If q_n is a sequence in $L(p;\mathcal{M},\omega)$ it will be called a generalized martingale adapted to the sequence $\mathcal{M}_n$ if $K_1 \circ \mathcal{E}_1^{\omega}(q_n) = q_1$ for $n > 1$. It will be called regular if there is a $q \in L(p;\mathcal{M},\omega)$ such that $q_n = K_n \mathcal{E}_n^{\omega}(q)$.

We get then the following results

<u>Theorem 2.</u> Let $\{q_n\} \subset L(p;\mathcal{M},\omega)$ be a martingale adapted to the sequence $\mathcal{M}_n$. Then the following conditions are equivalent:

I) $\{q_n\}$ is regular

II) $\{q_n\}$ converges in the norm of $L(p;\mathcal{M},\omega)$

III) $\sup \|q_n\|_{L(p;\mathcal{M},\omega)} < \infty$

<u>Theorem 3.</u> Let $\{q_n\} \subset L(1;\mathcal{M},\omega)$ be a martingale adapted to the sequence $\mathcal{M}_n$. Then the following conditions are equivalent

I) $\{q_n\}$ is regular

II) $\{q_n\}$ converges in $L(1;\mathcal{M},\omega)$

III) $\{q_n : n \in \mathbb{N}\}$ is relatively $(L(1), L(\infty))$ compact in $L(1;\mathcal{M},\omega)$

Bibliography

1) Accardi,L.,Cecchini, C. "Conditional expectations in von Neumann algebras and a theorem of Takesaki" J.Funct.Anal. 45 (1982)245-273;

2) Cecchini,C. "Non Commutative integration for states on von Neumann algebras" preprint to be pubblished in J.Op.Th.

3) Cecchini,C. "Non Commutative integration and K.M.S. spaces" Quantum probability and applications Proceedings, Heidelberg 1984 Lecture Notes in Mathematics 1136, Springer Verlag 136-142 (1985)

4) Cecchini,C., Petz,D., "Norm Convergence of generalized martingales in L_p spaces over von Neumann algebras" Acta Sci.Math.48 55-64(1985)

5) Connes, A., "On the spatial theory of von Neumann algebras" J.Funct. Anal. 35 , 153-164 (1980);

6) Hilsum, M., "Les espaces L^p d'une algèbre de von Neumann definies par la derivée spatiale" 40 151-169 (1981);

7) Sherstnev,A.N., "A general theory of measure and integration in von Neumann algebras (in Russian), Matematika 8 20-35 (1982);

8) Terp,M., "Interpolation spaces between a von Neumann algebra and its dual" J.Op.Th. 8 327-360 (1982);

9) Zoletarev,A.A., "L^p spaces on von Neumann algebras and interpolation" (in Russian) Matematika 8 36-43 (1982)

Prob. Theory and Math. Stat., Vol.1, pp. 331-356
Prohorov et al. (eds)
1986 VNU Science Press

ON OPTIMAL CONTROLS IN THE PROBLEM OF LOCALLY ABSOLUTELY CONTINUOUS CHANGE OF MEASURE (COMPACT SETS OF DECISIONS)

R.J.Chitashvili, M.G.Mania
Tbilisi Mathematical Institute,Tbilisi, USSR

Let $(\Omega, \mathcal{F}, F, P)$ be a probability space with the filtration $F=\{\mathcal{F}_t, 0\le t<\infty\}$ satisfying the usual conditions and $P^A=(P^a, a\in A)$ be a family of probability measures on $\mathcal{F}_\infty$ locally equivalent w.r.t. the measure P, i.e. $P_t^a \sim P_t$, $a\in A$, $0\le t<\infty$, where P_t^a, P_t are restrictions of the measures P^a, P on the σ-algebra $\mathcal{F}_t$. Let $M^A=(M^a, a\in A)$ denote a set of local martingales $M^a=(M(t,a), 0\le t<\infty)$ by means of which the elements of the set $R^A=(\rho^a, a\in A)$ of local densities $\rho^a=(\rho_t^a=dP_t^a/dP_t)$ $0\le t<\infty$ are represented as exponential martingales: $\rho^a=\mathcal{E}(M^a)$

The elements of the set A are interpreted as decisions. The introducing of the controls makes it possible for the decision to vary in time according to the accumulation of the information contained in the events from $(\mathcal{F}_t, t\ge 0)$. The set of probabilities corresponding to controls is generated by the operations of bifurcation, closure and convex closure.

A specific operation, which,in fact, determines the structure of the controlled process, is the closure of the class P^A with respect to bifurcation.

<u>Definition 1</u>. The class Q of probabilities on $\mathcal{F}_\infty$ is called closed with respect to bifurcation if for every elements P^1, P^2, P^3 from Q, every time moment $0\le t<\infty$ and every set $B\in\mathcal{F}_t$ there exists a proba-

bility $P^4 \in Q$ such that

$$P^4_t = P^1_t,\ P^4(c|\mathcal{F}_t)I_B = P^2(c|\mathcal{F}_t)I_B,\ c \in \mathcal{F}_\infty,\quad P^2 - \text{a.s.},$$

$$P^4(c|\mathcal{F}_t)I_{B^c} = P^3(c|\mathcal{F}_t)I_{B^c},\ c \in \mathcal{F}_\infty,\quad P^3 - \text{a.s.},$$

Let U_d denote a class of piecewise constant processes taking values in A such that for $u \in U_d$ there exist the (deterministic) moments $0 \le t_1 \le \dots \le t_n < \infty$, the sets $B_{ij} \in \mathcal{F}_{t_i}$ and the elements $a_{ij} \in A$ $(1 \le i \le n,$ $1 \le j \le m_n)$ such that

$$u_t(w) = a_{ij} \quad \text{for } t_i \le t < t_{i+1},\ w \in B_{ij}.$$

Lemma 1 ([1]). The minimal class of probabilities closed with respect to bifurcation and containing the class P^A is the set P_d^U of probabilities P^u, $u \in U_d$ and each of them is defined by the local density $\rho^u = \mathcal{E}(M^u)$ where

$$M_t^u = \sum_i [M(t \wedge t_{i+1}, u_{t_i}) - M(t \wedge t_i, u_{t_i})] \qquad (1)$$

(the symbol $\wedge$ denotes min).

In the paper [1] we describe a number of closures of the class $P_d^U = (P^u, u \in U_d)$ $(M_d^U = (M^u, u \in U_d)$, $R_d^U = (\rho^u, u \in U_d$ in the case of a finite set A and consider the basic aspects of the control problem (optimality principle, Bellman's equation, etc.), which were discussed in various investigations connected with the problem of optimal absolutely continuous change of measure.

In the present work some results of [1] are generalized for the case of compact sets of decisions A with a special attention to the notion of a stochastic curvilinear integral. This notion enables us to define the local martingales M^u for the processes more general then in [1].

Suppose that the following conditions are satisfied:

A) $(A, \mathcal{A})$ is a separable metric space with the metric τ and the Borel σ-algebra $\mathcal{A}$.

B) For every $a \in A$ a martingale M^a is locally square integrable: $M^a \in \mathcal{M}^2_{loc}(P)$.

C) There exists a predictable increasing locally integrable process $K(K \in \mathcal{A}^+_{loc})$ dominating the square characteristics of M^a, $a \in A$:

$$\langle M^a \rangle << K, \quad a \in A$$

d) The derivative $\varphi(a,b) = d\langle M^a, M^b\rangle / dK$ is a continuous function of (a,b) for almost every (ω, t) with respect to the measure μ^K where μ^K is the Dolean's measure of the process K (i.e. the measure defined on the predictable sets $C \subset \Omega \times [0,\infty[$, $C \in \mathcal{P}$, by the formula

$$\mu^K(C) = E\, I_C \circ K_\infty$$

Here and in the following we use the notation $X \circ K = \left(\int_0^t X_s\, dK_s\right)_{0 \le t < \infty}$ $\left(X \cdot M = \left(\int_0^t X_s\, dM_s\right)_{0 \le t < \infty}\right)$ for the Lebesgue-Stieltjes integrals (stochastic integrals with respect to martingales).

Let us introduce the classes $\mathcal{U}(\mathcal{U}_g)$ of all predictable processes taking values in A (in the space $Q(A)$ of all probabilities on A). It is convenient to denote the elements of $\mathcal{U}$, $\mathcal{U}_g$ by the identical symbols since every element $u \in \mathcal{U}$ may be considered as an element (degenerated) of $\mathcal{U}_g$. So we shall use the shortened notations: $u = (u_t)_{0 \le t < \infty} = (u_t(\omega))_{0 \le t < \infty}$ for $u \in \mathcal{U}$ and $u = u(c) = (u_t)_{0 \le t < \infty} = (u_t(\omega))_{0 \le t < \infty} = (u_t(\omega, c))_{0 \le t < \infty}$, $c \in \mathcal{A}$ for $u \in \mathcal{U}_g$.

§1. A CONSTRUCTION OF A STOCHASTIC CURVILINEAR INTEGRAL ON THE CLASS $\mathcal{V}$.

The stochastic curvilinear integral along the curve u with respect to the class $M^A = (M^\alpha, \alpha \in A)$ will be symbolically written as

$$M_t^u = \int_0^t M(ds, u_s)$$

and will be constructed analogously to the construction of a linear integral (see, e.g. Kunita and Watanabe [2]).

Let $H^2 = \{M \in \mathcal{M}_{loc} : \|M\|_2 = E^{1/2}(\sup_t |M_t|)^2 < \infty\}$ and let $\mathcal{L}^2(M^A)$ denote the minimal stable subspace H^2 containing the stochastic integrals $X \cdot M^\alpha$, where $X \in L^2(\langle M^\alpha \rangle) = \{f \in \mathcal{P} : E f^2 \circ \langle M^\alpha \rangle_\infty < \infty\}$. (The subspace H of the space H^2 is called stable if H is a linear closed subspace and $I_c M^\tau \in H$ for all $M \in H$, $c \in \mathcal{F}_c$ and stopping times τ [2].) We define $\mathcal{L}^2_{loc}(M^A)$ as a set of elements $m \in \mathcal{M}^2_{loc}$ for which there exists a localizing sequence of stopping times $(\tau_n, n \geq 1)$ such that $m^{\tau_n} = (m_{t \wedge \tau_n}, c \leq t < \infty) \in \mathcal{L}^2(M^A)$ for any $n \geq 1$.

Condition C) and the definition of $\mathcal{L}^2(M^A)$ imply that $\langle m \rangle << K$ for every element m of $\mathcal{L}^2(M^A)$. Hence, for the mutual characteristics we have

$$\langle M^\alpha, m \rangle << \langle m \rangle << K, \langle M^\alpha, m \rangle << \langle M^\alpha \rangle << K.$$

Denote by $H(\alpha, m)$ and $f'(\alpha, m)$ the modifications of the derivatives

$$d\langle M^\alpha, m \rangle / dK, \quad d\langle M^\alpha, m \rangle / d\langle m \rangle$$

which are μ^K-a.e. (and $\mu^{\langle m \rangle}$-a.e., respectively) continuous with respect to $a \in A$. Denote also

$$g(a,b) = \varphi(a,a) - 2\varphi(a,b) + \varphi(b,b) \quad , a, b \in A.$$

The symbol $f(u,m)$ for $u \in \mathcal{U}$ is used instead of the exact notation $(f_t(\omega, u_t, m), 0 \le t < \infty)$; the symbels $H(u,m)$, $\varphi(u,m)$, $g(u,m)$ are understood in a similar way.

Let

$$\mathcal{U}(K) = \{u \in \mathcal{U} : \varphi(u,u) \circ K_\infty \in \mathcal{A}^+\},$$

$$\mathcal{U}_{loc}(K) = \{u \in \mathcal{U} : \varphi(u,u) \circ K \in \mathcal{A}^+_{loc}\}.$$

<u>Theorem 1</u>. For every $u \in \mathcal{U}(K)$ there exists an unique element $M^u \in \mathcal{X}^2(M^A)$ (which will be called a stochastic curvilinear integral along the curve u with respect to the system M^A) such that for every $m \in \mathcal{X}^2(M^A)$

$$E\langle M^u, m\rangle_\infty = E f(u,m) \circ \langle m\rangle_\infty . \qquad (2)$$

Moreover,

$$\langle M^u\rangle = \varphi(u,u) \circ K = \sup_m f^2(u,m) \circ \langle m\rangle . \qquad (3)$$

The proof of the theorem is based on the following lemmas.

<u>Lemma 2</u>. For every $B \in \mathcal{P}$

a) $I_B \circ \langle M^a\rangle_t = \sup_m I_B f^2(a,m) \circ \langle m\rangle_t ,$

b) $I_B \circ \langle M^a - M^b\rangle = \sup_m I_B [f(a,m) - f(b,m)]^2 \circ \langle m\rangle$

Proof. Since for any $B \in \mathcal{P}$ the martingales $I_B \cdot M^a - I_B f(a,m) \cdot m$ and m are orthogonal we have

$$I_B \circ \langle M^a\rangle = \langle I_B \cdot M^a - I_B f(a,m) \cdot m\rangle + f^2(a,m) I_B \circ \langle m\rangle,$$

hence,

$$I_B \circ \langle M^a\rangle_t \geq \sup_m I_B f^2(a,m) \circ \langle m\rangle_t$$

On the other hand, substituting the martingales $m = I_B \cdot M^a$ in the equality $\langle m, M^a\rangle = f(a,m)\circ\langle m\rangle$ we obtain

$$I_B\circ\langle M^a\rangle = I_B f(a, I_B\cdot M^a)\circ\langle M^a\rangle = I_B f^2(a, M^a)\circ\langle M^a\rangle$$

which implies a). Equality b) is proved in a simiàar way.

Note that a) implies the equality

$$f(a,a) = \sup_{m\in\mathcal{L}^2(M^A)} f^2(a,m)\frac{d\langle m\rangle}{dK} \qquad \mu^K\text{- a.e.} \quad (4)$$

Lemma 3. For every $m\in\mathcal{L}^2(M^A)$ there exist a μ^K-a.e. continuous modification of the function $H(a,m)$ and $\mu^{\langle m\rangle}$-a.e. continuous modification of the function $f(a,m)$.

Proof. The proof of Lemma 1 shows that $\sup$ in the expressions a), b) is obtained when $\langle m\rangle \ll K$ and since all element of $\mathcal{L}^2(M^A)$ are characterized by this property we have for every $m\in\mathcal{L}^2(M^A)$

$$g(a,b) = f(a,a) - 2f(a,b) + f(b,b) \geq [f(a,m) - f(b,m)]^2\frac{d\langle m\rangle}{dK}, \ \mu^K\text{- a.s.} \quad (5)$$

Besides, for every $m\in\mathcal{L}^2(M^A)$

$$H(a,m) - H(b,m) = [f(a,m) - f(b,m)]\frac{d\langle m\rangle}{dK} \qquad \mu^K\text{- a.s.} \quad (6)$$

Taking into account that $\mu^{\langle m\rangle}\{(\omega,t): d\langle m\rangle/dK = 0\} = 0$ we see that under the condition D) relations (5) and (6) imply the existence of $\mu^{\langle m\rangle}$-a.e. (μ^K-a.e.) continuous modification of the function $f(a,m)$ $(H(a,m))$ for any $m\in\mathcal{L}^2(M^A)$ □

Proof of the Theorem 1. Consider the mapping

$$m \to C^u(m) = Ef(u,m)\circ\langle m\rangle_\infty = EH(u,m)\circ K_\infty$$

Let us show that $C^u(m)$ is a linear bounded functional on $\mathcal{L}^2(M^A)$.

Let $m, N \in \mathcal{L}^2(M^A)$. The linearity of the functional $C^a(m), a \in A$, immediately leads to

$$H(a, m+N) = H(a, m) + H(a, N) \qquad \mu^K\text{- a.e.}$$

and the continuity of $H(a,m)$ with respect to a for all $u \in U(K)$ gives

$$H_s(u_s, m+N) = H_s(u_s, m) + H_s(u_s, N) \qquad \mu^K\text{- a.e.}$$

which implies the linearity of $C^u(m)$.

Applying the Kunita-Watanabe's inequality with (4) in mind we have

$$C^u(m) = Ef(u,m)\circ\langle m\rangle_\infty \leq E^{1/2} f^2(u,m)\circ\langle m\rangle_\infty E^{1/2}\langle m\rangle_\infty \leq$$

$$\leq E^{1/2}\Big(\sup_{m\in\mathcal{L}^2(M^A)} f^2(u,m)\frac{d\langle m\rangle}{dK}\Big)\circ K_\infty E^{1/2}\langle m\rangle_\infty =$$

$$= E^{1/2} f(u,u)\circ K_\infty \| m\|_2$$

Consequently, $C^u(m)$ is a continuous functional on $\mathcal{L}^2(M^A)$ and by the Hilbert space properties there exists an unique element $M^u \in \mathcal{L}^2(M^A)$ such that for any $m \in \mathcal{L}^2(M^A)$ we have

$$E\langle M^u, m\rangle_\infty = Ef(u,m)\circ\langle m\rangle_\infty$$

The equality $\varphi(u,u)\circ K = \sup_m f^2(u,m)\circ\langle m\rangle$ follows from (4), and the equality $\langle M^u\rangle = \sup_m f^2(u,m)\circ\langle m\rangle$ is proved similarly to Lemma 1. □

The following assertion can be immediately verified.

<u>Proposition 1</u>. For any $u \in U(K)$ and stopping time τ

$$(M^u)^\tau = M^{u,\tau}$$

where $(\tau_n, n\geq 1)$ is a stochastic curvilinear integral with respect to a system of stopped martingales

$$M^{a,\tau}=\{M^a_{t\wedge\tau}, t\geq 0\}, a\in A$$

Hence we can define the stochastic curvilinear integral for every $u\in U_{loc}(K)$ as an unique element $M^u\in \mathcal{L}^2_{loc}(M^A)$ such that for all

$$(M^u)^{\tau_n}=M^{u,\tau_n}$$

where $\{\tau_n, n\geq 1)$ is a localized sequence of stopping times for $f(u,u)\circ K\in A^+_{loc}$

It is easy to show that in this case the equality

$$\langle M^u, m\rangle = f(u,m)\circ\langle m\rangle = H(u,m)\circ K$$

holds for all $m\in \mathcal{L}^2_{loc}(M^A)$.

Remark. The continuity of $\varphi(a,b)=d\langle M^a,M^b\rangle/dK$ enables us to define the expression $C^u(m)=Ef(u,m)\circ\langle m\rangle$ (and, hence, the element M^u) uniquely. Otherwise, depending on the choice of the modification of $f(a,m)$ the functional $C^u(m)$ may vary and M^u is defined ambiguously. Besides, condition D) provided the continuity of the integral M^u with respect to the curve u .

To simplify the subsequent discussions we introduce the boundedness condition

E) $\sup_a \varphi(a,a)\circ K_t<\infty$ P -a.s. $t\geq 0$

Under this condition there exists a sequence of stopping times $(\tau_n, n\geq 1)$ such that

$$\sup_a \varphi(a,a)\circ K_{\tau_n}\leq n \tag{7}$$

The definition of the curvilinear integral implies that under condition E) M^u is defined for all $u\in U$ and $\langle M^u\rangle_{\tau_n}\leq n$ for all $u\in U$. Besides, since $\langle M^u-M^v\rangle = g(u,v)\circ K$ $(g(a,b)=\varphi(a,a)-2\varphi(a,b)+\varphi(b,b))$ the continuity of $\varphi(a,b)$ implies that for any sequence $(u_i, i\geq 1)$

with $u_i \to u$ a.e. we have

$$E\langle M^{u^i} - M^u\rangle_{T_n} \to 0,\ n \geq 1$$

Furthermore, the continuity condition enables us to give a constructive definition of the curvilinear integral.

<u>Corollary 1</u>. The stochastic curvilinear integral is an unique element of $\mathcal{L}^2(M^A)$ having the following properties:

1) $I_{B\cap(u=a)} \cdot M^u = I_{B\cap(u=a)} \cdot M^a$ for any $B \in \mathcal{P},\ a \in A$;

2) $u_i \to u$ μ_K-a.e. implies $E\langle M^{u_i} - M^u\rangle_{T_n} \to 0, n \geq 1.$

Indeed, let us show the if $\tilde{M}^u$ does satisfy the condition of the corollary, then $\tilde{M}^u = M^u$.

On the class U_0 of countably-valued processes u taking values in the countable dense in A subset the martingale $\tilde{M}^u$ is uniquely defined by property 1)

$$\tilde{M}^u = \sum_{a\in A_0} I_{[u=a]} \cdot M^a,\quad u \in U_0 \tag{8}$$

It is immediately verified that satisfies the relation

$$\langle \tilde{M}^u, m\rangle = H(u,m)\circ K,\quad m \in \mathcal{L}^2(M^A), \tag{9}$$

which implies that $\tilde{M}^u = M^u$ for all $u \in U_0$.

On the other hand, by virtue of the separability of A for any $u \in U$ there exists a sequence $(u^i, i \geq 1) \in U_0$ such that $u^i \to u,\ i \to \infty$. In particular, the sequence $(u^i, i \geq 1) \in U_0$ may be constructed in the following manner.

Obviously, $\inf_j r(a_j, u_t(\omega)) = 0$ where $A_0 = (a_1, a_2, \dots)$. Hence, for a sequence u^i such that

$u_t^i(\omega) = a_{K_i(\omega,t)}$, where $K_i(\omega,t) = \inf(j : r(a_j, u_t(\omega)) \le \frac{1}{i})$,

the inequality $r(u_t^i(\omega), u_t(\omega)) \le \frac{1}{i}$ holds.

Hence, by condition 2) $E\langle M^{u^i} - M^u \rangle_{t_n} \to 0,\ n \ge 1$ and passing to the limit in equality (9) for $u = u^i$ we have

$$\langle m, \hat{M}^u \rangle = H(u,m) \circ K,\ m \in \mathcal{L}^2(M^A) \Rightarrow \tilde{M}^u = M^u \ \square$$

Inversely, the continuity condition D) appears to be necessary for the curvilinear integral to be continuous and uniquely defined.

Corollary 2. Let the conditions A), B), C), E) hold and let for every $u \in U$ there exists an unique element $M^u \in \mathcal{L}^2(M^A)$ satisfying conditions 1), 2) of Corollary 1. Then condition D) is satisfied.

Proof. Let a set $A_0 = (a_1, a_2, \dots)$ be dense in A . The function

$$g^*(a,b) = \lim_m \sup_{i:\, r(a_i,a) \le \frac{1}{m}} \ \sup_{j:\, r(a_j,b) \le \frac{1}{m}} g(a_i, a_j)$$

(where $g(a,b)$ is some separable modification of the derivative $d\langle M^a - M^b\rangle / dK$) is upper continuous a.e. with respect to .

It is easy to see that a.e. continuous modification of $g(a,b)$ exists iff

$$g_t^*(u_t, u_t),\ _{0 \le t < \infty} = 0 \qquad \mu^K\text{- a.e. for all } u \in U$$

Let there exists $u \in U$ such that

$$\mu^K\{(\omega,t) : g^*(u,u) > 0\} > 0$$

We can construct two sequences $u^m, v^m \in U_0$ such that

$u^m \to u,\ v^m \to u$ and $g(u^m, v^m) \to g^*(u,u)$ μ^K-a.e.

Put $u_t^m(\omega)=a_{i_m}(\omega,t)$, $v_t^m(\omega)=a_{j_m}(\omega,t)$ where the indices $i_m(\omega,t)$ and $j_m(\omega,t)$ are selected so that

$$r(a_{i_m(\omega,t)}, u_t(\omega)) \leq \frac{1}{m}, \; r(a_{j_m(\omega,t)}, u_t(\omega)) \leq \frac{1}{m}$$

and

$$g(a_{i_m(\omega,t)}, a_{j_m(\omega,t)}) \geq \sup_{i: r(a_i, u_t(\omega)) \leq \frac{1}{m}} \; \sup_{j: r(a_j, u_t(\omega)) \leq \frac{1}{m}} g(a_i, a_j) - \frac{1}{m}.$$

Now we have

$$\underline{\lim}_m E\langle M^{u^m} - M^{v^m}\rangle_{\tau_n} = \underline{\lim}_m Eg(u^m, v^m)\circ K_{\tau_n} \geq Eg^*(u,u)\circ K_{\tau_n} > 0$$

for some $n \geq 1$ which implies that the continuity condition for M^u with respect to u does not hold.

<u>Remark</u>. For the elements of U_0 the formula (8) implies immediately that for any moment τ

$$\Delta M_\tau^u = M(\tau, u_\tau) - M(\tau-, u_\tau) \qquad (10)$$

However, for the processes u of the general type the continuity condition D) does not provide the property (10), and it may be asserted only that there exists a sequence $(u^i, i \geq 1) \in U_c$ with $u^i \to u$ such that

$$\Delta M^u = \lim_{n\to\infty} [M(\tau, u_\tau^i) - M(\tau-, u_\tau^i)]$$

Property (10) can be guaranteed by an additional requirement of the continuity of ΔM^a with respect to a μ^c-a.e.

2. STOCHASTIC CURVILINEAR INTEGRAL ON THE CLASS U_g^r

Let $G(A)$ denote a family of all probability measures

on the Borel σ-algebra $\mathcal{A}$ of the subsets A. Assume that $Q(A)$ is provided with the Prokhorov's metric

$$r^Q(q,q') = \inf\{\varepsilon > 0 : q(c) \le q'(C^\varepsilon) + \varepsilon,\ q'(C) \le q(C^\varepsilon) + \varepsilon\},$$
$$q, q' \in Q(A),\ C^\varepsilon = \{a,\ r(a,c) < \varepsilon\},\ C \in \mathcal{A}$$

Let U_q (a class of generalized curves) be a class of predictable mappings of $\Omega \times R_+$ into $Q(A)$.

Suppose that conditions A)-E) hold.

At first we shall define the generalized stochastic curvilinear integral when $u \in U_q$ is independent of (ω,t) and $E \sup_a \varphi(a,a) \circ K_\infty < \infty$

Let $q \in Q(A)$. Consider the functional

$$m \to C^q(m) = E \int_A \langle M^q, m \rangle q(da)$$

By Fubini's theorem

$$C^q(m) = E\int_0^\infty \left(\int_A f(a,m) q(da)\right) d\langle m\rangle = E\int_0^\infty \left(\int_A H(a,m) q(da)\right) dK$$

The linearity of $C^q(m)$ is evident. Now we show that $C^q(m)$ is a bounded functional. Indeed,

$$C^q(m) = E\int_0^\infty \left(\int_A f(a,m) q(da)\right) d\langle m\rangle \le E^{1/2}\int_0^\infty \left(\int_A f(a,m) q(da)\right)^2 d\langle m\rangle \times$$
$$E^{1/2}\langle m\rangle_\infty \le E^{1/2}\int_A \left(\sup_M \int_0^\infty f^2(a,m)\, d\langle m\rangle\right) q(da)\, E^{1/2}\langle m\rangle_\infty \le$$
$$\le E^{1/2}\int_0^\infty \sup_a \varphi(a,a)\, dK\ \|m\|_2 . \qquad (11)$$

We define M^q as a unique element of $\mathcal{L}^2(M^A)$ for which

$$E\langle M^q, m\rangle_\infty = E\int_A \langle M^q, m\rangle q(da),\ m \in \mathcal{L}^2(M^A)$$

$$(\text{or } \langle M^q, m\rangle = \int_A \langle M^a, m\rangle q(da), m \in \mathcal{L}^2(M^A)) \qquad (12)$$

Let us find $\langle M^q, M^{q'}\rangle$ for $q, q' \in Q(A)$. By (12) we have

$$\langle M^q, M^{q'}\rangle = \int_A \langle M^a, M^{q'}\rangle q(da)$$

and

$$\langle M^q, M^{q'}\rangle = \int_A \langle M^a, M^{q'}\rangle q(da) = \int_A\int_A \langle M^a, M^b\rangle q'(db) q(da)$$

$$= \left(\int_A\int_A \varphi(a,b) q(da) q'(db)\right) \circ K$$

Denote $\varphi(q,q') = \int_A\int_A \varphi(a,b) q(da) q'(da)$

For convenience we preserve the notation for the expression $\int_A H_t(a,m) u_t(da)$ and in the case of generalized curves the notations $f(u,m)$, $f(u,v)$, $g(u,v)$, $u,v \in U_g$, will be understood in a similar way.

It is evident that the assumptions A)-E) guarantee that the following conditions hold:

A') $(Q(A), r^Q)$ is a separable metric space

B') For every $q \in Q(A)$ $M^q \in \mathcal{M}^2_{loc}(P)$

C') $\langle M^q\rangle \ll K$, $q \in Q(A)$

D') The derivative $\varphi(q,q') = d\langle M^q, M^{q'}\rangle / dK$ is continuous with respect to the couple (q,q') μ^K-a.e.

E') $\sup_q \varphi(q,q) \circ K_t \le \sup_a \varphi(a,a) \circ K_t < \infty$ P - a.s., $t \ge 0$

Therefore we may define the stochastic curvilinear integral for the generalized curves $u \in U_g$ similarly to the case $u \in U$) as an unique element of $\mathcal{L}^2_{loc}(M^A)$ for which

$$\langle M^u, m\rangle_t = \int_0^t \left(\int_A f_s(a,m) u_s(da)\right) d\langle m\rangle_s \equiv f(u,m) \circ \langle m\rangle$$

3. CLOSURES OF THE CLASS M_d^U

The stochastic curvilinear integral on the class which was constructed in the above paragraphs will be used here for the description of the strong, weak and convex closures of the class M_d^U .

Consider the sequence of moments $(\tau_n, n \geq 1)$ such that

$$\sup_a \varphi(a,a) \circ K^{\cdot}_{\tau_n} \leq n$$

Let $H^2_{loc}(A)$ denote a subspace $\mathcal{M}^2_{loc}$ of elements m localized by the sequence $(\tau_n, n \geq 1)$ and note that $H^2_{loc}(A)$ is a countably normed Hilbert space with a sequence of semi-norms (for $m \in H^2_{loc}(A)$)

$$\|m\|_n = E^{1/2} m^2_{\tau_n} = E^{1/2} \langle m \rangle_{\tau_n}, \quad n \geq 1$$

(Note that $\mathcal{L}^2_{loc}(M^A) \subset H^2_{loc}(A)$.)

Further consideration will be made under the strong condition.

A*) The set A is compact. Besides, we assume that the σ -algebra $\mathcal{F}_\infty$ is countably generated, which guarantees metrizability of the weak topology in $H^2_{loc}(A)$

For a predictable set $B \in \mathcal{P}$ we denote by $U_g(B)$ a class of predictable processes with values in A on the set B and in $Q(A)$ on B^c . We denote also by $U_g^f(B)$ the subsets of the elements of $U_g(B)$ for which a finite subset $A_1 \subset A$ exists such that $u_t(\omega) \in A_1$ if $(\omega,t) \in B$ and $u_t(\omega, A_1) = 1$ if $(\omega,t) \in B^c$.

<u>Lemma 4</u>.For every element of $u \in U_g(B)$ there exists

a sequence $(u^m, m \geq 1)$ of elements $U_g^f(B)$ such that

$$\|M^{u^m} - M^u\|_n \to 0,\ m \to \infty \quad \text{for all } n \geq 1.$$

Proof. The compactness of A implies that $Q(A)$ is compact and that there exists a finite ε -net consisting of the elements $q \in Q(A)$ with support in a finite subset of A. Besides, the functions $g(a,b) = \varphi(a,a) - 2\varphi(a,b) + \varphi(b,b)$ and $g(q,q') = \varphi(q,q) - 2\varphi(q,q') + \varphi(q',q')$ (where $\varphi(q,q') = \iint \varphi(a,b) q(da) q'(db)$) are uniformly continuous on A^{A} and $Q(A)$ a.e. with respect to the measure μ^k. This implies that the sequence

$$h_\varepsilon = \sup_{a,b:\, r(a,b) \leq \varepsilon} g(a,b) I_B + \sup_{q,q':\, r^Q(q,q') \leq \varepsilon} g(q,q') I_{B^c},\ \varepsilon > 0.$$

converges to zero as $\varepsilon \to 0$ a.e. with respect to the measure μ^K. Let u be an element of $U_g(B)$ and $A_\varepsilon, Q_\varepsilon(A)$ be the ε -nets of sets A and $Q(A)$, respectively. Let us define the element $u^\varepsilon \in U_g^f(B)$ by the relations

$$u_t^\varepsilon(\omega) = a_{i_\varepsilon(\omega)},\ \omega \in B;\ u_t^\varepsilon(\omega) = q_{j_\varepsilon(\omega)},\ \omega \in B^c,$$

where

$$i_\varepsilon(\omega) = \min\{k : r(a_k, u_t(\omega)) \leq \varepsilon\},\ A_\varepsilon = (a_1, \dots, a_{k_\varepsilon}),$$

$$j_\varepsilon(\omega) = \min\{k : r^Q(q_k, u_t(\omega,a)) \leq \varepsilon\},\ Q_\varepsilon = (q_1, \dots, q_{l_\varepsilon})$$

It is evident from the construction that $g(u^\varepsilon, u) \leq h_\varepsilon$ Hence

$$\|M^{u^\varepsilon} - M^u\|_n^2 = E\langle M^{u^\varepsilon} - M^u\rangle_{\tau_n} = E g(u^\varepsilon, u) \circ K_{\tau_n} \leq$$
$$\leq E h_\varepsilon \circ K_{\tau_n}$$

implies that $u^m = u^{\frac{1}{m}}$ satisfies the condition of the lemma.

Theorem 2. a) The class $M_g^U(B)=\{M^u, u\in U_g(B)\}$ is strongly closed.

b) The class M^u is a strong closure of the class M_a^U

c) The class M_g^U is a convex closure of the class M_d^U

d) The class $M_g^U(B^k)$ where B^k is a union of all atoms of the measure μ^k is a weak closure of the class M_d^U

Proof. Let $(u^i, i\geq 1)\in U_g(B)$ be a sequence such that M^{u^i} converges strongly to some element $M\in \mathcal{L}^2_{loc}(M^A)\subset H^2_{loc}(A)$. Since

$$\|M^{u^i}-M^{u^j}\|_n^2 = Eg(u^i,u^j)\circ K_{\tau_n} = g(n^i;n^j)\, I_{[[0,\tau_n]]}\circ \mu^k$$

we can choose a subsequence i' such that a.e. with respect to the measure μ^k

$$g(u^{i'}, u^{j'})\to 0,\quad i'\to\infty,\ j'\to\infty \tag{13}$$

It is convenient to preserve the notation $(u^i, i\geq 1)$ for the selected subsequence. We shall need an assertion from the following elementary lemma proved in [3].

Lemma 5. Let $f_n(x)$ be a sequence of measurable mappings of $(X, \mathcal{B})$ into a compact space Y . Then there exists a sequence $n_j(x), j\geq 1$, of measurable mappings of X into $N=\{1,2,\dots\}$ such that $n_j(x)\leq n_{j+1}(x), j\geq 1, \lim_{j\to\infty} n_j(x)=\infty$ and the sequence $(f_{n_j(x)}(x), j\geq 1)$ converges for all $x\in X$.

Let $i_j(\omega,t), j\geq 1$, be a sequence which exists for $(u^i, j\geq 1)$ by the lemma and $\hat{u}^j=(u_t^{i_j(\omega,t)}(\omega), 0\leq t<\infty)$ converges to some element $u\in U_g(B)$.

Obviously, M^{u^i} converges weakly to M and therefore $EH(u^i,m)\circ K_{\tau_n}=E\langle M^{u^i},m\rangle_{\tau_n}\to E\langle M,m\rangle_{\tau_n}, m\in H^2_{loc}(A)$

$m \in H^2_{loc}(A)$.

Relations (5), (6) and (13) imply that the sequence $H_t(u^i_t(w), m)$ converges μ^K-a.e. to some process $H_t(m), t \geq 0$. The the sequence

$$H(\tilde{u}^j, m), j \geq 1.$$

also converges to the same limit. But the latter, in view of the continuity of $H(a,m)$ with respect to α and of the choice of the sequence $\tilde{u}^j$ converges to $H(u,m)$ Hence, we have

$$E\langle M, m\rangle_{\tau_n} = EH(u,m)\circ K_{\tau_n} = E\langle M^u, m\rangle, m \in \mathcal{L}^2_{loc}(M^A)$$

and, consequently, $M = M^u$.

In order to prove b), c), d) we use the results of [1] according to which the assertions of the theorem hold for a finite set A. In particular, this, together with Lemma 4, implies the assertions b), c) and weak approximation of the elements of $M^v_g(B^k)$ by the elements M^u from M^u_d.

It remains to establish the weak closeness of the class $M^v_g(B^K)$. Let $(u^i, i \geq 1) \in V_g(B^K)$ be a sequence for which M^{u^i} converges weakly to some element $M \in H^2_{loc}(A)$. We can construct a sequence of elements $u^{i,\varepsilon} \in V^f_g(B^K)$ such that

$$\|M^{u^i} - M^{u^{i,\varepsilon}}\|^2_n \leq E h_\varepsilon \circ K_{\tau_n}, n \geq 1. \tag{14}$$

The proof of Lemma 4 shows that there exists a sequence $u^{i,\varepsilon}$ satisfying (14) and such that the supports of its elements lie in the same ε-net of the set A.

The weak compactness of M^v_g (which follows from the convexity, strong boundedness and strong closeness of this class) implies the existence of a subsequence i' such that $M^{u^{i'},\varepsilon}$ converges weakly to some element $M^{\varepsilon} \in M_g^{v}$. But the assertions of [1] imply that $M^{\varepsilon} \in M_g^{v}(B^{K})$. By (14) we have

$$\|M - M^{\varepsilon}\|_n \leq E h_{\varepsilon} \cdot K_{\tau_n}, \; n \geq 1, \; \varepsilon > 0,$$

and for the proof of $M \in M_g^{v}(B^{K})$ it remains to apply the strong closeness of the class $M_g^{v}(B^{K})$.

4. CLOSURES OF THE CLASS R_d^{v}.

In order to extend the assertions of Theorem 2 to the class of local densities $R_d^{v} = (\rho^u, u \in V_d)$, $R^{v} = (\rho^u, u \in V)$, $R_g^{u} = (\rho^u, u \in V_g)$, $R_g^{v}(B) = (\rho^u, u \in V_g(B))$ where ρ^u is an exponential martingale $\mathcal{E}(M^u)$, we use a theorem on the transition to the limit under the exponent sign in the space H^2_{loc}.

Theorem 3. ([1]). Let $M, M^i \in M^2_{loc}$ and for a stopping time τ we have $\langle M \rangle_{\tau} \leq c$, $\langle M^i \rangle_{\tau} \leq \alpha, i \geq 1$. Then the convergence

$$E \langle M^i - M \rangle_{\tau} \to 0, \quad i \to \infty$$

is true iff

$$E \left[\mathcal{E}_{\tau}(M^i) - \mathcal{E}(M) \right]^2 \to 0 \quad i \to \infty.$$

This assertion implies the equivalence of the strong convergence of $M^{u^i} \to M^u$ and the convergence $\mathcal{E}(M^{u^i}) \to \mathcal{E}(M^u)$ for the sequence $(M^{u^i}, i \geq 1) \in M_g^{v}$ in $H^2_{loc}(A)$ since for the localizing sequence $(\tau_n, n \geq 1)$ from (7) we have $\langle M^u \rangle_{\tau_n} \leq n$, $n \geq 1$.

Note also that the criterion of the convergence of

exponential martingales in the space H^1_{loc}, which was used in the proof of Theorem 3, had been obtained in [4].

<u>Theorem 4</u>. a) The class $R^v_g(B)$ is strongly closed for any $B \in \mathcal{P}$.

b) The class R^v is a strong closure of the class R^v_d

c) The class R^v_g is a convex closure of the class R^v_d

d) The class $R^v_g(B^K)$ is a weak closure of the class R^v_d (B^K is a union of all atoms of the measure μ^K).

<u>Proof</u>. Assertions a), b) are the immediate consequences of Theorem 2 and 3. We shall prove the convexity of the class R^v_g. Let $u^1, u^2 \in U_g$ and $\rho = \alpha\rho^{u^1} + (1-\alpha)\rho^{u^2}$

Using the expression of the differential of the exponential martingale

$$d\rho^u_t = \rho^u_{t-}\, dM^u_t, \; t \geq 0,$$

we have $d\rho_t = \rho_{t-}\, dM_t$, where the martingale M is given by the integral expression

$$M_t = \int_0^t \alpha \frac{\rho^{u^1}_{s-}}{\rho_{s-}} \cdot dM^{u^1}_s + \int_0^t (1-\alpha)\frac{\rho^{u^2}_{s-}}{\rho_{s-}}\, dM^{u^2}_s .$$

Now we can show that existence of $u \in U_g$ such that $M = M^u$. Consider the expression $\langle M, m\rangle$ for $m \in \mathcal{L}^2_{loc}(M^A)$. We have

$$\langle M,m\rangle = \int_0^t \alpha\, \rho^{u^1}_{s-}/\rho_{s-}\, d\langle M^{u^1}, m\rangle_s + \int_0^t (1-\alpha)\rho^{u^2}_{s-}/\rho_{s-}\, d\langle M^{u^2}, m\rangle_s =$$

$$= \int_0^t \Big[\alpha\rho^{u^1}_{s-}/\rho_{s-} \int_A H_s(a,m)\, u^1_s(da) + (1-\alpha)\rho^{u^2}_{s-}/\rho_{s-} \int_A H_s(a,m)\, u^2_s(da)\Big] dK_s =$$

$= \int_0^t \int_A H_s(a,m)\, u_s(da),$

where $u_t(c) = \alpha \rho_{t-}^{u^1}/\rho_{t-}\, u_t^1(c) + (1-\alpha)\rho_{t-}^{u^2}/\rho_{t-}\, u_t^2(c)$, $c \in \mathcal{A}$, and by the definition of the curvilinear integral with respect to the generalized curve we have $M = M^u$.

As in the proof of Theorem 2, by the assertions of Theorem 3, every element of R_g^v is strongly approximated by the sequence $\rho^{u^i}, u^i \in V_g^f$ and the latter we approximate strongly by the convex combinations of the elements of R_d^u (this possibility was shown in [1]). In a similar way we can prove the possibility of weak approximation of the elements of $R_g^v(B^K)$ by the elements of R_d^v and, finally, to prove the weak closeness of $R_g^v(B^K)$ we use the construction of the sequence $u^\varepsilon \in V_g^{df}(B^K)$ from the proof of Theorem 2 and the assertion of Theorem 3.

5. EXISTENCE OF THE OPTIMAL CONTROLS

The assertion of Theorem 4 can be regarded as a basis for a construction of a controlled process by means of a stochastic curvilinear integral proposed in [5]. If the elements of the classes $V_d, V, V_g(B^K), V_g$ constructed above are interpreted as simple (piecewise constant), usual, mixed and generalized controls respectively, then the corresponding classes of densities R_d^v, R^v, $R_g^v(B^K)$, R_g^v define the sets of measures $P_d^v, \dots, P_g^v$ which are successive extensions of the initial set P^A of measures corresponding to the decisions (constant controls) on the basis of the bifurcation operations, strong, weak and convex closure. It should be mentioned that the generalized controls are appeared not only due to the convex hull, but also due to the weak closure (in particular, due to the continuous dominating processes $K: P_g^{v} = P_g^{v}(B^K)$).

If the conditions A*), B), C), D) and E) are satisfied then in the problem of maximization of the linear functional

$$L(u) = L(\rho^u) = E^u \gamma = \int \gamma \, dP^u$$

where γ is some $\mathcal{F}_\infty$-measurable random variable we obtain easily.

Corollary 3. The class R_g^{v} is weakly compact. For the upper semi-continuous on R_g^{v} functional L there exists an optimal control in the class U_g .

The following propositions establish the criterion of upper semi-continuity of the functional L and the sufficiency of the class U .

For $u \in U_g$ we introduce the estimater S^u of the control u

$$S_t^u = E^u(\gamma \mid \mathcal{F}_t) \quad 0 \le t < \infty ;$$

let $S_t = \sup_u S_t^u$ denote the value process.

Assume that the following conditions hold:

F) $\sup_u E^u |\gamma| < \infty$

G) $E(S_{\tau_n})^2 < \infty$ for all $n \geq 1$.

Theorem 5. a) There exists a right-continuous and prossessing the left limits modification of the value process such that it is the minimal solution of the differential equation

$$dS_t = -\max_{a \in A} H_t(a, s)\, dK_t + dm_t, \quad m \in \mathcal{M}^2_{loc}(P),$$

in the class of P^u-uniformly integrable solutions with $S_\infty = \lim_{t \to \infty} S_t \geq \gamma$ P^u- a.s. for all $u \in U_g$ ($H(a, s) =$

$= d\langle M^a, s\rangle / dK = d\langle M^a, m\rangle / dK)$

b) The control $u \in \mathcal{V}_g$ is optimal iff

1) $H(u, s) = \max_{a \in A} H(a, s) \quad \mu^K$- a.e.

2) $S_\infty = \tau \quad P^u$-a.s.

c) The functional L is upper semi-continuous iff

$S_\infty = \tau \quad P^u$-a.s. for all $u \in \mathcal{V}_g$.

d) For any linear on $H^2_{loc}(A)$ continuous functional L there exists a random variable $\eta \in \mathcal{F}_\infty$ such that $L(\rho^u) = E^u \eta$ for all $u \in \mathcal{V}_g$ and conditions E), G) are satisfied.

Assertions a), b), c) are proved in [1] (without using the finiteness of the set A). We shall prove assertion d).

Consider the subspace $H^2_{loc}(A)$ for any $u \in \mathcal{V}_g$ and $n \geq 1$

$$H^2_n(u) = \{N \in H^2_{loc}(A) : N_t = E(\xi|\mathcal{F}_t), t < \tau_n; N_t = \xi(\rho^u_t / \rho^u_{\tau_n}), t \geq \tau_n, \xi \in \mathcal{F}_{\tau_n}, E\xi^2 < \infty\}$$

The functional L is continuous on the Hilbert space $H^2_n(u)$ with the norm $\|N\| = E^{1/2} N^2_{\tau_n}$ and, hence, there exists a random variable $X^u_n \in \mathcal{F}_{\tau_n}$ with $E(X^u_n)^2 < \infty$ such that

$$L(N) = E N_{\tau_n} X^u_n , \quad N \in H^2_n(u).$$

We can show that the sequence $(X^u_n, \mathcal{F}_{\tau_n})$ is the P^u-martingale. It is easy to verify that if $N \in H^2_{n-1}(u)$ then

$$N_t = E(\xi|\mathcal{F}_t) I_{\{t < \tau_n\}} + \xi(\rho^u_t | \rho^u_{\tau_n}) I_{\{t \geq \tau_n\}} ,$$

where $\xi = N_{\tau_{n-1}}(\rho^u_{\tau_n} | \rho^u_{\tau_{n-1}})$. Hence, $N \in H^2_n(u)$ and

$H^2_{n-1}(u) \subset H^2_n(u)$

So for $N \in H^2_{n-1}(u)$ we have

$$L(N) = E N_{\tau_{n-1}} X^u_{n-1} = E N_{\tau_{n-1}} (\rho^u_{\tau_n} \mid \rho^u_{\tau_{n-1}}) X^u_n .$$

If a martingale N is selected so that $N_{\tau_{n-1}} = \rho_{\tau_{n-1}} I_B$, $B \in \mathcal{F}_{\tau_{n-1}}$, then we obtain the equality

$$E^u I_B X^u_{n-1} = E I_B \rho^u_{\tau_{n-1}} X^u_{n-1} = E I_B \rho^u_{\tau_n} X^u_n = E I_B X^u_n$$

Because the set $B \in \mathcal{F}_{\tau_{n-1}}$ is arbitrary the sequence $(X^u_n, n \geq 1)$ is a martingale with respect to the measure P^u.

Now we can prove that the sequence $(X^u_n, n \geq 1)$ is uniformly integrable with respect to the measure P^v for all $v \in V_g$. Moreover, we show that

$$\sup_{u,v,n} E^v X^u_n I_{\{|X^u_n| > \lambda\}} \to 0 \qquad \text{as } \lambda \to \infty$$

Consider the set $C \subset H^2_{loc}(A)$:

$$C = \{h = h_{u,v,n,\lambda} \in H^2_{loc}(A) : h_{\tau_n} = \rho^v_{\tau_n} I_{\{|X^u_n| \geq \lambda\}}, n \geq 1, u, v \in V_g, \lambda \geq 1\}$$

It is easily seen that $E^v X^u_n I_{\{|X^u_n| \geq \lambda\}} = L(h_{u,v,n,\lambda})$. Since the set C is strongly bounded in $H^2_{loc}(A)$ and $h_{u,v,n,\lambda} \to 0$ as $\lambda \to \infty$ for every u, v, n relation (15) follows from the continuity of the functional L.

Denote $\eta^u = \overline{\lim_n} X^u_n$. By (15) we have $\lim_{n\to\infty} X^u_n = \eta^u$ P^u-a.s. and $L(\rho^u) = E^u \eta^u$. Fix some element $u^0 \in V_g$ and show that $L(\rho^u) = E^u \eta$, where $\eta = \eta^{u^0}$, for any $u \in V_g$.

Define the sequence $(u^n, n \geq 1)$:

$$u^n = u I_{[\![0,\tau_n]\!]} + u^0 I_{]\!]\tau_n,\infty[\![}\ , \quad n \geq 1.$$

Obviously, $\rho^{u^n} \to \rho^u$ as $n \to \infty$ (in $H^2_{loc}(A)$) and $\gamma^{u^n} = \lim_i X_i^{u^n} = \lim_i X_i^{u^0} = \gamma^{u^0}$. Hence (taking into account the p^u -uniform integrability of $(X_n^{u^0}, n \geq 1)$ for all $u \in V_g$) we have

$$L(\rho^u) = \lim_{n\to\infty} L(\rho^{u^n}) = \lim_{n\to\infty} E^{u^n} \rho^{u^n} = \lim_{n\to\infty} E^{u^n} E^{u^n}(\gamma^{u^0} | \mathcal{F}_{\tau_n}) =$$

$$= \lim_{n\to\infty} E^u E^{u^0}(\gamma^{u^0} | \mathcal{F}_{\tau_n}) = \lim_{n\to\infty} E^u X_n^{u^0} = E^u \gamma^{u^0}.$$

Hence, we obtain the representation $L(\rho^u) = E^u \eta$ of the functional L on R_g^V where η is an $\mathcal{F}_\infty$ - measurable random variable with $\sup_u E^u |\eta| < \infty$ (this follows from the weak compactness of the class R_g^V), $S_{\tau_n}^u = X_n^u$ and $E(S_{\tau_n}^u)^2 < \infty$ for all $u \in V_g$. Consequently, $E(S_{\tau_n})^2 < \infty$ since for every $n \geq 1$ and $\varepsilon > 0$ a control $u^{n,\varepsilon}$ exists such that $S_{\tau_n}^{u^{n,\varepsilon}} > \sup_u S_{\tau_n}^u - \varepsilon$ (see [1]).

Corollary 4. The extreme points of the set R_g^V belong to the class R^V . For the upper semi-continuous functional there exists optimal control in the class V (of usual controls).

The corollary immediately follows from the optimality criterion b) and assertions c), d) of Theorem 5, since the function $H_t(a, s.) (= H_t(a, m))$ is continuous with respect to a a.s. with respect to the measure μ^K and, hence, by Philippov's lemma ([6]) an element does exist such that

$$H(u^*, s) = \max_{a \in A} H(a, s)$$

Finally, it should be mentioned that the existence of

the optimal controls was shown by the compactification of the density family for controlled diffusion processes (the martingale M^{a} can be represented as an integral with respect to the Wiener process) in [6]; in [7] an infinite time interval for pointwise controlled processes is considered; the paper [8] is a kind of a survey, in particular, for the martingales M^{a} possessing an integral representation of the general form the conditions of the existence of optimal control are given; in [9] a dynamic programming method is applied to prove the existence ofthe optimal control in the case of diffusion processes (criterion b), Theorem 5). For the first time stochastic curvilinear integral was considered by Gikhman and Skorokhod (see [10]) in the case of uniformly in (u,t) continuous derivatives $\varphi(a,b) = d\langle M^{a}, M^{b}\rangle / dK$ and continuous processes K.

REFERENCES

1. Chitashvili R.J., Mania M.G. Optimal locally absolutely continuous change of measure. Stochastics (to appear).
2. Jacob J. Calcul stochastique et problemes de martingales, Lect.Notes in Math., N 714, Springer 1979.
3. Chitashvili R.J., Toronjadze T.A. On one-dimensional stochastic differential equations with unit diffusion coefficient. Stochastics, vol.4, p.281-315, 1981.
4. Kabanov Yu.M., Liptzer R.Sh., Shiryayev A.N. On variation distance for probability measures. Zeit für War (to appear).
5. Chitashvili R.J. Martingale ideology in theory of controoled processes, Lect.Notes in Math., 1982, N 1021, Springer, p.73-92.

6. Benes V.E. Existence of optimal stochastic control. law. SIAM J.of Control, 1971, v.9, p.446-72.
7. Kabanov Yu.M. On the solution existence in a countable process control problem (in Russian). Math. Sbornik 1982, 119 (161), N 3 (11), p.431-444.
8. El-Karoui N. Les aspects probabilists du controle stochastique. Lect.Notes in Math., 1979, N 876, p.74-284.
9. Davis M.H.A. On the existence of optimal policies in stochastic control. SIAM J. of Control, 1973, N 11, p.587-594.
10. Gikhman I.I., Skorokhod A.V. Stochastic differential equations and their application (in Russian). Kiev, "Naukova Dumka", 1982.

Prob. Theory and Math. Stat., Vol.1, pp. 357-361
Prohorov et al. (eds)
1986 VNU Science Press

NON-UNIFORM ESTIMATES OF THE REMAINDER TERM IN LIMIT THEOREMS WITH A STABLE LIMIT LAW

Gerd Christoph
Sektion Mathematik, Technische Universität Dresden,
Mommsenstr.13, DDR-8027 Dresden

1. Introduction and Results

Let $X_1, X_2, \dots$ be a sequence of i.i.d. random variables belonging to the domain of normal attraction of a stable law $G_{\alpha\beta}$, $0 < \alpha < 2$, $-1 \leqq \beta \leqq 1$ (The case $\alpha = 1$, $\beta \neq 0$ is excluded.). We define for $k \geqq 1$ and $r \geqq 0$

$$F_k(x) = P(X_1 + \dots + X_k < k^{1/\alpha} x), \quad H_k(x) = F_k(x) - G_{\alpha\beta}(x),$$

$$\nu_{k,r} = \int_{-\infty}^{\infty} |x|^r \, |dH_k(x)| \, , \quad \mu_{k,l} = \int_{-\infty}^{\infty} x^l \, dH_k(x), \; l = 0, 1, \dots, [r].$$

The density of $G_{\alpha\beta}(x)$ will be denoted by $p_{\alpha\beta}(x)$ and that of $F_k(x)$ by $p_k(x)$, if it exists. The symbol C denotes a positive constant which may stand for different values.

Theorem 1. Let $r > \alpha$ and let m be such an integer that $m < r \leqq m+1$. Suppose that there exist an integer $k \geqq 1$ that $\nu_{l,r} < \infty$, $\mu_{l,1} = \dots = \mu_{l,m} = 0$ for $k \leqq l \leqq 2k-1$ and an integer N such that $p_N(x)$ is bounded. Then

$$|p_n(x) - p_{\alpha\beta}(x)| \leqq C \, n^{-(r-\alpha)/\alpha} \, (|x| + 1)^{-[r]} \tag{1}$$

for sufficiently large n and every x , where C does not depend on n and x.

Theorem 2. Let the conditions of Theorem 1 be satisfied and, when r=m+1, additionally,

$$\int_{-\infty}^{\infty} |x|^{m+1} \ln(1+|x|) \, |dH_l(x)| < \infty \quad \text{for } k \leqq l \leqq 2k-1. \tag{2}$$

Then

$$|F_n(x) - G_{\alpha\beta}(x)| \leqq C \, n^{-(r-\alpha)/\alpha} \, (|x| + 1)^{-[r]} \tag{3}$$

for sufficiently large n and every x, where C does not

depend on n and x.

Usually non-uniform estimates have been investigated under the condition $\nu_{1,r} < \infty$ for $\alpha < r \leqq 1+\alpha$. See Paulauskas (1974). Uniform bounds similar to (3) with $r > 1+\alpha$ were proved in connection with asymptotic expansions under pseudomoment (or weaker) conditions in e.g. Zolotarev (1962), Mitalauskas and Statulevičius (1976) and Christoph (1982, 1985). The following examples show that $\nu_{k,r} < \infty$ may be satisfied for some $k > 1$, whereas $\nu_{1,r} = \infty$.

Example 1. If X_1 has the density $p(x) = \pi^{-1}(1-\cos x)x^{-2}$, then X_1 belongs to the domain of attraction of $G_{10}(x)$ and $\nu_{1,r} < \infty$ only for $r < 1$, $\nu_{2,r} < \infty$ only for $r < 2$ and $\nu_{k,r} < \infty$ only for $r < 3$, $\mu_{k,1} = 0$, $\mu_{k,2} = (k-1)/k$ for $k=3,4,\ldots$. Hence the conditions of the theorems are fulfilled with $r=2$ and $k=3$.

Example 2. Let $1/4 \leqq \alpha \leqq 1$. If X_1 has the charakteristic function

$$f(t) = \begin{cases} 1 - |t|^{\alpha} + \frac{1}{2}|t|^{2\alpha} - \frac{1}{6}|t|^{3\alpha} + \frac{1}{48}|t|^{4\alpha}, & \text{for } |t| < 2^{1/\alpha} \\ 0 & \text{, elsewhere} \end{cases}$$

then X_1 belongs to the domain of normal attraction of $G_{\alpha,0}(x)$ and $\nu_{1,r} < \infty$ only for $r < \alpha$. It can be easily checked that for $\alpha = 1/2$ the conditions of the theorems are fulfilled with $k = 3$ and $r < 5/2$ ($\mu_{3,1} = 0$, $\mu_{3,2} \neq 0$) and for $\alpha = 1$ with $k = 5$ and $r = 4$ ($\mu_{5,1} = 0$, $l = 1,2,3$).

Having new non-uniform results we can apply these to prove the following theorem concerning probabilities of large deviations with a stable limit law.

Theorem 3. Under the conditions of Theorem 2, as $n \to \infty$,

$$\frac{1 - F_n(b_n) + F_n(-b_n)}{1 - G_{\alpha\beta}(b_n) + G_{\alpha\beta}(-b_n)} = 1 + O(b_n^{-([r]-\alpha)}\, n^{-(r-\alpha)/\alpha})$$

for any sequence b_n of reals such that $b_n \uparrow \infty$.

2. Sketch of the Proofs

Lemma 1. Let Y_i, i=1,2, be two independent random variables belonging to the domain of normal attraction of $G(x) = G_{\alpha\beta}(x)$ with $V_i(x) = P(Y_i < x)$ and $H_i(x)=V_i(x) - G(x)$, i=1,2. Let

$$\int_{-\infty}^{\infty}|x|^r|dH_i(x)| < \infty \text{ and } \int_{-\infty}^{\infty}x^l dH_i(x) = 0 \text{ for } l=0,1,\dots,[r-\alpha],$$

i=1,2, then the same is true for $H_3(x)=V_1*V_2(x)-G^{2*}(x)$, where $*$ denotes the convolution sign.

To prove Lemma 1 we make use of $V_1*V_2 - G*G = H_1*G + H_2*G + H_1*H_2$ and $|d(G*H_i)(x)| \leq |(G'*H_i)(x)|\,dx$. For $|x| \geq 2A$ we split the convolution integral in $G'*H_i$ into two parts $|y| \leq x/2$ and $|y| \geq x/2$, and apply the Taylor expansion of $G'(x-y)$, $G'(z) \leq D$ and $|G^{(l)}(z)| \leq C|z|^{-l-\alpha}$ for $|z|>A$, where A, C, and D are certain positive constants.

This lemma is an answer to a question stated in Paulauskas (1974). Note that $\int_{-\infty}^{\infty}|x|^r|dH_3(x)| = \infty$, if $\int_{-\infty}^{\infty}x^l\, dH_i(x) \neq 0$ for own of i=1,2 and own of $l \in [1,r-\alpha]$

Proof of Theorem 1: Denote the characteristic functions of F_1, F_n, and $G_{\alpha\beta}$ by $f(t)$, $f_n(t)$, and $g(t)$, respectively, where $g(t) = \exp\{-|t|^\alpha(1 - i\beta \operatorname{sgn} t \tan(\alpha\pi/2)\}$. Put $[r] = s$ and $n = uk + v$, where $u = [n/k] - 1$ and $k \leq v \leq 2k-1$. It follows now from Lemma 1 that $\nu_{n,r} < \infty$ and $\mu_{n,l} = 0$, l=0,1,...,m. Hence, $f_n(t) - g(t)$ is differentiable s-times and, moreover, $(f_n(t) - g(t))^{(l)}\big|_{t=0} = 0$ for l=0,1,...,m. Now, by the inversion formula we have for sufficiently large n

$$|x|^s|H_n'(x)| \leq \frac{1}{2\pi}(I_1 + I_2),$$

where

$$I_1 = \int_{|t|<T}|(f_n(t)-g(t))^{(s)}|dt, \quad I_2 = \int_{|t|\geq T}(|f_n^{(s)}(t)| + |g^{(s)}(t)|)dt,$$

$T = \varepsilon n^{1/\alpha}$ and $\varepsilon > 0$ is a sufficiently small constant. To estimate I_1 we need for $|t| < \varepsilon n^{1/\alpha}$ and l=0,1,...,s

$$\left|(f_n(t)-g(t))^{(1)}\right| \leqq C\,|t|^{r-1}(1+|t|^{1\alpha})\, e^{-|t|^{\alpha}/2}\, n^{-(r-\alpha)/\alpha}, \quad (4)$$

where C is a positive constant not depending on n and t, which may be proved using

$$f_n(t)-g(t) = f_k^u\left(\frac{tk^{1/\alpha}}{n^{1/\alpha}}\right)\cdot f_v\left(\frac{tv^{1/\alpha}}{n^{1/\alpha}}\right) - g^{uk}\left(\frac{t}{n^{1/\alpha}}\right)\cdot g^{v}\left(\frac{t}{n^{1/\alpha}}\right),$$

$$\left|\left(f_w\left(\frac{tw^{1/\alpha}}{n^{1/\alpha}}\right)-g^w\left(\frac{t}{n^{1/\alpha}}\right)\right)^{(1)}\right| \leqq C\,|t|^{r-1}\, n^{-r/\alpha} \quad \text{for } w=k,v,$$

$$\left|g\left(\frac{t}{n^{1/\alpha}}\right)^{(1)}\right| \leqq C\left(\frac{|t|^{\alpha-1}}{n} + \frac{|t|^{1\alpha-1}}{n^{1}}\right) e^{-|t|^{\alpha}/n} \quad \text{for } t \neq 0$$

and Lemma 3,Chapter VI of Petrov (1975). It follows from (4) that $I_1 \leqq C\, n^{-(r-\alpha)/\alpha}$. The condition "$p_N$ is bounded" together with

$$f^n = f_k^u \cdot f_v\,, \quad (f_w)^{(s)} = (f_w - g^w)^{(s)} + (g^w)^{(s)}\,, \quad w=k,v,$$

and the above mentioned bounds imply $I_2 \leqq C\, e^{-cn}$, where $c > 0$ is a positive constant, and we derive the desired relationship (1).

<u>Lemma 2.</u> Under the conditions of Theorem 2 we have

$$|H_n(x)| \leqq C(1+|x|)^{-s}\left\{\int_{-M}^{M}(|a_0(t)|+|a_s(t)|)\frac{1}{|t|}dt + \frac{K}{M}\right\} \quad (5)$$

for every x and every fixed n (sufficiently large),where $M = \varepsilon n^{(r-\alpha)/\alpha}$, $a_w(t) = \int_{-\infty}^{\infty} e^{itx} d(x^w H_n(x))$, $w = 0,s$, and C and K are positive constants not depending on n and x.

<u>Proof:</u> It follows from $\nu_{ns} < \infty$ that $x^s H_n(x)$ is a function of bounded variation and $x^s H_n(x) \to 0$ as $|x| \to \infty$. Here n is fixed. In proving (5) we follow the proof of Osipov's lemma, Lemma 8,Chapter VI of Petrov(1975). Instead of certain estimations for F and G in Osipov's lemma now we make use of $H_n(x) = H_n(x_0) + (x-x_0)H_n'(y)$ and (1), where x_0 is that point with $\sup_x |x|^s |H_n(x)| = |x_0|^s |H_n(x_0)|$ and $y \in (\min(x_0,x),\max(x_0,x))$.

<u>Proof of Theorem 2:</u> It follows from Lemma 7,Chapter VI of Petrov (1975) that

$$(-it)^s \, a_s(t) = s! \sum_{l=0}^{s} \frac{(-t)^l}{l!} (f_n(t)-g(t))^{(l)} .$$

To estimate the integral on the right-hand side of (5) we apply (4) or, if $r = s = m+1$, a modification of (5) and (2) for $|t| \leq \varepsilon n^{1/\alpha}$, and similar bounds as in the local theorem for $|t| > \varepsilon n^{1/\alpha}$.

Proof of Theorem 3: The statement of Theorem 3 follows from (3) and $1 - G_{\alpha\beta}(x) + G_{\alpha\beta}(-x) \asymp |x|^{-\alpha}$ as $|x| \to \infty$.

References

Christoph,G.(1982). Asymptotic expansions in the case of stable limit law.II, Litov. matem. sb., XXII, Nr.2, 69-79 (Part I, same journal (1981), XXI, Nr.2, 87-100)

Christoph,G.(1985). Asymptotic expansions for sums of random variables and of reciprocals of these with a non-normal stable limit law. Math. Nachr. 122, 91-98

Mitalauskas,A. and Statulevičius,V.(1976). An asymptotic expansion in the case of stable approximating law. Litov. matem. sb. XVI, Nr.4, 149 - 166

Paulauskas,V.(1974). Uniform and non-uniform estimate of the remainder term in the limit theorem with a limiting stable law. Litov. matem. sb. ,XIV, Nr.4, 171 - 185

Petrov,V.V.(1975). Sums of independent random variables. Akademie-Verlag,Berlin

Zolotarev,V.(1962). An analogue of the Edgeworth-Cramér asymptotic expansion in the case of nearness to stable distribution laws. In: Abstacts VI All-Union-Conference on Probability Theory and Matem. Statistics, Vilnius, 49 - 50

Prob. Theory and Math. Stat., Vol.1, pp. 363-373
Prohorov et al. (eds)
1986 VNU Science Press

MULTIPLE STABLE STOCHASTIC INTEGRALS

by Z. Ciesielski

1.Introduction. During 1985 we celebrated the centennial of Alfred Haar and the author's lecture "The Haar system in probability" at the Fourth International Vilnius Conference was presented to pay tribute to the great Hungarian. Extended version of that lecture will be published in the Proceedings of the Alfred Haar Memorial Conference held in Budapest in August of 1985. One of the new results announced in that lecture concerns the construction of a multiple stochastic integral with respect to a stable with independent increments process. The aim of this note is to present in details the constructive proof of existence of such integrals. Partial result of this kind in dimension 2 was obtained by Szulga and Woyczyński in |8| with the help of tensor product Haar system. Their idea appeared to be fruitful after replacing the Haar system by more adequate bases as constructed in |3|. The existence and description of the multiple integrals in question atracted recently number of mathematicians (see e.g.|7|,|8|,|4| and |9|). Perhaps this functional and approximation theory approach will illuminate the structure of the multiple stable integrals ?

2.Preliminaries. For our purpose it is enough to consider the function spaces on d-dimensional cube $Q=I^d$, $d\geq 1$, $I=\langle 0,1\rangle$. We need to consider two families of Besov spaces (generalized Lipschitz classes) $B^s_{p,q}(Q)$, $\mathring{B}^s_{p,q}(Q)$, $s\in R:=(-\infty,\infty)$, $1\leq p,q\leq\infty$, defined in such a way that they are always separable. Similarly the Sobolev spaces $W^k_p(Q)$, $\mathring{W}^k_p(Q)$ are assumed to be separable for $k\in \underline{Z}:=(0,\pm 1, \ldots)$, $1\leq p\leq\infty$. In the notation used in |2| it means that we consider the spaces $\mathcal{C}B$, $\mathcal{C}\mathring{B}$, $\mathcal{C}W$, $\mathcal{C}\mathring{W}$ only, suppressing the $\mathcal{C}$. With this convention for given $s\in R$ we have for $l<s<r$, $l,r\in\underline{Z}$, $s=(1-\theta)l+\theta r$

$$(2.1)\qquad \begin{aligned} B^s_{p,q}(Q)&=(W^l_p(Q),W^r_p(Q))_{\theta,q}\ , \\ \mathring{B}_{p,q}(Q)&=(\mathring{W}^l_p(Q),\mathring{W}^r_p(Q))_{\theta,q}\ , \end{aligned}$$

i.e. the Besov spaces are the real interpolating spaces between Sobolev spaces (for details see |3|,|2|,|1|). In |3| we have constructed bases (functions) having all the properties listed below.

Let us fix the index of smootheness $m\in\underline{N}:=(0,1,\ldots)$. The usual scalar product in $L^2(Q)$ and in $L^2(I)$ is denoted by $(\cdot,\cdot)$. There is a system of functions $(f_i,\mathring{f}_j$, $i,j\geq -m)$ such that

(2.2) $f_i\in C^m(I)$, $f_i\in C^m(I)$,

(2.3) $(\mathring{f}_i,f_j) = \delta_{ij}$,

(2.4) $|f_n(t)| \leq C\ (m+n+1)^{\frac{1}{2}}\ q^{n\,\mathrm{dist}(t,t_n)}$,

(2.5) $|\mathring{f}_n(t)| \leq C\ (m+n+1)^{\frac{1}{2}}\ q^{n\,\mathrm{dist}(t,t_n)}$,

where $t_n=0$ for $n=-m,\ldots,1$ and $t_n=(2l-1)/2^{k+1}$ for $n=2^k+l$,

$1 \le l \le 2^k$, $k \ge 0$ and $0<q<1$. Now let

$$X_k := \operatorname{span}[\, f_n,\ n \le 2^k \,],\quad \mathring{X}_k := \operatorname{span}[\, \mathring{f}_n,\ n \le 2^k \,].$$

There is to each $k \in \underline{N}$ a system $(f_{k,i}, \mathring{f}_{k,j}, -m \le i, j \le 2^k)$ such that

(2.6) $X_k = \operatorname{span}[\, f_{k,i},\ i \le 2^k \,]$, $\mathring{X}_k = \operatorname{span}[\, \mathring{f}_{k,i}, i \le 2^k \,]$,

(2.7) $(\mathring{f}_{k,i}, f_{k,j}) = \delta_{i,j}$ for $i,j = -m,\dots,2^k$,

(2.8) $|f_{k,j}(t)| \le C\, 2^{\frac{1}{2}k}\, q^{2^k \operatorname{dist}(t,s)}$, $s = t_{k,j}$, $t \in I$,

(2.9) $|\mathring{f}_{k,j}(t)| \le C\, 2^{\frac{1}{2}k}\, q^{2^k \operatorname{dist}(t,s)}$, $s = t_{k,j}$, $t \in I$,

where $0<q<1$, $t_{k,j}=0$ for $-m \le j < 0$ and $t_{k,j} = j/2^k$, $j=0,\dots,2^k$. We are ready now to define the basic functions over Q. The set of indeces is now $N_m^d := (\underline{N}-m) \times \dots \times (\underline{N}-m)$ (d copies) and it is being decomposed as follows

$$N_m^d = \bigcup_0^\infty N_k\ ,\quad N_o = (\underline{n} = (n_1,\dots,n_d) \in N_m^d : n_i \le 1 \text{ for } i \in D),$$

and

$$N_k = \bigcup_{\substack{e \subset D\\ e \ne \emptyset}} N_{k,e},\quad N_{k,e} = (\underline{n} \in N_m^d : 2^k < n_i \le 2^{k+1} \text{ for } i \in e,\ n_i \le 2^k \text{ for } i \in D \setminus e),$$

where $D=(0,\dots,d)$. Define $f_{\underline{n}} = f_{n_1} \otimes \dots \otimes f_{n_d}$ for $\underline{n} \in N_o$ and

$$f_{\underline{n}} = \bigotimes_{i \in e} f_{n_i} \otimes \bigotimes_{i \notin e} f_{k,n_i} \text{ for } \underline{n} \in N_{k,e},\ e \subset D,\ e \ne \emptyset.$$

The $f_{\underline{n}}$ we define in the same way replacing f by $\mathring{f}$. It follows that

(2.10) $(\mathring{f}_{\underline{n}}, f_{\underline{n}'}) = \delta_{\underline{n},\underline{n}'}$ for $\underline{n}, \underline{n}' \in N_m^d$,

(2.11) $|f_{\underline{n}}(\underline{t})| \le C\, 2^{\frac{1}{2}dk}\, q^{2^k \operatorname{dist}(\underline{t},\underline{s})}$, $\underline{t} \in Q$, $\underline{s} = \underline{t}_{\underline{n}}$, $\underline{n} \in N_k$,

(2.12) $|\mathring{f}_{\underline{n}}(\underline{t})| \le C\, 2^{\frac{1}{2}dk}\, q^{2^k \operatorname{dist}(\underline{t},\underline{s})}$, $\underline{t} \in Q$, $\underline{s} = \underline{t}_{\underline{n}}$, $\underline{n} \in N_k$, where $0<q<1$ and for $\underline{n}\in N_{k,e}$ $t_{\underline{n}}=(t'_{n_1},\dots,t'_{n_d})$ with $t'_{n_i}=t_{n_i}$ for $i\in e$ and $t'_{n_i}=t_{k,n_i}$ for $i\notin e$.

The basic result in |3|, Section 10, needed here can now be formulated.

THEOREM A. The system $(f_{\underline{n}},\ \underline{n}\in N_m^d)$ is a Schauder basis in each of the spaces $W_p^k(Q)$, $B_{p,q}^s(Q)$ where $|k|\le m$, $k\in\underline{Z}$, $|s|<m$, $1\le p,q\le\infty$. Moreover, if

$$f=\sum_{\underline{n}} a_{\underline{n}} f_{\underline{n}}\ ,\quad a_{\underline{n}}=(\mathring{f}_{\underline{n}},f),$$

then

$$\| f \|_{B_{p,q}^s(Q)} \sim \Big(\sum_{k=0}^{\infty}\Big(2^{kdu}\Big(\sum_{\underline{n}\in N_k}|a_{\underline{n}}|^p\Big)^{1/p}\Big)^q\Big)^{1/q}$$

with $u=\frac{1}{2}+s/d-1/p$.

Remark. Theorem A has its dual version Theorem $\mathring{A}$. The latter one we obtain from Theorem A replacing B,W by $\mathring{B},\mathring{W}$ and interchanging $f_{\underline{n}}$ with $\mathring{f}_{\underline{n}}$.

LEMMA 2.13. Let $0<p\le\infty$. Then

$$\| f_n \|_{L^p(I)} \sim (n+m+1)^{\frac{1}{2}-1/p} \quad \text{for } n\ge -m,$$

$$\| f_{k,j} \|_{L^p(I)} \sim 2^{k(\frac{1}{2}-1/p)} \quad \text{for } -m\le j\le 2^k,$$

$$\| f_{\underline{n}} \|_{L^p(Q)} \sim 2^{kd(\frac{1}{2}-1/p)} \quad \text{for } \underline{n}\in N_k.$$

The same holds after replacing f by $\mathring{f}$.

Proof. The upper estimate for $\|f_n\|_{L^p(I)}$ and $\|\mathring{f}_n\|_{L^p(I)}$ follows immediately by (2.4) and (2.5). Let now $0<p<2$.

Since

$$1=(\mathring{f}_n,f_n)^2 \leq \|\mathring{f}_n\|^2_{L^2(I)} \|f_n\|^2_{L^2(I)} \leq C \|f_n\|^2_{L^2(I)}$$

$$\leq C \|f_n\|^{2-p}_{L^\infty(I)} \|f_n\|^p_{L^p(I)} \leq C (n+m+1)^{1-p/2} \|f_n\|^p_{L^p(I)},$$

we infer

$$\|f_n\|_{L^p(I)} \geq C (n+m+1)^{\frac{1}{2}-1/p},$$

and by similar argument

$$\|\mathring{f}_n\|_{L^p(I)} \geq C (n+m+1)^{\frac{1}{2}-1/p}.$$

For $p \geq 2$ the lower estimates one gets from the Hölder's inequality by the upper estimates for $1<p<2$. The proofs for $\mathring{f}_{k,j}$ and $\mathring{f}_{\underline{n}}$ are similar.

3.The main result. The probabilistic part of the proof follows the idea of J.Szulga and W.A.Woyczyński |8|. We are given the probability space $(\Omega,\mathfrak{A},P)$, a symmetric stable with exponent p, $0<p<2$, process $(X(t),t\in R_+)$ with independent increments. In particular

(3.1) $$E(e^{iuX(t)}) = e^{-t|u|^p}, \quad u\in R, \ t\in R_+.$$

For such stable processes J.Rosiński and W.A.Woyczyński developed in |5| Ito type stochastic integrals which in particular permit to define $X_1(f)=\int_I f(t)X(dt)$ for f in $L^p(I)$, $0<p<2$, and to state that $f \longrightarrow X(f)$ is a continuous mapping from $L^p(I)$ into $L^0(\Omega,\mathfrak{A})$, where the latter is the space of real functions on Ω $\mathfrak{A}$-measurable. It also follows that $X_1(f)$ has the same probability distribution as $X(t_0)$ with $t_0=\|f\|_p^p$.

Our aim is to extend the definition of $X(f)$ to multiple integrals. More precisely, we are going to find a Banach space $\langle \mathcal{F}(Q), \| \ \|_{\mathcal{F}} \rangle$ of functions over $Q=I^d$ and a mapping X_d such that

(3.2) $X_d : \mathcal{F}(Q) \dashrightarrow L^0(\Omega, \mathcal{O}\!\mathcal{L})$ is continuous,

(3.3) $X_d(F) = X_1(f^1)\ldots X_1(f^d)$ (equality with prob.1) whenever $F=f^1 \otimes \ldots \otimes f^d$, $F \in \mathcal{F}(Q)$ and $f^i \in L^p(I)$.

(3.4) $P(X_d(\cdot) \in \mathcal{F}^*(Q))=1$ i.e. there is $\Omega_0 \subset \Omega$ such that $P(\Omega_0)=1$ and for each $\omega \in \Omega_0$ the functional $X_d(\cdot)$ is in the dual space to $\mathcal{F}(Q)$.

Let now $f_{\underline{n}}$ be given as in Section 2 and let

(3.5)
$$X_d(f_{\underline{n}}) := X_1(f_{n_1})\ldots X_1(f_{n_d}) \text{ for } \underline{n} \in N_0,$$
$$X_d(f_{\underline{n}}) := \prod_{i \in e} X_1(f_{n_i}) \prod_{i \notin e} X_1(f_{k,n_i}) \text{ for } \underline{n} \in N_{k,e}.$$

LEMMA 3.6. Let $\delta>0$ and $c>0$ be given. Moreover, let $(Y_n)_1^\infty$ be a sequence of random variables such that

(3.6.1) $t^\delta P(|Y_n|>t) \leq c$ for $n,t>0$;

(3.6.2) $t^{\delta-1} E(|Y_n| I(|Y_n| \leq t)) \leq c$ for $n,t>0$.

Then for any sequence $(c_n)_1^\infty$ condition

(3.6.3) $\sum_1^\infty |c_n|^\delta < \infty$

implies

$$\sum_1^\infty |c_n Y_n| < \infty$$

with probability 1.

THEOREM 3.7. Let $m>s>d(d-1)/p$, $0<p<2$, $d\geq 2$. Then

$$(3.7.1)\qquad \sum_{\underline{n}\in N_m^d} \|\mathring{f}_n\|_{\mathring{B}^{-s}_{1,1}(Q)} |X_d(f_{\underline{n}})| <$$

with probability 1.

Proof. Let us introduce

$$Y_{\underline{n}} := X_d(f_{\underline{n}})/\| f_{\underline{n}}\|^p_{L^p(Q)}$$

and

$$c_{\underline{n}} := \|\mathring{f}_{\underline{n}}\|_{\mathring{B}^{-s}_{1,1}(Q)} \cdot \|f_{\underline{n}}\|^p_{L^p(Q)}$$

Clearly,

$$Y_{\underline{n}} = Y_{n_1} \cdots Y_{n_d}$$

where each Y_{n_i} has the same probability distribution as $X(1)$. However (3.1) implies

$$(3.8)\qquad P(|X(1)|>t)\leq c\, t^{-p} \quad \text{for } t>0,$$

whence

$$P(|Y_{\underline{n}}|>t) = P((|Y_{n_1}|\cdots|Y_{n_d}|)^{1/d}>t^{1/d})$$
$$\leq P(\max_i |Y_{n_i}|>t^{1/d}) \leq \sum_{i=1}^{d} P(|Y_{n_i}|>t^{1/d})$$
$$\leq dc\, t^{-p/d} \quad \text{for } t>0.$$

On the other hand, since $d\geq 2$, we get by (3.8)

$$E(|Y_{\underline{n}}|I(|Y_{\underline{n}}|\leq t)) = \int_0^t x\, dF_{|Y_{\underline{n}}|}(x) = tF_{|Y_{\underline{n}}|}(t) -$$

$$- \int_0^t F_{|Y_{\underline{n}}|}(x)dx = \int_0^t P(x\leq|Y_{\underline{n}}|<t)dx \leq \int_0^t P(|Y_{\underline{n}}|^{1/d}\geq x^{1/d})$$

$$\leq \int_0^t P(\max_i |Y_{n_i}| \geq x^{1/d})dx \leq cd \int_0^t x^{-p/d}dx \leq c't^{1-p/d}, \quad t>0.$$

Thus conditions (3.6.1) and (3.6.2) are satisfied with $\delta=p/d$. To check (3.6.3) we use Theorem $\mathring{A}$ and Lemma 2.13. The first implies

$$\|\mathring{f}_{\underline{n}}\|_{\mathring{B}^{-s}_{1,1}(Q)} \sim 2^{kd(-\frac{1}{2}-s/d)} \text{ for } \underline{n} \in N_k,$$

and the second

$$\|f_{\underline{n}}\|^p_{L^p(Q)} \sim 2^{kd(\frac{1}{2}-1/p)} \quad \text{for } \underline{n} \in N_k.$$

Consequently,

$$\sum_{\underline{n}\in N_k} |c_{\underline{n}}|^\delta \leq c\ 2^{k\delta(d(d-1)/p-s)},$$

and therefore

$$\sum_{\underline{n}\in N_m^d} |c_{\underline{n}}|^\delta < \infty.$$

Application of Lemma 3.6 gives

$$\sum_{\underline{n}\in N_m^d} |c_{\underline{n}} Y_{\underline{n}}| = \sum_{\underline{n}\in N_m^d} \|\mathring{f}_{\underline{n}}\|_{\mathring{B}^{-s}_{1,1}(Q)} |X_d(f_{\underline{n}})| < \infty$$

with probability 1.

We want now to find the space $\mathcal{F}(Q)$. Notice at first that $\mathring{B}^{-s}_{1,1}(Q) = (B^s_{\infty,\infty}(Q))$ (cf. |2|). On the other hand by Theorem $\mathring{A}$ the system $(\mathring{f}_{\underline{n}}, \underline{n}\ N^d_m)$ is unconditional basis in $\mathring{B}^{-s}_{1,1}(Q)$. Consequently the sums

(3.9) $$X_d(\cdot) = \sum_{\underline{n}\in N_m^d} X_d(f_{\underline{n}})\ \mathring{f}_{\underline{n}}$$

by Theorem 3.7 with probability 1 are in $\mathring{B}^{-s}_{1,1}(Q)$.

Clearly $\mathcal{F}(Q)=B^{s}_{\infty,\infty}(Q)$ with $s>d(d-1)/p$ satisfies the requirements (3.2) and (3.4). Now, for $F\in\mathcal{F}(Q)$ (3.9) implies the unconditional convergence of the series

$$X_d(F) = \sum_{\underline{n}\in N^d_m} (\mathring{f}_{\underline{n}},F)\, X_d(f_{\underline{n}}) \tag{3.10}$$

with probability 1. Introducing the partial sums operators

$$P^d_k F := \sum_{i=0}^{k} \sum_{\underline{n}\in N_i} (\mathring{f}_{\underline{n}},F)\, f_{\underline{n}}\ , \quad F\in L^1(Q)$$

and

$$P_k f = \sum_{n=-m}^{2^{k+1}} (\mathring{f}_n,f)\, f_n\ , \quad f\in L^1(I)$$

we find that

$$P^d_k(f^1\otimes \ldots \otimes f^d) = (P_k f^1)\otimes \ldots \otimes (P_k f^d),\ f^i\in L^1(I).$$

This and (3.5) imply

$$X_d(P^d_k F) = X_1(P_k f^1)\ldots X_1(P_k f^d)\ ,$$

where $F=f^1\otimes \ldots \otimes f^d$. Now, Theorem A implies that $P^d_k F \dashrightarrow F$ in $\mathcal{F}(Q)$ and for $F\in\mathcal{F}(Q)$, and therefore $X_d(P^d_k F) \dashrightarrow X_d(F)$ with probability 1. Since $F\in C(Q)$ it follows that $f^i\in C(I)$ and therefore $P_k f^i \dashrightarrow f^i$ in $C(I)$ whence in $L^p(I)$. Thus $X_i(P_k f^i) \dashrightarrow X_i(f^i)$ in $L^0(\Omega,\mathcal{O}\!\mathcal{L})$ and this completes the proof of (3.3).

<u>Corollary 3.11.</u> Let $m>s>d(d-1)/p$ and $\mathcal{F}(Q)=B^s_{\infty,\infty}(Q)$. Then (3.10) is a correct definition of the stochastic integral

$$X_d(F) =: \int_Q F(t_1,\dots,t_d)X(dt_1)\dots X(dt_d), \quad F\in\mathcal{F}(Q),$$

which has the properties (3.2),(3.3) and (3.4).

4. Concluding remarks. With a different method based on Theorem A and $\mathring{A}$ we can obtain stronger result. Namely, for given p, $0<p<2$, the s in Corollary 3.11 can be equal to d/p for $p\neq1$ and to $\varepsilon+d$, $\varepsilon>0$, for $p=1$. In case $1<p<2$ still stronger result is true i.e. one can show that the space $\mathcal{F}(Q)=B_{1,1}^{\frac{1}{p}d(p-1)}(Q)$ is good. The proofs will be given elsewhere.

Acknowledgement. The author is very much indebted to his colleague Tadeusz Figiel for instructive discussions.

References

|1| J.Bergh and J.Löfström, Interpolation Spaces, Springer-Verlag, Berlin-Heidelberg-New York 1976.

|2| Z.Ciesielski and T.Figiel, Spline approximation and Besov spaces on compact manifolds, Studia Math.75 (1982),13-35.

|3| Z.Ciesielski and T.Figiel, Spline bases in classical function spaces on compact C^∞manifolds, Studia Math.76(1983),Part I,1-58, Part II, 95-36.

|5| J.Rosiński and W.A.Woyczyński, On the stochastic integration with respect to the p-stable motion: inner clock, integrability of sample path, double and multiple integrals, Ann. Prob. 13(1985).

|6| J.Rosiński and W.A.Woyczyński, Multilinear forms in Pareto-like random variables and product random

measures, Colloquium Mathematicum (to appear).
|4| T.R.McConnell, Triple integration with respect to stable processes. Preprint.
|7| D.Surgailis, On the Multiple Stable Integral, Zeitschr. Wahrsch.70.4(1985), 621-632.
|8| J.Szulga and W.A.Woyczyński, Existence of a Double Random Integral with respect to Stable Measures, J. Multivariate Analysis 13:1(1983),193-201.
|9| Seminar Notes on Multiple Stochastic Integration, Polynomial Chaos and their Applications, 1985. Dept. Math. and Stat. Case Western Reserve Univ. Cleveland, Ohio.

Mathematical Institute
Polish Academy of Sciences
Branch in Gdańsk
ul. Abrahama 18
81-820 SOPOT, POLAND

Prob. Theory and Math. Stat., Vol.1, pp. 375-384
Prohorov et al. (eds)
1986 VNU Science Press

A GENERALIZATION OF ρ-TYPE SPACES

A.N.Čuprunov
Kazan Chemistry-Technological Insititute, Kazan, USSR

We introduce and study (F,F_1)-type spaces. Stable p-type spaces, as well as Ψ-type spaces introduced in [1], are particular cases of such spaces. More general definitions of space types are considered in [2]. Conditions for convergence of series of indpendent random elements and laws of large numbers are proved in (F,F_1)-type spaces. These results are used in the study of conditions of convergence of duble series of independent random variables and laws of large numbers for bilinear forms.

1. Let us recall a well-known generalization of the concept of Orlicz spaces. We introduce the following relation on the set of function $G: R^+ \to R^+$. $G_1 \overset{\circ}{\prec} G_2$, if there are numbers C, $x_0 > 0$ such that for each $x \in (0, x_0)$

$$G_1(x) \leq C G_2(x); \quad (1)$$

$G_1 \overset{\circ}{\sim} G_2$, if $G_1 \overset{\circ}{\prec} G_2$ and $G_1 \overset{\circ}{\succ} G_2$. Relation $\overset{\circ}{\prec}$ is an order relation, and relation $\overset{\circ}{\sim}$ is a relation of equivalence. If inequality (1) is fulfilled at every $x \in (x_0, \infty)$, then we get other order and equivalence relations, which will be denoted by $\overset{\infty}{\prec}$ and $\overset{\infty}{\sim}$. An **increasing function** $G \in \Sigma$ will be called an Orlicz function if $G(0) = 0 = \lim_{x \to 0+0} G(x)$ and if for G Δ_2-condition is fulfilled, i.e. there exist $C > 0$ such that for each $x \in R^+$, $G(2x) \leq CG(x)$. The Orlicz space $L^G(B)$ consists of Bochner measurable functions f defined on the measure space (Y, B, μ) with values in the Banach space B such that

$$\tau_G(f) = \int_Y G(\|f(x)\|)\, d\mu(x) < \infty.$$

The topology of the space $L^G(B)$ is defined by the quasinorm

$$\|f\|_G = \inf_{\lambda > 0} \lambda\left(1 + \tau\left(\frac{f}{\lambda}\right)\right), \quad f \in L^G(B).$$

Thus, the space $L^G(B)$ is a complete metric linear space. The Orlicz spaces of sequences in B will be denoted by $\ell^G(B)$. We shall consider Orlicz spaces with not necessary convex Orlicz functions G.

Let $L^0(B)$ be the space of random element with values in a Banach spaces B defined on a probability space $(\Omega, \mathcal{O}\!\mathcal{L}, P)$. We shall consider in $L^0(B)$ the opology of convergence in probability defined by the quasinorms

$$J_\alpha(\|X\|, P) = \inf\{a : P(\|X\| \geq a) \leq \alpha\}, \quad X \in L^0(B);$$

$0 < \alpha < 1$. If $B = R$ the spaces $L^G(B), \ell^G(B), L^0(B)$ will be denoted L^G, ℓ^G, L^0 respectively.

In what follows ξ_n, $n \in N$, is a sequence of independent symmetric identically distributed random variables, F is their common distribution function, $\bar{F}(x) = 1 - F(x)$, $x \in R^+$. Define the Orlicz functions $g_{F,p}$, $0 < p < \infty$, by

$$g_{F,p}(x) = \int_0^\infty \min(x^p t^p, 1)\, dF(t), \quad x \in R^+.$$

If $p = 2$ we write g_F instead of $g_{F,p}$.

We shall concider only symmetric random variables and their distribution functions.

Definition 1. Let F, F_1 be distribution functions. We call a Banach space B a space of (F,F_1)-type, if the convergence of the series $\sum_{n=1}^\infty g_{F_1}(\|x_n\|)$, $x_n \in B$

implies the almost sure convergence in B of the series $\sum_{n=1}^{\infty} \xi_n x_n$.

If distribution functions F, F_1 coincide then a Banach space B of (F, F_1) -type will be called a F-type space. A Banach space B of (F_r, F) -type, where F_r is the distribution function of a Rademacher random variable r (i.e. $P\{r=1\} = P\{r=-1\} = \frac{1}{2}$) will be called a space of r -type F.

The notions of type F and r -type F spaces are natural generalizations of the notons of Rademacher type p and stable type p spaces. The F type spaces were considered in [3]. The goal of this paper is the further study of the spaces of type F. We try in particular to find out for which distribution functions F, F_1 a Banach space B of r -type F_1 is of type F.

2. We give sufficient conditions for a subspace the space $L^G(B)$ to be of (F, F_1)-type and indicate some cases when these conditions are necessary as well.

Theorem 1. Let G be an Orlicz function such that

$$G(\sqrt{x+y}) \leq G(\sqrt{x}) + G(\sqrt{y}), \quad x, y \in R^+ \qquad (2)$$

and

$$G(xy) \leqslant C\, g_F(x)\, G(y), \quad x \in (0, x_0), \; y \in R^+ .$$

Then the space L^G is of r -type F.

Corollary. (1) Let $g_F \overset{\circ}{\succ} x^p$. Then the space L^p is of r -type F. (2) The space ℓ^p is of r -type F if and only if $g_F \overset{\circ}{\succ} x^p$.

Example 1. Any Orlicz function g_F satisfies condition (2). Thus if an Orlicz function g_F satisfies the Δ_3 -condition: there are numbers C, $x_0 > 0$ such that $g_F(xy) \leqslant C g_F(x) g_F(y)$, $x, y \in (0, x_0)$ then the Orlicz space ℓ^{g_F} is of r -type F.

The following theorem deals with an Orlicz space $L^G(B)$ of random elements.

Theorem 2. Let the Banach space B be of type (F,F_1) the Orlicz function G satisfy the condition $g_F(xy)\leq C g_F(x)G(y)$, $x\in(0,x_0)$, $y\in(x_1,\infty)$ and let B_1 be a subspace of $L^G(B)$ such that the topology of B_1 coinsides with the topology of convergence on probability. Then the space B_1 is of type (F,F_1).

Let ξ_1,ξ_2 be independent random variables with distribution functions F_1,F_2 respectively. Their product $\xi_1\cdot\xi_2$ is a symmetric random varialbe. Denote by F_3 the distribution fuction of $\xi_1\cdot\xi_2$.

Theorem 3. A Banach space B of r-type F_2 is a space of type (F_1,F_3).

From the theorem and Example 1 we have:

Corollary. Let the distribution function in the condition of theorem 3 satisfy the Δ_3-condition. Then

(1) The space $\ell^{g_{F_2}}$ is of (F_1,F_3)-type.

(2) If the space $\ell^{g_{F_2}}$ is of (F_1,F_4)-type, then $g_{F_4}\overset{c}{>} g_{F_3}$.

Item (2) of the corollary shows that the type (F_1,F_3) is a maximal element in the set of the types (F,F_4) of the space $\ell^{g_{F_2}}$.

Lemma 1. Let distribution function F_1,F_2 have the properties

$$\overline{\lim_{t\to\infty}}\frac{\bar{F}_1(t)}{\bar{F}_1(\alpha t)}=0,\ \left(\underline{\lim_{t\to 0+0}}\frac{g_{F_2,p}(\alpha t)}{g_{F_2,p}(t)}\right),\quad \alpha\to\infty,$$

and $g_{F_2}(xy)\leq\varphi(x)g_{F_2}(y)$, $x,y\in(0,x)$, $\varphi(x)\to 0$, $x\to 0$.
Then $g_{F_3,p}\overset{c}{\simeq} g_{F_1,p}\overset{c}{\simeq}\bar{F}_1(1/x)$.

From Lemma 1 with $p=2$ and Theorem 3 follows

Theorem 4. Let distribution functions F_1,F_2 satisfy the conditions of lemma 1 and the Banach space B be of r-type F_2. Then the space B is of type F_1.

In a private communication V.I.Tarieladze noted that a Banach space of type 2 is of type F for any distribution function F . The following statement complements this result.

Denote by Ψ_p , $0<p<\infty$ the class of distribution functions such that

$$\overline{\lim_{t\to\infty}} \frac{\bar{F}(t)}{\bar{F}(\alpha t)} = O(\alpha^p), \quad \alpha \to \infty .$$

Corollary 1. A Banach space B of Rademacher type p is of p -type for any distribution function $F\in\Psi_p$, $1<p<2$.

From Corollary 1 a well-known result of G.Pisier [4] follows: a Banach space B of Rademacher type p is of stable type $p-\varepsilon$ for any $\varepsilon>0$.

Corollary 2. Let $F\in\Psi_1$. Then any Banach space is of p -type.

It is shown in [5] that any Banach space is of a stable type p , $0<p<2$. Corollary 2 is a generalization of this statment.

Corollary 3. Let the Banach space B be of a stable type p , $0<p<2$. Then the space B is of F -type for any distribution function F belonging to the domain of attraction of a p -stable law.

Example 2. Let F be a distribution function such that $\bar{F}(x) = 1/x^2$, $x \geqslant 1$, Then $g_F \doteq x^2 \ln(1+\frac{1}{x})$. The Orlicz function g_F sutisfies the Δ_3 -condition. Therefore by Theorem 1 the space ℓ^{g_F} is of r -type F .

Since $\lim\limits_{t\to 0} \dfrac{g_F(\alpha t)}{g_F(t)} = \alpha^2$, $\alpha>0$ by Theorem 4 the space ℓ^{g_F} is of type F for any $F\in\Psi_2$. The space ℓ^{g_F} is not a type 2 space.

3. Let us introduce spaces of sequences in a Banach space B to be used in the study of spaces of (F,F_1)-type.

Denote by $\ell\{F,B\}$ the space consisting of such sequences $x_n\in B$ that the series $\sum\limits_1^\infty \xi_n x_n$ converges almost surely in B. The space $\ell\{F,B\}$ is embedded in $L^0(B)$ (a sequence $(x_n)\in\ell\{F,B\}$ is mapped into $\sum_{n=1}^{\infty}\xi_n x_n$) and the topology of $\ell\{F,B\}$ is defined as the pre-mage of the topology of the space $L^0(B)$. The space $\ell\{F,B\}$ is a complet metrrizable linear space [6].

A definition of the (F,F_1)-type can be given in terms of the space $\ell\{F,B\}$.

Proposition 1. A Banach space B is of type (F,F_1) if and only if $\ell^{g_{F_1}}(B)\in\ell\{F,B\}$. This embedding is continuous.

Corollary. Let B be a Banach space. The following conditions are equivalent:

(a) B is of type (F,F_1);

(b) for any $0<\alpha<1$ there exist $T=T(B,\alpha)$ such that for any finite sequence $x_1,\dots,x_n\in B$

$$J_\alpha(\|\sum_{i=1}^{n}\xi_i x_i\|,p)\leq T(B,\alpha)\|(x_i)_{i=1}^{n}\|g_{F_1}$$

The Corollary shows that the (F,F_1) type of a Banach space is an infraproperty. The following theorem is proved in the same way as Theorem 1 in [7], p.96.

Theorem 5. Let B_0 be a closed subspace of a Banach space B of type (F,F_1). Then the factorspace $B_1=\frac{B}{B_0}$ is of (F,F_1)-type and $T(B_1,\alpha)\leq T(B,\alpha)$, $0<\alpha<1$.

Let us describe properties of a space $\ell\{F,B\}$.

Theorem 6. Let F be a distribution function, $F\in\Psi_2$, B be a Banach space.

(1) The space $\ell\{F,B\}$ coincides with the space $\ell\{F_1,B\}$ if and only if $\bar{F}\simeq\bar{F}_1$.

(2) B is of type F if and only if $\ell\{F,B\}=\ell^{G}(B)$, where $G(x)=\bar{F}(1/x)$, $x\in R^+$.

We give conditions for the space $\ell(F,B)$ and $\ell^p(B)$, $0<p<2$ to coincide.

Corollary 1. The following conditions are equevalent:

(a) $\ell\{F, B\} = \ell^p(B)$;

(b) B is of the stable type p and there exist C_1, $C_2, x_0 > 0$ such that for each $x \in (x_0, \infty)$

$$C_1/x^p \leq \bar{F}(x) \leq C_2/x^p.$$

Let ξ_n^i, $n \in N$, $1 \leq i \leq m$, be mutually independent random variables such that the random variables ξ_n^i, $n \in N$ have a common distribution function F_i, $1 \leq i \leq m$. We define $\ell\{F_1, \dots, F_m\}$ as the space of such sequences $\alpha_{i_1,\dots,i_m}$ that the series $\sum_{i_1,\dots,i_m} \alpha_{i_1\dots i_m} \xi_{i_1}^1 \dots \xi_{i_m}^m$ converges in probability. The space $\ell\{F_1, \dots, F_m\}$ is embedded into L^0 (the sequence $(\alpha_{i_1\dots i_m}) \in \ell\{F_1, \dots, F_m\}$ is mapped into $\sum_{i_1\dots i_m} \alpha_{i_1\dots} \alpha_{i_m} \xi_{i_1}^1 \dots \xi_{i_m}^m$) and the topology of the space $\ell\{F_1, \dots, F_m\}$ is defined as the preimage of the topology in L^0.

<u>Corollary 2</u>. Let distributions functions F_1, F satisfy the conditions of lemma 1 and the Orlicz function g_F satisfy the Δ_3-condition. Then $\ell\{F_1, F\} = \ell^{g_{F_1}}(\ell^{g_F})$.

<u>Corollary 3</u>. Let F_i, $i \leq m$, $1 \leq i$, be the distribution functions of p_i-stable random variables and let $p_i < p_j$ if $i < j$. Then $\ell\{F_1, F_2, \dots, F_m\} = \ell^{p_1}(\ell^{p_2} \dots (\ell^{p_m}) \dots)$.

The following proposition shows that the space $\ell\{F, L^p\}$ has the structure of a space with mixed norm.

<u>Proposition 2.</u> (1) If a distribution function F has the second moment, then $\ell\{F, L^p\} = \ell^{g_{F,p}}(L^p) \cap L^p(\ell^2)$.

(2) If distribution function F has the p-th moment, then $\ell\{F, L^p\} = L^p(\ell^2)$.

The space $\ell\{F\}$ is isomorphic to a subspace of L^0. Consequently it does not contain a subspace isomorphic ot ℓ^q, $2 < q < \infty$, [7]. This is not true for spaces $\ell\{F, B\}$. Using Proposition 2 we can construct such a distribution function F that the space $\ell\{F, \ell^p\}$ contains a subspace isomorphic to ℓ^q, $2<q<p$.

4. Using Theorem 6 we can characterize (F,F_1)-type spaces in terms of the weak law of large numbers.

Let B be a Banach space, $x_k^n \in B$, $1 \le k \le n$, $n \in N$, $b_n \uparrow \infty$. Consider the relation of the following statements:

(a) the sequence of random elements $\frac{\sum_{k=1}^{n} \xi_k x_k^n}{b_n} \to 0$ as $n \to \infty$ in probability;

(b) $\sum_{k=1}^{n} g_{F_1}\left(\frac{\|x_k^n\|}{b_n}\right) \to 0$ as $n \to \infty$.

<u>Theorem 7</u>. (1) A Banach space B is of type (F,F_1) if and only if (a)$\Leftarrow$(b).

(2) Let $F \in \Psi_2$. A Banach space B has type F if and only if (a)$\Leftrightarrow$(b).

<u>Corollary</u>. Assume that the conditions of Corollary 2 Theorem 6 are satisfied. Let $\zeta_j^n = \sum_{i=1}^{\infty} \alpha_{ij}^n \xi_i^1$, $\alpha_i^n \in R$, $i,j,n \in N$, $b_n \uparrow \infty$. Then $b_n^{-1} \sum_{k=1}^{n} \zeta_k^n \xi_k^2 \to 0$ as $n \to \infty$ if and only if $\sum_{i,j}^{n} \alpha_{ij} e_{ij} \to 0$ in $\ell^{g_{F_2}}(\ell^{g_{F_1}})$ where $\{e_{ij}\}$ is a coordinate basis in the space $\ell^{g_{F_2}}(\ell^{g_{F_1}})$.

(F,F_1) type spaces can also be characterized in terms of the strong law of large numbers. Let b_n be a numerical sequence such that $b_n \uparrow \infty$, $n \to \infty$ and

$$\sup_{n \in N} b_n^{-1} b_{2n} < \infty .$$

<u>Theorem 8.</u> Let B be a Banach space. The following conditions are equivalent:

(a) B is of type (F,F_1);

(b) if $\sum_{n=1}^{\infty} g_{F_1}\left(\frac{\|x_n\|}{b_n}\right) < \infty$, $x_n \in B$, then $\frac{\sum_{k=1}^{n} x_k \xi_k}{b_n} \to 0$ as $n \to \infty$ almost shurely in B.

When the space B is of r-type F Theorem 8 can be sharpened as follows.

Theorem 9. Let X_n be a sequence of independent centered random elements with values in B and b_n be a numerical sequence from Theorem 8. The following conditions are equivalent:

(a) B is of r-type F ;

(b) if $\sum_{n=1}^{\infty} Eg_F(\|X_n\|) < \infty$, then the series $\sum_{n=1}^{\infty} X_n$ converges almost surely in B ;

(c) if $\sum_{n=1}^{\infty} Eg_F\left(\frac{\|X_n\|}{b_n}\right) < \infty$, then $\frac{\sum_{k=1}^{n} X_k}{b_n} \to 0$ as $n \to \infty$ almost surely in B ;

(d) if $\sum_{n=1}^{\infty} g_F\left(\frac{\|x_n\|}{b_n}\right) < \infty$, $x_n \in B$, then $\frac{\sum_{k=1}^{n} r_k x_k}{b_n} \to 0$ as $n \to \infty$ almost surely in B where $r_k, k \in N$ are Rademacher functions.

Theorem 10. Assume that $F \in \Psi_2$ has the p-th moment, $1 \le p \le 2$, and the Orlicz function g_F satisfies the Δ_3-condition. Let α_{ij}^{n}, $1 \le i, j \le n$, $n \in N$ be numbers such that $\sup_{i,j,n} |\alpha_{ij}^{n}| < \infty$. Then

$$\frac{1}{n^{2/p}} \sum_{i,j=1}^{n} \alpha_{ij}^{n} \xi_i \xi_j \to 0 \qquad \text{as} \quad n \to \infty$$

in probability.

REFERENCES

1. Ledoux M. Loi du Logarithme itere et type Ψ d'espaces de Banach. - Ann.sci.Univ.Clermont-Ferrund, 1981, N 19, p.207-220.
2. Lapreste T.-T. Charges, poids et mesures de Levy dans les espaces vectoriels localement convexes. - Seminaires de Probability de Strasbourg, X11, 1977-1978, p.41-71.

3. Чупрунов А.Н. Топологические признаки плотности цилиндрических мер. - Казань, КГУ. I44 с.
4. Pisier G. Une propriete du type p-stable. - Seminaire Maurey - Schwartz, 1973-1974, exp. n N 8.
5. Maurey B. Espaces de cotype p, $0<p<2$. - Seminaire Maurey - Schvatz, 1972-1973, exp.7.
6. Hoffman-Jørgensen J. Sums of independent Banach spaces valued random variables. - Preprint N 15 - Mat. inst., Aarhus univ., 1972, 89 p.
7. Муштари Д.Х. Вероятности и топологии в банаховых пространствах. - Рукопись деп. в ВИНИТИ, № 4264-79. 24I с.
8. Hoffman-Jørgensen J., Pisier G. The law of Large numbers and the central limit theorem in Banachspspaces. - Ann.Probab., v.4, N 4, p.587-599.

Prob. Theory and Math. Stat., Vol.1, pp. 385-404
Prohorov et al. (eds)
1986 VNU Science Press

SMOOTH MEASURES ON INFINITE DIMENSIONAL MANIFOLDS AND FORWARD KOLMOGOROV EQUATION

Yu.L.Dalecky
Kiev Polytechnik Institute, Kiev, U.S.S.R.

1. Measures as well as functions plays fundamental part in analysis on infinite dimensional manifolds. Thus it is a natural problem to construct and investigate those measures which are compatible with some additional structures (smooth or algebraic) on a manifold. Gaussian measures in linear spaces give a species of such measures. But this class is poor enough both from analytical point of view and from the point of view of mathematical physics of systems with infinite number of freedom degrees. An enrichment of this class may be construct by means of measures which are nonlinear transformations of Gaussian measures. In this report we consider some measure properties which are invariant with respect to smooth transformations and thus may be extend to measures defined on smooth manifolds.

Fundamental notions of differential calculus for measures defined in linear space had been introduced by S.Fomin and A. Skorochod (see [1,2,3]) and developed by some other authors (we are not going to give a historical review here and send a reader to comments and references in the book [4]). To deal with analysis on nonlinear manifolds one needs new invariant notions, connected with necessity to calculate derivatives of the measures along vector fields [5]. One of the motives of this activity was an attemp to extend Malliavin calculus to infinite dimensional manifolds (see [6]).

Stochastic differential Ito equations give possibility to construct random variables on smooth manifolds by means of special nonlinear transformations of a Gaussian white noise in a manifold model space. This procedure is described in [7] for infinite dimensional manifolds. Probability distributions of random values, constructed in this way are measures on manifolds and they may be treated as nonlinear images of Gaussian measures with respect to transformations possessing certain smooth properties. Compatibility of those measure properties with additional manifold structures is guaranteed by the corresponding coefficient properties of Ito equations. In particular, in the case of finite dimensional manifolds a smooth property of a measure is equivalent to existence of its smooth density with respect to a natural volume. For infinite dimensional manifolds there is no analogue of the volume measure and instead one say that a measure is smooth if it has derivatives along sufficiently large number of vector fields.

It is well known that Markov process transition probabilities are connected with solutions of Kolmogorov diffusion equations. A forward Kolmogorov equation (a Fokker-Plank equation) may be considered as an equation with respect to transition probability density in finite dimensional case. In infinite dimensional case the forward equation is essentialy a differential equation with respect to measure. This equation may be treated in a weak sense as a dual equation with respect to a backward Kolmogorov equation (which is an equation with respect to a function) (see [8]). Here formulate a method based on measure shift along trajectories of stochastic equation solutions, which gives a direct approach to construction a measure which is the solution of the forward Kolmogorov equation.

The following exposition is given in a review form. For the sake of simplicity we consider in detail the case of linear spaces, but at the end of the report there is a short description of possible extensions to the case of nonlinear manifolds.

2. Introduce some notions and notation needed below. By Banach space we mean a real separable Banach space. If B_1, B_2 are Banach space, then $L_n(B_1, B_2)$ is the Banach space of n-linear bounded mappings from B_1 into B_2; $L(B_1, B_2) = L_1(B_1, B_2)$; $L(B) = L(B, B)$, $B^* = L(B, \mathbb{R}^1)$, $\langle x, y\rangle_B$-pairing between $x \in B$ and $y \in B^*$ (the index B is omitted when no confusion can arise).

Let H be a Hilbert space with the inner product $(x, y)_H$. A mapping $\varphi \in L(H, B)$ will be called 2-absolutely summing, if there exists the constant C such that for any $x_1, \dots, x_n \in H$ the following inequality is true

$$\sum_j \| \varphi(x_j) \|_B^2 \leq C^2 \sup_{y \in H} \sum_j |(x_j, y)_H|^2.$$

Such mapping form the Banach space $L_{1,2}(H, B)$ with the norm $\sigma_2(\varphi) = \min C$. In the case, when B also is a Hilbert space, $L_{1,2}(H, B)$ consists of Hilbert-Schmidt operators and it is the Hilbert space with the inner product

$$\operatorname{Tr}_H \psi^* \varphi = \sum_k (\varphi(e_k), \psi(e_k))_B = \sum_k ((\psi^* \varphi) e_k, e_k)_H,$$

where $\{e_k\}$ is an ortonormal basis in H and the right-hand-sum is independent of basis choice. One can define the corresponding notions for the multilinear mapping in the same way.

We say that the Hilbert-Schmidt (HS) structure $(B; H)$ is defined in Banach space B, if some

Hilbert space H is densely embedded into B and the embedding $i: H \to B$ belongs to the class $L_{1,2}(H,B)$. In this case $i^*: B^* \to H^* = H$ and we obtain the tripple of densely embedded spaces $B^* \xrightarrow{i^*} H \to B$ with the relation $(h, i^* y)_H = \langle ih, y \rangle_B$ $(h \in H,\ y \in B^*)$.

Notice, that for $A \in L(B,H)$ we obtain $Ai \in L_{1,2}(H)$ and therefore operator $i^* A^* A i \in L(H)$ has the finite trace.

Let $f: U \to B_2$ be a mapping defined in some domain $U \subset B_1$. It will be called differentiable in $x_0 \in U$ along the linear space B densely embedded into B_1, if there exists $f'(x_0) h = \frac{d}{d\varepsilon} f(x_0 + \varepsilon h)\big|_{\varepsilon=0}$ for any $h \in B$ and $f'(x_0) \in L(B, B_2)$. If $f'(x)$ exists in the neighbourhood of x_0 and continuously depends on x, then $f(x_0+h) - f(x_0) = f'(x_0)h + o(\|h\|)$. It is possible to define the higher derivatives in the same way, for example $f''(x_0) \in L_2(B, B_2)$ such that $f(x_0+h) - f(x_0) = f'(x_0)h + \frac{1}{2} f''(x_0)(h,h) + o(\|h\|^2)$.

Denote $C_n(U, B, B_2)$ a Banach space of the continuously differentiable up to order n along B functions $f: U \to B_2$ with the norm $\|f\| = \sum_{k=1}^{n} \sup_x \|f^{(k)}(x)\|$ and $C_{n,2}(U, B, B_2)$ its subspace consisting of the functions with 2-absolutely summing derivatives with the norm $\|f\|_2^2 = \sum_{k=1}^{n} \sup_x \sigma_2^2 (f^{(k)}(x))$ (if B is the Hilbert space).

The mapping z of the Banach space B into itself belonging to the class C_n $(n \geq 1)$ will be called a vector field, the expression $\nabla_z \varphi = \langle z, \nabla \rangle \varphi = \langle z, \varphi' \rangle_B$ will be called the derivative of the function $\varphi \in C_1$ along the vector field z.

The vector field z generated in B the integral flow S_t^z $(t \in \mathbb{R}^1)$ of the smooth mappings $S_t: B \to B$, defined by the differential equation

$x'(t)=z(x(t))$, $x(0)=x$; $x(t)=S_t(x)$. The derivatives $S_t'(x)$ $(t\in\mathbb{R}^1)$ of these mappings satisfy the differential equations obtained through the formal differentiation of the original one for example

$$\frac{d}{dt}S_t'(x)=z'(x(t))S_t'(x)$$
.

3°. Let B a Banach space, $\mathcal{B}$ be a Borel σ-algebra of its subsets, $\mathfrak{M}_B$ be a Banach space of the real valued σ-additive bounded functions of sets in $\mathcal{B}$ with the norm $\|\mu\|=\operatorname{Var}\mu$ (such functions will be called measures below).

If α is μ-integrable function, then $\alpha\mu$ will be a measure absolutely continuous with respect to μ $(\alpha\mu<\mu)$ with density α . Denote $S\mu=\mu\circ S^{(-1)}$ the image of the measure μ with respect to a measurable mapping S .

A measure $\mu\in\mathfrak{M}_B$ will be called differentiable along the vector field z , if there exists the measure $\nabla_z\mu=-\frac{d}{dt}(S_t\mu)\big|_{t=0}$ - the derivative of the measure μ along the vector field z , defined in the following sence:

$$\int_B\varphi(x)(\nabla_z\mu)(dx)=-\lim_{t\to0}\int_B\varphi(x)\frac{(S_t\mu)-\mu}{t}(dx)=-\int_B(\nabla_z\varphi)(x)\mu(dx)$$

for any function $\varphi\in C_1(B,B,\mathbb{R})$. If in addition $\nabla_z\mu<\mu$ then the density $\rho_\mu(z;x)=\frac{(\nabla_z\mu)(dx)}{\mu(dx)}$ will be called a logarithmic derivative μ along z . Thus, a logarithmic derivative is definded by the relation (integration by part formula)

$$\int_B(\nabla_z\varphi)(x)\mu(dx)=-\int_B\varphi(x)\rho_\mu(z;x)\mu(dx) \qquad (1)$$

The next relations follows from (1):

$$\rho_\mu(\alpha z;x)=\alpha(x)\rho_\mu(z;x)+\nabla_z\alpha$$

and for $\alpha(x) \neq 0$ (μ - a.e.)

$$\rho_{\alpha\mu}(z;x) = \frac{1}{\alpha(x)} \rho_\mu(\alpha z; x) = \rho_\mu(z;x) + \nabla_z \ln \alpha(x).$$

These relations demonstrate that $\rho_\mu(z;x)$ possesses the properties of a first order differential operator with respect to z.

Let $(B; H)$ be a space equipped with an HS-structure. Consider a measure class $\mathcal{M}_1(B)$ consisting of measures possessing a logarithmic derivative along constant vector fields $z(x) \equiv h \in B^* \subset H$ representable in the form $\rho_\mu(h;x) = \langle \lambda(x), h \rangle_B$ (μ - a.e.), where $\lambda = \lambda_\mu : B \to B$ is a weakly measurable mapping which will be called a vector logarithmic derivative of the measure μ.

Let V_{B^*} be a space of the vector fields belonging to the class $C_1(B, B, B^*)$. For $\mu \in \mathcal{M}_1(B)$ and $z \in V_{B^*}$ the following relation holds

$$\rho_\mu(z;x) = \langle \lambda(x), z(x) \rangle_B + \operatorname{Tr} z'(x). \qquad (2)$$

It is possible to check this relation through the direct calculation in the finite dimensional case and then to extend it to the general case by means of limit procedure.

Then consider more restricted measure class $\mathcal{M}_2(B,H) \subset \mathcal{M}_1(B)$ consisting of measures whose vector logarithmic derivative $\lambda(x)$ is differentiable along H (μ - a.e.) and $\operatorname{vrai\,sup} \|\lambda'(x)\|_B < \infty$.

Let $\mathcal{H}(B, H, \mu)$ be a Hilbert space of the vector functions obtained by the complementation of $C_1(B, H, H)$ with respect to the norm

$$\sigma_{1,2}^{\mu}(z) = \left[\int_B \left\{ \|z(x)\|_H^2 + \operatorname{Tr} z'(x) z'(x)^* \right\} \mu(dx) \right]^{1/2}.$$

Theorem 1. (see [5]). Let $\mu \in \mathcal{M}_2(B,H)$ and $z \in \mathcal{H}(B, H, \mu)$. Then a measure μ is differentiable along a vector field z and possesses a logarithmic derivative $\rho_\mu \in \mathcal{L}_2(B,\mu)$ such that

$$\int_B |\rho_\mu(z;x)|^2 \mu(dx) = \int_B \{ \mathrm{Tr}_H\, z'(x)^2 - \langle \lambda'(x) z(x), z(x) \rangle_B \} \times \mu(dx). \quad (3)$$

<u>Proof</u>.

Let the vector field z be such that the function $\rho = \rho_\mu(z;x)$ is bounded on B along with its first derivative. It is easy to obtain that

$$\nabla_z \rho = (z, \nabla)\,[\langle \lambda(x), z(x) \rangle + \mathrm{div}\, z(x)] =$$

$$= \rho(\nabla_z z; x) + \langle \nabla_z \lambda(x), z(x) \rangle - \mathrm{Tr}\,[z'(x)]^2$$

and then

$$\int_B \rho^2(z;x)\mu(dx) = \int_B \rho(z;x)(\nabla_z \mu)(dx) =$$

$$= -\int_B \nabla_z \rho(z;x)\mu(dx) = \int_B \{ \mathrm{Tr}\, z'(x)^2 - \langle \lambda'(x) z(x), z(x) \rangle \} \mu(dx).$$

This relation yields the estimate

$$\int_B |\rho_\mu(z;x)|^2 \mu(dx) \leq \mathrm{const}\,[\sigma^\mu_{1,2}(z)]^2 \quad (4)$$

which demonstrates, that it is possible to extend the linear operation $\rho_\mu(z;x)$ to $\mathcal{H}(B,H,\mu)$ by continuity. Then it is possible to extend relations (1) and (3) to this place by means of limit procedure.

It is possible to estimate the integrals of the logarithmic derivative higher powers in such way. Transforming the integral

$$\int_B \rho^{2n}(z;x)\mu(dx) = \int_B \rho^{2n-1}(z;x)(\nabla_z \mu)(dx) =$$

$$= -(2n-1)\int_B \rho^{2n-2}(z;x)\nabla_z \rho(z;x)\mu(dx)$$

and by means of applying Holder and Minkovsky inequalities we obtain the estimate

$$\frac{1}{2n-1}\left\{ \int_B \rho^{2n}(z;x)\mu(dx) \right\}^{1/n} \leq \left\{ \int_B [\mathrm{tr}\, z'(x)^2 - \right.$$

$-\langle\lambda'(x)z(x),z(x)\rangle]^n\mu(dx)\}^{1/n}+\{\int_B \rho^n(\nabla_z z;x)\mu(dx)\}^{1/n}$.

As $z\in C_m(B,H)$ implies $\nabla_z z\in C_{m-1}(B,H)$ then it is possible to iterate this estimate. The following estimate was obtained in this way after the bulky enough calculations

$$\{\int_B \rho^{2k}(z;x)\mu(dx)\}^{1/2k} \le const\, \sigma^{\mu}_{2k,2k-1}(z), \qquad (5)$$

where

$$\sigma^{\mu}_{m,n}(z) = \sum_{j=0}^{n} \{\int_B \sigma_2^m(z^{(j)}(x))\mu(dx)\}^{1/m}.$$

Let $W^n_m(B,H,\mu)$ be a Banach space, obtained by completion of $C_m(B,H)$ with respect to norm $\sigma^{\mu}_{m,n}$.

Denote $\mathcal{M}_n(B,H)$ a measure class which consists of measures with vector logarithmic derivatives possessing the derivatives along H up to the order $n-1$, which are the uniformly bounded multilinéar mappings: $\max_{k\le n-1} \sup_x \|\lambda^{(k)}(x)\|_{L_k(H,H)} < \infty$

Consider the measures $\nabla_{z_n}\ldots\nabla_{z_1}\mu$ obtained by sequential differentiations of μ and let

$\rho_n(z_1,\ldots,z_n) = \frac{1}{n!}\sum(\nabla_{z_n}\ldots\nabla_{z_1}\mu)(dx)/\mu(dx)$, where the summation is taken over the all transpositions of the fields $z_1,\ldots,z_n$ be an order n symmetrize logarithmic derivative. It is possible to obtain the following results using the estimate (5).

<u>Theorem 2.</u> For $\mu\in\mathcal{M}_n(B,H)$ and $z_1,\ldots,z_n\in W^n_{2n}(B,H,\mu)$ $(n=2k)$ there exists $\rho(z_1,\ldots,z_n)\in\mathcal{L}_2(B,\mu)$.

4°. Consider the problem of a measure images smoothness corresponding to smooth space mapping. Let $f\in C_1(B,B_1)$ and z,z^f be the vector fields in B and B_1 respectively, f -compatible in the sense $z^f(f(x))=f'(x)z(x)$.

If μ is a measure, differentiable along z , then its image $\mu^f=\mu\circ f^{(-1)}$ will be differentiable along z^f and the following equality holds

$$\nabla_{z^f}(\mu^f) = (\nabla_z \mu)^f. \tag{6}$$

This equality demonstrates the invariance of the notion of measure derivative along a vector field with respect to smooth mapping. The relation (6) follows from the transformation sequence $\int_{B_1} \langle \varphi'(y), z^f(y)\rangle \mu^f(dy) =$

$$= \int_{B_1} \langle \varphi'(f(x)), f'(x)z(x)\rangle \mu(dx) = \int_B \langle (\varphi\circ f)'(x), z(x)\rangle \mu(dx) =$$
$$= -\int_B (\varphi\circ f)(x)(\nabla_z \mu)(dx) = \int_{B_1} \varphi(y)(\nabla_z \mu)^f(dy).$$

If f is the invertible mapping and the logarithmic derivative ρ_μ exists, then ρ_{μ^f} exists too and

$$\rho_{\mu^f}(z^f; y) = \rho_\mu(z; f^{(-1)}(y)).$$

In particular, for the constant directions the following relation is valid

$$\langle \lambda_{\mu^f}(f(x)), h\rangle = \langle \lambda_\mu(x), [f'(x)]^{-1}h\rangle + \mathrm{Tr}[f'(x)^{-1}h]'. \tag{7}$$

Under the hypotheses of theorem 1 it is possible (used the arguments used in [6] for the Gaussian measure μ).

Theorem 3. If $\mu \in \mathcal{M}_2(B, H)$, $z \in \mathcal{H}(B, H, \mu)$, $f \in C_1(B, B_1)$ then there exist the logarithmic derivative $\rho_{\mu^f}(z^f, y)$ belonging to the class $\mathcal{L}_2(B_1, \mu^f)$.

Proof.

For $\varphi \in C_1$ using (4) we obtain the estimate

$$\left|\int_{B_1} (\nabla_{z^f}\varphi)(y)\mu^f(dy)\right| = \left|\int_B (\varphi\circ f)(x)\rho_\mu(z; x)\mu(dx)\right| \leq$$
$$\leq \|\varphi\|_{\mathcal{L}_2} \sigma^\mu_{1,2}(z).$$

It follows from this that there exists the function $\beta \in \mathcal{L}_2$ such that

$$\int_{B_1} \nabla_{z^f}\varphi(y)\mu^f(dy) = -\int_{B_1} \varphi(y)\beta(y)\mu^f(dy)$$

and, obviously $\beta(y) = \rho_{\mu^{f}}(z^{f}, y)$

Notice, that it is possible to obtain analogous result for higher derivatives of the measure μ ,using theorem 2.

The obtained results give the way of construction examples of the differentiable measures in Banach spaces.

At first, it is known (see [2, 4]) that canonical Gaussian measure in the space B with HS-structure, defined by the characteristic functional $\int_B e^{i\langle x, y\rangle}\mu(dx) = e^{-\frac{1}{2}\|y\|_H^2}$ possesses the logarithmic derivative $\lambda(x) = -x$. One can construct more complicated examples combining variously the operation of smooth mappings and the changing of the given measure by the measure possessing the smooth density function with respect to the original measure.

5°. The notion of measure smoothness is closely connected with the quasi invariance notion that is the existence of a fairly rich collection of shifts such that the shifted measure is absolutely continuous with respect to the original one.

Let S_t $(t \in \mathbb{R}^1)$ be a mapping integral flow corresponding to the vector field $z \in V_{B^*}$ and $\mu_t = S_t\mu$, $\mu \in \mathfrak{M}_1(B)$. Let $\mu_t < \mu$ and the function $\beta(t, x) = \mu_t(dx)/\mu(dx)$ is differentiable with respect to t for $t = 0$ as element of $\mathcal{L}_1(B, \mu)$. Then it is easy to check the existence of the logarithmic derivative $\rho(z; x) = -\partial\beta(t, x)/\partial t\big|_{t=0}$. Notice, that it easy follows from the definition of β , that the following functional relation is valid

$$\beta(t+\tau, x) = \beta(t, S_{-\tau}x)\beta(\tau, x)$$

and the differential equation holds

$$\partial\beta(\tau,x)/\partial\tau = -\rho(z;S_{-\tau}x)\beta(\tau,x)$$

with solution of the form

$$\beta(t,x) = \exp\left\{-\int_0^t \rho(z;S_{-\tau}x)\,d\tau\right\}. \qquad (8)$$

Therefore, it is possible to obtain the function $\beta(t,x)$ if one know the logarithmic derivative $\rho(z;x)$ However, if one don't know apriori that $\mu_t<\mu$, we can't state the equality $\mu_t=\beta\mu$ which is equivalent to the relation

$$\int_B \varphi(x)\exp\left\{-\int_0^t \rho(z;S_{-\tau}x)\,d\tau\right\}\mu(dx)=\int_B \varphi(S_t x)\mu(dx) \qquad (9)$$

for the fairly rich collection of function φ .

A measure μ belonging to the class $\mathcal{M}_1(\mathcal{B})$ will be called analytical if $\lambda(x)$ be a (real) analytical function on the whole space B such that it is possible to expand $\lambda(x)$ into power series with infinite convergence radius.

<u>Theorem 4.</u> Let μ be an analytical measure and z be a vector field such that the function (8) is μ -integrable. Then $S_t\mu<\mu$ and $\beta(t,x)= = \frac{\mu_t(dx)}{\mu(dx)}$.

<u>Proof</u>.

It's sufficient to obtain the formula (9) for the bounded real analytical function φ . First let z be also a real-analytical vector field bounded along with its derivatives and such that the function $\rho(z,S_t x)$ is bounded and possesses the bounded derivatives. In this case the formula (9) is equivalent to the following relation

$$\int_B \varphi(x)\left\{\frac{d^k}{dt^k}e^{-\int_0^t \rho(z;S_{-\tau}x)d\tau}\right\}_{t=0}\mu(dx)=\int_B \frac{d^k}{dt^k}\varphi(S_t x)\Big|_{t=0}\mu(dx).$$

For $\kappa = 1$ it turns into the logarithmic derivative definition and for large κ it is possible to obtain this relation through the induction. It is not difficult to get rid of the circumscribed restrictions on the vector field by means of a limit procedure.

Corollary. Under the hypotheses of theorem for Higher derivatives $\nabla_z^{(\kappa)} \mu = \left(-\frac{d}{dt}\right)^{\kappa} (S_t \mu)\Big|_{t=0}$ the following relation holds

$$\nabla_z^{(\kappa)} \mu = \frac{d^{\kappa}}{dt^{\kappa}} \exp\left\{-\int_0^t \rho(z; S_{-\tau} x)\, d\tau\right\}\Big|_{t=0} \mu .$$

Example 1. Let $z(x) \equiv h \in B^*$, $S_\tau x = x + \tau h$.

Then $\rho(t, x) = \exp\left\{-\int_0^t < \lambda(x - \tau h), h > d\tau\right.$.

Example 2. Let $z(x) = Kx$, $K \in L(B, B^*)$, $S_t x = e^{Kt} x$.

Then

$$\rho(t, x) = \exp\left\{-\int_0^t < \lambda(e^{-K\tau} x), e^{-K\tau} x > d\tau - t \, Tr_H K\right\}.$$

(We assume λ is analytical and the obtained density function is μ -integrable). For the Gaussian measure the obtained formulae are well known.

6. Then we consider the aspects connected with the stochastic equation theory. First of all we describe the general construction following [6,7]. Let H_0 be a Hilbert space, $w(t)$, $(t \geq 0)$ -canonically connected with H_0 Wiener random process, whose trajectories a.e. belong to some extension $H_- \supset H_0$ with the HS -imbedding $H_0 \to H_-$, $\mathcal{F}$ - corresponding σ - algebra flow. Let (B, H) be a Banach space with HS-structure possessing a smooth norm $\|x\|_B^2 \in C_2$.

Consider a stochastic equation in B:

$$\xi(t) = x + \int_0^t a(\xi(s))\, ds + \int_0^t A(\xi(s))\, dw(s), \quad (10)$$

where $a: B \to B$, $A: B \to L_{1,2}(H_0, B)$ smooth functions with the bounded derivatives

Under those conditions the equation (10) has (up to stochastic equivalence) exactly one solution $\xi_x(t)$, which can be obtained by successive approximations with mean square convergence (i.e. with respect to the norm $\sup_t E\|\xi(t)\|_B^2$. In addition the mapping $x \mapsto \xi_x(t) = T_t(x)$ is mean square continuous differentiable. The derivative $T_t'(x)$ is described by the linear stochastic equation with the random coefficients (the first extension of (10))

$$T_t'(x)h = h + \int_0^t a'(\xi(s))T_s'(x)h\,ds + \int_0^t A'(\xi(s))(T_s'(x)h, dw(s)) \quad (11)$$

Along with the equation (10) we shall consider the "inverse type" equation

$$Y_t(y) = \eta_\tau(t) = y + \int_\tau^t \tilde{a}(\eta_\tau(s))ds + \int_\tau^t \tilde{A}(\eta_\tau(s))dw(s) \quad (12)$$
$$(t \le \tau)$$

with the same Wiener process $w(s)$, but with solutions, subordinate to the inverse σ-algebra flow $\mathcal{F}_{[t,\tau]}$. The linear equation for the derivatives of $Y_t'(y)h$ like (11) correspond to it.

In order to establish the connection between these mappings, generated by the equations (10) and (12) we use the following stochastic equation regularization. Let $\theta_\varepsilon(t)$ be the infinite differentiable function with the support $[-\varepsilon, \varepsilon]$ such that $|\varepsilon\theta_\varepsilon(t)| \le const$ $\int_{-1}^{1} \theta_\varepsilon(t)dt = 1$. Then a random function $w_\varepsilon(t) = \int_{-\infty}^{\infty} \theta_\varepsilon(t-s)w(s)ds$ is infinite differentiable (a.s.) and $w_\varepsilon'(t) = \int_{-\infty}^{\infty} \theta_\varepsilon(t-s)dw(s)$.

Consider an ordinary differential equation with the random coefficients in B:

$$\frac{d\xi_\varepsilon}{dt} = a_1(\xi_\varepsilon) + A_1(\xi_\varepsilon)G w_\varepsilon'(t), \quad \xi_\varepsilon(0) = x \quad (13)$$

where $G \in L_1(H_-, H_0), G|_{H_0} \in L_{1,2}(H_0)$. The equation (13) can be rewritten in the form $\xi_\varepsilon(t-\varepsilon) = x + \int_0^{t-\varepsilon} a_1(\xi_\varepsilon(s-\varepsilon))ds +$
$$+ \int_0^{t-\varepsilon} A_1(\xi_\varepsilon(s-\varepsilon))dw(s) + \frac{1}{2}\int_0^{t-\varepsilon} Tr\{A_1'(\xi_\varepsilon(s-\varepsilon)(A_1(\xi_\varepsilon(s-\varepsilon))\cdot,\cdot)\}\times$$
$$\times ds + \xi(t,\varepsilon,G),$$
where $\xi(t,\varepsilon,G)$ is a random variable with the properties
$$E\|\xi(t,\varepsilon,G)\|^2 \to 0 \quad (\varepsilon \to 0;\ \|Gy - y\| \to 0,\ \forall y).$$

These arguments permit to deduce the following approximation
$$E\|\xi(t) - \xi_\varepsilon(t)\|^2 \to 0, \tag{14}$$
where $\xi(t)$ is the solution of (10) with the coefficients $A = A_1;\ a = a_1 + \frac{1}{2} Tr\{A_1'(A_1\cdot,\cdot)\}$. The same arguments can be applied to the extended equation (11) and respectively to the differential mappings T_t', $T_{t\varepsilon}'$. Then notice, that the equation (13) can be inversed with respect to time. The solution $\eta_{\tau,\varepsilon}(t)$ of the following problem
$$\frac{d\eta_{\tau,\varepsilon}(t)}{dt} = -a_1(\eta_{\tau,\varepsilon}) - A_1(\eta_{\tau,\varepsilon})Gw_\varepsilon'(t),\quad \eta_{\tau,\varepsilon}(\tau) = y\ (t<\tau) \tag{15}$$
generates the mapping $y \mapsto Y_{0,\varepsilon}(y) = \eta_{\tau,\varepsilon}(0)$ inverse with respect to the mapping $x \mapsto T_{\tau,\varepsilon}(x) = \xi_\varepsilon(\tau)$, corresponding to the equation (13)
$$Y_{0\varepsilon} \circ T_{\tau,\varepsilon} = id,\quad Y_{0,\varepsilon}'(T_{\tau,\varepsilon}x)T_{\tau,\varepsilon}'(x) = I\ (\text{п.н.}) \tag{16}$$

On the other hand, the solutions of the equation (15) converge in the sense (14) to the solutions of the inverse type stochastic equation (12) with the coefficients
$$\tilde{a} = a_1 - \frac{1}{2} Tr\{A_1'(A_1\cdot,\cdot)\},\quad \tilde{A} = A_1$$

Thus $\mathcal{F}$ -measurable random mappings

$$T_{\tau,\varepsilon} : x \mapsto \xi_x(\tau) \quad ; \quad Y_0(y) : y \mapsto \eta_\tau(0)$$

possess the following properties. There exist the "smoothed" mappings $T_{\tau,\varepsilon}(x)$, $Y_{0,\varepsilon}(y)$ such that the following relation holds (16) and

$$M\|T_\tau(x) - T_{\tau,\varepsilon}(x)\|^2 \to 0 ; \quad M\|Y_0(y) - Y_{0,\varepsilon}(y)\|^2 \to 0; \qquad (17)$$
$$M\|T_\tau'(x) - T_{\tau,\varepsilon}'(x)\|^2 \to 0; \quad M\|Y_0'(y) - Y_{0,\varepsilon}'(y)\|^2 \to 0$$

($\varepsilon \to 0$, $G \to I$ in the strong sense).
Notice, that it is possible to construct the subsequences T_{τ,ε_n} Y_{0,ε_n} such,that the convergence in (17) is fulfilled with the probability 1, however the total measure set, for which such a convergence takes place, cen be depended on the starting value x, y. On the other hand it is possible to construct the subsequences such that for a given measure μ on B the corresponding relations converge with the probability 1 on the μ -total measure set of values x, y.

Now we consider the problem of the measure differentiation along the"stochastic vector field" associated with the equation (10). Let $\mu \in \mathfrak{M}_2(B,H)$ and $\mu_t = T_t \mu$ be the a.s.- defined random measure, inducing the random process in the space $\mathfrak{M}_B$.

The expression $d\mu_t = \beta_t dt + \nu_t dW(t)$, where β_t, ν_t are measures, depended on the parameter t, will be called the stochastic differential of this processs, if the following relations are valid (for $\varphi \in C_1(B,\mathbb{R}^1)$):

$$M\left\{\int_B [\varphi(T_t x) - \varphi(x)]\mu(dx) - \int_B \varphi(x)(d\mu_t)(dx)\right\} = o(t)$$

$$M\left\{\int_B [\varphi(T_t x) - \varphi(x)]\mu(dx) - \int_B \varphi(x)(d\mu_t)(dx)\right\}^2 = o(t)$$

Using formulae for the logarithmic derivative transformations and the limit relations (17) one can compute

$$\rho_{\mu_\tau}(T_\tau' z; y) = \langle \lambda_\mu(Y_0 y), a(Y_0 y)\rangle + Tr\, a'(Y_0 y) \quad (18)$$

Now let the coefficients of the stochastic equation (10) satisfy the conditions $a(x) \in B^*$, $A(x) \in L(H_0, B^*)$. Introduce for the operator function $C(x)$ the second order differential operator $(\nabla \otimes \nabla, C) =$

$$= \sum_{j,k} (e_j, \nabla)(e_k, \nabla)(C(x)e_j, e_k).$$

Using Ito formula one can compute the "logarithmic stocastic differential" for the random measure described above

$$\left.\frac{d\mu_t(dx)}{\mu_t(dx)}\right|_{t=0} = (\nabla + \lambda(x), A(x)\,dw) + \quad (19)$$
$$+ \left[\frac{1}{2}\left((\nabla + \lambda(x)) \otimes (\nabla + \lambda(x)), A^*(x)A(x)\right) - (\nabla + \lambda(x), a(x))\right] dt.$$

Consider the mapping in the measure space, obtained through the averaging of this random measure $\mu \mapsto U_t\mu = E\mu_t$. It is easy to see, that $U_t\mu(\Delta) = \int_B \mu(dx)\, P_t(x, \Delta)$ where $P_t(x, \Delta)$ is the transition probability of the Markov Process $\xi_x(t)$ and therefore the mapping U_t form the semigroup in the measure space $\mathcal{M}_B$. Using formula (19) one can compute the infinitezimal operator of this semigroup (more exactly, its restriction to $\mathcal{M}_2(B, H)$)

$$G\mu = \frac{1}{2}(\nabla \otimes \nabla, A^*(x)A(x)\mu) - (\nabla, a(x)\mu). \quad (20)$$

<u>Theorem 5</u>. Let $a \in C_1(B, B, B^*)$, $A \in C_1(B, B, L(H_0, B^*))$.

Under these conditions the Cauchy problem for the measure equation (a forward Kolmogorov equation, associated with the random process $\xi(t)$, satisfying the stochastic equation (10)

$$\frac{\partial\mu(t)}{\partial t} = G\mu(t) \quad ; \quad \mu(0)=\mu$$

is correct in the space $\mathcal{M}_{\mathcal{L}}(B,H)$ and its solution has the following form $\mu(t)=\mathcal{U}_t\mu=E\,T_t\mu$.

In [6] we indicated the other way of investigations the smoothness of the distributions of a random variables $\xi_x(t)$, described by the stochastic equations, which was based as well on the logarithmic derivative concept and the theorem 3 (more exactly, its variant for the Gaussian original measure). That way was connected with study the smoothness of the mapping $w'\mapsto\xi_x(t)$ of the "white noise trajectories" space into the space of values of the stochastic equation solution. The image of a canonycal Gaussian measure for this mapping is the probability distribution $\mu(t)(\Delta)=$ $=P\{\xi(t)\in\Delta\}$, in particular the transition probability $P_t(x,\Delta)$ for $\xi(0)=x$.

7°. In this section we shall briefly discuss some additional difficulties, arising if one is going to extend the above results to the case, when linear space is changed for the smooth nonlinear manifold.

Let X be a manifold belonging to the class C_n $(n\geq 2)$ modelled by a Banach space B with the smooth norm $\|x\|^2\in C_n$. On such a manifold there exists (see [9]) a unity partition permitting to reduce some constructions, in particular vector and tensor fields and connections to the local case. We shall assume that the model space possesses the HS-structure $(B;H)$ and respectively the analogous structure can be defined in X: the tangent bundle $\tau: TX\to X$ possesses the densely embedded Hilbert subbundle $\gamma: \mathcal{H}\to X$, $\gamma=\tau|_{\mathcal{H}}$, whose fibres induce the HS-structure $(T_xX, \mathcal{H}_x)$ for each $x\in X$.

By means unity partitions the problem of the measure differentiability can be reduced to local case. In order to consider this case it is necessary to obtain an invariant form of the formulas described above.

A local HS-manifold is a neighborhood of $\mathcal{U}$ a zero of a model space B. For the pair U, V of those neighborhoods the permissible mappings $f: U \to V$ belong to the class C_n and

$$f'(x) \in L(H), \dots, f^{(n)}(x) \in L_n(H).$$

Define a connection of a local manifolds by means of the bilinear mapping $\Gamma_x^{(u)}(y,z)$; $\Gamma_x^u : B \times B \to B$, smooth with respect to x and transforming under the mapping f in the following way

$$\Gamma_{f(x)}^{(V)}(f'(x)y, f'(x)z) = f'(x)\Gamma_x^{(U)}(y,z) - f''(x)(y,z).$$

The connection will be called compatible with HS-structure, if $\Gamma_x^{(u)}: H \times H \to H$. This property is invariant under the permissible mappings.

A covariant derivative of a vector field y along a vector field z is defined by formula

$\nabla_z y = y'(x)z(x) + \Gamma_x(z(x), y(x))$ and it is transformed according to the rule

$$\nabla_{f'(x)z} f'(x)y(x) = f'(x)\nabla_z y(x).$$

We shall consider the connections without torsion, i.e. with the symmetry property $\Gamma_x(z,y) = \Gamma_x(y,z)$.

Treelinear mapping

$$R_x(h,g,y) = \Gamma_x(h, \Gamma_x(g,y)) - \Gamma_x(g, \Gamma_x(h,y)) + \\ + \Gamma_x'(g)(h,y) - \Gamma_x'(h)(g,y)$$

will be called a curvature tensor for the given connection and its convolution - the bilinear functional

$\tilde{R}_x(g,y) = \sum_j (R_x(e_j, g, y), e_j)$, where $\{e_j\}$ is orthonormal basis in H, - will be called a Richitensor. We shall assme, that the Richi tensor exists and it is generated by a bounded operator $\tilde{R}_x$ in

$H: \widetilde{R}_x(g,y) = (\widetilde{R}_x g, y)$, $\sup\limits_x \|\widetilde{R}_x\| < \infty$.

As we indicated above, the described concept can be extend in the standard way to the nonlocal manifolds. In addition we obtain the following results.

The general definition (1) will be correct , if we take for $\nabla_z \varphi$ the derivative of the scalar function φ along the given vector field z , defined on the manifold X , assuming that the vector field induced the integral flow on the manifold (see [9]). Tt is possible to give an invariant description of a class $\mathcal{M}_2(B,H)$ of those measures on X for which theorem 1 would be valid after changing (3) by

$$\int\limits_X |\rho_\mu(z;x)|^2 \mu(dx) = \int\limits_X \left\{ \left([R_x - \nabla\lambda(x)] z, z \right) + Tr(\nabla z)^2 \right\} \times \mu(dx) \quad (21)$$

According to it the statement concerning the logarithmic derivative transformations under the manifold morphisms, in particular the analogy of theorem 3 are valid. Theorem 4 stating the connection between a logarithmic derivative and a density function of a measure shifted along the vector field trajectory, with respect to the original measure can be extended on the real-analytical manifold case.

We shall need more complicated constructions in order to consider the questions, connected with the stochastic equations. The description of those equations on the smooth manifold, the construction of its solutions and consideration of its differential properties one may see in [7] . Using this work results, we may extend theorem 5 to the considering situation. Notice that the formula (20) will be correct, if we take for ∇ corresponding operation, generating by the covariant derivative.

References

1. S.V.Fomin, Differentiable measures in linear spaces, Usp. mat nauk, 1968, 23, 1, 221-222.
2. V.I.Averbuch, O.G.Smoljanov, S.V.Fomin, Generalized functions and differential equations in linear topological spaces, Tr.MMO, 1971, 24, 133-174.
3. A.V.Skorochod, Integration in the Hilbert space, Moskva, Nauka, 1975, 231 p.
4. Yu.L.Dalecky, S.V.Fomin, Measures and differential equationes in infinite dimensional spaces, Moskva, Nauka, 1983, 384 p.
5. Yu.L.Dalecky, B.D.Marjanin, Smooth measures on infinite dimensional manifolds, DAN USSR,285,6, 1985.
6. Yu.L.Dalecky, Stochastic differential geometry, Usp. mat.nauk, 1983, 38, 3, 87-11 1.
7. Ya.I.Belopolskaja, Yu.L.Dalecky, Ito equations and differential geometry, Usp. mat. nauk, 1982, 37, 3, 95-142.
8. Yu.L.Dalecky, S.V.Fomin, Generalized measures in the Hilbert space and forward Kolmogorov equation, DAN USSR, 1972, 205, 4, 759-762.
9. S.Lang, Introduction to differentiable manifolds, New York, London, 1962.

Prob. Theory and Math. Stat., Vol.1, pp. 405-410
Prohorov et al. (eds)
1986 VNU Science Press

THE STRUCTURE OF DISTRIBUTIONS OF CONVEX FUNCTIONALS

Yu.A.Davydov
Leningrad State University, Leningrad, USSR

1. INTRODUCTION

Let X be a complete separable linear metric space, d be the invariant metric in X, P be a probability measure on Borel σ-algebra $\mathcal{B}_X$ of subsets of X, f be a continuous convex functional defined on X. If P is a Gaussian measure then in accordance with Tzirelson's result [1] the distribution Pf^{-1} of functional f has the following structure:

There is a number a_0 (possibly equals to $-\infty$) such that $Pf^{-1}([a_0,\infty))=1$, at the point a_0 (if $a_0 \neq -\infty$) Pf^{-1} can have an atom and on the semiline (a_0,∞) Pf^{-1} is absolutely continuous.

In this publication we investigate the structure of the distribution Pf^{-1} in nongaussian situation and demonstrate that the above qualitative picture is conserved under rather vast assumptions.

2. NOTATIONS

X - complete separable linear metric space,
d - invariant metric on X,
$\mathcal{B}_X$ - σ-algebra of Borel subsets of X,

P - probability measure on $\mathcal{B}_X$,
f - continuous convex functional defined on X ,
Pf^{-1} - distribution of functional f ,
$\operatorname{supp} P$ - topological support of the measure P ,
P_V - restriction of P on the σ-algebra $\mathcal{B}_V = \mathcal{B}_X \cap V$,
A^- - closure of the set A ,
$\operatorname{span} E$ - closed linear hull of the set E ,
λ - Lebesgue measure.

The relation $\mu \ll \nu$ means that the measure μ is absolutely continuous with respect to ν .

3. GROUPS OF ADMISSIBLE TRANSFORMATIONS

The following auxiliary notions will be subsequently needed. Let $V \in \mathcal{B}_X$ and let $G = (G_c)$, $c \in R^1$ be a group of bijection of V . Two points x, y will be called equivalent if there exists c such that $y = G_c x$. The correspondent equivalence classes will be called orbits. We shall say that a group G has smooth orbits if for all $x \in V$ there exists vector $L_x^G \in X$ such that that

$$d(x + cL_x^G, G_c x) = o(c), \quad c \to 0.$$

It is clear that L_x^G is the tangent vector for the curve $c \to G_c x$ at the point x .

<u>Definition 1.</u> The group G is admissible if

1) $P_V G_c^{-1} \ll P_V$ for a.a. $c \in R^1$;
2) orbits of G are smooth;
3) mapping $x \to L_x^G$ is continuous;
4) partition Γ of the set V generated by orbits of G is measurable.

The family of all admissible groups of transformations of V will be denoted by $\mathfrak{D}(V)$.

Example. Let G be the group of shifts of X in the direction defined by the vector $\ell \in X$:

$$G_c x = x + c\ell , \quad c \in R^1.$$

This group will be admissible iff for all $c \in R^1$ $PG_c^{-1} \ll P$ that is iff ℓ defines an admissible direction for the measure P .

Let us consider the family

$$\Pi(x) = \{ L_x^G \mid G \in \mathfrak{D}(V), V \ni x \}.$$

It is not difficult to see that $y \in \Pi(x)$ implies $cy \in \Pi(x)$ $\forall c \in R^1$.

Definition 2. We shall say that a measure P has a sufficient number of admissible groups if

$$\left(\Pi(x)\right)^- = \operatorname{span}(\operatorname{supp} P).$$

4. THE MAIN THEOREM

Let us put

$$a_0 = \inf_{x \in \operatorname{supp} P} f(x) , \quad a_1 = \sup_{x \in \operatorname{supp} P} f(x) .$$

Theorem 1. Let us suppose that P has a sufficient number of admissible groups. Let f be a convex locally Lipschitzian functional. Then the distribution Pf^{-1} is concentrated in the interval $[a_0, a_1]$, at the point a_0 (if $a_0 \neq -\infty$) Pf^{-1} can have an atom and in the interval $(a_0, a_1]$ the distribution Pf^{-1} is absolutely continuous. If $\operatorname{supp} P$ is connected and $Pf^{-1}(\{a_0\}) < 1$ then the density of Pf^{-1} is positive in almost all points of (a_0, a_1) .

Remark 1. If X is a Banach space then any continuous convex functional will be locally Lipschitzian.

Remark 2. Very simple examples demostrate that the distribution Pf^{-1} can be almost arbitrary if the condition of existence of sufficient number of admissible groups in omitted. For example, let $X = R^2$, $f(x) = \|x\|$ ($\|\cdot\|$ - Euclidean norm). Let P be a measure which is concentrated on the union of concentric circumferences. In this case Pf^{-1} can be identical to the distribution of arbitrary nonnegative random variable.

The proof of the theorem 1 which will be published in [5] is based essentially on the method of stratifications (see [6]).

5. APPLICATIONS

Gaussian processes. Let P be a gaussian measure. Without loss of generality we can suppose that average value of P is equal to zero and $\operatorname{supp} P = X$. It is well-known (see e.g. [2]) that gaussian measure has a sufficient number of admissible groups: we can confine ourselves only to the groups of shifts along admissible directions. The partition Γ corresponding to such a group is a partition of X in parallel lines and the conditional distributions for P are gaussian and nondegenerate. Due to the circumstance mentioned we obtain, in addition to the already known information, that the distribution Pf^{-1} has on the semiline (a_0, ∞) the density which is strictly positive.

Diffusion processes. Let $\{\xi(t)\}$, $t \in [0, T]$ be one-dimensional diffusion process defined by the equation:

$$d\xi(t) = a(t, \xi(t))dt + b(t, \xi(t))\,dw(t). \qquad (1)$$

Assume that the coefficients of this equation satisfy to the following condition: $a(t,x)$ is twice differentiable with respect to x, $b(t,x)$ is differentiable with respect to t and twice differentiable with respect to x, and all these derivatives are continuous with respect to (t,x). Assume also that (1) has a strong solution. In [3] it is proved that under these conditions the distribution $\mathcal{P}_\xi$ in $C[0,T]$ of the process ξ has a sufficient number of admissible groups. So the theorem 1 is applicable and we obtaine the following result:

<u>Theorem 2</u>. Let f be a convex continuous functional defined on $C[0,T]$. Then under the above conditions for distribution $\mathcal{P}_\xi f^{-1}$ the assertions of theorem 1 hold true.

<u>Processes with independent increments</u>. Consider now the case when P is the distribution in the Skorokhod space $D[0,1]$ of a stochastically continuous process $\{\xi(t)\}$, $t \in [0,1]$, having independent increments. Since the linear operations in $D[0,1]$ are not continuous the theorem 1 is not applicable directly. However it is not difficult to prove that under some additional assumptions about the process ξ there exists an analogue of the theorem 1. In particular the following result may be proved.

<u>Theorem 3.</u> Let ξ be a stable process (in the strong sense). Let P be its distribution in $D[0,1]$. Let f be a convex locally Lipschitzian functional. Then Pf^{-1} is concentrated on the semiline $[a_0, \infty)$, at the point a_0 Pf^{-1} can have an atom (if $a_0 > -\infty$) and on the semiline (a_0, ∞) the density of Pf^{-1} is strictly positive (if, of course, $Pf^{-1}(\{a_0\}) \neq 1$).

<u>Remark 3</u>. The existence of a sufficient number of admissible groups for P follows from the result of [4]. The strict positivity of the density $\frac{dPf^{-1}}{d\lambda}$ follows

from the fact that $\operatorname{supp} P$ coincides either with the subspace $D_0 = \{x \in D[0,1] \mid x(0) = 0\}$, or with the one of two cones

$$D_{\pm} = \{x \in D[0,1] \mid x(0) = 0,\ x(s) \gtrless 0,\ s \in [0,1]\}.$$

REFERENCES

1. Tzirelson B.S., The distribution density of maximum of gaussian process. - Theor.Probab.Appl., 1975, v.20, N 4, p.865-873.
2. Hui-Hsiung Kuo, Gaussian measures in Banach spaces.- Lect.Notes in Math., Springer-Verlag, Berlin, Heidelberg. New York, 1975.
3. Smorodina N.V., The absolute continuity of distributions of functionals of diffusion processes. - Uspekhi Mat.Nauk, 1982, v.37, N 6, p.185-192.
4. Lifschitz M.A., The method of stratifications for the processes with the independent increments. - LOMI seminares, 1983, v.130, p.109-121.
5. Davydov Yu.A., Distributions of convex functionals of random processes. - in the book: "Rings and modules. Limit theorems of the Probability theory", U.P., Leningrad State University (to appear in 1986).
6. Davydov Yu.A., Lifschitz M.A., The method of stratification in some probabilistic problems. - in the book: "Sum of science and technique", VINITI, Ser. Probab.Theor., Math.Stat., Kibern.Theor., v.22, Moscow 1984.

Prob. Theory and Math. Stat., Vol.1, pp. 411-418
Prohorov et al. (eds)
1986 VNU Science Press

WIENER GERMS APPLIED TO THE TAILS OF M-ESTIMATORS

Hermann Dinges
Fachbereich Mathematik, J.W. Goethe-Universität Frankfurt
Robert-Mayer-Str. 10, D-6000 Frankfurt, Fed. Rep. of Germany

1. **Introduction** : The main idea in my recent work on distributions is the definition of a Wiener germ given below. A Wiener germ is a family of distributions which converges to a δ-measure in a peculiar way. Sequences of distributions $L(X_n)$ which behave in some sense like $\mathcal{N}(\mu, \frac{1}{n}\sigma^2)$ turn up everywhere in stochastics. For many of these sequences there is also a theory of large deviations. The range in between is not so well studied; it is of practical importance if n is of moderate size. In order to understand this range we have to develop other types of asymptotics. (F. Hampel likes to call this goal "small sample asymptotics"; other people stress the methods by which the best results in this direction have been obtained hitherto; they say that we have to look closer at the saddle point approximations). Our definition of a Wiener germ stresses the type of mathematical results rather than the methods or the intended applications. We do not have a general method to prove that a sequence $L(X_n)$ follows a Wiener germ. However, in many cases we can rely on known results concerning sums of independent random variables. In this paper we shall not discuss any methods which go beyond this. (The application on M-estimators given below is quite in this line.) Rather than giving any outlook on further applications we shall show here that the tails of one-dimensional Wiener germs allow very nice approximations which seem to be new even for the classical case of sums of i.i.d. variables.

2. Main Definition

Definition : a) A d–dimensional Wiener germ is given by a triple

$$(K(x),\ H(x),\ \{S(\varepsilon,x) : \varepsilon\to 0\})\ .$$

K(x), H(x) and all the $S(\varepsilon,x)$ are functions on a fixed neighborhood U ($U \subseteq \mathbb{R}^d$) of a point x*, which is called the center of the Wiener germ. We stipulate

1) $K(x) \geq 0\ ,\quad K(x^*) = 0$
   K"(x) is positiv definite (|K"(x)| denotes the determinant)
   K(x) is four times continuously differentiable
2) $H(x) > 0\ ,\quad H(x^*) = 1$
   H(x) is twice continuously differentiable
3) $S(\varepsilon,x) = S(x) + O(\varepsilon)$ uniformly on compact subsets of U
   S(x) is continuously differentiable , S(x*) = (see below).

b) The Wiener germ given by our triple is the following family of densities $\{f_\varepsilon(x)dx : \varepsilon\to 0\}$

$$f_\varepsilon(x)dx = (2\pi\varepsilon)^{-\frac{d}{2}} \cdot \exp(-\varepsilon\ K(x)) \cdot \sqrt{|K''(x)|} \cdot H(x) \cdot \exp(\varepsilon \cdot S(\varepsilon,x))dx\ .$$

We stipulate that for every neighborhood U* of X*

$$\int_{U^*} f_\varepsilon(x)dx = 1 - O(\varepsilon^2)\ .$$

(This is a requirement on S(x*); S(x*) is determined by the derivatives of K(x) of order ≤ 4 and the derivatives of H(x) of order ≤ 2.)

c) The functions K(x), H(x), $S(\varepsilon,x)$ are uniquely determined by the Wiener germ

K(x) is called the entropy function

H(x) is called the modulating function

$D(x)dx = \sqrt{|K''(x)|} \cdot H(x)dx$ is called the modulating density of the Wiener germ.

The $S(\varepsilon,x)$ are called the correcting functions.

(Sometimes we stipulate that uniformly on every compact subset of U

$$S(\varepsilon,x) = S_0(x) + \varepsilon \cdot S_1(x) + O(\varepsilon^2)\ )\ .$$

Definition : We say that a sequence of random variables X_n follows along the sequence ε_n our Wiener germ if the density of X_n restricted to U coincides with $f_{\varepsilon_n}(x)dx$ for all n.

3. Sums of i.i.d. variables satisfying Cramér's condition

The famous theorem of H.E. Daniels (1954) provides us with lots of interesting examples.

Theorem (Daniels) : Let $Y_1, Y_2, \ldots$ i.i.d. with a density in $\mathbb{R}^d$ and

$$\psi(\vartheta) := \ln(E \exp(\vartheta Y)) < \infty$$

for all ϑ in some neighborhood V of the origin.
Then there exists S(x) continuous and

$$S(\tfrac{1}{n},x) = S(x) + O(\tfrac{1}{n})$$

on some neighborhood U of $x^* = EY$ such that the density of

$$X_n := \frac{1}{n}(Y_1 + \ldots + Y_n)$$

restricted to U has the form

$$(2\pi\frac{1}{n})^{-\frac{d}{2}} \cdot \exp(-n \cdot K(x)) \cdot \sqrt{|K''(x)|} \cdot \exp(\frac{1}{n} \cdot S(\frac{1}{n},x))dx.$$

Here K(x) is the Legendre transform of the cumulant generating function

$$K(x) := \sup\{\vartheta x - \psi(\vartheta) : \vartheta \in V\} \quad \text{for } x \in U .$$

Remark : In the language introduced above this theorem says that $L(\frac{1}{n}(Y_1 + \ldots + Y_n))$ follows along $\varepsilon_n = \frac{1}{n}$ a Wiener germ with entropy function K(x) and modulating function $H(x) \equiv 1$.

4. Beta-distributions

The Beta-distributions give us most simple examples of Wiener germs with a nontrivial modulating function. A random variable Z is Beta-distributed with parameter (α,β) if Z has the density

$$z^{\alpha-1}(1-z)^{\beta-1} \cdot dz \cdot \frac{1}{B(\alpha,\beta)} \quad \text{for } 0<z<1$$
$$= (2\pi\varepsilon)^{-\frac{1}{2}} \cdot \exp(-\frac{1}{\varepsilon}[\vartheta \ln \frac{\vartheta}{z} + (1-\vartheta) \ln \frac{1-\vartheta}{1-z}])$$
$$\cdot [z(1-z)]^{-1}[\vartheta(1-\vartheta)]^{\frac{1}{2}} \cdot \exp(\varepsilon \cdot S(\varepsilon,\vartheta))dz$$

where

$$\vartheta = \frac{\alpha}{\alpha+\beta}\ ;\quad \varepsilon = \frac{1}{\alpha+\beta}\ .$$

$\varepsilon \cdot S(\varepsilon,\vartheta)$ is independent of z and can be obtained by Stirling's formula.

This shows that for every fixed $\vartheta = \frac{\alpha}{\alpha+\beta} \in (0,1)$ the beta distributions with parameter $(\frac{1}{\varepsilon}\vartheta, \frac{1}{\varepsilon}(1-\vartheta))$ form a Wiener germ with center ϑ. Its entropy function is

$$K_\vartheta(z) = \vartheta \ln \frac{\vartheta}{z} + (1-\vartheta) \ln \frac{1-\vartheta}{1-z} \quad \text{for } z\in(0,1)\ .$$

We remark that for an Y which is Beta-distributed with parameter $(\frac{1}{\varepsilon}\vartheta, \frac{1}{\varepsilon}(1-\vartheta))$ we have

$$EY = \vartheta\ ,\quad \text{var } Y = \varepsilon\vartheta(1-\vartheta) \cdot (1+\varepsilon)^{-1}.$$

5. On the distributions of certain M-estimators

Theorem : Let $Y_1, Y_2, \ldots$ be i.i.d. E-valued random variables. Let $\Psi(y,t)$ be real valued on $E\times(t',t'')$ with

1) $\Psi(y,t)$ is antitone in (t',t'') for every fixed y.
2) $L(\Psi(Y,t))$ satisfies the conditions of Daniels' theorem quoted above for every fixed t.
3) $t \to E(\Psi(Y,t))$ maps (t',t'') diffeomorphically onto an interval (η',η'').

Then we have

a) For every fixed t the distributions of

$$X_n(t) = \frac{1}{n} \sum^{n} \Psi(Y_i,t)$$

follow (along $\frac{1}{n}$) a Wiener germ with some entropy function $K(t,x)$

$$K(t,x) = 0 \quad \Longleftrightarrow \quad x = \mathbb{E}(\Psi(Y,t)) .$$

b) For fixed $\eta \in (\eta', \eta'')$ consider

$$T_n^-(\eta) = \sup\{t : X_n(t) > \eta\}$$

$$T_n^+(\eta) = \inf\{t : X_n(t) < \eta\}$$

If we assume that $K(t,x)$ is four times continuously differentiable in t for fixed x then the distributions $L(T_n^\pm(\eta))$ follow (along $\frac{1}{n}$) a Wiener germ with the entropy function $K(t,\eta)$. The modulating density is

$$D(t,\eta)dt = \left(\frac{\partial^2}{\partial\eta^2} K(t,\eta)\right)^{\frac{1}{2}} \cdot \left(\frac{\partial}{\partial\eta} K(t,\eta)\right)^{-1} \cdot \left(\frac{\partial}{\partial t} K(t,\eta)\right) .$$

6. The tails of a one-dimensional Wiener germ

Theorem : Let $\{f_\varepsilon(x)dx : \varepsilon \to 0\}$ be a Wiener germ on U

$[x',x''] \subseteq U$, $x' < x^* < x''$ where x^* denotes the center.

Fix $c>0$ and consider the "distribution functions"

$$F_\varepsilon^-(x) := \int_{x'}^{x} f_\varepsilon(u)du \qquad \text{for} \quad x' \leq x \leq x^* + c\cdot\sqrt{\varepsilon}$$

$$F_\varepsilon^+(x) := 1 - \int_{x}^{x''} f_\varepsilon(u)du \qquad \text{for} \quad x^* - c\cdot\sqrt{\varepsilon} \leq x \leq x'' .$$

Assume that $\delta_\varepsilon = F_\varepsilon^+(x) - F_\varepsilon^-(x) = O(\varepsilon^3)$.

a) There exist $K_0(x)$, $K_1(x)$, $K_2(x)$ on $[x',x'']$ and

$K_3^-(\varepsilon,x)$ uniformly bounded on $[x'+\eta, x^*+c\cdot\sqrt{\varepsilon}]$

$K_3^+(\varepsilon,x)$ uniformly bounded on $[x^*-c\cdot\sqrt{\varepsilon}, x''-\eta]$

for every $\eta>0$ and every fixed c such that

$$\varepsilon\left[\frac{1}{2}\,\Phi^{-1}(F_\varepsilon^\pm(x))\right]^2 = K_0(x) + \varepsilon\cdot K_1(x) + \varepsilon^2\cdot K_2(x) + \varepsilon^3\cdot K_3^\pm(\varepsilon,x) .$$

b) $K_0(x)$ is in fact the entropy function of our Wiener germ. $K_1(x)$ is determined by the entropy function $K(x)$ and the modulating function $H(x)$

$$K_1(x) = -\ln H(x) + \frac{1}{2} \ln \frac{(K'(x))^2}{2K(x) \cdot K''(x)} .$$

Furthermore we have

$$K_2(x) = -S(x) + \frac{K_1'(x)}{K'(x)} - \frac{1}{2} \frac{K_1(x)}{K(x)} .$$

Remark :

$$\frac{1}{2} [\Phi^{-1}(p)]^2 \sim -\ln(p) \quad \text{for} \quad p \to 0$$

$$\sim -\ln(1-p) \text{ for } p \to 1 .$$

Our theorem tells us

$$\sqrt{\varepsilon}\, \frac{1}{2} [\Phi^{-1}(\Pr(x' \leqq X_\varepsilon \leqq x))]^2 \sim K_0 + \varepsilon K_1 .$$

Thus $-K_0$ is the rate function in the theory of large deviations.

7. Averages of i.i.d. random variables

We formulate a simplified version of the main theorem on the tails of sums of i.i.d. variables.

Theorem : Let $Y_1, Y_2, \ldots$ be i.i.d. with a density as in Daniels' theorem. $\psi(\vartheta)$ denotes the cumulant generating function. Put

$$k(\vartheta) := \vartheta \cdot \psi'(\vartheta) - \psi(\vartheta)$$

$$k_1(\vartheta) := -\frac{1}{2} \ln \frac{2k(\vartheta)}{\vartheta^2 \cdot \psi''(\vartheta)} .$$

Consider $x > EY$ and ϑ such that $\psi'(\vartheta) = x$. Put

$$z = \sqrt{n}\, \sqrt{2k(\vartheta)}$$

$$w = \sqrt{n}\, \vartheta \cdot \sqrt{\psi''(\vartheta)} .$$

Then there exist $k_2(n,\vartheta)$

$k_2(n,\vartheta) = k_2(\vartheta) + O(\frac{1}{n})$ uniformly in a fixed neighborhood of $\vartheta = 0$

such that

$$\Pr(\frac{1}{n}(Y_1+ \dots +Y_n) \geqq x)$$
$$= \Phi(-\sqrt{n}[2k(\vartheta) + \frac{1}{n} 2k_1(\vartheta) + \frac{1}{n^2} 2k_2(n,\vartheta)]^{\frac{1}{2}})$$
$$= \Phi(-[z^2 + 2\ln \frac{w}{z} + \frac{1}{n} 2k_2(n,\vartheta)]^{\frac{1}{2}}) \cdot$$

It might be interesting to compare this result with known results. For that we introduce the auxiliary function $q(z) := \frac{\varphi(z)}{\Phi(-z)}$, the reciprocal of Mill's ratio.

A) V.V. Petrov (1967) has given conditions under which
$$\Pr(\frac{1}{n}(Y_1+ \dots +Y_n)\geqq x) = \varphi(z) \frac{1}{w} [1+o(1)]$$
uniformly for $x \in [EY+\varepsilon, x'']$, $\varepsilon>0$, x'' suitably fixed.

B) T. Höglund (1979) has found the following estimate which covers small deviations at the same time
$$\Pr(\frac{1}{n}(Y_1+ \dots +Y_n)\geqq x) = \varphi(z) \frac{1}{q(w)} [1+o(1)]$$
uniformly for $x \in [EY, x'']$.

C) Lugannani & Rice recommend the approximation
$$\Pr(\frac{1}{n}(Y_1+ \dots +Y_n)\geqq x) = \Phi(-z) + \varphi(z)[\frac{1}{w} - \frac{1}{z}] .$$
We can show that B) and C) are not very different from the approximations which we get from the theorem above
$$\sim \Phi(-z - \frac{1}{z} \ln \frac{w}{z}) \quad \text{or} \quad \sim \Phi(- \sqrt{z^2+2\ln \frac{w}{z}} .$$
Our approach has the advantage that we go further by determining $k_2(\vartheta)$. Moreover we can handle Wiener germs which are more general than the averages of i.i.d. random variables.
In the general case we have
$$z = \sqrt{n} \cdot \sqrt{2K(x)} \, ; \quad w = \sqrt{n}\; K'(x)\; [K''(x)]^{-\frac{1}{2}} \cdot [H(x)]^{-1} .$$
We believe that our approximation formula gives numerical approximations for quite small n in many practically important cases.

References :

1) Alfers, D. & Dinges, H. (1984) : "A Normal Approximation for Beta and Gamma Tail Probabilities". Z. Wahrscheinlichkeitstheorie u. verw. Geb. 65 399–420.

2) Daniels, H.E. (1954) "Saddlepoint approximations in statistics". Ann. Math. Stat. 25, 631–50.

3) Daniels, H.E. (1983) "Saddlepoint approximations for estimating equations". Biometrika 70, 80–96.

4) Field, C.A. & Hampel, F.R. (1982) "Small sample asymptotic distributions of M–estimators of location". Biometrika 69, 29–46.

5) Höglund, T. (1979) "A unified formulation of the central limit theorem for small and large deviations from the mean." Z. Wahrscheinlichkeitstheorie u. verw. Geb. 49, 105–117.

6) Lugannani & Rice (1980) "Saddlepoint approximation for the distribution of the sum of independent random variables". Adv. Appl. Prob. 12, 479–90.

7) Petrov, V.V. "Sums of independent random variables". Berlin–Heidelberg–New York. Springer 1975 (in particular page 248)

Prob. Theory and Math. Stat., Vol.1, pp. 419-431
Prohorov et al. (eds)
1986 VNU Science Press

ON THE ASYMPTOTIC BEHAVIOUR OF THE LINEAR STOCHASTIC HEAT EQUATION SOLUTIONS

A.Ya.Dorogovtsev, S.D.Ivasishen, A.G.Kukush
Kiev State University, Kiev, U.S.S.R.

The linear heat equation with random sources given for space variable from bounded domain and on unbounded time interval is studied in present report. Various papers are devoted to the theory of similar and more general partial differential equations (see [1]). Thus the conditions of the existence and uniqueness of generalized and strong solutions of these equations are worked out in details. However the problem of the asymptotic behaviour of the solutions is practically unexplored in mathematical literature. We'll give the conditions under which both the solution and its derivative uniformly over space variable x tend to zero a.s. as time $t \to +\infty$. These results were announced in [2] . Note that the nonlinear case was also studied in [2] where the conditions of uniform tending to zero as $t \to +\infty$ of the solution were given, without consideration of the asymptotic properties of its derivative.

Let's consider the stochastic heat equation

$$\frac{\partial u}{\partial t}-\frac{\partial^2 u}{\partial x^2}=\sigma(u,t,x)\dot{w}(t),\ t\geq 0,\ 0\leq x\leq 1, \quad (1)$$

with initial condition

$$u(0,x)=\varphi(x),\ 0\leq x\leq 1, \quad (2)$$

and homogeneous boundary conditions

$$u(t,0)=u(t,1)=0,\ t\geq 0. \tag{3}$$

Here $\{w(t)\}$ is the standard Wiener process adapted to the flow of σ -algebras $\{\mathcal{F}_t\}$; the intensity function σ is a deterministic function, linear over u . There are two principally different ways to put mathematically correct problem (1)-(3). One of them is to replace the equation (1) by some stochastic integral equation containing the Green function [1]. The second one is to regard the equation (1) as an Ito equation with some infinite dimensional phase space and the drift operator $\partial^2/\partial x^2$ [3, 4].

Denote $H=L_2[0,1]$ with norm $\|\cdot\|$ and scalar product $(\cdot,\cdot)$; W_2^2 denotes Sobolev functional space on $[0,1]$,

$$H_+=\{f\in W_2^2 \mid f(0)=f(1)=0\}.$$

Together with the equation (1) consider unperturbed equation

$$\frac{\partial u}{\partial t}-\frac{\partial^2 u}{\partial x^2}=0,\ t\geq 0,\ 0\leq x\leq 1. \tag{4}$$

Denote $F(t,x),\ t\geq 0,\ 0\leq x\leq 1,$ the solution of the mixed boundary value problem (2)-(4); let G_0 be the Green function,

$$G_0(t-\tau,x,\xi)=\sum_{n=1}^{\infty} e^{-\lambda_n(t-\tau)}\varphi_n(x)\varphi_n(\xi),$$

$$0\leq\tau<t,\ 0\leq x,\xi\leq 1.$$

Here $\lambda_n=\pi^2n^2,\ \varphi_n(x)=\sqrt{2}\sin \pi n x,\ n\geq 1,\ 0\leq x\leq 1,$ are the eigen values and normalized eigen functions of the operator

$$A=-\frac{\partial^2}{\partial x^2}$$

in the space H .

DEFINITION 1. Random H -valued process $\{\tilde{u}_t\}$ which is $\mathcal{F}_t$ -measurable and satisfies a.s. the equality

$$\tilde{u}_t = \int_0^1 dw(\tau) \int_0^1 \sigma(\tilde{u}_\tau(\xi), \tau, \xi) G_0(t-\tau, x, \xi) d\xi + F(t,x) \tag{5}$$

is called the <u>generalized solution</u> of the problem (1) - (3). The relation (5) is understood for each t as the equality of functions from H ; under the sign of stochastic integral stands the random H -valued function.

Let's interpret (1) as an Ito equation. We'll assume the operator A to act on functions from the space $W_2^2 \subset H$. Introduce the function $\tilde{\sigma}(h,t) = \sigma(h(\cdot), t, \cdot)$, $h \in H$, $t \geq 0$, and consider in H the stochastic equation

$$du_t = -Au_t dt + \tilde{\sigma}(u_t, t) dw(t), \quad t \geq 0, \tag{6}$$

with initial condition

$$u_0 = \varphi \in H_+. \tag{7}$$

The solution of the problem (6), (7) is called the solution of the problem (1)-(3). More precisely, we'll give the following definition.

DEFINITION 2. Random process $\{u_t\}$ is called <u>strong solution</u> of the problem (1) - (3) if it is $\mathcal{F}_t$ -measurable for each t, with probability 1 $u_t \in H_+$, $t \geq 0$, and

$$u_t = u_0 - \int_0^t Au_s ds + \int_0^t \tilde{\sigma}(u_s, s) dw(s), \quad t \geq 0.$$

The questions of existence and uniqueness of the solutions of linear equations similar to (6) as well as conditions of the coincidence of generalized solution and strong solution were considered by V.V.Baklan [4] . We'll put some restrictions on the intensity function structure in order to apply the results of the mentioned paper. Namely we assume the function σ to have representation

$$\sigma(u,t,x)=u\,c(t,x)+d(t,x), \qquad u\in\mathbb{R},\ t\geqslant 0,\ 0\leqslant x\leqslant 1, \tag{8}$$

where $d(t,\cdot)\in H_+$ for each $t\geqslant 0$ and the derivative c''_{22} exists and is bounded over x . It's easy to verify that under these conditions for every fixed t the function $\tilde{\sigma}$ maps H_+ into itself.

LEMMA 1. Let $\varphi\in H_+$; let the functions c , c'_2 and c''_{22} be locally bounded over t ; for each $T>0$

$$\int_0^T \left\|\frac{\partial^2 d(t,\cdot)}{\partial x^2}\right\|^2 dt<\infty .$$

Then the problem (1) - (3) generalized solution exists, it is unique and coincides with the unique strong solution.

<u>Proof.</u> We'll show the equation (6) to be the particular case of the equations from [4]. For this purpose let's consider the Cauchy problem for fixed s

$$\frac{d}{dt}x_s(t)=-Ax_s(t),\quad 0\leqslant s\leqslant t\leqslant T,$$

$$x_s(s)=x_0\in H_+ .$$

We'll assume the operator A is defined on $H_+\subset H$. This operator is closed in H . The Cauchy problem is uniformly correct in the sense of [5] . Really, its solution

$$x_s(t)=\sum_{n=1}^{\infty} e^{-\lambda_n(t-s)}\varphi_n(x)\,(\varphi_n,x_0)$$

satisfies the following condition: the H -norm convergence $\{x_0^{(n)}\} \subset H_+$ to $\bar{0} \in H_+$ implies the convergence of the solutions corresponding to these initial conditions:

$$\sup_{s \le t \le T} \| x_s^{(n)}(t) \| \longrightarrow 0, \, n \to \infty.$$

This convergence follows from the inequality

$$\| x_s^{(n)}(t) \|^2 \le \| x_0^{(n)} \|^2, \; s \le t \le T, \, n \ge 1.$$

Let $U(t,s)$ be the evolution operator of the Cauchy problem

$$x_s(t) = U(t,s) x_0, \; s \le t.$$

Then it's easy to verify that this operator satisfies all the 5 conditions from [4] . We'll apply the theorem 1 [4] . For this case its conditions will be written in such a form:

$$\text{a)} \int_0^T \| \tilde{\sigma}(x_t, t) \|^2 dt + \int_0^T \left\| \frac{\partial^2}{\partial x^2} \tilde{\sigma}(x_t, t) \right\|^2 dt \le$$

$$\le K_T \left(1 + \int_0^T \| x_t \|^2 dt + \int_0^T \left\| \frac{\partial^2}{\partial x^2} x_t \right\|^2 dt \right);$$

$$\text{b)} \int_0^T \| \tilde{\sigma}(x_t, t) - \tilde{\sigma}(y_t, t) \|^2 dt + \int_0^T \left\| \frac{\partial^2}{\partial x^2} \tilde{\sigma}(x_t, t) - \right.$$

$$\left. - \frac{\partial^2}{\partial x^2} \tilde{\sigma}(y_t, t) \right\|^2 dt \le K_T \left(\int_0^T \| x_t - y_t \|^2 dt + \right.$$

$$\left. + \int_0^T \left\| \frac{\partial^2}{\partial x^2} (x_t - y_t) \right\|^2 dt \right).$$

Here $T>0$ is arbitrary, x_t and y_t are arbitrary measurable H_+ -valued functions, K_T doesn't depend upon the choice of x_t , y_t . The restrictions on the functions c and d easily imply the inequalities a) and b). Then the theorem 1 [4] implies the statement of the lemma.

REMARK 1. It follows from the proof of the theorem 1 [4] that for each $T>0$

$$\int_0^T E\|u_t\|^2 dt + \int_0^T E\|Au_t\|^2 dt < \infty .$$

We'll formulate now the main result of the report.

THEOREM. Let the following conditions be satisfied.

1. $\varphi \in H_+$.
2. $\sup_{t,x} [c^2(t,x) + 2(c_2'(t,x))^2] < 2\pi^2$,
   $\sup_{t,x} |c_{22}''(t,x)| < \infty$.
3. $\int_0^\infty \|d_{22}''(t,\cdot)\|^2 dt < \infty$.

Then the problem (1) - (3) strong solution $\{u_t\}$ exists and it is unique,

$$\int_0^\infty E\left\|\frac{\partial^2 u_t}{\partial x^2}\right\|^2 dt < \infty , \quad \sup_{t\geq 0} E\left\|\frac{\partial^2 u_t}{\partial x^2}\right\|^2 < \infty$$

and with probability 1

$$\left\|\frac{\partial^2 u_t}{\partial x^2}\right\|^2 \longrightarrow 0 , t \longrightarrow +\infty .$$

Before the straitforward proof of the theorem we'll formulate some lemmas. From now on we'll assume the theorem conditions are satisfied. Note that the lemma 1 implies the existence and uniqueness of $\{u_t\}$.

Let's denote $\sigma_\tau = \tilde{\sigma}(u_\tau, \tau), \tau \geq 0$.

LEMMA 2. For every $T, \varepsilon > 0$

$$\int_0^T E\|Au_t\|^2 dt \leq$$
$$\leq \frac{1+\varepsilon}{2\pi^2} \int_0^T E\|A\sigma_\tau\|^2 d\tau + \frac{1+\varepsilon^{-1}}{2\pi^2} \|A\varphi\|^2 .$$

Proof. According to lemma 1 $\{u_t\}$ coincides with the generalized solution. From (5) we obtain

$$Au_t = \sum_{n=1}^{\infty} \lambda_n \varphi_n \int_0^t e^{-\lambda_n(t-\tau)} (\varphi_n, \sigma_\tau) dw(\tau) +$$
$$+ \sum_{n=1}^{\infty} e^{-\lambda_n t} \lambda_n (\varphi_n, \varphi) \varphi_n .$$

The elementary inequality

$$\|a+b\|^2 \leq (1+\varepsilon)\|a\|^2 + (1+\varepsilon^{-1})\|b\|^2,$$
$$a, b \in H, \varepsilon > 0,$$

implies further the following estimate

$$\int_0^T E\|Au_t\|^2 dt \leq (1+\varepsilon) \sum_{n=1}^{\infty} \lambda_n^2 \int_0^T \int_0^t e^{-2\lambda_n(t-\tau)} \times$$
$$\times E(\varphi_n, \sigma_\tau)^2 d\tau dt + (1+\varepsilon^{-1}) \sum_{n=1}^{\infty} \lambda_n^2 (\varphi_n, \varphi)^2 \int_0^T e^{-2\lambda_n t} dt .$$

After changing the integration order we obtain the inequality

$$\int_0^T E\|Au_t\|^2 dt \leq$$
$$\leq (1+\varepsilon) \sum_{n=1}^{\infty} \lambda_n^2 \int_0^T \frac{1-e^{-2\lambda_n(T-\tau)}}{2\lambda_n} E(\varphi_n, \sigma_\tau)^2 d\tau +$$
$$+ (1+\varepsilon^{-1}) \sum_{n=1}^{\infty} \frac{1-e^{-2\lambda_n T}}{2\lambda_n} \lambda_n^2 (\varphi_n, \varphi)^2 .$$

The statement of the lemma follows now from the relations

$$\lambda_n \geq \pi^2 ; \quad \sum_{n=1}^{\infty} \lambda_n^2 (\varphi_n, h)^2 = \|Ah\|^2, \; h \in H_+ .$$

LEMMA 3. The following integral converges:

$$\int_0^{\infty} E\|u_t\|^2 dt < \infty .$$

Proof. As well as in the previous lemma we obtain from the equality (5) for each $T, \varepsilon > 0$

$$\int_0^T E\|u_t\|^2 dt \leq \frac{1+\varepsilon}{2\pi^2} \int_0^T E\|\sigma_\tau\|^2 d\tau + \frac{1+\varepsilon^{-1}}{2\pi^2} \|\varphi\|^2 .$$

Further

$$\|\sigma_\tau\|^2 \leq (1+\varepsilon) \| u_\tau c(\tau,\cdot)\|^2 + (1+\varepsilon^{-1}) \|d(\tau,\cdot)\|^2 ,$$

whence it follows that

$$\int_0^T E\|u_t\|^2 dt \leq \frac{(1+\varepsilon)^2}{2\pi^2} \sup_{t,x} c^2(t,x) \int_0^T E\|u_t\|^2 dt +$$

$$+ \frac{(1+\varepsilon)(1+\varepsilon^{-1})}{2\pi^2} \int_0^{\infty} \|d(\tau,\cdot)\|^2 d\tau .$$

The last improper integral converges because of the condition 3 of the theorem, since $d(\tau,\cdot)$ belongs to H_+ and the relation $\|h\|^2 \leq \|h''\|^2, \; h \in H_+$, holds true. Choosing ε to be small enough and collecting the integrals containing $E\|u_t\|^2$ to the L.H.S. of the inequality we obtain, with the help of remark 1, the statement of the lemma.

LEMMA 4. The following integral converges:

$$\int_0^\infty E\|Au_t\|^2 dt < \infty .$$

Proof. Let $h \in H_+$. Then for each $\varepsilon > 0$

$$\|h'\|^2 = -\int_0^1 h h'' dx \leq \|h\| \|h''\| \leq \frac{\varepsilon}{2}\|h''\|^2 + \frac{1}{2\varepsilon}\|h\|^2. \tag{9}$$

Let's estimate $\|A\sigma_\tau\|^2$. We have from the equality (8)

$$A\sigma_\tau = u''_\tau c + 2u'_\tau c'_2 + c''_{22} u + d''_{22} ,$$

where all the derivatives are taken over x. By means of the elementary inequality true for each $\varepsilon, \delta > 0$, $a, b, c \in H$

$$\|a+b+c\|^2 \leq (1+\varepsilon+\delta)\|a\|^2 + + (1+\frac{1}{\varepsilon}+\delta)\|b\|^2 + (2+\frac{2}{\delta})\|c\|^2,$$

we continue to estimate:

$$\|A\sigma_\tau\|^2 \leq (1+\varepsilon+\delta) \sup_{t,x} c^2(t,x)\|u''_\tau\|^2 + 4(1+\frac{1}{\varepsilon}+\delta) \times \times \sup_{t,x}(c'_2(t,x))^2 \|u'_\tau\|^2 + (2+\frac{2}{\delta})\|c''_{22}u_\tau + d''_{22}\|^2. \tag{10}$$

From this and from lemma 2 we have

$$\int_0^T E\|Au_t\|^2 dt \leq$$

$$\leq \frac{(1+\varepsilon)(1+\varepsilon+\delta)}{2\pi^2} \sup_{t,x} c^2(t,x) \int_0^T E\|Au_t\|^2 dt +$$

$$+\frac{1+\varepsilon}{2\pi^2}4\left(1+\frac{1}{\varepsilon}+\delta\right)\sup_{t,x}(c_2'(t,x))^2\int_0^T E\|u_t'\|^2dt+ \tag{11}$$
$$+\frac{1+\varepsilon}{2\pi^2}\left(2+\frac{2}{\delta}\right)\int_0^T E\|c_{22}''u_t+d_{22}''\|^2dt+$$
$$+\frac{1+\varepsilon^{-1}}{2\pi^2}\|A\varphi\|^2.$$

Further we estimate $\|u_t'\|^2$ by means of the inequality (9) and obtain in the R.H.S. of (11) the addend containing the integral

$$\int_0^T E\|Au_t\|^2dt$$

with the coefficient

$$\frac{(1+\varepsilon)(1+\varepsilon+\delta)}{2\pi^2}\sup_{t,x}c^2(t,x)+\frac{2(1+\varepsilon)(1+\varepsilon+\delta\varepsilon)}{2\pi^2}\sup_{t,x}(c_2'(t,x))^2.$$

This coefficient by the theorem conditions can be made less than 1 by means of suitable choice of ε and δ. Note also that by these conditions

$$\int_0^\infty E\|c_{22}''u_t+d_{22}''\|^2dt<\infty$$

and by lemma 3

$$\int_0^\infty E\|u_t\|^2dt<\infty.$$

The proof of the lemma can be completed in the same way as in lemma 3.

Proof of the theorem. We'll consider only the case $\varphi=\delta$ (slightly changing the further reasonings one can obtain the proof in general case by means of the lemmas proved before). The generalized solution definition implies

$$\|Au_t\|^2=\sum_{n=1}^{\infty}\lambda_n^2\Big[\int_0^t e^{\lambda_n(\tau-t)}(\varphi_n,\sigma_\tau)\,dw(\tau)\Big]^2.$$

From this

$$E\|Au_t\|^2=\sum_{n=1}^{\infty}\lambda_n^2\int_0^t e^{2\lambda_n(\tau-t)}E(\varphi_n,\sigma_\tau)^2\,d\tau\le$$

$$\le\int_0^t E\|A\sigma_\tau\|^2\,d\tau$$

and by the inequality (10), by the conditions of the theorem and by lemmas 3,4 we obtain the second statement of the theorem:

$$\sup_{t\ge 0}E\|Au_t\|^2<\infty.$$

Let's denote

$$\eta(t)=\|Au_t\|^2,\ t\ge 0.$$

By lemma 4 $\int_0^\infty E\eta(t)\,dt<\infty$ a.s., that is why it's sufficient to prove in order to obtain the convergence of $\{\eta(t)\}$ to zero that the random function $\{\eta(t)\}$ tends to some random variable η_∞ as $t\to+\infty$ a.s. Let's consider for $s<t$ the conditional mathematical expectation

$$E[\eta(t)\,|\,\mathcal{F}_s]\le$$

$$\le\sum_{n=1}^{\infty}\lambda_n^2\Big[\int_0^s e^{\lambda_n(\tau-s)}(\varphi_n,\sigma_\tau)\,dw(\tau)\Big]^2+$$

$$+\sum_{n=1}^{\infty}\lambda_n^2 E\Big[\int_s^t(\varphi_n,\sigma_\tau)^2\,d\tau\;\Big|\;\mathcal{F}_s\Big]=$$

$$=\eta(s)+E\Big[\int_s^t \|A\sigma_\tau\|^2 d\tau \,\Big|\, \mathcal{F}_s\Big].$$

From this we have that the process

$$\xi(t)=\eta(t)-\int_0^t \|A\sigma_\tau\|^2 d\tau,\ t\geqslant 0,$$

is a submartingale; moreover $\sup\limits_{t\geqslant 0} E|\xi(t)|<\infty$ by lemma 4. Applying the submartingales convergence theorem we obtain

$$\xi(t)\longrightarrow \xi_\infty,\ t\longrightarrow +\infty \quad \text{a.s.}$$

From this because of the convergence of the integral $\int_0^\infty \|A\sigma_\tau\|^2 d\tau$ we obtain the convergence as $t\longrightarrow +\infty$ with probability 1:

$$\eta(t)\longrightarrow \eta_\infty.$$

The theorem is completely proved.

COROLLARY. Under the conditions of the theorem as $t\longrightarrow +\infty$ a.s.

$$\max_{0\leqslant x\leqslant 1}|u(t,x)|+\max_{0\leqslant x\leqslant 1}|u'_x(t,x)|\longrightarrow 0.$$

References

I. Гихман И.И. Стохастические дифференциальные уравнения с частными производными. - В кн.: Качественные методы исследования нелинейных дифференциальных уравнений и нелинейных колебаний. Киев: Институт математики АН УССР, 1981, с.25-59.

2. Дороговцев А.Я., Ивасишен С.Д., Кукуш А.Г. Асимптотическое поведение решений уравнения теплопроводности с белым шумом в правой части. - Украинский математический журнал, 1985, 37, № I, с.13-20. ISSN 0041-6053.

3. Pardoux E. Equations aux dérivées partielles stochastiques non lineaires monotones. Etude de solutions fortes de type Ito. - Thése doct. Sci math. Univ. Paris Sud., 1975.

4. Баклан В.В. Об одном классе стохастических дифференциальных уравнений в гильбертовом пространстве. - В кн.: Теория случайных процессов. Республиканский межведомственный сборник. Киев, 1978, вып.6,с.10-15.

5. Крейн С.Г. Линейные дифференциальные уравнения в банаховом пространстве. - Москва: Наука, 1967. - 464 с.

Prob. Theory and Math. Stat., Vol.1, pp. 433-438
Prohorov et al. (eds)
1986 VNU Science Press

SOME UNIVERSAL DONSKER CLASSES OF FUNCTIONS

R. M. Dudley
Room 2-245, MIT
Cambridge, MA 02139, USA

Abstract. A class $\tilde{F}$ of measurable functions on a measurable space X is called a universal Donsker class if for every probability measure P on X, the centered and normalized empirical measures converge in law, with respect to uniform convergence over $\tilde{F}$, to the limiting Gaussian process. Some relations are stated between this property and several conditions: extensions of the Vapnik-Červonenkis combinatorial property and Pollard's integrability condition on Kolčinskii's entropy and other metric entropies. Proofs are to be published elsewhere.

Introduction. First, some definitions and previous results will be reviewed, mainly as in Dudley (1984). Let $(X,\tilde{A},P)$ be a probability space. For $f \in L^1(P)$ let $P(f) := \int f\,dP$. Let G_P be a Gaussian process indexed by $L^2(X,\tilde{A},P)$ with mean 0 and covariance $EG_P(f)G_P(g) = P(fg) - P(f)P(g)$ for all f and g. Let $\rho_P(f,g) := (E((G_P(f) - G_P(g))^2))^{1/2}$. A G_P process will be called coherent if each sample function $G_P(\cdot)(\omega)$ is bounded and uniformly ρ_P-continuous on $\tilde{F}$. A subset $\tilde{F}$ of $\tilde{L}^2$ will be called pregaussian or P-pregaussian (formerly called G_PBUC) iff a coherent G_P process exists on $\tilde{F}$. Let $(X^\infty,\tilde{A}^\infty,P^\infty)$ be the countable product of copies of $(X,\tilde{A},P)$, with coordinates $x(1)$, $x(2),\ldots$. Let $P_n := n^{-1}(\delta_{x(1)} + \ldots + \delta_{x(n)})$, the nth empirical measure for P. Let $\nu_n := n^{1/2}(P_n - P)$. Let $(\Omega,\tilde{S},Pr)$ be the product of $(X^\infty,\tilde{A}^\infty,P^\infty)$ and the unit interval

A.M.S. classifications: primary 60F17, 60F05. Key words and phrases: Central limit theorems, Vapnik-Červonenkis classes.

with Lebesgue measure. For any real-valued function G on $\tilde{F}$ let $\|G\|_{\tilde{F}} := \sup\{|G(f)|: f \in \tilde{F}\}$. Call $\tilde{F}$ a <u>functional Donsker class</u> for P iff on Ω, there exist independent coherent G_P processes $Y_1, Y_2, \ldots$, such that for each $\varepsilon > 0$,

$$\lim_{n\to\infty} \Pr^*\{n^{-1/2}\max_{k\leq n} \|k(P_k - P) - Y_1 - \ldots - Y_k\|_{\tilde{F}} > \varepsilon\} = 0.$$

$\tilde{F}$ is a <u>universal Donsker class</u> iff it is a functional Donsker class for every probability measure P on $\tilde{A}$.

A class $\tilde{C}$ of sets <u>shatters</u> a set F iff every subset A of F is the intersection of B with F for some B in $\tilde{C}$. Then $S(\tilde{C})$ is the supremum of cardinalities of finite sets shattered by $\tilde{C}$. $\tilde{C}$ is a <u>Vapnik-Červonenkis (VC) class</u> iff $S(\tilde{C}) < \infty$.

A measurable space $(S,\tilde{B})$ is <u>standard</u> iff $\tilde{B}$ is the Borel σ-algebra for some complete, separable metric on S. Then a measurable space $(Y,\tilde{U})$ is <u>Suslin</u> if $\tilde{U}$ is countably generated and separates points of Y and there is a measurable function from S onto Y. A set $\tilde{F}$ of measurable real functions on X will be called <u>image admissible Suslin via</u> $(Y,\tilde{U},T)$ iff $(X,\tilde{A})$ and $(Y,\tilde{U})$ are Suslin and T is a function from Y onto $\tilde{F}$ such that $T(y)(x)$ is jointly measurable in $\langle x,y\rangle$.

If $\tilde{F} = \{1_B: B \in \tilde{C}\}$, it is a universal Donsker class, or even pregaussian for all P, iff, up to measurability conditions, $\tilde{C}$ is a Vapnik-Červonenkis class (Durst and Dudley, 1981).

For any real function f let $\mathrm{diam}(f) := \sup f - \inf f$. Let $\tilde{F}$ be a class of nonnegative functions on X and $\tilde{C}$ a class of subsets of X. For each $f \in \tilde{F}$ and $t \in \mathbb{R}$, the set $\{x \in X: f(x) > t\}$ is a <u>major set</u> of $\tilde{F}$. $\tilde{F}$ is a <u>major class</u> for $\tilde{C}$ iff all its major sets are in $\tilde{C}$. If $\tilde{C}$ is a VC class then I call $\tilde{F}$ a VC major class (for $\tilde{C}$).

The subgraph of a function $f \geq 0$ will be defined as

sub(f) := $\{\langle x,t\rangle \in X \times \mathbb{R}:\ 0 \leq t \leq f(x)\}$.

If $\tilde{D}$ is a class of subsets of $X \times \mathbb{R}$, $\tilde{F}$ is a subgraph class for $\tilde{D}$ iff sub(f) $\in \tilde{D}$ for all $f \in \tilde{F}$. If $\tilde{D}$ is a VC class then I call $\tilde{F}$ a VC subgraph class (for $\tilde{D}$). VC subgraph classes have also been called "VC graph" classes (Alexander, 1984a,b) or "polynomial classes" (Pollard, 1984, 1985).

For a class $\tilde{F}$ of functions and $0 < M < \infty$, $H(\tilde{F},M)$ denotes M times the symmetric convex hull of $\tilde{F}$: the set of functions g such that for some $k < \infty$, $f_j \in \tilde{F}$, and real t_j, $j = 1,\ldots,k$,

$$\Sigma_{j=1}^{k} |t_j| \leq M \quad \text{and} \quad g(x) = \Sigma_{j=1}^{k} t_j f_j(x) \quad \text{for all } x.$$

Then $\bar{H}_s(\tilde{F},M)$ is the smallest class $\tilde{G}$ of functions including $H(\tilde{F},M)$ such that for any $g_n \in \tilde{G}$ with $g_n(x) \to g(x)$ as $n \to \infty$ for all x, we have $g \in \tilde{G}$. Say that $\tilde{H}$ is a hull class for $\tilde{F}$ if $\tilde{H}$ is included in $\bar{H}_s(\tilde{F},M)$ for some $M < \infty$. Then $\tilde{H}$ is a VC hull class if $\tilde{F} = \{1_C\colon C \in \tilde{C}\}$ and $S(\tilde{C}) < \infty$.

For any metric space (S,d) and $\varepsilon > 0$ let $D(\varepsilon,S) :=$ $D(\varepsilon,S,d) := \sup\{m:$ for some $x_1,\ldots,x_m \in S$, $d(x_i,x_j) > \varepsilon$ for $1 \leq i < j \leq m\}$. Given a law Q on $\tilde{A}$, $1 \leq p < \infty$ and $\varepsilon > 0$, $D^{(p)}(\varepsilon,\tilde{F},Q) := D(\varepsilon,\tilde{F},e_{Q,p})$ where $e_{Q,p}(f,g) := (\int|f-g|^p dQ)^{1/p}$. Let $D^{(p)}(\varepsilon,\tilde{F}) := \sup_Q D^{(p)}(\varepsilon,\tilde{F},Q)$ where the supremum is over all laws Q concentrated in finite sets (Kolčinskiĭ, 1981). Pollard (1982) worked with the entropy condition

(1) $\int_0^1 (\log D^{(2)}(\varepsilon,\tilde{F}))^{1/2} d\varepsilon < \infty$.

Let $D^{(\infty)}(\varepsilon,\tilde{F}) := D(\varepsilon,\tilde{F},d_\infty)$ where $d_\infty(f,g) := \sup|f-g|$.

Here are some other conditions to be treated:

(2) For some $r < \infty$, $D^{(2)}(\varepsilon,\tilde{F}) = O(\varepsilon^{-r})$ as $\varepsilon \downarrow 0$.

(3) $\tilde{F}$ is a sequence $\{f_j\}_{j\geq 1}$ with

$$\operatorname{diam}(f_j) = o((\log j)^{-1/2}) \text{ as } j \to \infty.$$

(4) $\int_0^1 (\log D^{(\infty)}(x,\tilde{F}))^{1/2}dx < \infty$.

Now the results on implications will be stated. Pollard's entropy condition (1) nearly characterizes universal Donsker classes, but does not, by parts h), s), t) and u) of the Theorem. For any universal Donsker class $\tilde{F}$, $\sup_{f\in\tilde{F}} \operatorname{diam}(f)$ is finite, and $\nu_n(f) \equiv \nu_n(f+c)$ for any constant c, so an assumption $0 \leq f \leq 1$, $f \in \tilde{F}$, amounts to an allowable rescaling.

<u>Theorem</u>. Let $\tilde{F}$ be a class of measurable functions f on $(X,\tilde{A})$ with $0 \leq f(x) \leq 1$ for all x, and $\tilde{C}$ a subclass of $\tilde{A}$.

a) Every major class for $\tilde{C}$ is also a hull class for $\tilde{C}$.

b) There exist VC hull classes $\tilde{F}$ which are not VC major.

c) (Pollard) For every VC subgraph class $\tilde{F}$, (2) and hence (1) hold.

d) (2) does not imply that $\tilde{F}$ is a VC subgraph class.

e) There exist classes $\tilde{F}$ satisfying (1) and not (2).

f) There exist VC major (thus VC hull) classes which do not satisfy (2), thus are not VC subgraph classes.

g) There are VC subgraph classes which are not VC major.

h) (Pollard) If $\tilde{F}$ satisfies (1), and is image admissible Suslin, it is a universal Donsker class.

i) (Paulauskas-Heinkel) If f_j are measurable functions on X

satisfying (3) then $\{f_j\}_{j\geq 1}$ is a universal Donsker class.

j) There are sequences $\{f_j\}$ satisfying (3) but not (1).

k) If $\tilde{F}$ is a functional Donsker class for P then so is $\bar{H}_s(\tilde{F},M)$ for any $M < \infty$.

m) (4) implies (1) (clearly).

n) (4) does not imply (2).

o) (4) fails for any infinite collection of indicators of sets.

p) (4) does not imply that $\tilde{F}$ is a VC major class.

q) There exist sequences $\tilde{F} = \{f_n\}_{n\geq 1}$ satisfying (2), (3) and (4), and VC subgraph classes $\tilde{G}$, such that $\tilde{F}$ and $\tilde{G}$ are not VC hull classes.

r) If $\tilde{G}$ is a uniformly bounded class of measurable functions such that for some $K < \infty$ and $0 < \lambda < \infty$, $D^{(2)}(\varepsilon,\tilde{G}) \leq K\varepsilon^{-\lambda}$ for $0 < \varepsilon \leq 1$, then for any $t > s = 2\lambda/(2+\lambda)$, there are constants C_1, C_2 with $D^{(2)}(\varepsilon,\bar{H}_s(\tilde{G})) \leq C_1\exp(C_2\varepsilon^{-t})$ for $0 < \varepsilon \leq 1$. Thus every VC hull class satisfies (1).

s) For any universal Donsker class $\tilde{F}$, $\varepsilon^2\log D^{(2)}(\varepsilon,\tilde{F})$ is bounded as $\varepsilon \downarrow 0$.

t) There exist universal Donsker classes $\tilde{F}$ with $\varepsilon^2\log D^{(2)}(\varepsilon,\tilde{F}) \to 0$ arbitrarily slowly as $\varepsilon \downarrow 0$.

u) There is a uniformly bounded class $\tilde{F}$ of measurable functions, which is not a universal Donsker class, such that

$$\log D^{(2)}(\varepsilon,\tilde{F}) \leq 3/(\varepsilon^2 L(1/\varepsilon)) \quad \text{as} \quad \varepsilon \downarrow 0,$$

so (by t)) no condition on $D^{(2)}$ characterizes the universal Donsker property (even aside from measurability).

Let a "VC subgraph hull class" be one included in $\bar{H}_s(\tilde{F},M)$ for some VC subgraph class $\tilde{F}$ and $M < \infty$, such as the "secondary VC graph difference classes" for which Alexander (1985, Theorem 2.2) proves extended Kiefer inequalities. Since (2) is known for VC subgraph classes, (1) holds for VC subgraph hull classes by r). I do not know if (4) implies that $\tilde{F}$ is a VC subgraph hull class.

ACKNOWLEDGMENTS. This research was partially supported by National Science Foundation Grant DMS 8506638. A journal referee of a longer version of the paper made helpful suggestions.

REFERENCES

Alexander, K. S. (1984a). Probability inequalities for empirical processes and a law of the iterated logarithm. Ann. Probability 12, 1041-1067.

____________(1984b). The central limit theorem for empirical processes on Vapnik-Červonenkis classes (preprint).

____________(1985). Rates of growth for weighted empirical processes. In: Proc. Berkeley Conf. in honor of Jerzy Neyman and Jack Kiefer II, LeCam, L., and Olshen, R. (Eds.). Wadsworth, Monterey, CA, pp. 475-493.

Assouad, Patrice (1983). Densité et dimension. Ann. Inst. Fourier (Grenoble) 33, 233-282.

Dudley, R. M. (1984). A Course on Empirical Processes. In: Ecole d'été de probabilités de St.-Flour XII (1982), Hennequin, P. L. (Ed.). Lecture Notes in Math. (Springer) 1097, 1-142.

Durst, Mark, and R. M. Dudley (1981). Empirical processes, Vapnik-Chervonenkis classes and Poisson processes. Prob. Math. Statist. (Wrocław) 1 #2, 109-115.

Heinkel, B. (1983). Majorizing measures and limit theorems for c_o-valued random variables. In: Probability in Banach Spaces IV, Beck, A. and Jacobs, K. (Eds.). Lecture Notes in Math. (Springer) 990, 136-149.

Kolčinskiĭ, V. I. (1981). On the central limit theorem for empirical measures. Theor. Probability Math. Statist. 24, 71-82 = Teor. Verojatnost. i Mat. Statist. 24, 63-75.

Paulauskas, V. (1980). On the central limit theorem in the Banach space c_o. Doklady Akad. Nauk SSSR 254, 286-288 = Sov. Math. Doklady 22, 349-351.

Pollard, D. (1982). A central limit theorem for empirical processes. J. Austral. Math. Soc. (Ser. A) 33, 235-248.

_________(1984). Convergence of Stochastic Processes. Springer, New York.

_________(May, 1985). New ways to prove central limit theorems (preprint).

Prob. Theory and Math. Stat., Vol.1, pp. 439-465
Prohorov et al. (eds)
1986 VNU Science Press

DUPLICATES IN MIXED SEQUENCES AND A FREQUENCY DUPLICATION PRINCIPLE. METHODS AND APPLICATIONS

A.T. Fomenko

Moscow State University, 117234 Moscow, USSR

1. This article is an immediate continuation of a series of the author's papers [6-21], Here we give the results of a new research carried out in 1983-1985 on the basis of the frequency damping principle,formulated and varified by the author in [11-15] against considerable experimental material. A new method of ststistical processing of quantitative textual information of the narrative character is introduced. The method makes it possible to discover the chronologically correct order of separate textual fragments, and recognize duplicates and fragments describing the same historical events.Its numerous versions are based on the analysis of frequency of using proper names and parallel passages by chroniclers, etc.

The "frequency damping" and "frequency duplication" principles have been formulated by the author in [9]. They are, actually, a reformulation in modern terms, of the important general idea of loss and distortion of information with time, expressed by acad. N.A. Morozov who extended this rule not only to a "verbal tradition",but also to a "written memory" ([5],v.4,pp.621,809;v.6,793).

It should be noted that none of our methods claims for universality, since we analyze multilayer chronicles which are often subjectively tinged.

For brevity, we call a textual fragment, describing the events of one generation a "chapter-generation". We shall assume that the average length of one generation is that of the rule of authentic personages fixed in the surviving chronicles. While analizing J. Blair's chronological tables [2] it was calculated (see, the author's [7]) that it was equal to 17.1 years. When working with the authentic texts, we sometimes face natural difficulties if we want to distinguish "chapter-generations" from them; therefore, we confine ourselves only to their appoximate partition. Let text X describe the events in a sufficiently long time interval (A, B), with, at least, several generations of personages succeeding each other. Suppose that it can be partitioned into chapters-generations $X(T)$, each describing the events of one generation numbered T in the $X(T)$ enumeration which was fixed in the text. How correctly are these chapters numbered, or ordered, in text X? And if this enumeration is lost or doubtful, then how can it be restored, i.e., $X(T)$ correctly arranged in time respect to each other?

Now let us formulate a model: name = personage. If the time interval described by the author is sufficiently large, say, tens or hundreds of years, then, as it was verified by analysing a large set of concrete texts (see [11,17]), different characters usually have different names in X. In other words, the number of different historical figures possessing the same name in one text is negligibly small as compared to that of all textual personages. This statement turned out to be true for all several hundred texts investigated by the author, and describing medieval Rome, Greece, Germany, Russia, etc.

Now we formulate the frequency damping principle characterizing the chronologically correct order of chapters-generations inside text X. On numbering them correctly, and while passing from the description of the events of one generation to another, the author of the chronicle also replaces the acting personage, viz., he says noting of those from the generations prior to the one numbered T_0, since they have not yet been "born"; then, in describing T_0, he speaks most of the historical characters just at this time, because the events described are related to them; finally, while coming to the description of the subsequent generations, the chronicler mentions the earlier personages still more rarely, for he describes new events whose protagonists supplant the dead.

Thus, each generation gives birth to its own, new, unique and mostly inimitable personages; if the generations are replaced, then these figures are replaced, too. In spite of its outward simplicity, the principle is far from being obvious, and turns out to be extremely useful in creating a statistical dating method. Since personages are almost uniquely determined by their names (name= =personage), we shall study a complete set of all the textual names.

Consider a group of names first appearing in a chapter-generation T_0, and denote the quantity of their occurences in the chapter $X(T_0)$ by $K(T_0, T_0)$. Then we count how many times they were mentioned in the chapter $X(T)$ numbered T. We designate the obtained value by $K(T_0, T)$. If the same name is repeated several times, i.e., it possesses a certain multiplicity, then all the mentions are counted. All $K(T_0, T)$ with variable T_0 and T are naturally organized into a square matrix $K\{T\}$ of the order $n \times n$, where n is the number of chapters. We call it a square (name) frequency matrix.

The author and G.V. Nosovsky have recently found a new modification of the procedure for discovering dependent texts on the frequency damping principle basis. G.N. Nosovsky has computerized the whole work.

2. First consider a chronological list of secular or church rulers. Normally, each ruler had several names. We shall assume that al the names of a certain ruler are

listed consecutively in the appropriate place of the list, and that there are no separation signs between the names of neighbouring personages (in time). Our methods and the obtained results do not depend on the way of name ordering. Order the list with respect to the average year of the ruling period, and denote it by $X=\{a_1,a_2,\dots,a_N\}$. We assume a decomposition of the list X into chapters $X_1, X_2, \dots, X_n$ given (ibid.). Denote by $I=\{u_1,u_2,\dots,u_m\}$, $m \leq N$, the set of different names in X, and the name of the i-th entry of X by $u(a_i)$, $u(a_i) \in I$.

Definition 1. We call an integer $\rho(a_i,a_j)=|r-s|$ the <u>scattering of two entries</u> $a_i, a_j \in X$, $a_i \in X_s$, $a_j \in X_r$.

Definition 2. We shall say that two names $u_\ell, u_k \in I$ are <u>of the same age</u>, and denote the fact by $u_\ell \approx u_k$, if their first occurences are in one chapter of X.

Definition 3. We shall say that two names $u_\ell, u_k \in I$ are <u>conjugate</u>, and denote the fact by $u_\ell \sim u_k$ if there exists a chapter X_p in X, containing both.

If two entries a_i and a_j from the list X are conjugate (or of the same age) as two names from I, then we shall also call them conjugate (resp. of the same age), and employ the corresponding notation.

Consider a finite scheme (Ω, Σ, P) of equiprobabilistic sampling with replacement of two elements from X. Thus, $\Omega = X \times X$, $\Sigma = 2^\Omega$, $P(\omega)=\frac{1}{N^2}$ for any $\omega \in \Omega$.

We shall denote the first selected element by $a_{(1)}$, and the second one by $a_{(2)}$. Consider the scattering of the pair $a_{(1)}, a_{(2)}$

$$\xi_1(\omega)=(a_{(1)}, a_{(2)}). \qquad (1)$$

It is a random variable defined on Ω.

Assume that the events $A=\{\omega : a_{(1)} \approx a_{(2)}\}$ and $B=\{\omega : a_{(1)} \sim a_{(2)}\}$ are non-zero, and $P(A) \neq 0$, $P(B) \neq 0$. Consider the conditional probabilities P_A and P_B on Ω, viz.,

$$P_A(C)=\frac{P(AC)}{P(A)}, \quad P_B(C)=\frac{P(BC)}{P(B)}, \quad \forall C \in \Sigma .$$

Denote by f_1, f_2 and f_3, respectively, the distributions of the random variable ξ_1 relative to the probabilities P, P_A and P_B, viz.,

$$f_1(k)=P(\xi_1=k),$$

$$f_2(k)=P_A(\xi_1=k), \quad f_3(k)=P_B(\xi_1=k),$$

what means the three random variables ξ_1, ξ_2, ξ_3; $\xi_1(\omega)=$ $=\xi_2(\omega)=\xi_3(\omega)$, are considered on three different probability spaces (Ω,Σ,P), (Ω,Σ,P_A) and (Ω,Σ,P_B) with the respective distributions f_1, f_2 and f_3.

In the sequel, we shall also use the term "frequency histogram" for the distribution of discrete random variables defined on a finite scheme.

We shall call the frequency histograms of random variables of type ξ_2 and ξ_3 (i.e., the conditional distributions of the random variable ξ_1 under a certain "locally" determined condition), the related name scattering frequency histograms, meaning the "relation" in the sense of this condition. We shall call the histogram $f_1(k)=P(\xi_1=k)$ simply a <u>name sacttering frequency histogram</u>.

3. Now we try to investigate the structure of list X by comparing the distribution of the random variable ξ_1 with ξ_2 and ξ_3. In particular, natural ideas of how the rulers' names should be arranged chronologically "correctly" lead us to the following statement.

(A) The condition $u_i \sim u_j$ (or $u_i \approx u_j$) imposed on the names u_i, u_j from I does not influence the details of the mutual disposition of u_i, u_j with respect to the whole of X.

It is clear that statement (A) is closely related to the frequency damping principle: as a matter of fact, we assume that "local" relations in the chronologically correct list must not statistically lead to any global relations.

By means of ξ_1, ξ_2, ξ_3 (A) can be made more precise as follows.

(B) The random variables ξ_1, ξ_2, ξ_3 constructed by the chronologically correct list should be distributed in a similar way. In other words, the distribution of ξ_1 should not depend either on the event A or B.

Remark. It is clear that a certain divergence of the distribution of ξ_1 from ξ_2 (or ξ_3) will arise even in the case when (A) is valid, just because of the finiteness of the scheme. However, here we consider sufficiently long lists containing about 300 to 600 entries, so we shall neglect their finiteness.

Assume now that the chronological list X under investigation contains some duplicates, with the system $S_1, S_2, \dots, S_m$ of the most frequent (typical)shifts among them. We do not suppose that X is divided into disjoint duplicate systems, for those from different groups may overlap (cf. concept of "fibered chronicle" in [12].

Under this assumption, the distribution of the random variable ξ_1 is naturally dependent on the condition (event) A (and B). In fact, if two names u_i and u_j occured in chapter X_1 (were "born" there), then we should also expect them to be found among the duplicates of X_1. Thus, the value of the scattering of any two entries in the list X containing them will more often be close to zero, and to the shifts typical of the given duplicate system, than that of an arbitrary pair of names from I. Therefore, the histograms f_2 and f_3 will contain (in contrast with f_1) splashes near to the origin and the values of the shifts $S_1, S_2, \dots, S_m$.

Consider the problem of the verification of (B). We shall see that the distribution of ξ_1 has the same form in all the cases. Consequently, a problem arises how to verify the hypothesis that the distribution of ξ_2 (ξ_3) is close to a certain given one. As it has already been stressed, a certain deviation of f_1 from f_2 (f_3) may be due to the finitiness of the scheme. It is then natural to specify the problem as follows.

Consider the histogram f_2 (or f_3) as an empirical distribution on the set $\{0, 1, 2, \ldots, n-1\}$, constructed from a finite sample from the universe, and verify the hypothesis H_0 whether the general distribution coincides with that on the set $\{0, 1, \ldots, n-1\}$. We check H_0 by the test χ^2; H_0 is rejected in the considered examples. By the universe, we understand here a scheme constructed from a certain unknown extension of the list in question. The number of chapters is regarded invariable, and the chapter volume increasing rapidly. Thus, we can include into the extended list all the names of the relatives, courtiers, etc.; in the case of a narrative source, we include all personages active in the country at that time. Hence, the general population is constructed from all sorts of data both from the preserved and lost sources. What was constructed from a known list can then be regarded as a finite from a very large, practically "infinite" universe. This statement is rather general in the considered problems (see [11], where a similar situation arises). We assume that the available sample contains information just about the general distribution of the random variables considered in the above sense. In other words, any feasible way of selecting personages from a sufficiently long composite chronicle does not affect the distribution of the related name scatterings. In fact, such a choice is usually of "local" character, whereas the scattering distribution is a global characteristic, and stable to local perturbations. The values of the statistic χ^2 and of the actual level of the test α are given under the graphs.

4. It is easy to calculate that, in the case of a uniformly dense list $X=(a_1, a_2, \dots, a_N)$ such that all the chapters X_i, $i=1,2,\dots,n$, contain the same number p of entries, the histogram $f_1(j)=P(\xi_1=j)$ linearly decreases on the set $j\in\{0, 1, \dots, n-1\}$, and $f_1(j)=0$ for $j<0$ and $j\geqslant n$. In fact, ξ_1 takes the value j in $(n-j)\cdot p^2$ cases out of N^2 possible ($|\Omega| = N^2$), since there exist $n-j$ ways of fixing the chapter with the first name; the second chapter is fixed uniquely in accordance with the first one and number j, whereas the set of their name pairs with scattering j is of power p^2. Thus,

$$f_1(j)=P(\xi_1=j)=\frac{n-j}{N^2}\cdot p^2, \qquad 0\leqslant j\leqslant n-1.$$

In the sequel, we shall always make the list under consideration dense sufficiently uniformly, i.e., the histogram $f_1(j)$ linear with respect to j on the set $\{0, 1, \dots, n-1\}$. E.g., computations show that this condition is immediately fulfilled to a very high accuracy for the list of popes' names.

We can determine the histogram $f_2(j)=P_A(\xi_2=j)$ by means of the square name matrix $K_{n\times n}$ corresponding to the given list. Viz., the following formula is valid

$$f_2(j)=\begin{cases}\frac{1}{N^2}\sum_{i=1}^{n} K(i,s+j)\cdot K(i,s), & 0\leqslant j\leqslant n-1\\ 0 & \text{otherwise,}\end{cases} \tag{2}$$

where $K(\cdot,\cdot)$ are the elements of $K_{n\times n}$.

Formula (2) directly follows from the definition of the random variable ξ_2 and $K(i,s)$ being the total of the multiple names from the set of those "born" in X_i, which fell into chapter X_s.

The square matrix is insufficient for the construction of f_3. Therefore, we have to resort to a rectangular name matrix supplying complete information regarding chapters of the list.

It is constructed by ordering the names as they first appear in X, writing them out in a vertical column, and then entering all their mention frequencies. It is clear that the square matrix is derived from the obtained rectangular one.

5. Now we discuss the construction of the histograms f_1 and f_2, related to the well-known popes' name and nationality list Π and H dated from A.D. 50 (Apostle Peter) until the present day (see, e.g. [2-4]). Their peculiarity is, the names or nationalities have no explicit succession. Consequently, there are every grounds to believe that statement (B) is fulfilled if the above lists are chronologically correct. Computations show that this really happens for part of the list from the XIIIth c. A.D. till now days. Note that if we do assume the existence of a succession, then hypothetically correct chronology can only explain the splash near to the origin on the histogram f_2 and f_3 (see Item 6).

We divided Π and H into 10-year long chapters the list length being N=293, number of chapters n=190, and the number of different names k=87. We made use of the rectangular and square matrices constructed from Π and H by A.A. Makarov.

By direct computation it was established that the histogram $f_1(j)$ for Π and H is, to a very high accuracy, a linearly decreasing function for $j = 0, 1, \ldots, n-1$, and zero for other j (see the form of f_2 in Figs. 1,2); on the axis of abscissas, the values of scatterings were recalculated into years.

It can be seen that f_2 for Π possesses a series of sharp splashes (which are absent in the XIII-XXth cc. A.D.). According to the above argument, we can single out the following shift groups for f_2 and Π, viz.,

(i) through 40-50 and (doubling it) 80-100 years,

(ii) through 300 and 330-350 years,

(iii) the group of 11 consective shifts separated by a 100 years: through 400,480,580,670,760,940, 1,500, 1,140, 1,230 years

and

(iv) through 1,400 years.

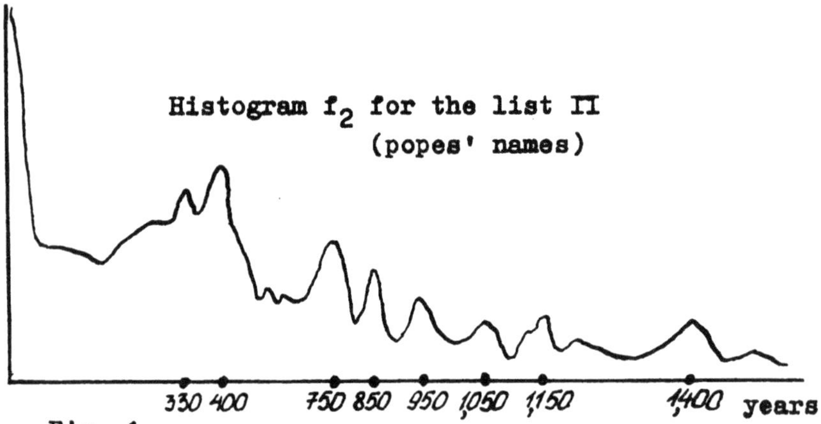

Fig. 1

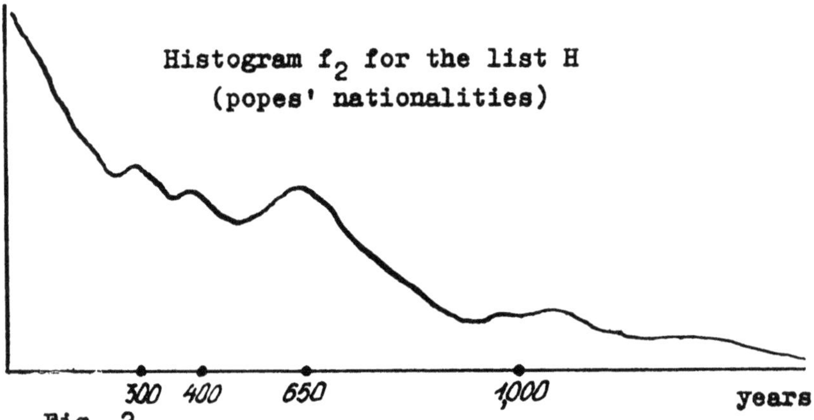

Fig. 2

Besides, we observe an exceptionally sharp, four-fold splash near to the origin. The shifts through 330, 400,760,850,960, 1,050, 1,150 and 1,400 years are also explicit.

The historgram f_2 for H supplies much less information, and contains two sharp splashes about the origin and 600-640 years as well as two weaker splashes about 330 and 450 years, which can probably be explained by the popes' nationalities having been determined quite recently and indirectly.

6. Now consider a narrative source divided into chapters-generations $X_1, X_2, \ldots, X_n$. Let us write out of each chapter all personal names with their multiplicities, and indicate their numbers. Note that we mean here personal and incomplete names, i.e., we break each complete composite name into separate ones. Then we obtain the same name collection divided into chapters as the one considered in Item 2. The related names and those of the same age, whereas the random variables ξ_1, ξ_2, ξ_3 and their frequency histograms f_1, f_2, f_3 will be defined in a similar way. However, there exists a natural dependence of the name set in X_i on that in $X_{i \pm l}$ (for small l) in a narrative source. We call this dependence a damping succession.

The existence of a damping succession in a narrative source leads to the necessity of specifying statement (A), and altering (B). In fact, if the names u_i and u_j got into some chapter X_m, then even a local relation leads to a statistically strong relation (u_i and u_j often occur in X_m and neighbouring chapters), what implies splashes near to the origin on the frequency histograms f_2 and f_3. Therefore, the statement (B) is replaced in a narrative source by the following.

(C) If the chronology of a source with the damping succession is correct, and its histogram $f_1(j) = P(\xi_1 = j)$ on its support is linearly recreasing, then f_2 and f_3 monotonically decrease on $0, 1, \ldots, n-1$ (if they are different from zero).

In order to apply (C) to the problem of verification of the narrative source chronology, it is necessary for $f_1(j) = P(\xi_1 = j)$ to decrease linearly on the support, what is, however, incorrect in the general case. As a matter of fact, narrative sources are often composite, and their chapter volume can be sharply non-uniform, of which the Bible is a good example. It is easy to see that the name scattering histogram f_1 will then have splashes for the values of the distances between detailed chapters with a large number of names.

The study of scatterings between source detailed chapters may also be useful for determining the duplicate shifts. The biblical histogram f_1 consists of a series of strong splashes, viz., 0,420, 650, 1,050, 1,300 and 1,600-1,800 years. The recalculation into years was performed, assuming 17 years for one chapter-generation.

To apply statement (C) to a narrative source, we assume that the number of mentions of a person involved in a historical event is directly proportional to the length of a chronicle describing it. With this assumption in mind, we can norm the source by dividing the multiplicity of a name occuring in a chapter by the total number of the names mentioned. For simplicity, all fractions will be reduced to the common denominator. We shall speak of a normed source or a normed source matrix in the sequel, implying the above norming procedure.

It is clear that a normed source is uniformly dense relative to the chapters; consequently, the histogram f_1 is on the support a linearly decreasing function (see Item 4).

7. Now we describe the results related to the normed lists B and M (see Item 6) of the biblical names and parallel passages, or repetitions,[11,12] (for their separation into chapters-generations, see ibid.). Containing tens of thousands of elements and several thousand different names, they were divided into n=218 chapters (see the form of the frequency histograms f_2 in Figs. 3,4). Both graphs possess sharp splashes in the interval

$0 \leqslant j \leqslant n-1$. We indicate the values of scattering in years and chapters (in parentheses), assuming each is 17 years long.

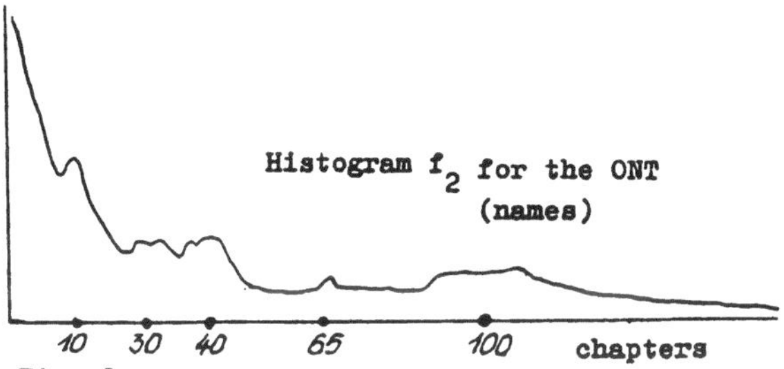

Fig. 3

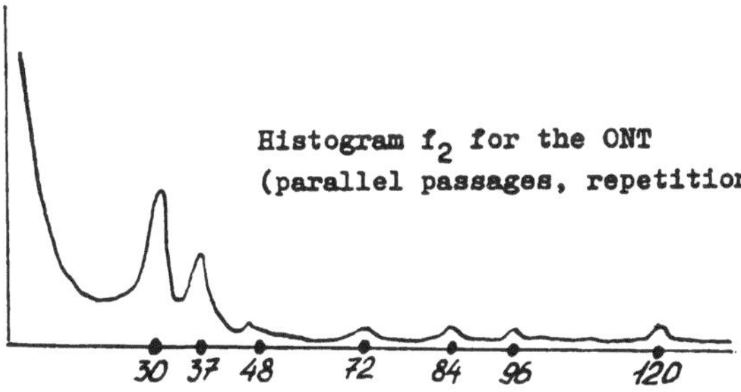

Fig. 4

We distinguish the following shifts: 170(10), 330 (19), 410(24), 500(29-32), 650-700(36-41), 1,100(65), 1,250(73), 1,500-1,700(92-102), in B, and 500(30), 650 (37), 1,250(72), 1,550(96), 2,000(120) in M, the first two being the strongest.

Their values in years have been found indirectly by recomputing the generations, and are, therefore, less exact than for the lists with a natural annual scale (e.g. popes).

8. We have already stressed in Item 3 that the basic statement (A) means that local relations must not lead to global ones in a chronologically correct list.

Two local conditions for a pair of names have been considered. Now, some more requirements.

Take the probabilistic scheme from Item 2. Let C be a certain subset of the list chapters, viz., $C = \{X_{i_1}, \ldots, X_{i_\ell}\}$. We shall say that two names u_i, u_j from C are of the same age if they were "born" together in one of its chapters. We shall call u_i and u_j conjugate in C ($u_i \overset{C}{\approx} u_j$) if they have been mentioned in one of its chapters, and write $a_i \overset{C}{\approx} a_j$, or $a_i \overset{C}{\sim} a_j$, if the corresponding relation is valid for the two entries in X as for the names from I.

Defining the events

$$A_C = \{\omega : a_{(1)} \overset{C}{\approx} a_{(2)}\},\; B_C = \{\omega : a_{(1)} \overset{C}{\sim} a_{(2)}\},\; \omega = (a_{(1)}, a_{(2)}),$$

we consider the frequency histograms for the names related in C as in Item 2, viz.,

$$f_2^C(j) = P_{A_C}(\xi_1 = j) = P(\xi_2^C = j),$$

$$f_3^C(j) = P_{B_C}(\xi_1 = j) = P(\xi_3^C = j),$$

where the random variables ξ_2 and ξ_3 are defined on the probability spaces $(\Omega, \Sigma, P_{A_C})$ and $(\Omega, \Sigma, P_{B_C})$, respectively;

$$\xi_1(\omega) = \xi_2^C(\omega) = \xi_3^C(\omega), \quad \omega \in \Omega.$$

By means of f_2^C, f_3^C, f_2 and f_3 we can also determine the shifts between the duplicates in chronologically incorrect lists. However, the shifts, typical to the system of duplicates of chapters in C, can be only found from f_2^C and f_3^C in the same way as those described in Item 3, whereas the duplicates themselves may not belong to C. By investigating f_2^C or f_3^C for different C, we can study the shift structure more in detail (certain examples of f_2^C for the popes' lists Π, their nationalities' H, biblical names' and biblical parallel passages' M are shown in Figs. 5-9).

9. Here, we discuss the problem, simulating the formation of an incorrect chronology mechanism by the exa-

mple of card shuffling. Both procedures are unknown, but both admit natural assumptions. Nothing prevents us from assuming that a deck is shuffled in the same manner as the duplicate in chronologically incorrect lists. Note that the problem is not a strict statement well-posed, it restates the initial one in the language of simpler patterns, and serves as the principal basis for working out the methods under consideration.

Suppose originally there were several decks of cards, identical in composition and (unknown) order P_0. Assume that the cards were then put in one large deck K, and shuffled, obtaining a new order P_1. Suppose that the "traces" of the initial order P_0 are retained in K, i.e., the shuffling is "incomplete", and that the number of the original decks (and their volumes) is unknown,

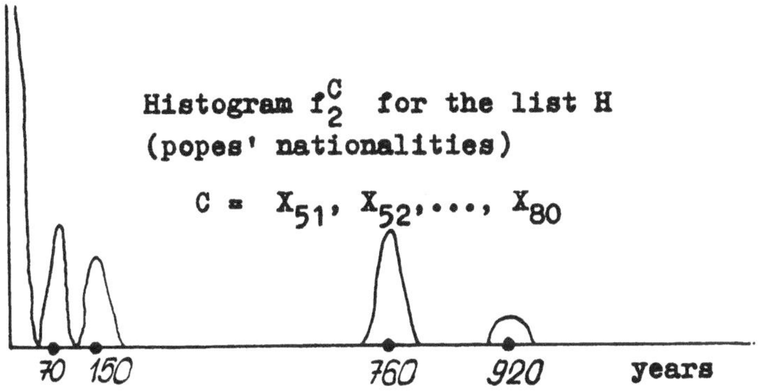

Fig. 5

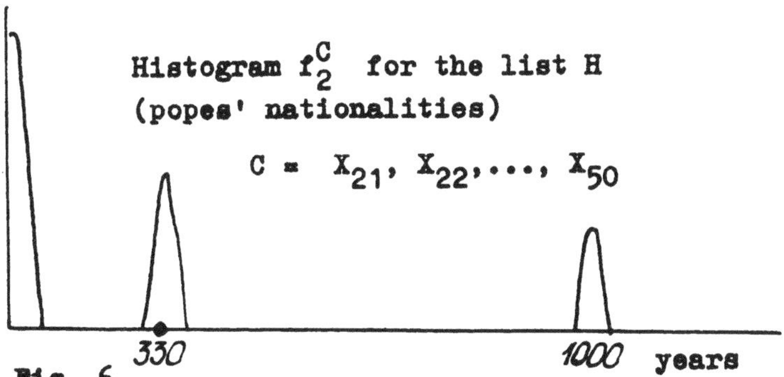

Fig. 6

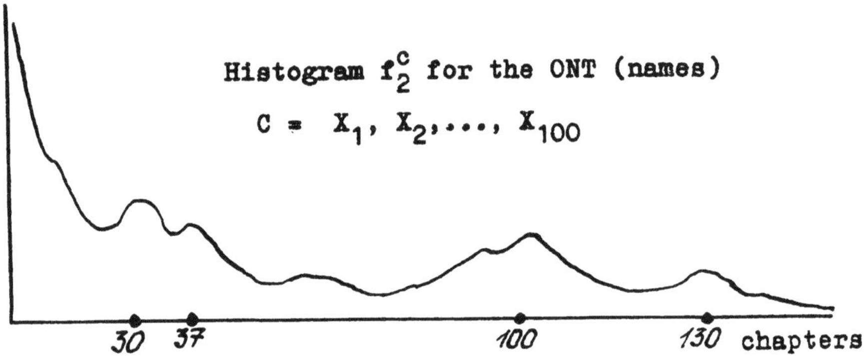

Fig. 7

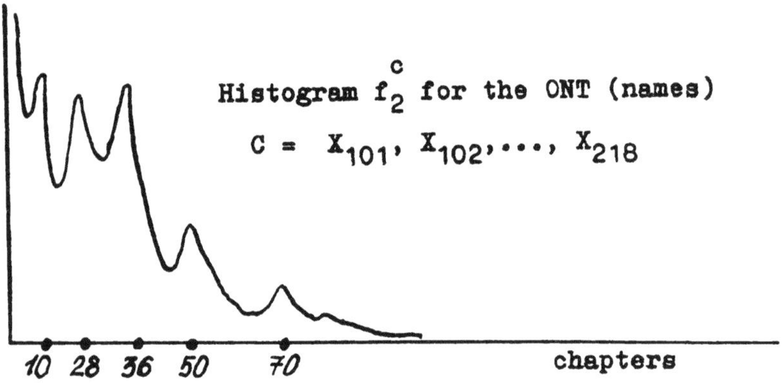

Fig. 8

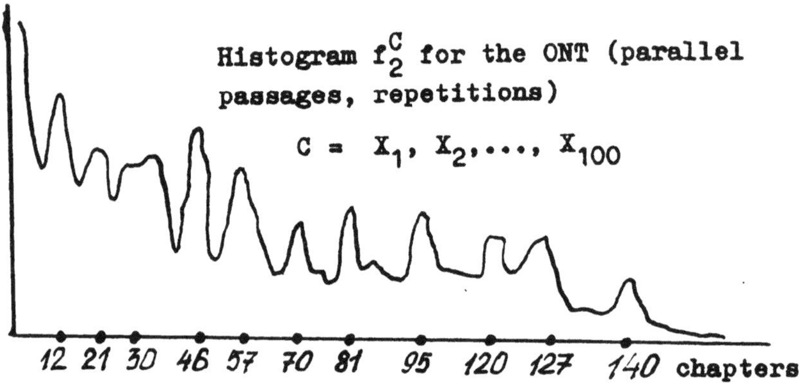

Fig. 9

assuming it cosiderably less than the volumes. How can we learn for a certain P_0 whether or not the deck K of order P_1 was obtained by the same method, and what was the initial order P_0 ?

The natural approach rests on the search of similar pieces in K. The more similar pieces have been found, the more sure the assertion is that this or other piece preserves the influence of P_0. Thus, we can attempt to restore P_0 piecewise. Besides, investigating in K the mutual disposition of similar pieces, we can determine whether or not the order P_1 is obtained, on the basis of inserting several decks of order P_0, somehow shifted relative to each other, as it is always done in shuffling cards and also find the shift values. We should, therefore, construct the frequency histogram for the "distances" between the similar pieces in K and see if there are typical distances. If there are such values, and the histogram does possess sharp splashes, then they can be naturally regarded as the shifts between the portions influenced by P_0.

The simplest piece in K is two consecutive cards. If K was, really, obtained by means of the described mechanism, then we can expect a considerable number of nearby cards in the final deck to be neighbouring also in the original ones. Therefore, the frequency histogram for scattering between the cards, which were placed side by side in K at least once, should make splashes around the values of typical shifts between the "duplicates".

The argument of this sort leads to the study of the frequency histograms for names related in chronological lists. Analogously, we can also model the methods considered below.

10. Now we turn to the relation matrix constructed from a given chronological list. We shall employ the notation from Item 2.

By means of the related name frequency histograms in Items 2-9, we did verify the hypothesis about the existence of duplicates in a chronological list, and determine the values of typical shifts among them, but did not find out which parts of the list exactly were duplicates. Recall that in accordance with the concept of a fibered chronicle, two parts of a list are regarded as duplicates if they contain fibers repeating each other [11],[12].

Now we turn to the card deck problem. We call two parts of the final deck K duplicates if they contain cards numbered identically or similarly before shuffling the original deck. Thus, two parts Δ_1 and Δ_2 of K are regarded as duplicates if they contain such subsets $A \subset \Delta_1$ and $B \subset \Delta_2$ that the cards from A and B were originally among the copies of the same sufficiently small, connected piece Δ of the original deck. Note that Δ_1 and Δ_2 may contain no identical cards at all, since it is possible that $A \cap B = \emptyset$. However, in shuffling incompletely, there must be copies of Δ, distributed in K with certain cards from A and B not far each from other. In the case where Δ_1 and Δ_2 contain fragments arising from the common propotype of Δ, that means there increases a probability of two cards from Δ_1 and Δ_2, respectively, being close somewhere in K. This fact can be used for making the concept of "similarity" of pieces in K more precise and for introducing a measure relation among them based on the quantity of such pairs of cards.

Now we present detailed reasoning for long chronological lists. Let there be a list X which may contain errors, omissions and/or duplicates. We denote by Y an unknown original list on which X is based. Thus, Y is an imaginary list containing complete data of a certain sort (say, about the rulers' names) for a long historical time interval I_Y. Let I_Y be described by a number of chroniclers, each making his own short list Z according to the contemporary events. Denote by $\{Z_i\}$ a set of them, forming a certain covering of Y, assumed to be sufficiently dense (with large multiplicity), and consisting of somehow dispersed and, possibly, erroneous pieces, i.e. each of Z_i mentioning not all the rulers' names or not everyone of the personages. Besides,errors and gaps could occur in rewriting and compiling, which will be assumed for simplicity, to be intrinsic to Z_i from a scratch.

The resulting chronology in its contemporary form was a certain new sewing of Z_i, so that the known list X was obtained. Consider two intervals Δ_1 and Δ_2 in X. Let us try to determine whether or not there is a pair Z_i, Z_j in X, which were related to one period in Y, and sewed to Δ_1 and Δ_2, respectively. As well as in the example with the cards, we conclude that if there is such a pair, then the probability that the names from Δ_1 and Δ_2 will be proximal somewhere in X increases, on account of a third, "sewing" chronicle Z_l (see for the detailed mathematical treatment in [22]).

11. Here and in the next items we consider the consequences of the study of the relation matrix for the popes' lists, beginning from Peter until 1950, and Roman emperors' from the IVth c.A.D. (Western Roman Empire) until the Holy Roman Empire in 962-1254 and the Habsburgs' Empire in 1273-1619 A.D. The emperors' list is extended up to 1700 A.D. (according to Leopold). To make the discussion of the results independent of the above argument, we recall the basic ideas of the method [22].

The so-called relation matrix is constructed from a large chronological list of rulers' names, for which each pair of connected fixed-length pieces (neighbourhoods) is associated with the number (relation), so that the following conditions are fulfilled, viz., in the case when the given list contains no duplicates, and consists of a random (in a certain sense) name sequence, the mean value of the relation does not depend on the choice of the numbers of the neighbourhood pairs, and, in the case when the list does contain duplicates, the relation of the pairs possessing duplicating fibers, is, on the average, greater than for those without such fibers.

By means of the study in the matrix of the distribution frequency histograms (for a fixed number of common names in the neighbourhoods), the theresholds were defined, separating the essential relations (the conclusion being that the neighbourhoods do contain duplicating fibers) and inessential relations (with the conclusion that the neighbourhoods are independent).

The essential relation matrix (or, simply, relation matrix), obtained in such a way provides for a decomposition of the list into duplicate systems (meanwhile, different systems can intersect, i.e., certain parts of the list possibly "fiber").

Note briefly certain overlappings determined by the constructed matrices. All of them are completely consistent with the GCD decomposition, obtained by the author ([6]-[22]).

(1) Three sharp splashes of heresies in church history, viz., the IVth c., "heresy age"; XII-XIIIth cc., the Albigenses, war with the heretics, establishment of the Inquisition; and heresies of the XV-XVI cc., strengthening of the Inquisition, putting witches to death, religious wars (see Fig. 10).

(2) Church schism in 1054 A.D.; irrevocable separation of the church in 1204 A.D. (see Fig. 10).

(3) Three papal elections under Henry III around

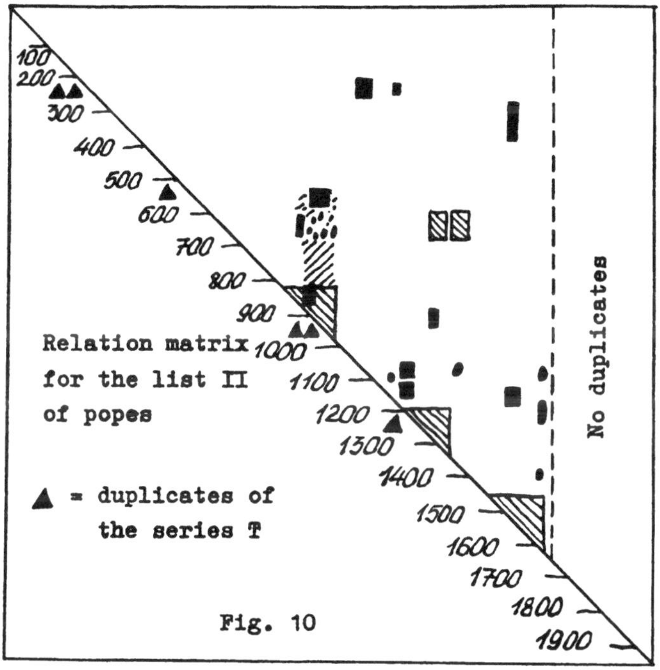

Fig. 10

1050 A.D.; the three popes in 1378-1417 A.D. during the great schism (see Fig, 10).

(4) Roman Empire in 753-523 B.C.; Roman Empire in 82 B.C. - 217 A.D. (see Fig. 11).

(5) Roman Empire in 82 B.C. - 217 A.D.; Roman Empire in 270-526 A.D.; Holy Empire in 962-1254 A.D. (see Fig. 11).

(6) Carolingians in 681-887 A.D., the Holy Empire in 962-1254 A.D. and the Habsburgs' Empire in 1273-1619 A.D. (see Fig. 11).

(7) Overlapping of the limits of the two Roman republics (Fig. 11).

(8) Principal originals from the Habsburgs' Empire in the XVIth c., 1500-1530 and c. 1600 A.D. (see Fig. 11).

Besides, the author's conclusion regarding the existence of authentic consecutive chronology after 900 A.D.

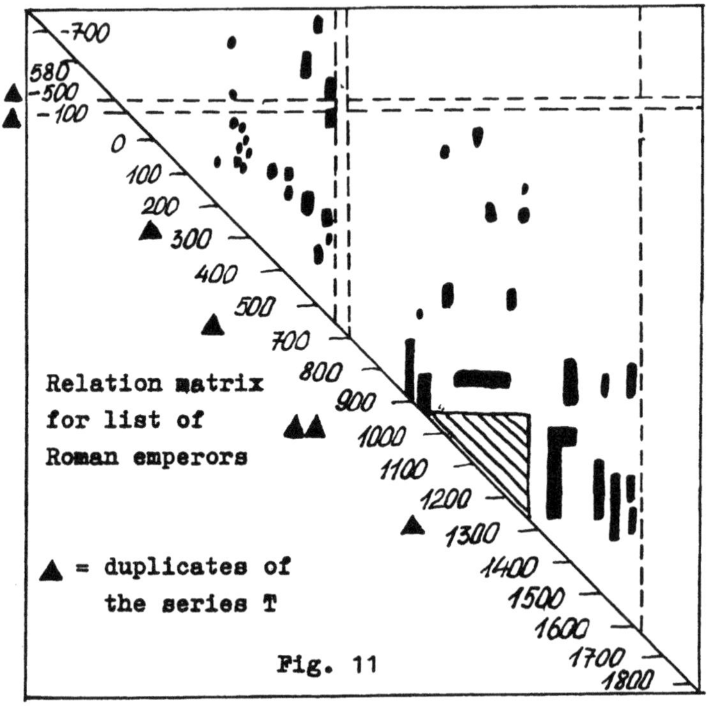

Fig. 11

and reliable chronology from the end of the XIIIth c. A.D. is entirely confirmed (see Figs. 10, 11).

The popes' list in 50-1950 A.D. contains 87 different names of the total number of 293, their maximum multiplicity being 21. The essential relation matrix is shown in Fig. 10. Note immediately that the interval 1700-1950 A.D. contains no duplicates, whereas 1600-1700 A.D. is comparatively weakly related only to certain epochs after 1100 A.D. (the only stronger relation to the interval 1180-1260 A.D.).

The duplicate system list is given below in accordance with Fig. 10. The arrowheads denote relations with the (key) time interval which is the first in each line, whereas the dotted ones mark comparatively weak relations. The interval time limits are accurate up to a century (see the details in Fig. 10). The right-hand column contains the generated shifts (cf. those distinguished from the lists considered in Item 5).

No.	Duplicate system (cc.)	Shifts (yrs.)
1	II ←---- XV-XVI	1,300-1,400
2	II,III ←—— XII	1,000
3	III,IV ←---- XI, ←—— XVI	700, 1,200
4	VI,VII ←—— X	280-330
5	VI_2,VII ←— XIII_2: ←----- XIV,XV_1	660, 760
6	XI ←---XIII; ←---- XVII_1	300, 600
7	XII,XIII ←—— XVII_1	500
8	XIII ←--- XIV	170
9	XIV ←---- XVII_1	330

A Roman numerals indicates a century, and a subscript shows half a century.

12. A list of Roman emperors' names and the related conclusions. The list was constructed from the names of Roman emperors (after the separation of the Western Roman Empire) from Romulus (753 B.C.) up to Leopold I (c. 1700 A.D.). If several names of one emperors are known, then they all are written out one by one. The ordering was made by the average year of ruling, the list contains no separation signs between the names of consecutive emperors, its length is 555, the number of different names 193, and the maximal name multiplicity is 40 (see the essential relation matrix in Fig. 11). Similarly to the previous item, we also list the distinguished duplicate systems and their generated shifts. The Roman numerals denote centuries, whereas the subscripts 1 or 2 designate the first or second half of the century, respectively.

This matrix determines the decomposition of the emperors' list into duplicate series which coincide with those found by the author on investigating dynastic parallels [8]. E.g., the overlapping of the two Roman Empires in 82 B.C. - 217 A.D. and 270-526 A.D. is reflected in the diagonal parallel to the principal (see Fig. 11). It means that these duplicates overlap nearly without distortions. The T duplicate series of the GTR - war is also explicit.

No.	Duplicate system (cc.)	Shifts (yrs.)
1	VII B.C. ←--- III A.D.; ←— V A.D.	150, 1,050, 1,200
2	VI B.C., I A.D. ←— III A.D.; ←— VII	750, 1,050
3	I-III A.D. ←— III-VI A.D.	250-300, 950-1,050
4	I, III ←— XI, XIII	1,000-1,050
5	IV, V ←— VI	100-200
6	VI ←— X; ←— XIII	500, 700
7	VIII, IX ←— c. 900 A.D.	100
8	IX ←— XI_2, XII ←— c. 1400A.D. ←— c. 1500 A.D. ←— c. 1600 A.D.	200-400
9	X-XIII ←— XIII	300
10	XI ←— c. 1350 A.D.	300
11	XII, XIII ←— c. 1350 A.D.; ←— (1500-1550 A.D.) ←— (1600-1625 A.D.)	200-400

13. The comparison of the obtained results with the GCD decomposition constructed by the author.

The present study was carried out by the author and G. Nosovsky in 1983-1985 with the purpose of independent verification of the GCD (e.g., see [8], [11] or a more detailed treatment in [12], [19]).

Finally, two circumstances were made clear.

(1) Ancient and medieval historical data possess explicit statistical regularities appearing earlier than the XIIIth c. A.D. It is possible that this can be easily explained by the hypothetical existence of chronological duplicates in the form given by the author (ibid.).

(2) If the hypothesis regarding the existence of chronological duplicates is accepted, then the interpretation of the obtained results leads to the same results as those derived by the author earlier (viz.,the GCD decomposition). It is essential that the GCD decomposition was originally constructed from the dynamic rule durations. If we proceed from the dynastic name set, the same conclusions derived supply an independent confirmation of the author's hypothesis.

Remark. It is important that the above anomalous splashes on the histograms f_2 and f_3 are not accidental. This is confirmed by the following experiment. We have subjected the names' Π, H, ONT and Roman emperors' lists to random mixing (shuffling). Then we calculated the histograms f_1, f_2, f_3. It is remarkable that they all tend into linear functions. Thus, discovered a priori non-random fibered structure in the original lists, generated by the three shifts through c. 330, 720 and 1,050 years (approximately).

REFERENCES

1. Bickerman, E. Chronology of the ancient world. Thames & Hudson, London, 1968.
2. Blair, J. Blair's Chronological and Historical Tables, from the Creation to the present time, etc. G. Bell & Sons, London, 1882.
3. Gregorovius, F. History of the city of Rome in the Middle Ages. G. Bell & Sons, London, 1900-1909.
4. Lozinsky, S.G. History of Papacy. Ogiz, Moscow, 1934 (Russian).
5. Morozov, N.A. Christ. Gosizdat, Moscow-Leningrad, 19261932 (Russian).
6. Fomenko, A.T. Computation of second derivative of moon's elongation and statistical regularities in distribution of certain astronomical data. "Operations Research and ACS", v. 20, Kiev University Press, Kiev, 1982, pp. 98-113 (Russian).
7. Fomenko, A.T. Certain statistical regularities of information density distribution in text with scale. "Semiotika i informatika", Vyp. 15, VINITI, Moscow, 1980, pp. 99-124 (Russian).
8. Fomenko, A.T. A method of duplicate recognition and

some applications. DAN SSSR, v. 258, 6 (1981), pp.1326-1333 (Russian).

9. Fomenko, A.T. Informative functions and related statistical regularities. Abstracts of Communications of the Third International Vilnius Conference on Probability Theory and Mathematical Statistics. Inst.of Mathematics and Cybern.of ASc. LSSR, v. 2, 1981, pp. 211-212 (Russian).

10. Fomenko, A.T. On properties of second derivative of moon'c elongation and related statistical regularities. "Problems of Computational and Applied Mathematics", vyp. 63, AN USSR, Tashkent, 1981, pp. 136-150 (Russian).

11. Fomenko, A.T. New empirico-statistical methods in ordering texts and applications to dating problems. DAN SSSR, v. 268, 6 (1983), pp. 1322-1327 (Russian).

12. Fomenko, A.T. New Experimental and Statistical Methods for Dating Events of Ancient History and Applications to Global Chronology of Ancient World and Middle Ages. Preprint N B 07201 (dated Dec.9, 1981), Moscow, 1981, pp. 1-100. (English translation is catalogoued in the British Library, Depart. of printed books, Cup. 918-87).

13. Fomenko, A.T. On geometry of distribution of integral points in hyperregions. "Trudy seminara po vekronomu i tenzornomu analizu", vyp. 21, 1983, Moscow Univ.Press, Moscow, pp. 106-152 (Russian).

14. Fomenko, A.T. New methods of dating texts chronologically correctly and applications to problems of dating ancient history events. "Operations Research and ACS", vyp. 21, Kiev Univ.Press, Kiev,1983, pp. 40-59 (Russian).

15. Fomenko, A.T. Methods of statistical investigation of investigating parallelisms in chronographic texts and global chronological diagram. "Operations Research and ACS", vyp. 22, Kiev Univ.Press, Kiev, 1983, pp. 29-40 (Russian).

16. Fomenko, A.T. The jump of the second derivative of the moon's elongation. "Celestial Mechanics", v. 25, 1 (1981), pp. 33-40.

17. Fomenko, A.T. Global Chronological Diagram. "Khimiya i Zhizn'", 9, 1983, pp. 85-92 (Russian).

18. Fedorov, V.V. and Fomenko, A.T. Statistical estimation of chronological proximity of historical texts. "Stability Problems of Stochastic Models". System Research Institute, Moscow, 1983, pp. 101-107 (Russian).

19. Fomenko, A.T. New empirical and statistical methods of discovering parallels and dating duplicates. "Stability Problems of Stochastic Models". Systems Research Institute, Moscow, 1984, pp. 154-177 (Russian).

20. Fomenko, A.T. Informative functions and related statistical regularities. "Statistics. Probability. Economics", Nauka, Moscow, 1985, pp. 335-342 (Russian).
21. Fomenko, A.T. Duplicates in Mixed Sequences and Frequency Duplication Principle. Abstracts of Communications of the Fourth International Vilnius Conference on Probability Theory and Mathematical Statistics. Institute of Mathem.and Cybernetics of ASc LSSR, v. 3, 1985, pp. 246-248 (Russian).
22. Nosovsky, G.V. and Fomenko, A.T. The calculation of initial structures in mixed sequences. "Trudy seminara po vektornomu i tenzornomu analizu", vyp. 22, Moscow Univ.Press, Moscow, 1985 , pp. 119-131 (Russian).
23. Newton I. The chronology of ancient kingdoms amended, etc. J. Tonson, London, 1728.

Prob. Theory and Math. Stat., Vol.1, pp. 467-481
Prohorov et al. (eds)
1986 VNU Science Press

BOOTSTRAPPING EMPIRICAL MEASURES INDEXED BY VAPNIK-CHERVONENKIS CLASSES OF SETS

Peter Gaenssler
Mathematical Institute, University of Munich, Theresienstrasse 39,
D-8000 Munich 2, W.-Germany

1. INTRODUCTION

Let ξ_i, $i \in \mathbb{N} = \{1,2,\ldots\}$, be i.i.d. real valued random variables with law μ (defined on the Borel σ-field $\mathcal{B} = \mathbb{B}$ in $X = \mathbb{R}$) and let μ_n be the empirical measure based on $\xi_1,\ldots,\xi_n$, i.e.

$$\mu_n(B) := n^{-1} \sum_{i=1}^{n} 1_B(\xi_i), \quad B \in \mathcal{B},$$

where

$$1_B(x) := \begin{cases} 1, & \text{if } x \in B \\ 0, & \text{if } x \in \complement B \end{cases}, \quad x \in X.$$

Then, concerning the asymptotic behaviour of μ_n, two basic classical results state that

(1.1) $$\sup_{t\in\mathbb{R}} |\mu_n((-\infty,t]) - \mu((-\infty,t])| \to 0 \quad \text{a.s.} \quad \text{as} \quad n \to \infty$$

(GLIVENKO-CANTELLI THEOREM), and

(1.2) $$(n^{1/2}(\mu_n((-\infty,t]) - \mu((-\infty,t])))_{t\in\mathbb{R}} \xrightarrow{\mathcal{L}} \mathbb{G}_\mu$$

(DONSKER'S FUNCTIONAL CLT),

where $\mathbb{G}_\mu = (G_\mu(t))_{t\in\mathbb{R}}$ is a mean-zero Gaussian process with covariance structure given by

$$\operatorname{cov}(G_\mu(s),G_\mu(t)) = \mu((-\infty,s] \cap (-\infty,t]) - \mu((-\infty,s])\mu((-\infty,t]), \quad s,t \in \mathbb{R}.$$

In case of random elements ξ_i in arbitrary sample spaces $(X,\mathcal{B})$ it is natural to ask whether the same results hold true for certain subclasses $\mathcal{C}$ of $\mathcal{B}$ which play the role of $\mathcal{C} = \{(-\infty,t] : t \in \mathbb{R}\}$ in the just mentioned classical situation of the real line $X = \mathbb{R}$.

It is known that (under certain additional measurability conditions)

(1.1) and (1.2) remain true in arbitrary sample spaces $(X,\mathcal{B})$ for so-called Vapnik-Chervonenkis classes $\mathcal{C} \subset \mathcal{B}$ having the following characteristic property (in common with the just mentioned class of all half rays in $X = \mathbb{R}$):
There exists some $s \in \mathbb{N}$ such that no s-element subset of X is shattered by $\mathcal{C}$, where, given any finite subset F of X, F is said to be shattered by $\mathcal{C}$, if for every $F' \subset F$ there exists a $C \in \mathcal{C}$ such that $F' = F \cap C$.
So, given a sequence of i.i.d. random elements ξ_i with law μ on $\mathcal{B}$ in an arbitrary sample space $(X,\mathcal{B})$ and a Vapnik-Chervonenkis class $\mathcal{C} \subset \mathcal{B}$ (assumed here to be countable in order to avoid measurability problems), the following results extend (1.1) and (1.2) to general sample spaces (where μ_n is defined as before):

(1.3) $\quad \|\mu_n - \mu\|_{\mathcal{C}} := \sup_{C\in\mathcal{C}} |\mu_n(C) - \mu(C)| \to 0$ a.s. as $n \to \infty$,

and

(1.4) $\quad (n^{1/2}(\mu_n(C) - \mu(C)))_{C\in\mathcal{C}} \overset{\mathcal{L}}{\to} \mathbb{G}_\mu$,

where $\mathbb{G}_\mu = (G_\mu(C))_{C\in\mathcal{C}}$ is a mean-zero Gaussian process (indexed by $\mathcal{C}$) with covariance structure given by
$\operatorname{cov}(G_\mu(C),G_\mu(D)) = \mu(C \cap D) - \mu(C)\mu(D)$, $C,D \in \mathcal{C}$,
and where $\overset{\mathcal{L}}{\to}$ denotes weak convergence (convergence in law) as studied in Gaenssler (1983)(called $\mathcal{L}_b$-convergence there).

The present paper is concerned with a method of proof being appropriate not only for proving (1.3) and (1.4) but also for proving their so-called bootstrapped versions, where e.g. the bootstrapped version of (1.4) states that for any $\underline{x} = (x_1,x_2,\ldots) \in X^{\mathbb{N}}$ satisfying (1.3) (i.e. for which $\|\mu_n^{\mathbf{x}} - \mu\|_{\mathcal{C}} \to 0$ as $n \to \infty$, where $\mu_n^{\mathbf{x}}(C) = n^{-1} \sum_{i=1}^{n} 1_C(x_i)$) one has

$(m^{1/2}(\mu_{m,n}(C) - \mu_n^{\mathbf{x}}(C)))_{C\in\mathcal{C}} \overset{\mathcal{L}}{\to} \mathbb{G}_\mu$ if $m = m(n) \to \infty$ as $n \to \infty$, where $\mu_{m,n}(C) := m^{-1} \sum_{i=1}^{m} 1_C(\xi_i^*)$, the ξ_i^*'s being i.i.d. random elements in $(X,\mathcal{B})$ with law $\mu_n^{\mathbf{x}}$ on $\mathcal{B}$.

2. VAPNIK-CHERVONENKIS CLASSES OF SETS

Concerning the Glivenko-Cantelli statement (1.3) it is easily seen that it cannot hold in general for any μ if the corresponding class $\mathcal{C}$ is too rich in a certain sense which will become apparent in a moment.

Let e.g. $\mathcal{C} := \{C \subset X : |C| < \infty\}$ and let μ be a nonatomic measure on a σ-field $\mathcal{B}$ in X containing all singletons (where nonatomic means that $\mu(\{x\}) = 0$ for all $x \in X$); then $\|\mu_n - \mu\|_{\mathcal{C}} = 1$ for all $n \in \mathbb{N}$.

To mention another example, let

$$\mathcal{C} := \{C \subset \mathbb{R}^k, \ C \text{ a convex Borel set}\},$$

and let μ be the uniform distribution on the unit sphere S_1 in $X = \mathbb{R}^k$, $k \geq 2$; then again $\|\mu_n - \mu\|_{\mathcal{C}} = 1$ a.s. for all $n \in \mathbb{N}$, since, given any sample $x_1,\ldots,x_n$ (belonging a.s. to S_1), there exists a $C \in \mathcal{C}$ contained in the unit ball of $\mathbb{R}^k$ such that $C \cap S_1 = \{x_1,\ldots,x_n\}$, whence $\mu_n(C) = 1$ whereas $\mu(C) = 0$.

Both examples show that the underlying classes $\mathcal{C}$ are too rich in the following sense:

(2.1) For every $n \in \mathbb{N}$ there exists an $F \subset X$ with $|F| = n$ such that $|\{F \cap C : C \in \mathcal{C}\}| = 2^n$.

(As to the second example, take any $F = \{x_1,\ldots,x_n\} \subset S_1$ observing that for any subsample $\{x_{i_1},\ldots,x_{i_k}\} \subset F$ one has $\{x_{i_1},\ldots,x_{i_k}\} = F \cap C$ for C being the convex hull of $\{x_{i_1},\ldots,x_{i_k}\}$.)

A class $\mathcal{C}$ of sets not satisfying (2.1) will be called a Vapnik-Chervonenkis class (VCC), i.e.

$\mathcal{C} \subset \mathcal{B}$ in $X = (X,\mathcal{B})$ is a VCC iff there exists some $s \in \mathbb{N}$ such that no s-element subset is shattered by $\mathcal{C}$, which means that for every $F \subset X$ with $|F| = s$ one has $|\{F \cap C : C \in \mathcal{C}\}| < 2^s$.

2.2 Examples of VCC's (cf. Wenocur and Dudley (1981) and Gaenssler (1983)).

a) $X = \mathbb{R}^k$, $k \geq 1$, $\mathcal{C} := \{(-\infty,\underline{t}] : \underline{t} \in \mathbb{R}^k\}$, $s = k + 1$.

k=1 :

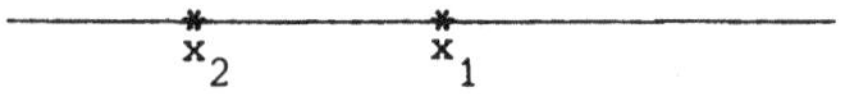

$F = \{x_1,x_2\} \Rightarrow F' := \{x_1\} \neq F \cap C$ for all $C = (-\infty,t]$.

b) $X = \mathbb{R}^k$, $k \geq 1$, $\mathcal{C} := \times_{i=1}^{k} [s_i,t_i] : -\infty < s_i \leq t_i < \infty\}$, $s = 2k + 1$.

c) $X = \mathbb{R}^k$, $k \geq 1$, $\mathcal{C} := \{B \subset \mathbb{R}^k : B \text{ a closed Euclidean ball}\}$,

$s = k + 2$.

Other examples of VVC's can be obtained from the following Lemma (cf. Dudley (1978), Theorem (7.2), and Pollard (1984), II. Lemma 18):

<u>2.3 Lemma.</u> Let $\mathcal{G}$ be any m-dimensional vector space of real functions on a set X ; then $\mathcal{C} := \{\{g \geq 0\} : g \in \mathcal{G}\}$ is a VCC.

It follows immediately from the definition that any subclass of a VCC is again a VCC. On the other hand, as shown by Dudley (1978), Proposition (7.12), one has:

<u>2.4 Lemma.</u> For any VCC $\mathcal{C}$ and any fixed $k < \infty$, the class $\mathfrak{A}_k(\mathcal{C}) := \cup\{\mathfrak{A}(C_1,\ldots,C_k) : C_1,\ldots,C_k \in \mathcal{C}\}$ is a VCC, where $\mathfrak{A}(C_1,\ldots,C_k)$ denotes the algebra generated by $C_1,\ldots,C_k$.

This implies especially that the classes $\{C \Delta D : C,D \in \mathcal{C}\}$, $\{C\setminus D : C,D \in \mathcal{C}\}$ and $\{C \cap D : C,D \in \mathcal{C}\}$ are VCC's if $\mathcal{C}$ is a VCC.

Next, concerning again the Glivenko-Cantelli statement (1.3), it is known through the work of Vapnik and Chervonenkis (1971) that for any VCC $\mathcal{C}$ (1.3) holds true provided that some additional measurability conditions are satisfied (cf. Theorem 2 in Gaenssler (1983) and Example (19) there which is due to Durst and Dudley (1981)).

As mentioned in the introduction we are going to restrict ourselves to countable VCC's in order to avoid measurability problems at all. That countable classes $\mathcal{C}$ are not necessarily VCC's can be seen by the following example:

Let $(X,\mathcal{B}) = (\mathbb{R},\mathbb{B})$, $J_1 := \{[a,b] : a < b,\ a,b \in \mathbb{Q}\}$, where $\mathbb{Q}$ denotes the rational numbers in $\mathbb{R}$; let $J(n) := \bigcup_{i=1}^{n} J_i$ with $J_i \equiv J_1$. Then $\mathcal{C} := \bigcup_{n\in\mathbb{N}} J(n)$ is a countable subclass of the Borel σ-field $\mathbb{B}$ in $\mathbb{R}$ for which (2.1) holds true.

3. THE FUNDAMENTAL VAPNIK-CHERVONENKIS INEQUALITY AND SOME OF ITS CONSEQUENCES

Let $X = (X,\mathcal{B})$ be an arbitrary sample space and let

$(\xi_{ni})_{n\in\mathbb{N},1\leq i\leq i(n)}$ be an array of row-wise i.i.d. random elements in $(X,\mathcal{B})$ defined on some basic p-space $(\Omega,\mathcal{F},\mathbb{P})$. Let $\mathcal{C}\subset\mathcal{B}$ be a VCC assumed here to be countable in order to avoid measurability problems. Given any $n\in\mathbb{N}$, let

$$\mu_{i(n),n}(C) := \frac{1}{i(n)}\sum_{i=1}^{i(n)} 1_C(\xi_{ni}),\ C\in\mathcal{C},$$

and

$$\|\mu_{i(n),n}-\nu_n\|_{\mathcal{C}} := \sup_{C\in\mathcal{C}} |\mu_{i(n),n}(C)-\nu_n(C)|,$$

where $\nu_n := \mathcal{L}\{\xi_{ni}\}$, $i = 1,\dots,i(n)$.

Then the following inequality due to Vapnik and Chervonenkis (1971) is basic (cf. Lemma 10 in Gaenssler (1983)):

3.1 Lemma (VC-Inequality). For any $\varepsilon > 0$ and $n\in\mathbb{N}$ one has

$$(3.2)\quad \mathbb{P}(\|\mu_{i(n),n}-\nu_n\|_{\mathcal{C}}\geq\varepsilon) \leq \left(1-\frac{1}{\varepsilon^2 i(n)}\right)^{-1}\cdot[2i(n)]^v\cdot 2\exp\left(-\frac{\varepsilon^2 i(n)}{8}\right),$$

where $v := V(\mathcal{C})$ is the so-called Vapnik-Chervonenkis index pertaining to the VCC $\mathcal{C}$ (i.e. v equals the least $s\geq 1$ such that no s-element subset of X is shattered by $\mathcal{C}$).

As a corollary we obtain the following Glivenko-Cantelli results (cf. (1.3)):

3.3 Lemma.

(i) $\|\mu_{i(n),n}-\nu_n\|_{\mathcal{C}}\xrightarrow{\mathbb{P}} 0$ if $i(n)\to\infty$ as $n\to\infty$;

(ii) $\|\mu_{i(n),n}-\nu_n\|_{\mathcal{C}}\to 0$ $\mathbb{P}$-a.s. if $\frac{i(n)}{\ln n}\to\infty$ as $n\to\infty$.

Proof. (i) is an immediate consequence of (3.2).
As to (ii), it suffices to show that

$$(*)\quad \limsup_{n\to\infty}\|\mu_{i(n),n}-\nu_n\|_{\mathcal{C}}\leq\varepsilon\ \ \mathbb{P}\text{-a.s. for every } \varepsilon>0;$$

according to the Borel-Cantelli Lemma, (*) is implied by

$$(**)\quad \sum_{n\in\mathbb{N}}\mathbb{P}(\|\mu_{i(n),n}-\nu_n\|_{\mathcal{C}}>\varepsilon)<\infty \text{ for every } \varepsilon>0,$$

whence the proof will be concluded by showing that (**) holds true. According to the VC-Inequality (3.2) this amounts to show that for every $\varepsilon > 0$

$$\sum_{n\in\mathbb{N}} [i(n)]^v \exp(-\frac{\varepsilon^2 i(n)}{8}) < \infty :$$

But this follows at once, since by assumption $\frac{i(n)}{\ln n} \to \infty$ as $n \to \infty$, whence for large enough n,

$$[i(n)]^v \exp(-\frac{\varepsilon^2 i(n)}{8}) = \exp[v \ln i(n) - \frac{\varepsilon^2 i(n)}{8}] =$$

$$\exp[i(n)(\frac{v \ln i(n)}{i(n)} - \frac{\varepsilon^2}{8})] \leq \exp(-\frac{\varepsilon^2 i(n)}{16}) = \exp(-\frac{i(n)}{\ln n}\frac{\varepsilon^2}{16} \ln n) =$$

$$n^{-\frac{i(n)}{\ln n}\frac{\varepsilon^2}{16}} \leq n^{-2} . \ \square$$

Taking $\xi_{ni} = \xi_i^*$, $i = 1,\dots,m = m(n)$, the ξ_i^* being i.i.d. random elements in $(X,\mathcal{B})$ with law $\mu_n^{\mathbf{x}}$ on $\mathcal{B}$ (cf. section 1), Lemma 3.3 yields the following "bootstrapped" Glivenko-Cantelli results:

<u>3.4 Lemma.</u> For any $\underline{x} = (x_1,x_2,\dots) \in X^{\mathbb{N}}$ one has

(i) $\|\mu_{m,n} - \mu_n^{\mathbf{x}}\|_{\mathcal{C}} \xrightarrow{\mathbb{P}} 0$ if $m(n) \to \infty$ as $n \to \infty$,

and

(ii) $\|\mu_{m,n} - \mu_n^{\mathbf{x}}\|_{\mathcal{C}} \to 0$ $\mathbb{P}$-a.s. if $\frac{m(n)}{\ln n} \to \infty$ as $n \to \infty$,

where $\mu_{m,n}$ denotes the empirical measure based on $\xi_1^*,\dots,\xi_{m(n)}^*$.

4. WEAK CONVERGENCE OF BOOTSTRAPPED EMPIRICAL $\mathcal{C}$-PROCESSES INDEXED BY VAPNIK-CHERVONENKIS CLASSES $\mathcal{C}$ OF SETS

Let again $X = (X,\mathcal{B})$ be an arbitrary sample space and let $(\xi_i)_{i\in\mathbb{N}}$ be a sequence of i.i.d. random elements in $(X,\mathcal{B})$ with distribution μ on $\mathcal{B}$ and defined on some basic p-space $(\Omega,F,\mathbb{P})$. Let $\mathcal{C} \subset \mathcal{B}$ be a VCC assumed again to be countable in order to avoid measurability problems. Extensions to uncountable VCC's under appropriate measurability conditions can be e.g. achieved along the lines of Gaenssler (1983).

Given $\underline{x} = (x_1,x_2,\dots) \in X^{\mathbb{N}}$, $x_i = \xi_i(\omega)$, $\omega \in \Omega$, let for each $n \in \mathbb{N}$ $\mu_n^{\mathbf{x}}$ be the empirical measure based on $x_1,\dots,x_n$. Then it follows from Lemma 3.3 (ii) (applied with $\xi_{ni} = \xi_i$, $i = 1,\dots,i(n) = n$) in connection with Lemma 2.4 that

(4.1) $\|\mu_n^{\mathbf{x}} - \mu\|_{\mathfrak{A}_k(\mathcal{C})} := \sup_{E\in\mathfrak{A}_k(\mathcal{C})} |\mu_n^{\mathbf{x}}(E) - \mu(E)| \to 0$ as $n \to \infty$

for almost every sequence $\underline{x} = (x_1, x_2, \ldots)$ of realizations x_i of ξ_i .

Now, let $\underline{x} = (x_1, x_2, \ldots) \in X^{\mathbb{N}}$ be any sequence satisfying (4.1) and being fixed in what follows. Define the

BOOTSTRAPPED EMPIRICAL $\mathcal{C}$-PROCESS

$$\beta_{m,n} = (\beta_{m,n}(C))_{C \in \mathcal{C}}$$

by

(4.2) $\beta_{m,n}(C) := m^{1/2}(\mu_{m,n}(C) - \mu_n^{\underline{x}}(C)),\ C \in \mathcal{C}$,

where $\mu_{m,n}$ denotes the empirical measure based on $\xi_1^*, \ldots, \xi_m^*$, $m = m(n)$, the ξ_i^*'s being i.i.d. random elements in $(X, \mathcal{B})$ with law $\mu_n^{\underline{x}}$ on $\mathcal{B}$ and defined on the same basic p-space $(\Omega, \mathcal{F}, \mathbb{P})$ (properly enlarged).

Based on the results of section 3 we are going to show that Pollard's (1981) method of proving (1.4) can also be applied to show that empirical processes indexed by Vapnik-Chervonenkis classes $\mathcal{C}$ of sets can be bootstrapped , that is the following theorem holds:

<u>4.3 Theorem.</u> $\beta_{m,n} \xrightarrow{\mathcal{L}} \mathbb{G}_\mu$ if $m = m(n) \to \infty$ as $n \to \infty$, where $\mathbb{G}_\mu = (G_\mu(C))_{C \in \mathcal{C}}$ is a mean-zero Gaussian process with covariance structure given by

$\operatorname{cov}(G_\mu(C_1), G_\mu(C_2)) = \mu(C_1 \cap C_2) - \mu(C_1)\mu(C_2),\ C_i \in \mathcal{C},\ i = 1,2,$ $\mathbb{G}_\mu$ having all its sample paths in the space of bounded real-valued and uniformly d_μ-continuous functions on $\mathcal{C}$, where d_μ is the pseudo-metric defined on $\mathcal{C}$ by $d_\mu(C_1, C_2) := \mu(C_1 \Delta C_2)$.

<u>4.4 Remarks.</u> Here $\xrightarrow{\mathcal{L}}$ denotes weak convergence (convergence in law) as studied in Gaenssler (1983) (called $\mathcal{L}_b$-convergence there), where in the present situation the $\beta_{m,n}$'s are viewed as random elements in $(\ell^\infty(\mathcal{C}), \mathcal{B}_b(\ell^\infty(\mathcal{C})))$, $\ell^\infty(\mathcal{C})$ being the space of all bounded real-valued functions on $\mathcal{C}$ equipped with the supremum-norm $\|\cdot\|_{\mathcal{C}}$ and $\mathcal{B}_b(\ell^\infty(\mathcal{C}))$ being the σ-field in $\ell^\infty(\mathcal{C})$ generated by the open $\|\cdot\|_{\mathcal{C}}$-balls (which coincides in the present situation of countable $\mathcal{C}$'s with the σ-field generated by the coordinate projections Π_C from $\ell^\infty(\mathcal{C})$ onto $\mathbb{R}$, defined by $\Pi_C(f) := f(C)$ for $f \in \ell^\infty(\mathcal{C})$).

Note also that the space of all bounded real-valued and uniformly d_μ-continuous functions on a VCC $\mathcal{C}$ is a separable and closed subspace of $(\ell^\infty(\mathcal{C}), \|\cdot\|_{\mathcal{C}})$ (cf. Dudley (1978) (7.13) and Gaenssler (1983), (51)).

Proof of theorem 4.3. Let $(\sigma_i)_{i\in\mathbb{N}}$ be i.i.d. random variables defined also on our basic p-space $(\Omega,\mathcal{F},\mathbb{P})$ such that $\mathbb{P}(\sigma_i = 1) = \frac{1}{2} = \mathbb{P}(\sigma_i = -1)$ and where we assume that the sequence $(\sigma_i)_{i\in\mathbb{N}}$ is independent of $(\xi_i^*)_{1\le i\le 2m}$ for each $m \in \mathbb{N}$, the ξ_i^*, $1 \le i \le 2m$, being i.i.d. with law $\mu_n^{\mathbf{x}}$ on $\mathcal{B}$.

Set $\mu_{2m,n}^{+}(B) := \frac{1}{N_{2m}^{+}} \sum_{i=1}^{2m} \sigma_i^{+} 1_B(\xi_i^*)$, $B \in \mathcal{B}$,

where $N_{2m}^{+} := |\{i \in \{1,\ldots,2m\} : \sigma_i = 1\}|$,

and let

$$\beta_{2m,n}^{+} = (\beta_{2m,n}^{+}(C))_{C\in\mathcal{C}}$$

be defined by

$$\beta_{2m,n}^{+}(C) := (2m)^{1/2}(\mu_{2m,n}^{+}(C) - \mu_n^{\mathbf{x}}(C)),\ C \in \mathcal{C},$$

viewed also as random elements in $(\ell^\infty(\mathcal{C}), \mathcal{B}_b(\ell^\infty(\mathcal{C})))$.

Then, in the first step we are going to show:

(4.5) $\beta_{2m,n}^{+} \xrightarrow{\mathcal{L}} \sqrt{2}\,\mathbb{G}_\mu$ implies $\beta_{m,n} \xrightarrow{\mathcal{L}} \mathbb{G}_\mu$ if $m = m(n) \to \infty$ as $n \to \infty$.

For this, given any $k > 0$, let

$$A_m := \{|N_{2m}^{+} - m| > k\sqrt{m}\},$$

noticing that $\mathbb{P}(A_m) \le 1/2k^2$ by Chebyshev's inequality. Due to our assumption that $(\sigma_i)_{i\in\mathbb{N}}$ is independent of $(\xi_i^*)_{1\le i\le 2m}$, it follows easily that

(a) $\mathcal{L}\{\beta_{2m,n}^{+}\} = \mathcal{L}\{(\frac{2m}{N_{2m}^{+}})^{1/2}\beta_{N_{2m}^{+},n}\}$.

Furthermore, for any $\varepsilon > 0$,

$$\mathbb{P}(\{\|(\tfrac{N_{2m}^{+}}{m})^{1/2}\beta_{N_{2m}^{+},n} - \beta_{m,n}\|_{\mathcal{C}} \ge \varepsilon\} \cap \complement A_m) =$$

$$\sum_{\{i\le 2m: |i-m|\le k\sqrt{m}\}} \mathbb{P}(\{\|(\tfrac{N_{2m}^{+}}{m})^{1/2}\beta_{N_{2m}^{+},n} - \beta_{m,n}\|_{\mathcal{C}} \ge \varepsilon\} \cap$$

$$\{|N_{2m}^{+} - m| \le k\sqrt{m}\} \cap \{N_{2m}^{+} = i\}) \le$$

$$\sum_{\{i\le 2m: |i-m|\le k\sqrt{m}\}} \mathbb{P}(\|(\tfrac{i}{m})^{1/2}\beta_{i,n} - \beta_{m,n}\|_{\mathcal{C}} \ge \varepsilon)\cdot\mathbb{P}(N_{2m}^{+} = i),$$ where

$$\mathbb{P}(\|(\tfrac{i}{m})^{1/2}\beta_{i,n} - \beta_{m,n}\|_{\mathcal{C}} \ge \varepsilon) =$$

$$\mathbb{P}(\tfrac{1}{\sqrt{m}}\|i(\mu_{i,n} - \mu_n^{\mathbf{x}}) - m(\mu_{m,n} - \mu_n^{\mathbf{x}})\|_{\mathcal{C}} \ge \varepsilon) =$$

$$\mathbb{P}\left(\frac{|i-m|}{\sqrt{m}}\,\|\mu_{|i-m|,n}-\mu_n^{\underline{x}}\|_{\mathcal{C}}\geq\varepsilon\right)\leq$$

$$\left(1-\frac{|i-m|}{\varepsilon^2 m}\right)^{-1}\cdot[2|i-m|]^{v}\cdot 2\exp\left(-\frac{\varepsilon^2 m}{8|i-m|}\right)$$

by the Vapnik-Chervonenkis inequality (3.2). Therefore

$$\sum_{\{i\leq 2m:\,|i-m|\leq k\sqrt{m}\}}\mathbb{P}\left(\left\|\left(\tfrac{i}{m}\right)^{1/2}\beta_{i,n}-\beta_{m,n}\right\|_{\mathcal{C}}\geq\varepsilon\right)\cdot\mathbb{P}(N_{2m}^{+}=i)\leq$$

$$\max_{\{i\leq 2m:\,|i-m|\leq k\sqrt{m}\}}\left(1-\frac{|i-m|}{\varepsilon^2 m}\right)^{-1}\cdot[2|i-m|]^{v}\cdot 2\exp\left(-\frac{\varepsilon^2 m}{8|i-m|}\right)\leq$$

$$\left(1-\frac{k}{\varepsilon^2\sqrt{m}}\right)^{-1}\cdot[2k\sqrt{m}]^{v}\cdot 2\exp\left(-\frac{\varepsilon^2\sqrt{m}}{8k}\right)\to 0\quad\text{as}\quad m\to\infty\,.$$

This shows that

$$\limsup_{n\to\infty}\mathbb{P}\left(\left\|\left(\frac{N_{2m}^{+}}{m}\right)^{1/2}\beta_{N_{2m}^{+},n}-\beta_{m,n}\right\|_{\mathcal{C}}\geq\varepsilon\right)\leq\frac{1}{2k^2}\quad\text{for each}\quad k>0,$$

implying that for each $\varepsilon > 0$

(b) $\mathbb{P}\left(\left\|\left(\frac{N_{2m}^{+}}{m}\right)^{1/2}\beta_{N_{2m}^{+},n}-\beta_{m,n}\right\|_{\mathcal{C}}\geq\varepsilon\right)\to 0$ if $m = m(n)\to\infty$ as $n\to\infty$.

Now, (b) readily implies that for each $\varepsilon > 0$

(c) $\mathbb{P}\left(\left\|\left(\frac{m}{N_{2m}^{+}}\right)^{1/2}\beta_{N_{2m}^{+},n}-\frac{m}{N_{2m}^{+}}\,\beta_{m,n}\right\|_{\mathcal{C}}\geq\varepsilon\right)\to 0$ if $m = m(n)\to\infty$ as $n\to\infty$.

In fact, let $B_m := \{\frac{N_{2m}^{+}}{m}\leq\frac{1}{2}\}$ noticing that $\mathbb{P}(B_m)\to 0$ since $\frac{N_{2m}^{+}}{m}\to 1$ $\mathbb{P}$-a.s.; then, for each $\varepsilon > 0$

$$\mathbb{P}\left(\left\|\left(\frac{m}{N_{2m}^{+}}\right)^{1/2}\beta_{N_{2m}^{+},n}-\frac{m}{N_{2m}^{+}}\,\beta_{m,n}\right\|_{\mathcal{C}}\geq\varepsilon\right)=$$

$$\mathbb{P}\left(\left\|\left(\frac{N_{2m}^{+}}{m}\right)^{1/2}\beta_{N_{2m}^{+},n}-\beta_{m,n}\right\|_{\mathcal{C}}\geq\frac{\varepsilon N_{2m}^{+}}{m}\right)\leq$$

$$\mathbb{P}\left(\left\|\left(\frac{N_{2m}^{+}}{m}\right)^{1/2}\beta_{N_{2m}^{+},n}-\beta_{m,n}\right\|_{\mathcal{C}}\geq\varepsilon/2\right)+\mathbb{P}(B_m)\to 0\quad\text{if}$$

$m = m(n)\to\infty$ as $n\to\infty$ according to (b).

The assertion in (4.5) follows now from (a) and (c):

$$\frac{1}{\sqrt{2}}\,\beta_{2m,n}^{+}\xrightarrow{\mathcal{L}}\mathbb{G}_\mu\ \underset{(a)}{\Rightarrow}\ \left(\frac{m}{N_{2m}^{+}}\right)^{1/2}\beta_{N_{2m}^{+},n}\xrightarrow{\mathcal{L}}\mathbb{G}_\mu\ \underset{(c)}{\Rightarrow}$$

$$\frac{m}{N_{2m}^{+}}\,\beta_{m,n}\xrightarrow{\mathcal{L}}\mathbb{G}_\mu\ \underset{\left(\frac{m}{N_{2m}^{+}}\to 1\text{ a.s.}\right)}{\Rightarrow}\ \beta_{m,n}\xrightarrow{\mathcal{L}}\mathbb{G}_\mu\,.$$

According to (4.5) it remains to show

(4.6) $\beta_{2m,n}^{+}\xrightarrow{\mathcal{L}}\sqrt{2}\,\mathbb{G}_\mu$ if $m = m(n)\to\infty$ as $n\to\infty$.

As to (4.6) it is easy to show weak convergence of the corresponding finite-dimensional distributions (using the assumption that $\underline{x}$ satisfies (4.1)). Secondly, introducing for any $f\in\ell^\infty(\mathcal{C})$, $\delta > 0$, and a p-measure ν on $\mathcal{B}$, the modulus $w_f(\delta;\nu)$ of continuity, defined by

$$w_f(\delta;\nu) := \sup\{|f(C)-f(D)| : C,D \in \mathcal{C} \text{ and } \nu(C \Delta D) < \delta\},$$

it remains to verify the following equicontinuity condition (cf. Theorem B in Gaenssler (1983)):

(4.7) For any $\varepsilon,\eta > 0$ there exists a $\delta = \delta(\varepsilon,\eta) > 0$ such that

$$\limsup_{n\to\infty} \mathbb{P}(w_{\beta^+_{m,n}}(\delta;\mu) > \varepsilon) < \eta.$$

Making again use of the assumption that $\underline{x}$ satisfies (4.1), (4.7) is implied by

(4.8) For any $\varepsilon_1,\eta_1 > 0$ there exists a $\delta_1 = \delta_1(\varepsilon_1,\eta_1) > 0$ such that

$$\limsup_{n\to\infty} \mathbb{P}(w_{\beta^+_{m,n}}(\delta_1;\mu_n^{\mathbf{x}}) > \varepsilon_1) < \eta_1 .$$

To prove (4.8) one can proceed as follows:

Let $\mu^o_{m,n}$ be the symmetrized empirical measure based on $\xi^*_1,\ldots,\xi^*_m$, defined by

$$\mu^o_{m,n}(B) := \frac{1}{m}\sum_{i=1}^{m} \sigma_i 1_B(\xi^*_i),\quad B \in \mathcal{B} .$$

Then, as in Pollard (1981), Lemma (3.3), one obtains

(4.9) Given any $\delta > 0$ and $\varepsilon > 0$ such that $\varepsilon < \frac{1}{4}\delta$, and given any countable subclass $\mathcal{C}$ of $\mathcal{B}$, one has for all $n \in \mathbb{N}$ with $m = m(n) \geq \frac{2\delta}{\varepsilon^2}$

$$m^{-1}\frac{\delta^2}{\varepsilon^2} + \mathbb{P}(w_{\mu^o_{m,n}}(\delta;\mu_n^{\mathbf{x}}) > \varepsilon) \geq \frac{1}{2}\mathbb{P}(w_{\beta^+_{m,n}}(\delta;\mu_n^{\mathbf{x}}) > 8m^{1/2}\varepsilon)$$

which, together with 3.4(i) (assuming w.l.o.g. that $C \Delta D \in \mathcal{C}$ for any $C,D \in \mathcal{C}$) allows for an easy proof of

(4.10) Suppose that for any $\varepsilon,\eta > 0$ there exists a $\delta = \delta(\varepsilon,\eta) > 0$ such that

$$\limsup_{n\to\infty} \mathbb{P}(w_{\mu^o_{m,n}}(2\delta;\mu_{m,n}) > \frac{1}{8}\varepsilon m^{-1/2}) < \eta ;$$

then (4.8) holds true.

Thus, along these lines the proof of theorem 4.3 has reduced to prove the following statement:

(4.11) Let $\mathcal{C}$ be a countable VCC with index $v = V(\mathcal{C})$; then, for any $\varepsilon,\eta > 0$ there exists a $\delta = \delta(\varepsilon,\eta,v) > 0$ and there exists an $n_o = n_o(\varepsilon,\eta,v) \in \mathbb{N}$ such that for all $n \geq n_o$ and $m = m(n)$

$$\mathbb{P}(w_{\mu^o_{m,n}}(2\delta;\mu_{m,n}) > 5\varepsilon m^{-1/2}) < \eta .$$

The proof of (4.11) necessitates only minor modifications compared with Pollard's (1981) method employed to prove (1.4).

To keep the present paper selfcontained we are going to present here the remaining steps.

Proof of (4.11) (cf. Pollard (1981), Lemma (4.6)):

a) The proof of (4.11) makes use of the following fact (cf. Pollard (1981), Lemma (2.5)):

(*) There exist constants $A = A(v)$ and $B = B(v)$ such that for any $\mathcal{E} \subset \mathcal{C}$, $x^* \in X^m$ and $\lambda \in (0,1]$, one has $|\mathcal{E}| \leq A\cdot\lambda^{-B}$ if $|(E \,\Delta\, E') \cap \{x^*\}| \geq m\lambda$ for any $E,E' \in \mathcal{E}$ with $E \neq E'$.

Note also that for any $C,D \in \mathcal{C}$

(+) $|\mu^o_{m,n}(C) - \mu^o_{m,n}(D)| \leq \mu_{m,n}(C \,\Delta\, D)$, and

$|\mu^o_{m,n}(C) - \mu^o_{m,n}(D)| \leq |\mu^o_{m,n}(C\setminus D)| + |\mu^o_{m,n}(D\setminus C)|$.

b) Now, given any $\varepsilon,\eta > 0$ and a countable VCC $\mathcal{C} = \{C_i : i \in \mathbb{N}\}$, let for $j \in \mathbb{N}$

$$\delta_j := 2^{-j} \quad\text{and}\quad \varepsilon_j := (96\cdot B\cdot j\cdot\delta_{j-1})^{1/2} ;$$

choose $s_o \in \mathbb{N}$ such that

$$\sum_{j=s_o+1}^{\infty} \varepsilon_j < \varepsilon , \qquad \sum_{j=s_o+1}^{\infty} 4A^2 \exp(-2Bj) < \eta/3$$

and such that

$$a(s) := 4A^2 \exp(2Bs - \frac{\varepsilon^2 2^s}{64}) < \eta/6 \quad \text{for all} \quad s \geq s_o$$

(where A and B are the constants occuring in (*)).

Choose $n_o \in \mathbb{N}$ such that $\varepsilon m_o^{-1/2} < 2^{s_o}$, $m_o = m(n_o)$, and such that

$$4A^2 \exp(2Bs_o - \frac{\varepsilon^2/4}{\varepsilon m^{-1/2} + 16\; 2^{-s_o}}) \leq a(s_o) + \eta/6$$

for all $n \geq n_o$, $m = m(n)$.

Let $\delta = \delta(\varepsilon,\eta,v) := \delta_{s_o-1}$; then we claim that for all $n \geq n_o$ and $m = m(n)$

(++) $\mathbb{P}(w_{\mu^o_{m,n}}(2\delta;\mu_{m,n}) > 5\varepsilon m^{-1/2} | \xi^*_1,\ldots,\xi^*_m) < \eta$

which yields the assertion of (4.11) by integrating out over $\mu_{m,n}$.

c) To prove (++), let $n \in \mathbb{N}$ with $n \geq n_o$; choose $k \in \mathbb{N}$ such that $\delta_k \leq \varepsilon m^{-1/2} < \delta_{k-1}$, whence $k > s := s_o$.

Given any $x^* \in X^m$ and $j \in \mathbb{N}$ we define recursively a sequence $(b_i(x^*,j))_{i\in\mathbb{N}}$ by $b_1(x^*,j) := 1$ and, for $i > 1$,

$$b_i(x^*,j) := \begin{cases} \inf N_i(x^*,j), & \text{if } N_i(x^*,j) \neq \emptyset \\ b_{i-1}, & \text{if } N_i(x^*,j) = \emptyset , \end{cases}$$

where, for $i > 1$,

$$N_i(x^*,j) := \{n \in \mathbb{N} : n > b_{i-1}(x^*,j) \text{ and } |\{x^*\} \cap (C_n \Delta C_{b_r(x^*,j)})| \geq m\delta_j \ \forall r \in \{1,\dots,i-1\}\} ;$$

furthermore, given any $x^* \in X^m$ and $j \in \mathbb{N}$, let

$$\mathcal{C}(x^*,j) := \{C_{b_i(x^*,j)} : i \in \mathbb{N}\}$$

(i.e. $\mathcal{C}(x^*,j)$ is a maximal subclass of $\mathcal{C}$ such that

$\nu_m^{x^*}(C' \Delta C'') \geq \delta_j$ for each distinct pair C', C'' in $\mathcal{C}(x^*,j)$,

where $\nu_m^{x^*}$ denotes the empirical measure based on x^*).

It follows from (*) that

(γ_o) $|\mathcal{C}(x^*,j)| \leq A\cdot\delta_j^{-B} = A\ 2^{jB} \leq A \exp(jB)$

for any $x^* \in X^m$ and $j \in \mathbb{N}$.

Also, due to the maximality of $\mathcal{C}(x^*,j)$, one has that for any $C \in \mathcal{C}$ there exists an $i \in \mathbb{N}$ such that

$$m\nu_m^{x^*}(C \Delta C_{b_i(x^*,j)}) = |\{x^*\} \cap (C \Delta C_{b_i(x^*,j)})| < m\delta_j .$$

d) "CHAINING ARGUMENT":

Given $j \in \{s,\dots,k\}$ ($s := s_o$) and $x^* \in X^m$, let

$$K(x^*,j) : \mathcal{C} \to \mathcal{C}(x^*,j)$$

be defined by

$K(x^*,j)(C) := C_{b_i(x^*,j)}$ iff

$\nu_m^{x^*}(C \Delta C_{b_i(x^*,j)}) < \delta_j$ and $\nu_m^{x^*}(C \Delta C_{b_r(x^*,j)}) \geq \delta_j\ \forall r \in \{1,\dots,i-1\}$.

Next, given any $x^* \in X^m$ and $k \in \mathbb{N}$ such that $\delta_k \leq \varepsilon m^{-1/2} < \delta_{k-1}$, let

$T(x^*,k) := K(x^*,k)$, and

$T(x^*,j) := K(x^*,j)\circ\dots\circ K(x^*,k)$ for $j \in \{s,\dots,k-1\}$.

Then one has

(γ_1) $\nu_m^{x^*}(T(x^*,j)(C) \Delta T(x^*,j-1)(C)) < \delta_{j-1}$

$\forall x^* \in X^m$, $j \in \{s+1,\dots,k\}$ and $C \in \mathcal{C}$

and

(γ_2) $\nu_m^{x^*}(T(x^*,k)(C)\ \Delta\ C) < \delta_k$

$\forall x^* \in X^m$ and $C \in \mathcal{C}$.

Now, let $\xi^* := (\xi_1^*,\ldots,\xi_m^*)$; then, by (γ_2) and the choice of k it follows that

(γ_3) $\mathbb{P}(\sup_{C\in\mathcal{C}} |\mu_{m,n}^o(C) - \mu_{m,n}^o(T(\xi^*,k)(C)| > \varepsilon m^{-1/2}|\xi^*) = 0$.

In what follows, let for $j \in \{s,\ldots,k\}$ and $C \in \mathcal{C}$

$T_j(C) := T(\xi^*,j)(C)$.

Then we are going to show

(α) $\mathbb{P}(\sup_{C\in\mathcal{C}} |\mu_{m,n}^o(C) - \mu_{m,n}^o(T_s(C))| > 2\varepsilon m^{-1/2}|\xi^*) < \eta/3$

and

(β) $\mathbb{P}(\sup\{|\mu_{m,n}^o(T_s(C)) - \mu_{m,n}^o(T_s(D))| : C,D \in \mathcal{C}$ s.t.

$\mu_{m,n}(C\ \Delta\ D) < 2\delta\} > \varepsilon m^{-1/2}|\xi^*) < \eta/3$.

The proof of (α) and (β) makes use of the following fact (cf. Pollard (1981), (2.1)):

(**) Given a countable subclass $\mathcal{E}$ of $\mathcal{B}$, $\delta > 0$ and $x^* \in X^m$ such that $\nu_m^{x^*}(E) \le \delta$ for each $E \in \mathcal{E}$, then

$$\mathbb{P}(\sup_{E\in\mathcal{E}} |(\nu_m^{x^*})^o(E)| > \varepsilon) \le 2|\mathcal{E}|\exp\left(\frac{-n\varepsilon^2}{2\varepsilon+2\delta}\right) ,$$

where $(\nu_m^{x^*})^o(E) := \frac{1}{m}\sum_{i=1}^m \sigma_i 1_E(x_i^*)$, $x^* = (x_1^*,\ldots,x_m^*)$.

Proof of (α):

$$\mathbb{P}(\sup_{C\in\mathcal{C}}|\mu_{m,n}^o(T_k(C)) - \mu_{m,n}^o(T_s(C))| > \varepsilon m^{-1/2}|\xi^*) \le$$

$$\mathbb{P}(\sup_{C\in\mathcal{C}} \sum_{j=s+1}^k |\mu_{m,n}^o(T_j(C)) - \mu_{m,n}^o(T_{j-1}(C))| > m^{-1/2}\sum_{j=s+1}^k \varepsilon_j|\xi^*) \underset{((+)\text{ in a)})}{\le}$$

$$\sum_{j=s+1}^k [\mathbb{P}(\sup_{C\in\mathcal{C}}|\mu_{m,n}^o(T_j(C)\setminus T_{j-1}(C))| > \tfrac{1}{2}m^{-1/2}\varepsilon_j|\xi^*) +$$

$$\mathbb{P}(\sup_{C\in\mathcal{C}}|\mu_{m,n}^o(T_{j-1}(C)\setminus T_j(C))| > \tfrac{1}{2}m^{-1/2}\varepsilon_j|\xi^*)] \le$$

(by independence of (ξ_i^*) and (σ_i))

$$\sum_{j=s+1}^k [\max_{x^*\in X^m}\mathbb{P}(\sup_{C\in\mathcal{C}}|(\nu_m^{x^*})^o(T(x^*,j)(C)\setminus T(x^*,j-1)(C))| > \tfrac{1}{2}m^{-1/2}\varepsilon_j) +$$

$$\max_{x^*\in X^m}\mathbb{P}(\sup_{C\in\mathcal{C}}|(\nu_m^{x^*})^o(T(x^*,j-1)(C)\setminus T(x^*,j)(C))| > \tfrac{1}{2}m^{-1/2}\varepsilon_j)]$$

$$\underset{((\gamma_1),(\gamma_o),(**))}{\le} \sum_{j=s+1}^k 4A^2\exp(2Bj)\exp\left(\frac{-\varepsilon_j^2/4}{m^{-1/2}\varepsilon_j+2\delta_{j-1}}\right) \le$$

$(\varepsilon_j^2 := 96Bj\delta_{j-1}$ and $m^{-1/2}\varepsilon_j < m^{-1/2}\varepsilon < \delta_{k-1} \le 4\delta_{j-1}$ $\forall j \in \{s,\ldots,k\})$

$$\sum_{j=s+1}^{k} 4A^2 \exp(2Bj - \frac{24Bj\delta_{j-1}}{6\delta_{j-1}}) = \sum_{j=s+1}^{k} 4A^2 \exp(-2Bj) < \eta/3 \,. \quad \text{(cf. b))}$$

This, together with (γ_3), implies (α).

Proof of (β): First, it is easy to see that

(***) $\forall x^* \in X^m$ and $\forall C,D \in \mathcal{C}$ one has that $\nu_m^{x^*}(C \,\Delta\, D) < 2\delta$ implies $\nu_m^{x^*}(T(x^*,s)(C) \,\Delta\, T(x^*,s)(D)) < 8\delta_s$.

It follows that

$$\mathbb{P}(\sup\{|\mu_{m,n}^{o}(T_s(C)) - \mu_{m,n}^{o}(T_s(D))| : C,D \in \mathcal{C} \text{ s.t. } \mu_{m,n}(C \,\Delta\, D) < 2\delta\} > \varepsilon m^{-1/2} | \xi^*) \underset{((+)\text{ in a}))}{\leq}$$

$$2\mathbb{P}(\sup\{|\mu_{m,n}^{o}(T_s(C)\setminus T_s(D))| : C,D \in \mathcal{C} \text{ s.t. } \mu_{m,n}(C \,\Delta\, D) < 2\delta\} > \frac{1}{2}\varepsilon m^{-1/2} | \xi^*) \leq$$

$$2 \max_{x^*\in X^m} \mathbb{P}(\sup|(\nu_m^{x^*})^{o}(T(x^*,s)(C)\setminus T(x^*,s)(D))| : C,D \in \mathcal{C} \text{ s.t. } \nu_m^{x^*}(C \,\Delta\, D) < 2\delta\} > \frac{1}{2}\varepsilon m^{-1/2}) \underset{(***)}{\leq}$$

$$2 \max_{x^*\in X^m} \mathbb{P}(\sup\{|(\nu_m^{x^*})^{o}(C'\setminus D')| : C',D' \in \mathcal{C}(x^*,s) \text{ s.t. } \nu_m^{x^*}(C' \,\Delta\, D') < 8\delta_s\} > \frac{1}{2}\varepsilon m^{-1/2})$$

$$\underset{((\gamma_o),(**))}{\leq} 4A^2 \exp(2Bs) \exp(\frac{-\varepsilon^2/4}{\varepsilon m^{-1/2} + 16\delta_s}) \underset{\text{(cf. b))}}{\leq} a(s) + \eta/6 < \eta/3 \,,$$

which proves (β) .

Now, (α) and (β) immediately imply (++) in part b) which concludes the proof of (4.11) and therefore theorem 4.3 is proved. □

ACKNOWLEDGEMENT. I am grateful to W. Schneemeier for some helpful comments.

REFERENCES

Dudley, R.M. (1978). Central limit theorems for empirical measures. Ann. Probability 6, 899-929. (Correction, ibid. 7 (1979), 909-911).

Durst, M. and Dudley, R.M. (1981). Empirical processes, Vapnik-Červonenkis classes and Poisson processes. Prob. Math. Statist. (Wrocław) 1, no. 2, 109-115.

Gaenssler, P. (1983). Empirical Processes. IMS Lecture Notes - Monograph Series, Vol. 3.

REFERENCES

Pollard, D. (1981). Limit theorems for empirical processes. Z. Wahrscheinlichkeitstheorie verw. Gebiete 57, 181-195.

Pollard, D. (1984). Convergence of Stochastic Processes. Springer, New York.

Vapnik, V.N. and Chervonenkis, A.Ya. (1971). On the uniform convergence of relative frequencies of events to their probabilities. Theor. Probability Appls. 16, 264-280 = Teor. Verojatnost. i Primenen. 16, 264-279.

Wenocur, R.S. and Dudley, R.M. (1981). Some special Vapnik-Cervonenkis classes. Discrete Math. 33, 313-318.

Prob. Theory and Math. Stat., Vol.1, pp. 483-487
Prohorov et al. (eds)
1986 VNU Science Press

A REMARK ON THE CENTRAL LIMIT THEOREM FOR RANDOM MEASURES AND PROCESSES *

Evarist Giné and Joel Zinn
Texas A&M University
College Station, Texas 77843

Let $(T, \mathcal{T})$ be a measurable space, let $S = \mathcal{M}(T)$ be the set of measures of finite total variation on $(T, \mathcal{T})$, let $\|\mu\| = \sup_{A \in \mathcal{T}} |\mu(A)|$ denote the finite total variation norm of $\mu \in \mathcal{M}(T)$, and $|\mu|$ the corresponding total variation measure. Let $\mathcal{C} \subset \mathcal{T}$ be a *Vapnik-Červonenkis Class* (see e.g. [1]) and let $\mathcal{F}_\mathcal{C}$ be the collection of evaluation functions

$$\mathcal{F}_\mathcal{C} = \{f_C : \mathcal{M}(T) \to \mathbf{R}, \quad f_C(\mu) = \mu(C), \quad \mu \in \mathcal{M}(T), \quad C \in \mathcal{C}\}.$$

Finally, let $\mathcal{S}$ be a σ-algebra on $S = \mathcal{M}(T)$ for which the functions in $\mathcal{F}_\mathcal{C}$ are measurable, and let P be a probability measure on $(S, \mathcal{S})$. We will assume the "required measurability conditions" on $\mathcal{F}_\mathcal{C}$; for Theorem 1 below this means that $\mathcal{F}$ is NLDM(P) and $\mathcal{G} := \{(f - g)^2 : f, g \in {}_0\mathcal{F}_\mathcal{C}\}$ is NLSM(P), and for Theorem 2, just that $\mathcal{F}_\mathcal{C}$ is NDM(P) (see [2], Definition 2.3).

Examples of measurable $\mathcal{F}_\mathcal{C}$. (1) If $\mathcal{C}$ is countable then $\mathcal{F}_\mathcal{C}$ satisfies the required measurability conditions.

(2) Let $T = [0,1]^d$ and let $X(t)$, $t \in T$, be a separable process on T such that $\mathrm{EX}^2(t) < \infty$ for all $t \in [0,1]^d$. For each rectangle $(s,t] := \{(x_1, \ldots, x_d) : s_i < x_i \leq t_i, \ i = 1, \ldots, d\}$, where $s = (s_1, \ldots, s_d)$, $t = (t_1, \ldots, t_d)$ and $s_i \leq t_i$ for all $i \leq d$, define

$$\begin{aligned}\mu(\omega)(s,t] &:= \Delta X(\omega)(s,t] \\ &:= \sum_{\{\tau_i = s_i \text{ or } t_i \text{ for } i=1,\ldots,d\}} (-1)^{\#\{\tau_i = s_i\}} X((\tau_1, \ldots, \tau_d), \omega).\end{aligned}$$

---

* Research supported by grants from the National Science Foundation, USA

Define also $\mathcal{C} = \{[0,t],\ 0 \le t_i \le 1,\ i = 1,\dots,\}$, $\mathcal{S} = \sigma(\mathcal{F}_\mathcal{C})$, the σ-algebra of $\mathcal{M}(T)$ generated by the functions in $\mathcal{F}_\mathcal{C}$, and let P be the law of the measure $\mu(\omega)$. Then $\mathcal{F}_\mathcal{C}$ satisfies the required measurability conditions due to the separability of X.

With $(S, \mathcal{S})$, $\mathcal{F}_\mathcal{C}$ and P as described in the first paragraph of this note, one has the following two results.

Theorem 1. *(Pisier [4]). If $\int_\mathcal{M}^* \|\mu\|^2\, dP(\mu) < \infty$ then $\mathcal{F}_\mathcal{C}$ is a functional P-Donsker class.*

Theorem 1 is not stated in the present form in [4], but it is actually proved there. In fact Pisier proves much more, in particular, that, for $\mathcal{C}$ countable, the inclusion map $j : (\mathcal{M}, \|\cdot\|) \to (\mathcal{M}, \|\cdot\|_\mathcal{C})$ is of type 2 if and only if $\mathcal{C}$ is a Vapnik-Cervonenkis class, where $\|\mu\|_\mathcal{C} := \sup_{C \in \mathcal{C}} |\mu(C)|$.

Theorem 2. *If $P\{\|\mu\|_\mathcal{C} \le \lambda\} = 1$ for some $\lambda < \infty$, then the conditions*

(i) $\mathcal{F}_\mathcal{C}$ is P-pregaussian

and

(ii) $\lim_{\mu \to \infty} u^2 P^*\{\ln_+ \|\mu\| > u\} = 0$,

imply that $\mathcal{F}_\mathcal{C}$ is a functional P-Donsker class.

Note the difference between the conditions on the total variation norm in both theorems. Compare Theorem 2 with Theorem 7.3 in [2].

The proofs of Theorems 1 and 2 are identical except that for the first one we invoke Theorem 8.10 in [2] whereas the second one is based on Theorem 5.4 (or alternatively 5.1) in [2]. We recall these theorems for the reader's convenience. The first one asserts that

(1) If $\mathcal{F}$ satisfies the required measurability conditions and its envelope function is in $\mathcal{L}_2(P)$, then the condition

$$\lim_{\delta \downarrow 0} \limsup_n E\left(1 \wedge \int_0^\delta (\ln N_{n,2}(\epsilon, \mathcal{F}))^{1/2}\, d\epsilon\right) = 0$$

implies that $\mathcal{F}$ is a functional P-Donsker class.

The second one is as follows:

(2) If $\mathcal{F}$ satisfies the required measurability conditions and is a uniformly bounded class of functions, then the conditions:

(i) $\mathcal{F}$ is P-pregaussian and

(ii) $\lim_{n\to\infty} E^*\left(1 \wedge \int_0^{n^{-1/4}} (\ln N_{n,2}(\epsilon, \mathcal{F}))^{1/2}\, d\epsilon\right) = 0$

imply that $\mathcal{F}$ is a functional P-Donsker class.

Here $N_{n,2}(\epsilon, \mathcal{F})$ is the covering number of $\mathcal{F}$ for the distance $d_{n,2}^2(f,g) = \sum_{i=1}^n (f(\mu_i) - g(\mu_i))^2/n$ where μ_i are i.i.d. S-valued r.v.'s with law P, defined as coordinates on $(S^{\mathbf{N}}, \mathcal{S}^{\mathbf{N}})$ (see [2], Section 2 for the exact framework).

The other main ingredient in the proofs of Theorems 1 and 2 is the following important lemma of Dudley [1]:

(3) If $\mathcal{C} \subset \mathcal{T}$ is a Vapnik-Červonenkis class and Q is a probability measure on $(T, \mathcal{T})$, then

$$N(\epsilon, \mathcal{C}, \|\cdot\|_{L_2(Q)}) \leq K\left[\left(\frac{1}{\epsilon}\right)^k \vee 1\right]$$

for some numbers $K, k \in (0,\infty)$ depending only on $\mathcal{C}$ (and not on Q).

Proof of Theorems 1 and 2. Let μ_i be i.i.d. with law P. Then, essentially as in [4],

$$\begin{aligned} d_{n,2}^2(f_A, f_B) &= \sum_{i=1}^n (\mu_i(A) - \mu_i(B))^2/n \\ &\leq \sum_{i=1}^n (|\mu_i|(A \Delta B))^2/n \\ &\leq \left(\frac{\sum_{i=1}^n \|\mu_i\|^2}{n}\right) Q_n(A \Delta B) \end{aligned}$$

with $Q_n = \sum_{i=1}^n \|\mu_i\| \, |\mu_i| / \sum_{i=1}^n \|\mu_i\|^2$. Since Q_n is a (random) probability measure, Dudley's lemma (3) gives

$$\begin{aligned} N_{n,2}(\epsilon, \mathcal{F}, P) &\leq N\left(\frac{\epsilon}{\sqrt{\sum_{i=1}^n \|\mu_i\|^2/n}}, \mathcal{C}, \|\cdot\|_{L_2(Q_n)}\right) \\ &\leq K\left[\left(\frac{\sqrt{\sum_{i=1}^n \|\mu_i\|^2/n}}{\epsilon}\right)^k \vee 1\right]. \end{aligned} \tag{4}$$

Now, using this estimate for the random entropy respectively in (1) and in (2) gives immediately Theorems 1 and 2. Let us just check it on (2). Let $a_n = \sqrt{\sum_{i=1}^n \|\mu_i\|^2/n}$, and let us use $\lesssim$ or $\simeq$ if multiplicative constants and/or lower order summands are ignored. Then

$$\int_0^{n^{-1/4}} \sqrt{\ln N_{n,2}(\epsilon, \mathcal{F}_{\mathcal{C}})}\, d\epsilon \lesssim \int_0^{n^{-1/4}} \sqrt{\ln_+ \frac{a_n}{\epsilon}}\, d\epsilon$$

$$= a_n \int_{n^{1/4}a_n}^{\infty} \frac{\sqrt{\ln_+ u}}{u^2}\, du \simeq \sqrt{\frac{\ln_+ n^{1/4}a_n}{n^{1/2}}} + a_n I(n^{1/4}a_n \leq 1)$$

Now the second summand tends to zero (since it is bounded by $n^{-1/4}$), and so does the first (in pr^*) if $\frac{\ln_+ n^{1/4} a_n}{n^{1/2}} \to 0$ in pr^*. So, this is what is needed for the entropy condition (2ii) to hold. But for n large enough depending on $\epsilon > 0$,

$$\Pr^*\left\{\ln_+ \sqrt{\sum_{i=1}^{n} \|\mu_i\|^2/n^{1/2}} > 2\epsilon n^{1/2}\right\} \leq n \Pr^*\{\ln_+ \|\mu_1\| > \epsilon n^{1/2}\}.$$

and this tends to 0 as $n \to \infty$ for all $\epsilon > 0$ if condition (ii) of Theorem 2 holds. □

Here is an interesting corollary of Theorem 2 about the CLT in $C[0,1]^d$ (or $D[0,1]^d$) under conditions on the incremental moments of the process (see [2], Theorem 7.4 and examples 7.5 for $d = 1$, and compare to Ibragimov's result [3]). The boundedness condition is essential.

Proposition 3. *Let $T = [0,1]^d$ and let $X(t)$, $t \in T$, be a stochastic process on T with all its trajectories continuous (or $X(t)$ stochastically continuous and with its trajectories in $D[0,1]^d$)) and uniformly bounded, i.e. $\sup_{t\in T} |X(t,\omega)| \leq c$ a.s. for some $c < \infty$ Then if for some $K < \infty$*

(5) $$E|X(t) - X(s)| \leq K|t - s|^d \quad s,t \in [0,1]^d$$

(where $|\cdot|$ also denotes norm in $\mathbf{R}^d$), the process X satisfies the CLT in $C(T)$ (in $D[0,1]^d$ with the uniform norm).

Proof. If $\mathcal{C} = \{[0,t] : 0 \leq t_i \leq 1,\ i = 1,\dots,d\}$, which is a Vapnik-Červonenkis class, then $\mathcal{F}_C$ satisfies the required measurability conditions with respect to the law P of the measure defined by the process $\mu(s,t] = \Delta X(s,t]$, as in the example above. To show that $X \in$ CLT (in $C(T)$ or $D(T)$, whichever the case is) it is enough to prove that $\mathcal{F}_C$ is a functional P-Donsker class (by definition and observing that there is no loss of generality in assuming $X(t) = 0$ if at least one coordinate of t equals zero). We apply Theorem 2. We note that

$$\|\mu\|_{\mathcal{C}} = \sup_{(s,t]} |\Delta X(s,t]| \leq 2^d \sup_t |X(t)| \leq 2^d c < \infty .$$

Let $t_n(k_1,\dots k_d) = (k_1 2^{-n}, k_2 2^{-n}, \dots, k_d 2^{-n})$. $0 \leq k_i \leq 2^n$. Then, by (5)

$$E\|\mu\| \lesssim \lim_{n\to\infty} \sum_{k_i=1, i\leq d}^{2^n} E|\mu(t_n(k_1-1,\dots,k_d-1), t_n(k_1,\dots,k_d)]|$$

$$\leq \lim_{n\to\infty} \sum_{k_i=1, i\leq d}^{2^n} 2^{d-1} 2^{-nd} = 2^{d-1} < \infty .$$

So Theorem 2 applies and yields the result. □

The following is another direct consequence of Theorem 2 (whose proof we omit since it is similar to the previous one).

Proposition 4. *Let $X(t)$ be a process on $[0,1]^d$ with continuous uniformly bounded trajectories. Assume there exists $0 < r \le d$ and a real random variable L such that $u^2 \Pr\{\ln_+ L > u\} \to 0$ as $u \to \infty$ for which*

$$|X(t,\omega) - X(s,\omega)| \le L(\omega)\|t-s\|^r, \quad \omega \in \Omega, \ s,t \in [0,1]^d .$$

Then $X \in$ CLT in $C[0,1]^d$.

This is a multiparameter generalization of Theorem 7.6 in [2].

We finish this note with two remarks:

1. W. T. Rhee [5] has proved that for $d = 1$ Proposition 3 can not be improved in the sense that for each convex monotonically increasing positive function ϕ, $\phi(0) = 0$, $\phi(|s|)/|s| \to 0$ as $s \to 0$, there are bounded continuous processes on $[0,1]$ which do not satisfy the CLT but such that $E\phi(|X(t) - X(s)|) \le |t-s|$.

2. Simple modifications of the counterexample at the end of Strassen and Dudley [6] show that for every $0 < \delta < 1$ there are probability measures P on $\mathcal{M}([0,1])$ such that, with $\mathcal{C} = \{[0,t], 0 \le t \le 1\}$,

   1) $P\{\|\mu\|_{\mathcal{C}} \le 1\} = 1$,

   2) μ is P-pregaussian,

   3) $u^{1-\delta} P\{\ln_+ \|\mu\| > u\} \to 0$ as $u \to \infty$, but

   4) $\mathcal{F}_{\mathcal{C}}$ is not P-Donsker.

This shows that one cannot replace u^2 by $u^{1-\delta}$ in Theorem 2, condition (ii), for any $\delta > 0$. It is plausible that the best power in condition (ii) is not 2 but $1+\delta$ or even 1 – this would give a sharp result e.g. in the CLT for the empirical characteristic function.

References

1. Dudley, R. M. (1978). Central limit theorems for empirical measures. Ann. Probability 6, 899-929.
2. Giné, E. and Zinn, J. (1984). Some limit theorems for empirical processes. Ann. Probability 12, 929-989.
3. Ibragimov, I. (1979). Sur la regularité des trajectoires des fonctions aléatoires. C. R. Acad. Sc. Paris A 289, 545-547.
4. Pisier, G. (1984). Remarques sur les classes de Vapnik-Červonenkis. Ann. Inst. H. Poincaré, Prob. et Stat. 20, 287-298.
5. Rhee, W. T. (1985). Central limit theorem and increment conditions. To appear.
6. Strassen, V. and Dudley, R. (1969). The central limit theorem and ϵ-entropy. Lecture Notes in Math. 89, 224-231. Springer, Berlin.

Prob. Theory and Math. Stat., Vol.1, pp. 489-507
Prohorov et al. (eds)
1986 VNU Science Press

ELLIPTIC LAW AND ELEMENTS OF G-ANALYSIS

V.L.Girko
Kiev State University, Kiev, USSR

Some questions of statistical analysis of multidimensional observations are considered as well as the main limit theorem of the spectral theory random matrices is proved in this report. The mentioned theorem is a logical conclusion of the following statements: in 1958 E.Wigner in connection with the consideration of mathematical models for energetic levels of heavy atomic nuclei proved a theorem which was called "the semicircular law": if elements ξ_{ij}, $i \geqslant j$ of random symmetric matrices $\Xi_n = (\xi_{ij} n^{-1/2})$ are independent, symmetric, $\mathcal{D}\xi_{ij} = 1$, $i,j = 1,2,\ldots$ and for any integer $k \geqslant 0$ $M\xi_{ij}^k \leqslant A_k$, then $\lim_{n\to\infty} M\mu_n(x) = \mu(x)$,
where $\mu_n(x) = n^{-1}\sum_{k=1}^n \chi(\lambda_k < x)$, $\mu'(x) = (2\pi)^{-1}\sqrt{4-x^2}\,\chi(x^2<4)$, λ_k are the eigen-values of the matrix Ξ_n; $\chi(A)$ - is the indicator of the set A. Later on this statement was extended by U.Grenader [11] L.Arnold [19], M.L.Mehta [21], L.A.Pastur [17] , F.A.Berezin. The author in the book (1975)[7] proved the statement which generalizes the results of the indicated authors.

Theorem 1. (Semicircular law). Let for every $n=1,2,\ldots$ random elements $\xi_{pl}^{(n)}$, $p \geqslant l$, $p,l=\overline{1,n}$ of symmetric real matrices Ξ_n be independent and given on a common probability space, $M\xi_{pl}^{(n)} = 0$, $\mathcal{D}\xi_{pl}^{(n)} = \sigma^2 n^{-1}$, $0 < \sigma^2 < \infty$. Then $\lim_{n\to\infty} \mu_n(x\sigma) = \mu(x)$ a.s. if and only if the Lindeberg condition is held for every $\tau > 0$

$$\lim_{n\to\infty} n^{-1}\sum_{p,l=1}^n M\chi(|\xi_{pl}^{(n)}| > \tau)|\xi_{pl}^{(n)}|^2 = 0. \tag{1}$$

The sufficiency of the condition (1) in the case of convergence $\mu_n(x) \to \mu(x)$ in probability was established by L.A. Pastur in his paper [17].

Later investigations, in particular, were directed to find the spectral limit

$$\nu_n(x,y) = n^{-1} \sum_{k=1}^{n} \chi(\operatorname{Re}\lambda_k < x, \operatorname{Jm}\lambda_k < y)$$

of random matrices Ξ_n, for which the condition $\xi_{ij} = \xi_{ji}$ does not hold while the other restrictions on the elements of Ξ_n are the same as in theorem 1. The application of the method of demonstration used in the theorem 1 to the solution of this problem was not successful as the perturbation formulae for the resolvents of nonselfadjoint matrices Ξ_n on which the proofs of the semicircular law are founded in the works [7, 17] turned out to be unsuitable for the finding of the limit $\lim_{n\to\infty} M\nu_n(x,y)$. Nevertheless the limit spectral function was found in one particular case for the complex random matrices whose real and imaginary parts are distributed by the standard normal law. D.Ginibre [20] proved that the distribution density function of eigen-values λ_k of matrices H_n, arranged in increasing order of their modeles is equal to

$$c \exp\left\{-\frac{1}{2}\sum_{p=1}^{n} |z_p|^2\right\} \prod_{p\neq l} |z_p - z_l|^2, \quad |z_1| > \ldots > |z_n|,$$

where $c > 0$ is a normalizing constant. Another proof of this formula is contained in the authors's paper [6].

Using this formula, M.L.Meta [21] proved for such matrices that $\lim_{n\to\infty} M\nu_n(x\sqrt{n}, y\sqrt{n}) = \nu(x,y)$, where

$$\frac{\partial^2 \nu(x,y)}{\partial x \partial y} = \frac{1}{2}\pi^{-1}\chi(x^2 + y^2 \leqslant 2).$$

In order to prove this statement in a more general case the author offered to use V-transformation with the help of which the proof of limit theorems for the functions $\nu_n(x,y)$ was reduced to the proof of limit theorems for spectral functions of some Hermitean random matrices.

Theorem 2. (Circular law). Let for every value n random elements $\xi_{pl}^{(n)}, p,l=\overline{1,n}$ of the complex matrix $\Xi_n = (\xi_{pl} n^{-1/2})$ be independent, $M\xi_{pl}=0$, $M|\xi_{pl}|^2=\sigma^2$, $0<\sigma^2<\infty$ and the values $Re\,\xi_{kl}$, $Im\,\xi_{kl}$ have densities $p_{kl}(x)$, $q_{kl}(x)$, satisfying the condition: for some $\beta>1$

$$\sup_n \sup_{k,l=\overline{1,n}} \int [p_{kl}^{\beta}(x) + q_{kl}^{\beta}(x)]\,dx < \infty, \quad (2)$$

for some $\delta>0$

$$\sup_n \sup_{k,l=\overline{1,n}} M|\xi_{pl}^{(n)}\sqrt{n}|^{2+\delta} < \infty. \quad (3)$$

Then for any x,y $\operatorname{plim}_{n\to\infty} \nu_n(x,y) = \nu(x,y)$.

So two laws were proved: the semicircular law (theorem 1) for symmetric random matrices and the circular law (theorem 2) for nonsymmetric random matrices with independent elements. One problem remained unsolved that is the problem of finding the limit spectral function in the case when the vectors (ξ_{pl},ξ_{lp}), $p\geq l$, $p,l=\overline{1,n}$ are independent, $M\xi_{pl}\xi_{lp}=\rho$, but the other constraints on the elements ξ_{pl} are the same as in the theorem 2. It turned out that with the help of V--transformation for such matrices one can find the density of the limit spectral function which is equal to some constant value on the domain whose boundary is an ellipse.

1. A theorem of the type of the strong law of large numbers for normalized spectral functions of nonselfadjoint random matrices.

In this paragraph theorem 1 is generalized. This theorem was proved by the author in paper [3] for complex random matrices

$$H_n = (\xi_{pl}^{(n)} n^{-1/2}), \quad p,l=\overline{1,n}.$$

Theorem 3. Let for every value n -random vectors $(\xi_{pl}^{(n)}, \xi_{lp}^{(n)}), l \geq p, \ l,p=\overline{1,n}$ be stochastically independent and given on a common probability space, $M\xi_{pl}^{(n)}=0, M|\xi_{pl}^{(n)}|^2=\sigma_{pl}^2, \ \sup_n \sup_{p,l=\overline{1,n}} \sigma_{pl}^2<\infty,$ (n)

Imaginary and real parts of random elements $\xi_{pl}, \xi_{lp}^{(n)}$ have the joint probability densities $p_{kl}^{(n)}(x_i, i=\overline{1,4})$ satisfying the condition:

for some $\beta>1$

$$\sup_n \sup_{k,l=\overline{1,n}} \int p_{kl}^{\beta}(x_i, i=\overline{1,4}) \Pi_{i=1}^4 dx_i<\infty,$$

$$\lim_{h\to\infty} \underline{\lim}_{n\to\infty} M \iint_{|x|<h, |y|<h} d\nu_n(x,y)=1.$$

Then for almost all x and y with respect to the Lebesque measure on the plane with probability 1

$$\lim_{n\to\infty} [\nu_n(x,y)-M\nu_n(x,y)]=0. \tag{6}$$

The proof is founded on the use of spectral functions V- transformations

$$m_n(p,q,c,d)=\frac{p^2+q^2}{4iq\pi n}\int_{-c}^{c}\int_{-d}^{d}\frac{\partial}{\partial s}\ln|\det(\bar{I}z-H_n)|^2 e^{itp+isq}\,dt\,ds,$$

for which the condition (5) implies

$$\lim_{c\to\infty}\lim_{d\to\infty}\sup_n M\left|\iint \exp(ipx+iqy)d\nu_n(x,y)-m_n(p,q,c,d)\right|=0.$$

2. Regularized V- transformation for spectral functions.

Theorem 4. Let the conditions of theorem 3 be hold. Then for finite z and $\alpha_n>0$ such that

$$\lim_{n\to\infty}\alpha_n^{\varepsilon n}=0, \quad 0<\varepsilon<(\beta-1)(2\beta)^{-1} \tag{9}$$

$$\lim_{n\to\infty} n^{-1}\Big[Mm_n(p,q,c,d)-\frac{p^2+q^2}{4iq\pi n}\int_{-c}^{c}\int_{-d}^{d}\frac{\partial}{\partial s}M\ln\det\big[\bar{I}\alpha_n+ \tag{10}$$

$$+(\bar{I}z-H_n)(\bar{I}z-H_n)^*\gamma\big]\exp(itp+isq)\,dt\,ds\Big]=0,$$

where γ is a random variable with the distribution density e^{-x}, $x\geq 0$. The expression in the right hand side of this difference will be called the regularized V-transformation of the spectral function.

Proof. Using the proof of the theorem 2 in the paper [3] we have $n^{-1}|\ln\det[(Iz-H_n)(Iz-H_n)^*\gamma]-$

$$-\ln\det[I\alpha_n + (Iz-H_n)(Iz-H_n^*)]| \leq n^{-1}\sum_{k=1}^{n} M\ln[1+\alpha_n x_k^{-1}], \quad (11)$$

where $x_k \geq |(h_k, \bar{\xi}_k)|$, ξ_k is the k-th row of the matrix $Iz-H_n$, h_k is some random vector of dimension n not depending on the components of the vector $\bar{\xi}_k$, but depending on the other elements of the matrix H_n and such that $(h_k, \bar{h}_k)=1$.

Using (11) in a similar way as in the proof of theorem 3 we obtain (10). Theorem 4 is proved.

3. The estimate of rate of convergence of spectral function Styltyes' Transformations to the limit function.

Let $\mu(x,z) = n^{-1}\sum_{k=1}^{n} \chi(\lambda_k(z) < x)$, where $\lambda_k(z)$ are the eigen values of matrix $(Iz-H_n)(Iz-H_n)^*$.

Denote $H_n = (\xi_{pl}^{(n)})$ the quadratic complex matrices of n-th order. Using the proof of the semicircular law in the paper we obtain the following statement.

Theorem 5. If for every value $n=1,2,\ldots$ random vectors

$$(\xi_{lp}^{(n)}, \xi_{pl}^{(n)}), \; p \geq l \;; \; p,l=\overline{1,n}$$

are stochastically independent, given on a common probability space, $M|\xi_{pl}^{(n)}|^2 = n^{-1}$, $M\xi_{pl}^{(n)}\xi_{lp}^{(n)} = n^{-1}\rho$, $l \neq p$ and the Lindeberg condition (1) holds, then for almost all x and z with probability 1 $\lim_{n\to\infty}\mu_n(x,z) = \mu(x,z)$, where $\mu(x,z)$ is the distribution function in x under fixed z whose Styltyes transform satisfies the equation

$$\frac{|\rho|}{m(\theta)(1+m(\theta))} - \frac{(bt-as)^2}{(1+m(\theta)-|\rho|m(\theta))^2} - \frac{(at+sb)^2}{(1+m(\theta)+|\rho|m(\theta))^2} = \rho\theta, \quad (12)$$

where $\alpha = \mathrm{Re}\sqrt{\rho}$, $\beta = \mathrm{Im}\sqrt{\rho}$, $z = t+is$, $\theta > 0$, $\sqrt{\rho}$ - is the main value of the square root. The solution of the equation (12) exists and is unique in the class of analytical functions $m(\theta)$, $\theta > 0$.

If in addition to the enumerated conditions instead of the condition (1) for some $\delta > 0$

$$\sup_n \sup_{p,l=\overline{1,n}} M|\xi_{pl}^{(n)}\sqrt{n}|^{2+\delta} < \infty, \tag{13}$$

then

$$\left|\int_0^\infty (1+i\theta x)^{-1} d\mu_n(x,z) - \int_0^\infty (1+i\theta x)^{-1} d\mu(x,z)\right| \leq c_3(|\theta| + \theta^2 + |\theta|^3) n^{-\gamma}, \tag{14}$$

where $0 < \gamma < 1$.

Theorem 6. Let $\tilde{\mu}_n(x)$ be the normalized spectral function of the random matrix $(Iz - H_n)(Iz - H_n)^*$, satisfying the conditions of the theorem 5, the condition (13) holds and $\tilde{\mu}(x)$ is a spectral function whose Styltyes transform satisfies the equation (13). Then as $|z|^2 < c$ where γ is a random variable with the distribution density e^{-x}, $x > 0$

$$\sup_x |M\tilde{\mu}_n(x\gamma^{-1}) - M\tilde{\mu}(x\gamma^{-1})| \leq cn^{-\alpha}, \quad 0 < \alpha < 1.$$

4. The Elliptic Law.

Theorem 7. Let for every value $n = 1, 2, \dots$ random vectors $(\xi_{lp}^{(n)}, \xi_{pl}^{(n)})$, $p \geq l$, $p, l = \overline{1,n}$ be stochastically independent, given in a common probability space

$$M\xi_{pl}^{(n)} = 0,\ M|\xi_{pl}^{(n)}|^2 = n^{-1},\ M\xi_{pl}^{(n)}\xi_{lp}^{(n)} = n^{-1}\rho;\ l \neq p,\ |\rho| < 1$$

imaginary and real parts of random elements $\xi_{pl}^{(n)}$, $\xi_{lp}^{(n)}$ of the random matrix $H_n = (\xi_{pl}^{(n)})_{p,l=1}^{n}$ have joint distribution densities $p_{kl}^{(n)}(x_i, i = \overline{1,4})$, satisfying the condition (1) and let the condition (2) be true. Then for any x and y with probability 1

$$\lim_{n\to\infty} \nu_n(x,y) = \lambda(x,y), \quad \text{where} \quad \frac{\partial^2 \lambda(x,y)}{\partial x \partial y} = \pi^{-1}[1-(\alpha^2+$$

$$+b^2)^2]^{-1}\chi\left[(bx-ay)^2(1-a^2-b^2)^{-2}(a^2+b^2)^{-1}+(ax+by)^2\times\right.$$
$$\left.\times(1+a^2+b^2)^{-2}(a^2+b^2)^{-1}<1\right],\quad a=\mathrm{Re}\sqrt{\rho},\quad b=\mathrm{Jm}\sqrt{\rho}.$$

Proof. Using theorem 3 we get that with probability 1 for almost all x and y $\lim_{n\to\infty}[\nu_n(x,y)-M\nu_n(x,y)]=0$. By theorems 4-6

$$\iint e^{ipx+iqx}\,dM\nu_n(x,y)=-\frac{q^2+p^2}{4iq\pi}\iiint\left[\int_{\alpha}^{\infty}\frac{\partial}{\partial s}m(\theta,z)\,d\theta\right]e^{itp+iqs}\,dt\,ds+o(1),\quad(15)$$

where $\alpha_n>0$ and $\lim_{n\to\infty}\alpha_n^{\varepsilon}n=0$, $0<\varepsilon<(\beta-1)(2\beta)^{-1}$, $q\neq 0$, $m(\theta,z)$ is the solution of the equation (13).

From the equation (12) we find

$$\frac{\partial m}{\partial s}=\left[(-2bat+2a^2s)(1+m-|\rho|m)^{-2}-(2abt+2b^2s)(1+m+|\rho|m)^{-2}\right]\times$$
$$\times\left[-|\rho|(1+2m)m^{-2}(1+m)^{-2}+2(bt-as)^2(1-|\rho|)(1+m-|\rho|m)^{-3}+\right.\quad(16)$$
$$\left.+2(at+sb)^2(1+|\rho|)(1+m+|\rho|m)^{-3}\right]^{-1}.$$

In the integral (15) we make a substitution

$$\theta=x^{-1}(1+x)^{-1}-\frac{(bt-as)^2}{|\rho|}(1+x-|\rho|x)^{-2}-\frac{(at+sb)^2}{|\rho|}(1-x+|\rho|x)^{-2}.$$

After such a substitution due to the equation (16) the integral (15) will have the form

$$\iint e^{ipx+iqx}\,dM\nu_n(x,y)=\frac{q^2+p^2}{4iq\pi}\iiint\left[\int_0^{m(\alpha_n,z)}\left\{-\frac{2bat+2a^2s}{|\rho|}(1+x-\right.\right.\quad(17)$$
$$\left.\left.-|\rho|x)^{-2}+\frac{2bat+2b^2s}{|\rho|}(1+x+|\rho|x)^{-2}\right\}dx\right]e^{itp+isq}\,dt\,ds+o(1).$$

Find $m(0,z)$. If $m(0,z)<\infty$ for some values of z then we get

$$m(0,z)=\Delta(1-\Delta(1-|\rho|))^{-1},\quad \Delta=\frac{1}{2\pi}\int_{-2}^{2}\frac{\sqrt{4-x^2}\,dx}{(t-ax)^2+(s-bx)^2}.$$

Since $m(0,t)>0$ this equality is valid as $\Delta\leq(1-|\rho|)^{-1}$. If $\Delta>(1-|\rho|)^{-1}$ then $m(0,z)=\infty$.

Therefore (17) is equal to

$$\iint e^{ipx+iqx}\,dM\nu_n(x,y)=(q^2+p^2)(4iq\pi)^{-1}\iint_{\Delta<(1-|\rho|)^{-1}}\left\{(2a^2s-\right.$$
$$\left.-2bat)|\rho|^{-1}\Delta+(2b^2s+2bat)|\rho|^{-1}\Delta(1+2\Delta|\rho|)^{-1}\right\}\times$$

$$\times e^{itp+isq}dt\,ds + \iint\limits_{\Delta>(1-|\rho|)^{-1}} \{(2a^2s-2bat)|\rho|^{-1}(1-|\rho|)^{-1} +$$

$$+(2b^2s+2bat)|\rho|^{-1}(1+|\rho|)^{-1}\}\, e^{itp+isq}dt\,ds\Big],+o(1)$$

where $p(x,y)$ is the density of the limit spectral function.

Hence one can easily conclude that

$$\iint p(x,y)(t+is-x-iy)^{-1}dx\,dy = \frac{a^2s-bat}{|\rho|(1-|\rho|)} + \frac{b^2s+bat}{|\rho|(1+|\rho|)} -$$

$$-i\left\{\frac{b^2t-abs}{|\rho|(1-|\rho|)} + \frac{a^2t+bas}{|\rho|(1+|\rho|)}\right\},$$

as $\Delta > (1-|\rho|)^{-1}$.

The same formula yields the statement of the theorem 7. Since the integral of the density $p(x,y)$ over the domain $\Delta(x,y)>(1-|\rho|)^{-1}$ is equal to 1, $p(x,y)\equiv 0$ as $\Delta(x,y)<(1-|\rho|)^{-1}$. Theorem 7 is proved.

One can easily see that the formula (15) implies, in particular, the semicircular (under $\rho \to 1$) and circular (under $\rho \to 0$) laws.

5. Stiltyes' transform of the covariance matrix normalized spectral function G_2-estimate.

The consistent and asymptotically normal estimation method of some covariance matrices functions $\varphi(R_{m_n})$ under large dimensions m_n of observed vectors is offered in this paragraph. Earlier, in most cases the estimates $\varphi(\hat{R}_{m_n})$, where $\hat{R}_{m_n}$ is the emperical covariance matrix required so many observations n, that there have been serious doubts about the usefulness of multivariate statistical analysis for the solution of practical problems at large m_n. These doubts were partly dispersed after the works [1,12-15] were published, in which corrections to the estimates $\varphi(\hat{R}_{m_n})$ were found under A.N.Kolmogorov condition: $\lim_{n\to\infty} m_n n^{-1} = c,\quad 0<c<\infty$. After many years of research it seemed that under Kolmogorov conditions consistent and asymptotically normal estimates of functions $\varphi(R_{m_n})$ do not exist. However, due to the developed matrix spectral theory one succeeded to establish that under G -condition: $\overline{\lim}\, f(m_n) n^{-1} < \infty$

$$\operatorname{plim}_{n\to\infty} \left[\varphi(\hat{R}_{m_n}) - \psi(R_{m_n})\right] = 0, \tag{18}$$

where ψ is some known measurable function of the matrix R_{m_n} elements. This is the principal statement making up the basis of G -analysis of observations of large dimensions. For the proof of (18) one use as a rule the limit theorems for sums of martingale differences and perturbation formulae for random matrices resolvents which form the theoretical foundation of G -analysis.

Using the equation (18) one can find such a measurable function $G(\hat{R}_{m_n})$ (G - estimate), that

$$\left[G(\hat{R}_{m_n}) - \varphi(R_{m_n})\right] c_n^{1/2} \Rightarrow N(0,1),$$

where c_n is a sequence of numbers. It is shown in the paper that the error of G -estimates of some functions $\varphi(R_{m_n})$ has the order $(m_n n)^{-1/2}$, while the estimates $\varphi(\hat{R}_{m_n})$ have an error equivalent to $m_n n^{-1/2}$.

Consider the main problem of G -analysis - the estimation of Stiltyes transforms of normalized spectral functions $\mu_{m_n}(x) = m_n^{-1} \sum_{k=1}^{m_n} \chi(\lambda_k < x)$ of covariance matrices R_{m_n} by the observations over random vector ξ with covariance matrix R_{m_n}, where λ_k are eigenvalues of R_{m_n} matrix. Note, that many analytical functions of covariance matrices which are used in multivariate statistical analysis can be expressed through spectral functions $\mu_{m_n}(x)$. For example, $m_n^{-1}\,\mathrm{Sp}\, f(R_{m_n}) = \int_0^\infty f(x)\,d\mu_{m_n}(x)$, where f is an analytical function.

The expession $\varphi(t,R_{m_n}) = \int_0^\infty (1+tx)^{-1} d\mu_{m_n}(x) = m_n^{-1}\,\mathrm{Sp}(I+tR_{m_n})^{-1}$ is called the Styltyes transform of the function $\mu_{m_n}(x)$. The G - estimate of Styltyes' transform $\varphi(t,R_{m_n})$ is by definition the following expression $G_2(t,\hat{R}) = \varphi(\hat{\theta}_n(t),\hat{R})$, where $\hat{\theta}_n(t)$ is the solution of the equation

$$\theta\left(1-\frac{m_n}{n-1}+\frac{m_n}{n-1}\varphi(\theta,\hat{R}_{m_n})\right)=t,\quad t\geqslant 0. \tag{19}$$

It is obvious that the positive solution of the equation (19) exists and is unique as $t\geqslant 0$.

Theorem 8. Let independent observations $x_1,\dots,x_n$ over m_n - dimensional random vector ξ be given, let G -condition be fulfilled:

$$\overline{\lim}_{n\to\infty} m_n n^{-1} < 1, \tag{20}$$

$$0 < c_1 \leqslant \lambda_i \leqslant c_2 < \infty,\quad i=\overline{1,m_n}, \tag{21}$$

let the components of the vector $\eta_k = (\eta_{1k},\dots,\eta_{m_n k}) = R^{-1/2}(\xi - M\xi)$ be independent and

$$\sup_n \sup_{k=\overline{1,n}} \sup_{i=\overline{1,m_n}} M|\eta_{ik}|^{4+\delta} < \infty,\quad \delta>0. \tag{22}$$

Then

$$\lim_{n\to\infty} P\left\{\left[G_2-\varphi(t,R_{m_n})\right]\sqrt{(n-1)m_n}\; a_n(t)+c_n(t)<x\right\}=\frac{1}{\sqrt{2\pi}}\int_{-\infty}^{x} e^{-y^2/2}\,dy, \tag{23}$$

as $t>0$, where $a_n(t) = q_1^{-1}(t)\,q_2(t)$,

$$q_1(t) = 1-\frac{M\hat{\varphi}'(\theta_n)\,k_n\theta_n}{1-k_n+k_n\theta_n M\hat{\varphi}'(\theta_n)+k_n M\hat{\varphi}(\theta_n)}, \tag{24}$$

$$q_2(t) = \left[m_n^{-1}\,\mathrm{Sp}A^2 + 3m_n^{-1}\sum_{i=1}^{m_n} a_{ii}^2\right]^{-1/2},\quad A=(a_{ij})=$$

$$=\left[1+\frac{\theta_n}{n-1} M\,\mathrm{Sp}\,SR_{m_n}\right]^{-1}\theta_n S^2 R_{m_n} + \theta_n^2 S R_{m_n} M\,\frac{\mathrm{Sp}\,S^2 R_{m_n}}{n-1}\times$$

$$\times\left[1+\frac{\theta_n}{n-1} M \operatorname{Sp} S R_{m_n}\right]^{-2}, \quad S=(I+\theta_n Q)^{-1}, \quad \kappa_n=\frac{m_n}{n-1},$$

θ_n is the solution of the equation

$$\theta_n\left(1-\kappa_n+\kappa_n M \hat{\varphi}(\theta_n)\right)=t, \tag{25}$$

$$\hat{\varphi}(\theta_n)=m_n^{-1} \operatorname{Sp} S, \quad Q=(n-1)^{-1} \sum_{k=1}^{n}(x_k-Mx_k)(x_k-Mx_k)',$$

$$c_n(t)=-a_n(t) \frac{\theta_n}{\sqrt{(n-1)m_n}} M(S\nu,\nu)\left[1-\frac{\theta_n}{n-1} M(S\nu,\nu)\right]^{-1} q_1(t)+$$

$$+\sqrt{\frac{n-1}{m_n}} \sum_{p=1}^{m_n}\left\{-\left[1+\lambda_p t\right]^{-3} M \varepsilon_{2p}^2+\left[1+\lambda_p t\right]^{-2} M \varepsilon_{2p}\right\}\Big],$$

$$\nu=\sum_{i=1}^{n} n^{-1/2}(x_i-Mx_i), \quad \varepsilon_{2p}=\theta_n \sum_{l=1}^{n}(b_{pl}^2-\lambda_p)- \tag{26}$$

$$-\theta_n^2\left\{(B_p'\Gamma_p B_p b_p, b_p)-\frac{\lambda_p}{n-1} \operatorname{Sp} R_p B_p B_p'\right\}+$$

$$+\theta_n \lambda_p\left[m_n^{-1} \operatorname{Sp} \Gamma_p - M\hat{\varphi}(\theta_n)\right],$$

B_p is a matrix obtained by deleting of the p-th row of the matrix $B=(\xi_1,\xi_2,\ldots,\xi_n)(n-1)^{-1/2}$, $\xi_i=\sqrt{\Lambda_{m_n}}\nu_i$,

$$\nu_k=HR_{m_n}^{-1/2}(\xi_k-M\xi_k), \quad \Lambda_{m_n}=\operatorname{diag}(\lambda_i, i=\overline{1,m_n}),$$

H is a matrix of R_{m_n} eigenvectors, $\Gamma_p=(I+\theta_n B_p B_p')^{-1}$, b_p is the row-vector of the matrix B, function $a_n(t)$ and $c_n(t)$ satisfy the inequality

$$\sup\left[|a_n(t)|+|c_n(t)|\right]<\infty, \tag{27}$$

as $0<t<c<\infty$.

Lemma 1.

$$G_2(t,\hat{R}_{m_n})-\varphi(t,R_{m_n})=\left[\hat{\varphi}(\theta_n)-M\hat{\varphi}(\theta_n)\right]c_n+ \tag{28}$$

$$+\left[M\hat{\varphi}(\theta_n)-\varphi(t,R_{m_n})\right]+\varepsilon_n,$$

$$C_n = 1 - \hat{\varphi}'(\theta_n) K_n \theta_n \left[1 - K_n + K_n \hat{\theta}_n \hat{\varphi}'(\theta_n) + K_n \hat{\varphi}(\theta_n)\right]^{-1}, \quad 0 \leq \alpha \leq 1,$$

$$\varepsilon_n = \frac{\frac{1}{2}\hat{\varphi}'(\theta_n) K_n \hat{\theta}_n \hat{\varphi}''(\theta_n + \alpha(\hat{\theta}_n - \theta_n))(\theta_n - \hat{\theta}_n)^2}{1 - K_n + K_n \hat{\theta}_n \hat{\varphi}'(\theta_n) + K_n \hat{\varphi}(\theta_n)} +$$

$$+ \frac{1}{2}\varphi''(\theta_n + \alpha(\hat{\theta}_n - \theta_n))(\hat{\theta}_n - \theta_n)^2. \tag{29}$$

Proof. Expanding the function $\hat{\varphi}$ into Taylor's series and using the equation $\theta_n(1 - K_n + K_n M\hat{\varphi}(\theta_n)) = t$, $\hat{\theta}_n(1 - K_n + K_n \hat{\varphi}(\hat{\theta}_n)) = t$, we have $\hat{\theta}_n(1 - K_n) + \hat{\theta}_n K_n \times$
$\times [\hat{\varphi}(\theta_n) + \hat{\varphi}'(\theta_n)(\hat{\theta}_n - \theta_n) + \frac{1}{2}\hat{\varphi}''(\theta_n + \alpha(\hat{\theta}_n - \theta_n))](\hat{\theta}_n - \theta_n)^2] = t,$
$\theta_n(1 - K_n) + \theta_n K_n M\hat{\varphi}(\theta_n) = t$.

Subtracting these two equations we obtain

$$\hat{\theta}_n - \theta_n = -\frac{1}{2}\frac{\theta_n K_n[\hat{\varphi}(\theta_n) - M\hat{\varphi}(\theta_n)] + \hat{\theta}_n K_n \hat{\varphi}''(\hat{\theta}_n + \alpha(\hat{\theta}_n - \theta_n))}{1 - K_n + K_n \hat{\varphi}(\theta_n) + \hat{\theta}_n K_n \hat{\varphi}'(\theta_n)}(\theta_n - \hat{\theta}_n)^2. \tag{30}$$

Obviously

$$G_2(t, \hat{R}_{m_n}) - \varphi(t, R_{m_n}) = \hat{\varphi}'(\theta_n)(\hat{\theta}_n - \theta_n) + \frac{1}{2}\hat{\varphi}''(\theta_n + \tag{31}$$

$$+ \alpha(\hat{\theta}_n - \theta_n))(\hat{\theta}_n - \theta_n)^2 + \hat{\varphi}(\theta_n) - \varphi(t, R_{m_n}).$$

The equalities (30) (33) entail the formula 28. Lemma 1 is proved.

Lemma 2 . Let non-degenerate matrices $A_n = (a_{ij})_{i,j=1}^n$ be given. Then [6, p. 54, p. 56]

$$\det A_n = \det A_{n-1}(a_{11} - (A_{n-1}^{-1} a_1, b_1)), \quad a_1 = (a_{12}, \dots, a_{1n}), \tag{32}$$

$$b_1 = (a_{21}, \dots, a_{n1}), \quad Sp(I + itA_n)^{-1} - Sp(I + itA_{n-1})^{-1} =$$

$$= -t\frac{d}{dt}\ln\left[1 + ita_{11} + t^2((I + itA_{n-1})^{-1} a_1, b_1)\right].$$

Let x_K be the m_n-dimensional column-vectors and the matrix $R_{m_n}=\sum_{K=1}^{n} x_K x_K'$, $n>m_n$ nondegenerate. Then [6,p.54]

$$(\underline{I}+R_{m_n}+x_{n+1}x_{n+1}')^{-1}-(\underline{I}+R_{m_n})^{-1}=-\frac{(\underline{I}+R_{m_n})^{-1}x_{n+1}x_{n+1}'(\underline{I}+R_{m_n})^{-1}}{1+((\underline{I}+R_{m_n})^{-1}x_{n+1},x_{n+1})}. \quad (33)$$

Lemma 3.

$$[M\hat{\varphi}(\theta_n)-\varphi(t,R_{m_n})]\sqrt{(n-1)m_n}=\sqrt{\frac{n-1}{m_n}}\sum_{p=1}^{m_n}\{-[1+\lambda_p t]^{-3}\times \quad (34)$$
$$\times M\varepsilon_{2p}^2+[1+\lambda_p t]^{-2}M\varepsilon_{2p}\}+o(1).$$

Besides

$$M\varepsilon_{2p}^2\leqslant\frac{c_1}{n-1},\ |M\varepsilon_{2p}|\leqslant\frac{c_2}{n-1},\ c_i<\infty,\ i=\overline{1,2}. \quad (35)$$

Proof. Since $t=\theta_n(1-K_n+K_n M\hat{\varphi}(\theta_n))$, then

$$\varphi(t,R_{m_n})=m_n^{-1}\sum_{p=1}^{m_n}[1+\lambda_p\theta_n(1-K_n+K_n M\hat{\varphi}(\theta_n))]^{-1}. \quad (36)$$

Evidently $Sp(\underline{I}+\theta_n Q)^{-1}=Sp(\underline{I}+\theta_n BB')^{-1}$.

Therefore, using the formula (32), we have (37)

$$M\hat{\varphi}(\theta_n)=m_n^{-1}\sum_{K=1}^{m_n}M[1+\theta_n(b_K,b_K)-\theta_n^2(B_K'\Gamma_K B_K b_K,b_K)]^{-1}= \quad (37)$$

$$=m_n^{-1}\sum_{p=1}^{m_n}M[1+\theta_n\lambda_p-\theta_n^2\lambda_p(n-1)^{-1}Sp\,\Gamma_p B_p B_p'+\varepsilon_{1p}]^{-1},$$

where

$$\varepsilon_{1p}=\theta_n\left(\sum_{l=1}^{n}b_{pl}^2-\lambda_p\right)-\theta_n^2\left\{(B_p'\Gamma_p B_p b_p,b_p)-\frac{\lambda_p}{n-1}Sp\,\Gamma_p B_p B_p'\right\}$$

From equation (32) the following emerges

$$M\hat{\varphi}(\theta_n)=m_n^{-1}\sum_{p=1}^{m_n}M[1+\theta_n\lambda_p[1-K_n+K_n M\hat{\varphi}(\theta_n)]+\varepsilon_{2p}]^{-1}, \quad (38)$$

where

$$\varepsilon_{2p}=\theta_n\left(\sum_{l=1}^{n}b_{pl}^2-\lambda_p\right)-\theta_n^2\{(B_p'\Gamma_p B_p b_p,b_p)-$$
$$-\frac{\lambda_p}{n-1}Sp\,\Gamma_p B_p B_p'\}+\theta_n\lambda_p\left[(n-1)^{-1}Sp\,\Gamma_p-\frac{m_n}{n-1}M\hat{\varphi}(\theta_n)\right].$$

It is easy to verify that

$$M\varepsilon_{1K}=0,\quad M\varepsilon_{1K}^2\leqslant\frac{c}{n-1}(\theta_n^2+\theta_n^4). \tag{39}$$

Using the equation (32), we obtain

$$\delta_n=(n-1)^{-1}Sp\Gamma_p-\frac{m_n}{n-1}\hat{\varphi}(\theta_n)-(n-1)^{-1}\theta_n\frac{d}{d\theta_n}\ln[1+\theta_n(b_p,b_p)-$$

$$-\theta_n^2(B_p'\Gamma_pB_pb_p,b_p)=-\frac{1}{n-1}\theta_n\frac{d}{d\theta_n}\ln\left[1+\theta_n\left(\left(I+\theta_nB_p'B_p\right)^{-1}b_p,b_p\right)\right].$$

It is easy to derive form this equation that

$$|\delta_n|\leqslant(n-1)^{-1}c. \tag{40}$$

Obviously $\hat{\varphi}(\theta_n)-M\hat{\varphi}(\theta_n)=\sum_{K=1}^n\gamma_K$,

where $\gamma_K=M_{K-1}\hat{\varphi}(\theta_n)-M_K\hat{\varphi}(\theta_n)$, M_K is a conditional expectation with respect to the minimal σ -algebra generated by $\xi_{K+1},\dots,\xi_n$. The variables γ_K are martingale difference, hence, using (40) and (41), we obtain

$$M[\hat{\varphi}(\theta_n)-M\hat{\varphi}(\theta_n)]^2=\sum_{K=1}^nM\gamma_K^2\leqslant n\,m_n^{-2}c, \tag{41}$$

$$M|\varepsilon_{2p}|^{2+\delta}(n-1)^{2+\delta/2}\leqslant c. \tag{42}$$

Using (39)-(42) from this equation and taking into account that $M\varepsilon_{2K}=\theta_n\lambda_KM\delta_n$ we obtain (34) and (35).

Lemma 3 is proved.

We shall need the following statement.

Lemma 4. [4 p.92] If for some $\delta>0$

$$\lim_{n\to\infty}\sum_{K=1}^nM|\gamma_K\sqrt{(n-1)m_n}\,a_n|^{2+\delta}=0, \tag{43}$$

$$\lim_{n\to\infty}\sum_{K=1}^nM|M_K\gamma_K^2-M\gamma_K^2|(n-1)m_n\,a_n^2=0, \tag{44}$$

where

$$a_n^{-2}=\sum_{K=1}^nM\gamma_K^2(n-1)m_n,$$

then

$$\lim_{n\to\infty}\left[a_n^{-2}-m_n^{-1}SpA^2-3m_n^{-1}\sum_{i=1}^{m_n}a_{ii}^2\right]=0,$$

$$\lim_{n\to\infty} P\left[\left(\hat{\varphi}(\theta_n)-M\hat{\varphi}(\theta_n)\right)\sqrt{(n-1)m_n}\; a_n < x\right] = \frac{1}{\sqrt{2\pi}}\int_{-\infty}^{x} e^{-y^2/2}\,dy. \quad (45)$$

We verify (43) and (44).
By virtue of (33) we have

$$\left[\hat{\varphi}(\theta_n)-M\hat{\varphi}(\theta_n)\right]\sqrt{(n-1)m_n} = -\sum_{k=1}^{n-1}\Big\{M_{k-1}\theta_n(S_k^2\xi_k,\xi_k)(n-1)^{-1}\times$$

$$\times\left[1+\theta_n(S_k\xi_k,\xi_k)(n-1)^{-1}\right]^{-1} - M_k\,\theta_n(S_k^2\xi_k,\xi_k)(n-1)^{-1}\left[1+\theta_n\times\right. \quad (46)$$

$$\left.\times(S_k\xi_k,\xi_k)(n-1)^{-1}\right]^{-1}\Big\}\sqrt{\frac{(n-1)}{m_n}} = -\frac{1}{\sqrt{n}}\sum_{k=1}^{n}\left(M_{k-1}\delta_k - M_k\delta_k\right),$$

where

$$S_k = \left(I+\theta_n(n-1)^{-1}\sum_{p\neq k}\xi_p\xi_p'\right)^{-1},$$

$$\delta_k = \frac{\left[\theta_n(S_k^2\xi_k,\xi_k)-\theta_n\,Sp\,S_k^2\Lambda_{m_n}\right]\frac{1}{\sqrt{m_n}}}{1+\theta_n(S_k\xi_k,\xi_k)(n-1)^{-1}} + \theta_n\frac{Sp\,S_k^2\Lambda_{m_n}}{n-1}\times$$

$$\times\frac{\left[\theta_n(S_k\xi_k,\xi_k)-\theta_n\,Sp\,S_k\Lambda_{m_n}\right]\frac{1}{\sqrt{m_n}}}{\left[1+\theta_n(n-1)^{-1}(S_k\xi_k,\xi_k)\right]\left[1+\theta_n(n-1)^{-1}Sp\,S_k\Lambda_{m_n}\right]}.$$

The equality (46) shows that (43) holds. One can easily see, that $\operatorname{plim}_{n\to\infty}\left[(S_k\xi_k,\xi_k)-M\,Sp\,S\Lambda_{m_n}\right](n-1)^{-1}=0,$

$$\operatorname*{plim}_{n\to\infty}\left[Sp\,S_k^2\Lambda_{m_n}-M\,Sp\,S^2\Lambda_{m_n}\right](n-1)^{-1}=0. \quad (47)$$

and the condition (44) can be replaced by the following one

$$\lim_{n\to\infty} m_n(n-1)a_n^2\sum_{k=1}^{n} M\left|M_k(M_{k-1}\delta_k)^2-M(M_{k-1}\delta_k)^2\right|=0. \quad (48)$$

We present $M_{k-1}\delta_k$ in the form

$$M_{k-1}\delta_k = m_n^{-1/2}\left[(B_k\xi_k,\xi_k)-Sp\,B_k\Lambda_{m_n}\right],$$

where

$$B_K = M_{K-1}\frac{\theta_n S_K^2}{1+\theta_n(S_K\xi_K,\xi_K)(n-1)^{-1}} + M_{K-1}\frac{\theta_n^2 S_K \dfrac{Sp\, S_K^2 \Lambda_{m_n}}{n-1}}{\left[1+\frac{\theta_n}{n-1}(S_K\xi_K,\xi_K)\right]\left[1+\frac{\theta_n}{n-1}Sp\, S_K R_m\right]}.$$

Obviously

$$M_K(M_{K-1}\delta_K)^2 = m_n^{-1}\, Sp(B\Lambda_{m_n}) + 3m_n^{-1}\sum_{i=1}^{m_n}\left[(B\Lambda_{m_n})_{ii}\right]^2. \quad (49)$$

We note that owing to the formula (33)

$$S - MS = \sum_{p=1}^{n}(M_{p-1}S - M_p S) = -\frac{1}{n-1}\sum_{p=1}^{n}\Big\{M_{p-1}\theta_n S_p\xi_p\xi_p' S_p \times$$

$$\times\left[1+\frac{\theta_n}{n-1}(S_p\xi_p,\xi_p)\right]^{-1} - M_p\theta_n S_p\xi_p\xi_p' S_p\left[1+\frac{\theta_n}{n-1}(S_p\xi_p,\xi_p)\right]^{-1}\Big\}.$$

Using this relation it is easy to prove that the condition (48) holds and also

$$M(M_{K-1}\delta_K)^2 = M\delta_K^2 + o(1) = \alpha_n^{-2} + o(1).$$

Hence (45) is fulfilled.

Thus lemma 4 is valid.

Lemma 5. Under the conditions of theorem 2

$$\operatorname*{plim}_{n\to\infty}\Big\{\left[G_2(t,\hat{R}_{m_n}) - \hat{\varphi}(\theta_n)\right]\sqrt{(n-1)m_n} - \theta_n(n-1)^{-1/2}m_n^{-1/2}\times \quad (50)$$

$$\times M(S^2 z,z)\left[1-\frac{\theta_n}{n-1}M(Sz,z)\right]^{-1}\Big\} = 0.$$

Proof. According to (33) we obtain

$$\left[m_n^{-1}\, Sp(I+\theta_n\hat{R}_{m_n})^{-1} - m_n^{-1}\, Sp\, S\right]\sqrt{(n-1)m_n} = \frac{\theta_n}{\sqrt{(n-1)m_n}}\times \quad (51)$$

$$\times(S^2 z,z)\frac{n-1}{n}\left[1-\frac{\theta_n}{n-1}(Sz,z)\frac{n-1}{n}\right]^{-1}.$$

We shall again make use of difference martingales sums

$$\frac{1}{n-1}(Sz,z) - \frac{1}{n-1}M(Sz,z) = \frac{1}{n-1}\sum_{K=1}^{n}\gamma_K, \quad (52)$$

where $\gamma_K = M_{K-1}(Sz,z) - M_K(Sz,z),$

M_k - conditional expectation with respect to $\xi_{k+1},\dots,\xi_n$.

Denote $\eta_k = \sum_{i\neq k}\xi_i n^{-1/2}$.

It is evident that $\gamma_k = M_{k-1}\Delta_k - M_k\Delta_k$, $\Delta_k = (S\eta_k,\eta_k) - (S_k\eta_k,\eta_k) + \frac{2}{\sqrt{n}}((S-S_k)\xi_k,\eta_k) + \frac{2}{\sqrt{n}}(S_k\xi_k,\eta_k) + \frac{1}{n}(S\xi_k,\xi_k)$.

Owing to the formula (33) we have for γ_k

$$\gamma_k = M_{k-1}\varkappa_k - M_k\varkappa_k,$$

where $\varkappa_k = \Big[\frac{\theta_n}{n-1}(S_k\xi_k,\eta_k)^2 + \frac{2\theta_n}{\sqrt{n}\,(n-1)}(S_k\xi_k,\xi_k)(S_k\xi_k,\eta_k)\Big]\times$
$\times\Big[1+\frac{\theta_n}{n-1}(S_k\xi_k,\xi_k)\Big]^{-1} + \frac{2}{\sqrt{n}}(S_k\xi_k,\eta_k) + \frac{1}{n}(S\xi_k,\xi_k)$.

Then

$$M\Big(\frac{1}{n-1}\sum_{k=1}^{n}\gamma_k\Big)^2 = (n-1)^{-2}\sum_{k=1}^{n}M\gamma_k^2 \leq \frac{cn}{(n-1)^2}\sup_k\big[\theta_n(n-1)^{-2}\times$$
$$\times M(S_k\xi_k,\eta_k)^4 + n^{-1}M(S_k\xi_k,\eta_k)^2 + n^{-2}M(S\xi_k,\xi_k)^2\big]. \tag{53}$$

Owing to the conditions of theorem 8 it is easy to verify

$$M(S_k\xi_k,\eta_k)^4 \leq c\,M(\eta_k,\eta_k)^2 \leq c_1 m_n^2,\quad M(S_k\xi_k,\eta_k)^2 \leq c_2 m_n,$$

$$M(S\xi_k,\xi_k)^2 \leq c\,M(\xi_k,\xi_k)^2 \leq c_3 m_n^2.$$

Using these inequalities we obtain

$$M\Big((n-1)^{-1}\sum_{k=1}^{n}\gamma_k\Big)^2 \leq cn^{-1} + o(1).$$

Hence, owing to (52)

$$\operatorname{plim}_{n\to\infty}\big[(S\eta,\eta) - M(S\eta,\eta)\big](n-1)^{-1} = 0. \tag{54}$$

Similarly

$$\frac{1}{\sqrt{(n-1)m_n}}\Big[(S^2\eta,\eta) - M(S^2\eta,\eta)\Big] = \sum_{k=1}^{n}\Big(\theta_n\frac{\partial}{\partial\theta_n}\gamma_k + \gamma_k\Big)\frac{1}{\sqrt{(n-1)m_n}}.$$

Having made the same simple transformations we obtain that this difference also tends to zero in probability.

Thus, using (54) and (51), we obtain (50).

It should be noted that in the general case

$$\lim_{n\to\infty} (n-1)^{-1} \left| M(S\vec{z},\vec{z}) - M\, Sp\, SR_{m_n} \right| > 0.$$

We shall show that the expression $\rho_n = \left(1 - \frac{\theta_n}{n-1} M(S\vec{z},\vec{z})\right)^{-1}$ is bounded. It is evident that owing to the formula (35)

$$1 - \frac{\theta_n}{n-1}(S\vec{z},\vec{z}) = \det\left[\bar{I} + \theta_n (Q - (n-1)^{-1}\vec{z},\vec{z}\,')\right] \det\left[\bar{I} + \theta_n Q\right]^{-1} =$$

$$= \left[1 + \theta_n (n-1)^{-1} \left(\left(\bar{I} + \theta_n (Q - \vec{z}\vec{z}\,'(n-1)^{-1})\right)^{-1} \vec{z},\vec{z}\right)\right]^{-1}.$$

From this equation we obtain

$$|\rho_n| \leq 1 + c\theta_n + o(1). \tag{55}$$

Lemma 6 is proved.

Using lemma 5 and 6 it is easy to establish that (27) is correct, $\operatorname{plim}_{n\to\infty} |\theta_n - \hat{\theta}_n| = 0$, $\operatorname{plim}_{n\to\infty} |\hat{\varphi}(\theta_n) - M\hat{\varphi}(\theta_n)| = 0$, $\operatorname{plim}_{n\to\infty} |\hat{\varphi}'(\theta_n) - M\hat{\varphi}'(\theta_n)| = 0$, $\operatorname{plim}_{n\to\infty} |\varepsilon_n \sqrt{m_n(n-1)}| = 0$. Therefore, using lemma 1,3,4,5 and formulae (29), (31), (55) we obtain the statement of the theorem 8.

References

I. Айвазян С.А., Енюков И.С.,Мешалкин Л.Д. Прикладная статистика, Справочное издание,М.,Финансы и статистика, 1985, 487.

2. Березин Ф.А. Несколько замечаний о распределении Вигнера.- Теоретическая и математическая физика.,1973, т.17, в.3, с.305-317.

3. Гирко В.Л. Круговой закон.- Теория вероятн. и ее применение,1984,т.29,в.4.

4. Гирко В.Л. Предельные теоремы для функций случайных величин.Киев,Вища школа,1983,208 с.

5. Гирко В.Л. - преобразование.-ДАН УССР,1982, №3,с.5-6.

6. Гирко В.Л. Теория случайных детерминантов.Киев, Вища школа,1980,368 с.

7. Гирко В.Л. Случайные матрицы.Киев,Вища школа,1975, 448 с.

8. Гирко В.Л. Предельные теоремы общего вида для спектральных функций случайных матриц. -Теория вероятн. и ее применение, 1977, т.22, в.1,с.160-163.

9. Гирко В.Л. Эллиптический закон.- Теория вероятн. и ее применение,1985, т.30,в.4.

10. Гирко В.Л. Центральная предельная теорема для случайных детерминантов,Теория вероятностей и ее применения,т.26,вып.3,с.532-542.

11. Гренандер У. Вероятности на алгебраических структурах. М.:Мир,1965,273 с.

12. Деев А.Д. Асимптотические разложения распределений статистик дискриминантного анализа "Статистические методы классификации", М.,Изд-во МГУ, вып.31, 1972,с.6-51.

13. Деев А.Д. Представление статистик дискриминантного анализа и асимптотические разложения при размерности пространства,сравнимой с объемом выборок, ДАН СССР,1970,195, № 4, с.759-762.

14. Заруцкий В.И. О классификации нормальных векторов простой структуры зависимостей в пространстве большей размерности.Теория вероятн. и ее примен.,1978, т.23, вып.2,с.473-475.

15. Мешалкин Л.Д. Сердобольский В.И., Ошибки при классификации многомерных наблюдений.Теория вероятн., 1978,т.23, вып.4, с.772-781.

16. Мусхелишвили Н.И.,Некоторые основные задачи математической теории упругости.М.: Наука,1966,707 с.

17. Пастур Л.А. О спектре случайных матриц.- Теоретическая и математическая физика.,1972, т.10, в.1, с.102-112.

18. Петров В.В. Суммы независимых случайных величин. М.:Наука, 1972, 414 с.

19. Arnold L. On the asymptotic distribution of the eigenvalues of random matrices. - 1. Math.Anal. Appl.; 1960, v.20, p.262-268.

20. Ginibre 1. Statistical ensembles of complex, quaternion and real matrices. - 1.Math.Phys., 1965, v.6, p,440-452.

21. Mehta M.L. Random matrices and the statistical theory of energy levels. - Acad. Press, New York, 1967, 260 p.

22. Wigner E.P. On the distribution of the roots of certain symmetric matrices. - Ann. Math., 1958, v.67, 2, p.325-327.

Prob. Theory and Math. Stat., Vol.1, pp. 509-512
Prohorov et al. (eds)
1986 VNU Science Press

MATHEMATICAL ASPECTS ON THE VARIATION OF AIR POLLUTANT CONCENTRATIONS

Jan Grandell
The Royal Institute of Technology, Stockholm, Sweden

We shall consider mathematical models for describing the time variability of the concentration of pollutants in the atmosphere. This note is based on Grandell (1985), which appered after the conference.

Let $c(t)$ (kg/m^3) be the concentration of the material at time t. The value of $c(t)$ fluctuates over time. We shall consider models where this fluctuation is due to variation in the source, i.e., in the rate of emission into the atmosphere, and the sink, i,e., the rate of removal. This means that the material enters and leaves the atmosphere with varying rates. More precisely we consider a stationary stochastic process $c(t)$ decribing the concentration of some kind of aerosol particles or some trace gas. Assume that

$$c'(t) = Q(t) - c(t)\lambda(t)$$

where $Q(t)$ (kg/m^3h) and $\lambda(t)$ (h^{-1}) are stationary processes called the <u>source strength</u> and the <u>sink intensity</u>. This simple physical model was first proposed by Gibbs and Slinn (1973), and we shall essentially restrict ourselves to that model. For particles in the size range 0.1 - 1 μm and water-soluble gases we interpret the sink intensity as the precipitation scavenging intensity. This means that the sink is related to precipitation (rain or snow). Let the stationary process $R(t)$ (kg/m^2h) decribe the <u>precipitation intensity</u>. In this case we shall sometimes assume that $\lambda(t) = aR(t)$ where a (m^2/kg) is a "parameter" depending on the type and the size of the particles or the kind of the gas. In many applications the value of a ranges from 0.1 to 1.

Define the <u>precipitation process</u> $h(t)$ (kg/m^2), the <u>sink process</u> $\Lambda(t)$ and the <u>source process</u> $q(t)$ (kg/m^2) by

$$h(t) = \int_0^t R(\tau)d\tau, \quad \Lambda(t) = \int_0^t \lambda(\tau)d\tau \text{ and } q(t) = \int_0^t Q(\tau)d\tau$$

respectively.

Thus

$$c(t) = \int_{-\infty}^{t} e^{-(\Lambda(t) - \Lambda(s))} dq(s)$$

and this is our general definition which is valid for and sink and source processes; i.e., also for processes which do not have an intensity representation.

The <u>residence time</u> T is the time spent in the atmosphere by a single particle, which enters the atmosphere independently of the sink process. Then we have

$$G(t) = \Pr\{T < t\} = E(e^{-\Lambda(t)}).$$

Put $c_0 = E(c(t))$, $\lambda_0 = E(\Lambda(1))$ and $Q_0 = E(q(1))$. Assume now that $\Lambda(t)$ and $q(s)$ are independent. Then

$$c_0 = \int_{-\infty}^{t} G(t - s)\, Q_0 ds = Q_0 \int_0^{\infty} G(s)\, ds = Q_0 E(T)$$

which is quite natural. In this case

$$c_0 \geq Q_0/\lambda_0$$

which implies that the mean concentration is systematically underestimated if the random variation in the sink process is disregarded.

In general, i.e., when $\Lambda(t)$ and $q(s)$ may be dependent, we have

$$c_0 \geq \exp\{E[\log(Q(0))]\}/\lambda_0$$

provided the source is described by an intensity model. In the same way as Q_0 corresponds to an arithmetic mean $\exp\{E[\log(Q(0))]\}$ corresponds to a geometric mean. We observe that if $\Pr\{Q(0) = 0\} > 0$ then this reduces to the triviality $c_0 \geq 0$.

Consider now "long-lived" particles. Hence we assume that $\Lambda(t) = ah(t)$ for some small value of a. Assume further that $q(t) = ag(t)$ for some non-decreasing process g(t) with stationary increments and that h(t) and g(s) are independent. Put $R_0 = E(h(1))$ and $\gamma_0 = E(g(1))$,

$$\Gamma_h = \lim_{t\to\infty} Var(h(t))/t \quad \text{and} \quad \Gamma_g = \lim_{t\to\infty} Var(g(t))/t.$$

Under general assumptions we have

$$c_0 \approx \frac{\gamma_0}{R_0}\left(1 + \frac{a\Gamma_h}{2R_0}\right) \text{ and } Var(c(0)) \approx \frac{a}{2R_0}\left(\frac{\gamma_0^2\Gamma_h}{R_0^2} + \Gamma_g\right) \qquad (1)$$

for small values of a.

Assume further that the source is deterministic; i.e., $g(t) \equiv \gamma_0 t$. Define the normalized concentration process $\bar{c}_a$ by

$$\bar{c}_a(t) = \left(c(t/a) - \gamma_0/R_0\right)/\sqrt{a}.$$

Let X be a sationary normal process with $E(X(t)) = 0$ and covariance function

$$r(\tau) = \frac{\gamma_0^2\Gamma_h}{2R_0^3} e^{-R_0|\tau|};$$

i.e., X is an Ornstein-Uhlenbeck process. Then, again under general assumptions,

$$\bar{c}_a \overset{d}{\to} X \text{ as } a \to 0 \qquad (2)$$

where $\overset{d}{\to}$ means "convergence in distribution". This limit theorem leads to the approximation

$$c(0) \overset{d}{\approx} \gamma_0\left(R_0^{-1} + W\cdot\{a\Gamma_h/2R_0^3\}^{1/2}\right)$$

where W is a normally distributed random variable with $E(W) = 0$ and $Var(W) = 1$. The underlying normalization is based on the crude approximation $c_0 \approx \gamma_0/R_0$. If, instead, (1) is used, we are led to

$$c(0) \overset{d}{\approx} \gamma_0(\{R_0^{-1} + a\Gamma_h/(2R_0^2)\} + W\cdot\{a\Gamma_h/2R_0^3\}^{1/2}).$$

Since $\{a\Gamma_h/(2R_0^2)\}/\sqrt{a} \to 0$ as $a \to 0$ obviously (2) still holds if $\bar{c}_a$ is correspondingly modified. Numerical comparisons indicate that the choice of normalization is very important for the validity of the approximation.

As an advertisement - or a warning - we end this note with a short summary of Grandell (1985). After an introduction and a section about basic probability we formulate in section 3 the model for c(t) and discuss its interpretation. In section 4 we consider the mean concentration and its relation to residence times. In section 5 we consider the variance of c(t) when the sink and the source are independent. Section 6 is devoted to the approximation proposed by Gibbs and Slinn (1973). In section 7 we consider precipitation scavenging from a somewhat more practical point of view. In section 8 we consider some "stochastic process properties" of c(t) and approximations of the distribution of the concentration. In one of the appendices models with dependent sink and source are considered.

REFERENCES

Gibbs A. G. and Slinn W. G. N. (1973). Fluctuations in trace gas concentrations in the troposphere. J. Geophys. Res. 78. 574 - 576.

Grandell J. (1985). Stochastic models of air pollutant consentration. Lecture Notes in Statistics 30. Springer-Verlag. Berlin.

Prob. Theory and Math. Stat., Vol.1, pp. 513-522
Prohorov et al. (eds)
1986 VNU Science Press

WEAK SOLUTIONS OF THE STOCHASTIC EVOLUTION EQUATIONS AND INVARIANCE PRINCIPLES

B. Grigelionis, R. Mikulevičius
Vilnius

INTRODUCTION

In the case of the classical stochastic Ito's equations in R^m weak solutions are usually defined as a probality measures on the Skorokhod space $D_{[0,\infty)}(R^m)$, which are characterized as the solutions to the corresponding martingale problems, formulated in the terms of the initial distributions and the coefficients of equations. The canonical process $X_t(\hat{\omega})=\hat{\omega}(t)$, $t\geq 0$, $\hat{\omega}\in D_{[0,\infty)}(R^m)$ is the semimartingale having the defined triplet of the predictable characteristics with respect to such probability measures and the canonical filtration of σ-algebras. Combining this characterization with the conveniet criteria of relative compactness of the sets of probability measures we can easily obtain general invariance principles on weak convergence of the semimartingales to the weak solutions of Ito's stochastic equations (see, e. g., Stroock, Varadhan (1979); Gikhman, Skorokhod (1982); Grigelionis, Mikulevičius (1981); Papanicolaou et al. (1977) and bibliography therein).

However in the case of stochastic equations with random coefficients or more general driving processes the above definition of weak solution fails. It is natural to consider the stochastic basis $(\Omega, \mathcal{F}, \mathbb{F}, P)$, where coefficiens and driving processes are defined, and to define a weak solution as a probability measure $\bar{P}$ on $\bar{\Omega}=\Omega\times D_{[0,\infty)}(R^m)$ such that the Ω-marginal $\bar{P}^{\Omega}$ of $\bar{P}$ equals to P and the canonical process $X_t(\omega, \hat{\omega})=\hat{\omega}(t)$, $t\geq 0$, $\omega\in\Omega$, $\hat{\omega}\in D_{[0,\infty)}(R^m)$ satisfies the equation with the obvious extensions to $\bar{\Omega}$. This definition corresponds to the notion of a good solution-measure in Jacod, Memin (1981). In the classical situation the $D_{[0,\infty)}(R^m)$ – marginal of a weak solution $\bar{P}$ coincides with a solution of the corresponding martingale problem.

The aim of this paper is to discuss a notion of weak solution to the stochastic evolution equation on a rigged Hilbert space with the Frechet valued driving processes, its connection to the theory of the infinite dimensional semimartingales and the corresponding invariance principles (see, e. g., Viot (1976); Kozlov (1978); Vishik et al. (1979); Grigelionis, Mikulevičius (1985 a, b); Dawson (1975, 1981, 1985); Gallavotti, Jona-Lasinio (1975); Martin-Lof (1976); Holley, Stroock (1976, 1978); Tanaka, Hitsuda (1981); Kotelenez (1982); Bouc, Pardoux (1983); Hitsuda (1983); Gorostiza (1983, 1985); Lavrent'ev (1983); Morrow (1983); Metivier (1984); Mikulevičius (1985 b); Shiga, Tanaka (1985), Joffe, Metivier (1986) and bibliography therein).

1. WEAK SOLUTIONS AND SEMIMARTINGALES

Let us consider a triplet $V\subset H\equiv H'\subset V'$, where V is a separable reflexive Banach space, which is a dense subset of a Hilbert space H and an imbedding $V\to H$ is continuous, V' is the topological dual space of V, and a stochastic basis $(\Omega, \mathcal{F}, \mathbb{F}, P)$. We shall use the standard terminology and notations of stochastic calculus from Jacod (1979), Metivier (1982). Besides for a Banach space E we shall denote $\mathcal{V}(E, P, \mathbb{F})$ a class of E-valued cadlag $\mathbb{F}$-adapted processes having P – a. e. locally finite variation, $\mathcal{A}_{loc}(E, P, \mathbb{F})$ a subclass of processes with the locally integrable variation, $\mathcal{M}_{loc}(E, P, \mathbb{F})$ a class of E-valued $(P, \mathbb{F})$ – local martingales, $\mathcal{M}^c_{loc}(E, P, \mathbb{F})$ a subclass of E-valued continuous local martingales, $\mathcal{M}^2_{loc}(E, P, \mathbb{F})$ a subclass of E-valued locally square integrable local martingales and $\mathcal{S}(E, P, \mathbb{F})$ a class of E-valued semimartingales.

We shall define a weak solutions and discuss the properties of the following stochastic evolution equation:

$$X_t = X_0 + \int_0^t A(s, X_{s-})dN_s + \sum_{j=1}^{\infty} \int_0^t B^j(s, X_{s-})dM^j_s, \quad t \geq 0, \tag{1}$$

where $X_0 \in H$, N is a $\mathbb{F}$-predictable increasing process, $X_{t-} \in V$ dNdP– a. e., $A: R_+ \times \Omega \times V \to V'$ is $(\mathcal{B}(V'), \mathcal{P}(\mathbb{F}) \otimes \mathcal{B}(V))$ – measurable, $\int_0^t |A(s, X_{s-})|_{V'} dN_s < \infty$ P – a. e. for each $t > 0$, $M^j \in \mathcal{M}^2_{loc}(P, \mathbb{F})$, $\langle M^j, M^k \rangle_t = \delta_{jk} \int_0^t Q^j_s dN_s$, $t \geq 0$, j, k=1, 2, ..., the $(P, \mathbb{F})$ – compensator of the jump measure of $M = (M^j, j \geq 1)$ is $\Pi([0, t] \times \Gamma) = \int_0^t \pi(s, \Gamma)dN_s$, $t \geq 0$, $\Gamma \in \mathcal{B}(R^{\infty} \setminus \{0\})$, $B^j: R_+ \times \Omega \times V \to H$ is $(\mathcal{B}(H), \mathcal{P}(\mathbb{F}) \otimes \mathcal{B}(V))$ – measurable, $j \geq 1$, and $\sum_{j=1}^{\infty} \int_0^t |B^j(s, X_{s-})|^2_H Q^j_s dN_s < \infty$ P – a. e. for each $t > 0$.

Remark that the cases of stochastic evolution equations with the driving local martingale M, taking values in the more general Frechet spaces, can be reduced to the case of the space R^{∞} (see Grigelionis, Mikulevičius (1983)).

Let $D_{[0,\infty)}(V'_w)$ be the Skorokhod space of cadlag functions $\hat{\omega}(\cdot): R_+ \to V'$, considering the weak topology in V', $D = \{\mathcal{D}_t, t \geq 0\}$ be the canonical filtration on $D_{[0,\infty)}(V'_w)$, $\mathcal{D} = \bigvee_{t \geq 0} \mathcal{D}_t$, $\bar{\Omega} = \Omega \times D_{[0,\infty)}(V'_w)$, $\bar{\mathbb{F}} = \{\bar{\mathcal{F}}_t, t \geq 0\}$, $\bar{\mathcal{F}}_t = \bigcap_{s > t} \mathcal{F}_s \otimes \mathcal{D}_s$, $\bar{\mathcal{F}} = \mathcal{F} \otimes \mathcal{D}$, $X_t(\omega, \hat{\omega}) = \hat{\omega}(t)$, $t \geq 0$, $(\omega, \hat{\omega}) \in \bar{\Omega}$.

Definition 1 (cf. Jacod, Memin (1981); Grigelionis, Mikulevičius (1985 b)). A probability measure $\bar{P}$ on $(\bar{\Omega}, \bar{\mathcal{F}})$ is called a weak solution of the stochastic evolution equation (1), if $\bar{P}^{\Omega} = P$, $M^j \in \mathcal{M}^2_{loc}(\bar{P}, \bar{\mathbb{F}})$, $\langle M^j, M^k \rangle_t = \delta_{jk} \int_0^t Q^j_s dN_s$, $t \geq 0$, j, k=1, 2, ..., the $(\bar{P}, \bar{\mathbb{F}})$ – compensator of the jump measure of $M = (M^j, j \geq 1)$ is $\int_0^t \pi(s, \Gamma)dN_s$, $t \geq 0$, $\Gamma \in \mathcal{B}(R^{\infty} \setminus \{0\})$ and X statisfies (1), considering the natural extensions to $\bar{\Omega}$ of functions, defined only on Ω.

The following martingale characterization of the weak solution $\bar{P}$ to (1) plays an important role. Let $\mathcal{Y} = V' \times R^{\infty}$, $Y = \{Y_t = (X_t - X_0, M_t), t \geq 0\}$, p^Y be the jump measure of Y, Π^Y be the $(\bar{P}, \bar{\mathbb{F}})$ – compensator of p^Y. For the sake of simplicity assuming that N is continuous in t, from (1) we shall have that

$$\Pi^Y([0, t] \times \hat{\Gamma}) = \int_0^t \pi^Y(s, \hat{\Gamma})dN_s, \; t \geq 0, \quad \hat{\Gamma} \in \mathcal{B}(\mathcal{Y} \setminus \{0\}),$$

where

$$\pi^Y(s, \hat{\Gamma}) = \int_{R^{\infty} \setminus \{0\}} \chi_{\hat{\Gamma}}\Big(\sum_{j=1}^{\infty} B^j(s, X_{s-}) z_j, z\Big) \pi(s, dz).$$

Let $\langle v', v \rangle$, $v \in V$, $v' \in V'$, be the canonical bilinear form, $\langle y' | y \rangle = \langle v', v \rangle + \sum_{j=1}^{\infty} x_j x'_j$,

$y = (v', x) \in \mathcal{Y}$, $y' = (v, x') \in \mathcal{Y}'$, $v \in V$, $x' \in R^{\infty}$

and has only finite number of non-zero components,

$$M^{y'}_t = e^{i\langle y' | Y_t \rangle} - \int_0^t e^{i\langle y' | Y_s \rangle} \varphi_{y'}(s, Y_s)dN_s, \; t \geq 0,$$

where

$$\varphi_{y'}(s, Y_s) = \int_{\mathcal{Y} \setminus \{0\}} [e^{i\langle y' | y \rangle} - 1 - i\langle y' | y \rangle] \, \pi^Y(s, dy) + i\langle A(s, X_{s-}), v \rangle -$$

$$- \frac{1}{2} \sum_{j=1}^{\infty} [\langle B^j(s, X_{s-}), v \rangle + 2\langle B^j(s, X_{s-}), v \rangle x'_j + (x'_j)^2] \, Q^j_s.$$

For any Hilbert space $\hat{H}$ and $(t, \omega)\in R_+ \times \Omega$ denoting $L_t^2(\hat{H}^\infty)$ a Hilbert space of sequences $\hat{h}=(\hat{h}_1, \hat{h}_2, \dots)$ with the norm

$$|\hat{h}|_t=\left(\sum_{j=1}^{\infty} |\hat{h}_j|_{\hat{H}}^2 Q_t^j\right)^{1/2},$$

remark that $dN\,dP$ – a. e. $B(t, X_{t-})=(B^j(t, X_{t-}), j\geq 1)\in L_t^2(H^\infty)$.

The following statment holds.

Theorem 1 (cf. Jacod, Memin (1981); Grigelionis, Mikulevičius (1985 b)). A probability measure $\bar{P}$ on $(\Omega, \mathcal{F})$ is a weak solution of the stochastic evolution equation (1) if $\bar{P}^{\Omega}=P$, $X_{t-}\in V$ $dN\,d\bar{P}$ – a. e., for each $t>0$ $\bar{P}$ – a. e.

$$\int_0^t |A(s, X_{s-})|_{V'} dN_s<\infty, \quad \int_0^t |B(s, X_{s-})|_s^2 dN_s<\infty,$$

and for each $y'\in\mathcal{Y}'$ $M^{y'}\in\mathcal{M}_{loc}(\bar{P}, \bar{\mathbb{F}})$.

P r o o f. Necessity part of the theorem we obviously obtain applying the Ito's formula to the one-dimensional semimartingales $e^{i\langle y'|Y_t\rangle}$, $t\geq 0$, $y'\in\mathcal{Y}'$. Following the proof of the theorem 1 in Grigelionis, Mikulevičius (1981), we find that for each $v\in V$ and $t\geq 0$ the equality

$$\langle X_t, v\rangle=\langle X_0, v\rangle+\int_0^t \langle A(s, X_{s-}), v\rangle dN_s+\sum_{j=1}^{\infty}\langle B^j(s, X_{s-}), v\rangle dM_s^j$$

holds, which implies the sufficiently part of the theorem.

Note that the existence, uniqueness, and stability properties of weak solutions were discussed in Grigelionis, Mikulevicius (1985 b), using coercitivity and monotonicity assumptions. We shall not discuss these problems now, noting only that the assumption VI of the theorem 4 in Grigelionis, Mikulevičius (1985 b) can be weakened taking $\varepsilon_0=0$.

If $\bar{P}$ is a weak solution to (1), then the canonical process X_t is evidently decomposable to the sum of a V' – valued process A_t of $\bar{P}$ – a. e. locally finite variation, $A_0=0$, and a H – valued $(\bar{P}, \bar{\mathbb{F}})$ – semimartingale Z_t. However these properties are too weak in order to develope the satisfactory theory of the characterization of such processes in the terms of the suitable predictable characteristics and to obtain the invariance principles in the analogy with the case of the finite dimensional semimartingales. For two normed spaces $E_1\supseteq E_2$ and $e\in E_1$ denoting $|e|_{E_2}$ the norm of e in E_2, if $e\in E_2$, and $=\infty$, if $e\bar{\in} E_2$, the following integrability property often holds: for each $t>0$ $\bar{P}$ – a. e. $\int_0^t |X_s|_V d|A|_s<\infty$, where $|A|_t=\int_0^t|dA_s|_{V'}$. Under this assumption the process X is $\bar{P}$ – a. e. H – valued cadlag in the strong topology of H and the following canonical decomposition is true (see Grigelionis, Mikulevičius (1985 a):

$$X_t=X_0+\alpha_t^h+X_t^c+\int_0^t\int_{H\setminus\{0\}} h(x)q(du, dx)+\int_0^t\int_{H\setminus\{0\}}(x-h(x))p(du, dx),$$

where $h: H\to H$ is a continuous bounded function, $h(x)=x$ for $|x|_H\leq\frac{1}{2}$ and $=0$ for $|x|_H\geq 1$, $X^c\in\mathcal{M}_{loc}^c(H, \bar{P}, \bar{\mathbb{F}})$, $p(du, dx)$ is the jump measure of X, $\Pi(du, dx)$ is the $(\bar{P}, \bar{\mathbb{F}})$ – compensator of p, $q(du, dx)=p(du, dx)-\Pi(du, dx)$, $\alpha^h\in\mathcal{A}_{loc}(V', \bar{P}, \bar{\mathbb{F}})\cap\mathcal{P}(\bar{\mathbb{F}})$, $\Delta\alpha_t^h=\int_{H\setminus\{0\}} h(x)\Pi(\{t\}\times dx)$, $t\geq 0$, and there exist $C\in\mathcal{V}(V', \bar{P}, \bar{\mathbb{F}})$, $S\in\mathcal{S}(H, \bar{P}, \bar{\mathbb{F}})$ such that $\bar{P}$ – a. e. for each $t>0$ $\int_0^t|X_s|_V d|C|_s<\infty$ and $\alpha^h=C+S$. The class of such processes, defind on some stochastic basis $(\tilde{\Omega}, \tilde{\mathcal{F}}, \tilde{\mathbb{F}}, \tilde{P})$ we shall denote $\mathcal{S}(V, H, \tilde{P}, \tilde{\mathbb{F}})$. Moreover under above integrability assumption the important Ito's formula holds (see Grigelionis, Mikulevičius (1985 a); Krylov, Rozovskii (1979); Gyondy, Krylov (1982)):

$$|X_t|_H^2=|X_0|_H^2+2\int_0^t X_s dA_s+2\int_0^t A_{s-}dZ_s+|Z_t|_H^2-\sum_{s\leq t}|\Delta A_s|_H^2, \quad t\geq 0.$$

The triplet (α^h, B, Π), where B is defined by means of the equality

$<(X^c, v)_H>_t = (v, B_t v)_H$, $t \geq 0$, $v \in V$,

is called the triplet of the $(\overline{P}, \overline{\mathbb{F}})$ – predictable characteristics of the process X and plays an analogous role as in the case of the finite dimensional semimartingales.

2. INVARIANCE PRINCIPLES

Let we are given a sequence of stochastic bases $(\Omega, \mathcal{G}, \mathbb{F}^n, P)$, $n \geq 1$, such that $\mathcal{F} \subseteq \mathcal{G}$, $\mathcal{F}_t \subseteq \bigcap_{n \geq 1} \mathcal{F}_{nt}$, $t \geq 0$, and a sequence $\{X^n, n \geq 1\}$, $X^n \in \mathcal{S}(V, H, \mathbb{F}^n, P)$ with the triplets of the $(P, \mathbb{F}^n)$ – predictable characteristics $(\alpha^{h,(n)}, B^{(n)}, \Pi^{(n)})$, $n \geq 1$. Assume that the paths of the processes X^n, $n \geq 1$, are contained in the subspace $\mathfrak{X} \subseteq D_{[0,\infty)}(V'_w)$, equipped with the completely regular topology τ, which is not weaker then the topology induced by the Skorokhod's $\mathcal{J}_1$ – topology in $D_{[0,\infty)}(V'_w)$ and such that each τ – continuous function is $\mathfrak{D} \cap \mathfrak{X}$ – measurable. Let $\tilde{\Omega} = \Omega \times \mathfrak{X}$, $\tilde{\mathcal{F}} = \mathcal{F} \otimes \mathfrak{B}(\mathfrak{X})$, $M_{mc}(\tilde{\Omega}, P)_H$ be a space of all probability measures μ on $\tilde{\mathcal{F}}$ such that $\mu^{\Omega} = P$ with the topology generated by the mappings $\mu \to \mu(f) = \int_{\tilde{\Omega}} f(\omega, \hat{\omega})\, \mu(d\omega, d\hat{\omega})$ for all bounded $\tilde{\mathcal{F}}$ – measurable functions, which are τ – continuous for each fixed $\omega \in \Omega$, $M_c(\mathfrak{X})$ be a space of all probability measures on $\mathfrak{B}(\mathfrak{X})$ with the topology of weak convergence. Let $\tilde{\mathbb{F}} = \{\tilde{\mathcal{F}}_t = \tilde{\mathcal{F}}_t \cap \tilde{\Omega}, t \geq 0\}$, ν be a probability measure on $\tilde{\mathcal{F}}_0$, $X_t(\omega, \hat{\omega}) = \hat{\omega}(t)$, $(\omega, \hat{\omega}) \in \tilde{\Omega}$, $\mathcal{S}(\nu, \tilde{\alpha}^h, \tilde{B}, \tilde{\Pi})$ be a class of all probability measures $\mu \in M_{mc}(\tilde{\Omega}, P)$ such that $X \in \mathcal{S}(V, H, \tilde{\mathbb{F}}, \mu)$ with the triplet $(\tilde{\alpha}^h, \tilde{B}, \tilde{\Pi})$ of the $(\tilde{\mathbb{F}}, \mu)$ – predictable characteristics and $\mu|_0 = \nu$.

Let us consider the seguence of measures $\mu_n(d\omega, d\hat{\omega}) = P(d\omega)\, \varepsilon_{X^n(\omega)}(d\hat{\omega})$ on $(\tilde{\Omega}, \tilde{\mathcal{F}})$, where ε_a is a Dirac measure, $\mu_n^{(0)} = \mu_n|\tilde{\mathcal{F}}_0$, $n \geq 1$.

Denote

$$\tilde{\Gamma}_v^h(t, \omega, \hat{\omega}) = (v, \tilde{B}_t(\omega, \hat{\omega})v)_H + \int_{H \setminus \{0\}} (h(x), v)_H^2\, \tilde{\Pi}(\omega, \hat{\omega}, [0, t] \times dx) -$$

$$- \sum_{s \leq t} \int_{H \setminus \{0\}} (h(x), v)_H \tilde{\Pi}(\omega, \hat{\omega}, \{s\} \times dx),$$

$$\Gamma_v^{h,(n)}(t, \omega) = (v, B_t^{(n)}(\omega)v)_H + \int_{H \setminus \{0\}} (h(x), v)_H^2 \Pi^{(n)}(\omega, [0, t] \times dx) -$$

$$- \sum_{s \leq t} \int_{H \setminus \{0\}} (h(x), v)_H \Pi^{(n)}(\omega, \{s\} \times dx), \quad t \geq 0, \ v \in V,$$

$$\tilde{\Pi}_t(\omega, \hat{\omega}, \varphi) = \int_{H \setminus \{0\}} \varphi(x) \tilde{\Pi}(\omega, \hat{\omega}, [0, t] \times dx),$$

$$\Pi_t^{(n)}(\omega, \varphi) = \int_{H \setminus \{0\}} \varphi(x) \Pi^{(n)}(\omega, [0, t] \times dx), \quad t \geq 0,$$

$\varphi \in \hat{C}^+ (H \setminus \{0\})$,

where $\hat{C}^+ (H \setminus \{0\})$ is a space of nonnegative continuous bounded functions which are equal to zero in the neighbourhood of zero, $\mathcal{T}(\mathbb{F}^n)$ a class of $\mathbb{F}^n$ – stopping times, $\Sigma(T)$ a class of sequences $(S_n, n \geq 1)$, where $S_n = S_n(T) \in \mathcal{T}(\mathbb{F}^n)$, $S_n \leq T$ and $\lim_{n \to \infty} P\{S_n < T\} = 0$, $T > 0$.

Theorem 2 (cf. Grigelionis, Mikulevičius (1985 a)). Assume that:

1) the sequence $\{\mu_n^{\mathfrak{X}}, n \geq 1\}$ is tight in $M_c(\mathfrak{X})$, i. e. for each $\varepsilon > 0$ there exist a compact set $K_\varepsilon \subset \mathfrak{X}$ such that $\mu_n^{\mathfrak{X}}(K_\varepsilon) \geq 1 - \varepsilon$, $n \geq 1$;

2) for some subclass of sets $\tilde{A} \in \tilde{\mathcal{F}}_0$, defining ν, $\lim_{n \to \infty} \mu_n^{(0)}(\tilde{A}) = \nu(\tilde{A})$;

3) for each $t>0$, $v\in V$, $\varphi\in\hat{C}^+(H\backslash\{0\})$ and P – a. e. ω $<\tilde{\alpha}_t^h(\omega,\cdot), v>$, $(v, \tilde{B}_t(\omega,\cdot)v)_H$ and $\tilde{\Pi}_t(\omega,\cdot,\varphi)$ are τ – continuous in $\hat{\omega}$;

4) for each $v\in V$, $\varphi\in\hat{C}^+(H\backslash\{0\})$ and $T>0$ there exists a sequence $(S_n, n\geq 1)\in\Sigma(T)$ such that $\{\Gamma_v^{h,(n)}(S_n)+\Pi_{S_n}^{(n)}(\varphi), n\geq 1\}$ is uniformly integrable;

5) for each $\varepsilon>0$, $t>0$, $v\in V$ and $\varphi\in\hat{C}^+(H\backslash\{0\})$ $\lim_{n\to\infty} P\{|<\alpha_t^{h,(n)}-\tilde{\alpha}_t^h(X^n), v>|+|\Gamma_v^{h,(n)}(t)-$ $-\tilde{\Gamma}_v^h(t, X^n)|+|\Pi_t^{(n)}(\varphi)-\tilde{\Pi}_t(X^n,\varphi)|>\varepsilon\}=0$.

Then $\{\mu_n, n\geq 1\}$ is relative compact in $M_{mc}(\tilde{\Omega}, P)$ and all limiting points are contained in $\mathcal{S}(\nu, \tilde{\alpha}^h, \tilde{B}, \tilde{\Pi})$.

P r o o f of the theorem is analoguous to the proof of theorem 1 in Grigelionis, Mikulevičius (1985 a), if we note that the tightness of $\{\mu_n^{\mathfrak{X}}, n\geq 1\}$ in $M_c(\mathfrak{X})$ is sufficient for the relative compactness of $\{\mu_n, n\geq 1\}$ in $M_{mc}(\tilde{\Omega}, P)$.

Remark that, if $\tilde{\Pi}\equiv 0$, then $\tilde{\alpha}_t^h(\omega,\hat{\omega})\equiv\tilde{\alpha}_t(\omega,\hat{\omega})$ does not depend on h. In this situation the following theorem is often useful. Let $X_t^n=X_0^n+A_t^n+M_t^n$, $t\geq 0$, $n\geq 1$, where $A^n\in\mathcal{V}(V', P, \mathbb{F}^n)$, $M^n\in\mathcal{M}_{loc}^2(H, P, \mathbb{F}^n)$ and $X_\cdot^n\in\mathfrak{X}$ P – a. e., $n\geq 1$.

Theorem 3. Assume that:

1) the assumption 1) – 2) of the theorem 2 are satisfied;

2) for each $v\in V$ P – a. e. ω the mappings $x\to(<\alpha_\cdot^h(\omega, x), v>, (v, \tilde{B}_\cdot(\omega, x)v)_H)$ from $\mathfrak{X}$ to $D_{[0,\infty)}(R^2)$ are continuous;

3) for each $\varepsilon>0$, $t>0$ and $v\in V$ $\lim_{n\to\infty} P\{[\sup_{s\leq t}|<A_s^n-\tilde{\alpha}_s(X^n), v>|+\sup_{s\leq t}|\Delta<X_s^n, v>|+|(\tilde{B}_t(X^n)v,v)_H-$ $-<(M^n, v)_H>_t|]>\varepsilon\}=0$;

4) for each $v\in V$ and $T>0$ there exists a sequence $\{S_n, n\geq 1\}\in\Sigma(T)$ such that for each $T_n\leq S_n$, $T_n\in\mathcal{T}(\mathbb{F}_n)$, $\{\Delta<A_{T_n}^n, v>, n\geq 1\}$ is uniformly integrable.

Then the sequence $\{\mu_n, n\geq 1\}$ is relative compact in $M_{mc}(\tilde{\Omega}, P)$ and all limiting points are contained in $\mathcal{S}(\nu, \tilde{\alpha}, \tilde{B}, 0)$.

P r o o f. It follows from 1) – 2) that the sequence $\{\mu_n, n\geq 1\}$ is relative compact in $M_{mc}(\tilde{\Omega}, P)$. Let $\bar{P}$ be a limiting point. Now fix $v\in V$, $T>0$ and consider $(S_n, n\geq 1)\in\Sigma(T)$, satisfying 4). Let $T_r=\inf\{t: <X_t-\tilde{\alpha}_t, v>>r\}$, $\bar{P}\{T_r=T_{r-}\}=1$.

Denote

$$M_u^{nvr}=<X^n_{u\wedge T_r(X^n)\wedge S_n}-A^n_{u\wedge T_r(X^n)\wedge S_n}, v>,$$

$$Y_u^{vr}=<X_{u\wedge T_r}-\tilde{\alpha}_{u\wedge T_r}, v>,$$

$$U_n=\inf\{t: |M_t^{nvr}-Y_t^{vr}(X^n)|>1\}.$$

Let f be a $\tilde{\mathcal{F}}_s$ – measurable, bounded, $\bar{P}$ – a. s. continuous function. It is obvious that for each $\varepsilon>0$

$$\lim_{n\to\infty} P\{\sup_{u\leq T}|M_u^{nvr}-Y_u^{vr}(X^n)|>\varepsilon\}=0$$

and there exists a subsequence $\{n_k, k\geq 1\}$ such that

$$\bar{P}[(Y_t^{vr}-Y_s^{vr})f]=\lim_{k\to\infty} E[M^{n_k vr}_{t\wedge U_{n_k}}-M^{n_k vr}_{s\wedge U_{n_k}})f(X^n)]=0,$$

i. e. $<X-\tilde{\alpha}, v>\in\mathcal{M}^c_{loc}(\bar{P}, \tilde{\mathbb{F}})$.

Analoguously we check that for each $v \in V$ $\langle X_\cdot - \tilde{\alpha}_\cdot, v\rangle^2 - (\tilde{B}_\cdot v, v)_H \in \mathcal{M}^c_{loc}(\bar{P}, \bar{\mathbb{F}})$. Therefore $\bar{P} \in \mathcal{S}(\nu, \tilde{\alpha}, \tilde{B}, 0)$.

Let as consider now some special cases. Assume that $\mathcal{X} = D_{[0,\infty)}(H) \cap L^2_{loc,w}(V, N)$, where N is a nonrandom continuous increasing function, $L^2_{loc,w}(V, N) = \{\hat{\omega}(\cdot) : \int_0^t |\hat{\omega}(s)|^2_V dN_s < \infty, \ t > 0\}$, and the space $\mathcal{X}$ is considered to be endowed with the supremum of $\mathcal{J}_1$ – topology and the topology of weak convergence in $L^2_{loc,w}(V, N)$. The sufficient conditions for the tightness of $\{\mu_n, n \geq 1\}$ are given in this case by the theorem 4 in Grigelionis, Mikulevičius (1985 a). For example, the main limit theorems in the papers Holley, Strook (1978) and Gorostiza (1983) can be proved under the same assumptions in the space $\mathcal{X} = D_{[0,\infty)}(H^{(-d-3)}) \cap L^2_{loc,w}(H^{(-d-2)}, dt)$ instead of the space $D_{[0,\infty)}(\mathcal{S}'(R^d))$.

In the case when $\mathcal{X} = D_{[0,\infty)}(H_w) \cap L^2_{loc,w}(V, N) \cap L^2_{loc}(H, N)$, endowed with the supremum topology, where H_w denotes Hilbert space H with the weak topology, $L^2_{loc}(H, N) = \{\hat{\omega}(\cdot) : \int_0^t |\hat{\omega}(s)|^2_H dN_s < \infty, \ t > 0\}$ with the topology of strong convergence, using lemma in Grigelionis, Mikulevičius (1985 b) and the know Aldous–Rebolledo criterion, we can prove that the analoguous result to the theorem 4 in Grigelionis, Mikulevičius (1985 a) holds.

Denote $\tilde{\Sigma}(T) = \{(T_{n,j})_{n,j\geq 1} : T_{n,j} \in \mathcal{T}(\mathbb{F}^n), T_{n,j} \leq T, \lim_{j\to\infty} \overline{\lim_{n\to\infty}} P\{T_{n,j} < T\} = 0\}, T > 0$.

Theorem 4. Let the following assumptions hold:

1) $\{\mu_n^{(0)\mathcal{X}}, n \geq 1\}$ is tight in $M_c(H)$;

2) for each $T > 0$ and $K > 0$ there exist $(T_{n,j}) \in \tilde{\Sigma}(T)$, $\lambda_{n,j}, A_{n,j} \in \mathcal{A}^+(P, \mathbb{F}^n) \cap \mathcal{P}(\mathbb{F}^n)$ such that for each $\varepsilon > 0$

$$\lim_{l\to\infty} \overline{\lim_{n\to\infty}} P\{\Pi^{(n)}([0,T] \times \{x : |x|_H > l\}) > \varepsilon\} = 0,$$

$$\sup_{n,j} [\lambda_{n,j}(\infty) + A_{n,j}(\infty)] \leq \text{const},$$

$$\overline{\lim_{n\to\infty}} P\{A_{n,j}(\infty) > \varepsilon\} = 0.$$

and on the set $]\!]0, T_{n,j}]\!]$ for some $\varepsilon_0 > 0$

$$2\langle d\alpha_s^{h,(n)}, X^n_{s-}\rangle + 2(X^n_{s-}, \int_{1<|x|_H\leq K} x\Pi^{(n)}(ds, dx))_H + d\langle X^c_n\rangle_s + \int_{|x|_H\leq K} |x|^2_H \Pi^{(n)}(ds, dx) +$$

$$+ \varepsilon_0 |X^n_{s-}|^2_V dN_s \leq |X^n_{s-}|^2_H d\lambda_{n,j}(s) + d A_{n,j}(s);$$

3) for each $T_n \in \mathcal{T}(\mathbb{F}_n)$, $T_n \leq T$, $T > 0$, $\delta_n \downarrow 0$, $\varepsilon > 0$ and some orthonormal basis $\{e_j, j \geq 1\}$ of H, consisting of the elements from V,

$$\lim_{n\to\infty} P\{|\alpha^{h,(n)}_{T_n+\delta_n} - \alpha^{h,(n)}_{T_n}|_{V'} + \sum_{j=1}^{\infty} |\Gamma^{h,(n)}_{l_j}(T_n + \delta_n) - \Gamma^{h,(n)}_{e_j}(T_n)| +$$

$$+ \Pi^{(n)}((T_n, T_n + \delta_n) \times \{x : |x|_H > 1\}) > \varepsilon\} = 0.$$

Then $\{\mu_n^{\mathcal{X}}, n \geq 1\}$ is tight in $M_c(\mathcal{X})$.

Finally we shall consider the case when $\mathcal{X} = D_{[0,\infty)}(H_w)$ with the Skorokhod $\mathcal{J}_1$ – topology. Let $a : \Omega \times [0,\infty) \times V \to V'$, $B : \Omega \times [0,\infty) \times V \to L_+(H)$ be $\mathcal{F} \otimes \mathcal{B}([0,\infty) \times V)$ – measurable functions, where $L_+(H)$ denotes the space of positive symmetric linear mappings of H to H with the nuclear norm $|\cdot|_1$, $\tilde{V} \subset V$ be a dense subspace of V,

$\varphi_t^v(x)=i\langle a_t(x), v\rangle-\frac{1}{2}\langle v, B_t(x)v\rangle_H$, $v\in\tilde{V}$,

$x\in V$, N be a continuous $\mathcal{F}$ – measurable increasing process, ν be a probability measure on $\mathcal{F}\otimes\mathcal{B}(H_w)$. Define $\mathcal{S}(\nu, a, B, N)=\{\bar{P}: \bar{P}\in M_{mc}(\tilde{\Omega}, P)$, $X_t\in V$ $dN\,d\bar{P}$ – a. e., $\int_0^t|\varphi_s^v(X_s)|dN_s<\infty$ $\bar{P}$ – a. e. $t\geq0$, $v\in\tilde{V}$, and $e^{i\langle X_\cdot, v\rangle}-\int_0^{\cdot} e^{i\langle X_s, v\rangle}\varphi_s^v(X_s)dN_s\in\mathcal{M}_{loc}^c(\bar{P}, \tilde{\mathbb{F}})$, $\varepsilon_{X_0}d\bar{P}=d\nu\}$.

Let $\{X^n, n\geq1\}$ be a sequence of V' – valued $(P, \mathbb{F}^n)$ – semimartingales such that $\varepsilon_{X_0^n}dP=d\nu_n$ and for each $v\in\tilde{V}$ $e^{i\langle X_\cdot^n, v\rangle}-\Phi^{vn}\in\mathcal{M}_{loc}(P, \mathbb{F}^n)$, where Φ^{vn} is a process of locally finite variation. Denote V_{wH} the space V with the supremum topology of the weak topology and the topology induced by the strong topology of H.

Theorem 5 (Mikulevičius (1985 b)). Let the imbedding $V\to H$ be compact. Assume that:

1) $\nu^n\to\nu$ as $n\to\infty$;

2) the mappings $(t, x)\to\varphi_t^v(\omega, x)$ are continuous on $[0, \infty)\times V_{wH}$ for P – a. e. ω and each $v\in\tilde{V}$;

3) for each $T>0$, $v\in\tilde{V}$ and some $\mathcal{F}$ – measurable processes N^n the following is true:

a) there exists $(S_n, n\geq1)\in\Sigma(T)$ such that

$$\sup_{n\geq1} E[\sup_{t\leq S_n}|X_t^n|_H^2+\int_0^{S_n}|X_s^n|_V^2dN_s^n]<\infty$$

and for each $T_n\leq S_n$, $T_n\in\mathcal{T}(\mathbb{F}^n)$ $\{\Delta\Phi_{T_n}^{vn}, n\geq1\}$ is uniformly integrable;

b) N_T is bounded, for each $\varepsilon>0$

$$\lim_{n\to\infty} P\{|N_T^n-N_T|+\sup_{t\leq T}|\Phi_t^{vn}-\int_0^t e^{i\langle X_s^n, v\rangle}\varphi_s^v(X_s^n)dN_s^n|>\varepsilon\}=0$$

c) $|a_t(x)|_{V'}\leq K(1+|x|_V)$, $|B_t(x)|_1^{1+\delta}\leq K(1+|x|_V^2)$ for each $t\leq T$ and some $\delta>0$, $K>0$.

Then the sequence $\{\mu_n, n\geq1\}$ is relative compact in $M_{mc}(\tilde{\Omega}, P)$, all limiting points $\bar{P}$ are contained in $\mathcal{S}(\nu, a, B, N)$ and $\int_0^t|X_s|_V^2dN_s<\infty$ $\bar{P}$ – a. e. for each $t>0$.

P r o o f. We can assume that $\sup_{n\geq1} N_\infty^n+N_\infty$ is bounded and for each $t\in[0, \infty]$ $N_t^n\to N_t$ as $n\to\infty$ in probability. From the assumption 3 a) it follows that for each $t\geq0$

$$\lim_{l\to\infty}\sup_{n\geq1} P\{\sup_{s\leq t}|X_s^n|_H^2+\int_0^t|X_s^n|_V^2dN_s^n>l\}=0. \tag{2}$$

Let $v\in\tilde{V}$, $T>1$ be fixed,

$$\tilde{\Phi}_t^{v,n}=\int_0^t e^{i\langle X_s^n, v\rangle}\varphi_s^v(X_s^n)dN_s^n,$$

$$T_n=\inf\{t: |\Phi_t^{v,n}-\tilde{\Phi}_t^{vn}|>1\}.$$

$$\hat{\Phi}_t^{v,n}=\Phi_{t\wedge T_n\wedge S_n}^{vn}, \quad \check{\Phi}_t^{vn}=\tilde{\Phi}_{t\wedge T_n\wedge S_n}^{vn},$$

where $(S_n, n\geq1)\in\Sigma(T)$ and satisfies 3 a).

According to the assumption 3) the sequence

$\{\sup_{t\geq0}(|\hat{\Phi}_t^{vn}|+|\check{\Phi}_t^{vn}|, n\geq1\}$ is uniformly integrable.

Let $U_n\in\mathcal{T}(\mathbb{F}^n)$, $\delta_n\downarrow0$, $U_n\leq T-1$,

$$|E\exp\{i\langle X_{(U_n+\delta_n)\wedge T_n\wedge S_n}^n-X_{U_n\wedge T_n\wedge S_n}^n, v\rangle-1|=$$

$$=|E\exp\{-i<X^n_{U_n}\wedge T_n\wedge S_n, v>[\hat{\Phi}^{vn}_{U_n+\delta_n}-\hat{\Phi}^{vn}_{U_n}]|=c_n.$$

From the assumption 3) it follows that $c_n\to 0$ as $n\to\infty$, i. e. for each $v\in\tilde{V}$

$$\lim_{n\to\infty} E[\exp\{i<X^n_{U_n+\delta_n}-X^n_{U_n}, v>-1]=0$$

and thus for each $v\in\tilde{V}$

$$<X^n_{U_n+\delta_n}-X^n_{U_n}, v>\to 0 \text{ as } n\to\infty \tag{3}$$

in probability.

Therefore $\{\mu_n, n\geq 1\}$ is relative compact in $M_{mc}(\tilde{\Omega}, P)$ (see Mikulevičius (1985 a)).

Let $\bar{P}$ be a limiting point, $\hat{\Omega}=\tilde{\Omega}\times R_+\times V_{wH}$, $\hat{\mathfrak{F}}=\tilde{\mathfrak{F}}\otimes\mathfrak{B}(R_+\times V_{wH})$, $\hat{\hat{\Omega}}=\tilde{\Omega}\times R_+\times H$, $\hat{\hat{\mathfrak{F}}}=\tilde{\mathfrak{F}}\otimes\mathfrak{B}(R_+\times H)$. Denote $M^+_{mc}(\hat{\Omega},\hat{\mathfrak{F}}, P)$ $(M^+_{mc}(\hat{\hat{\Omega}},\hat{\hat{\mathfrak{F}}}, P))$ the space of all finite measures on $(\hat{\Omega},\hat{\mathfrak{F}})$ $(\hat{\hat{\Omega}},\hat{\hat{\mathfrak{F}}})$, the Ω – marginals of which are absolute continuous with respect to the measure P, endowed with the topology generated by mappings $\mu\to\mu(f)$ for $\hat{\mathfrak{F}}$ $(\hat{\hat{\mathfrak{F}}})$ – measurable bounded functions such that $f(\omega,\cdot)$ are continuous for each $\omega\in\Omega$. Define the measures $\{\hat{\mu}_n, n\geq 1\}$, $\hat{\mu}$ on $\hat{\mathfrak{F}}$ by means of the equalities:

$$\hat{\mu}_n(f)=E[\int_0^\infty f(X^n, s, X^n_s)dN^n_s], \quad f\in\hat{\mathfrak{F}}^+,$$

$$\hat{\mu}(f)=\bar{P}[\int_0^\infty f(X, s, X_s)dN_s], \quad f\in\hat{\mathfrak{F}}^+.$$

From the assumption 3 a), the relation (3), the boundedness of $\sup_{n\geq 1} N^n_\infty$ and the compactness of the imbedding $V\to H$ it follows that the sequence $\{\hat{\mu}_n, n\geq 1\}$ is relative compact in $M^+_{mc}(\hat{\Omega},\hat{\mathfrak{F}}, P)$ (see Mikulevičius (1985 a)). Then it is easy to check that $\hat{\mu}$ is the limiting point of $\{\hat{\mu}_n, n\geq 1\}$ and because $x\to|x|_V=\sup_{\substack{|y|_{V'}\leq 1\\ y\in V'}}|<y, x>|$ is semicontinuous from below in H_w, we have that $\bar{P}[\int_0^\infty |X_s|^2_V dN_s]<\infty$.

Define the measures $\{\hat{\hat{\mu}}_n, n\geq 1\}$, $\hat{\hat{\mu}}$ on $\hat{\hat{\mathfrak{F}}}$ by means of the equalities:

$$\hat{\hat{\mu}}_n(f)=E[\int_0^\infty f(X^n, s, X^n_s)dN^n_s], \quad f\in\hat{\hat{\mathfrak{F}}}^+,$$

$$\hat{\hat{\mu}}(f)=\bar{P}[\int_0^\infty f(X, s, X_s)dN_s], \quad f\in\hat{\hat{\mathfrak{F}}}^+.$$

From the above assumptions it follows again that the sequence $\{\hat{\hat{\mu}}_n, n\geq 1\}$ is relative compact in $M^+_{mc}(\hat{\hat{\Omega}},\hat{\hat{\mathfrak{F}}}, P)$. It is easy to check that $\hat{\hat{\mu}}$ is the limiting point of $\{\hat{\hat{\mu}}_n, n\geq 1\}$. Let now $s\leq t\leq T$, f be $\mathfrak{F}\otimes\mathfrak{D}_s$ – measurable, bounded and $\bar{P}$ – a. e. continuous in x. Then from the above statements there exists a subsequence $\{n_k, k\geq 1\}$ such that as $n\to\infty$

$$0=E\{[\exp\{i<X^n_{T_n}\wedge S_n\wedge t, v>\}-\exp\{i<X^n_{T_n}\wedge S_n\wedge s, v>-(\hat{\Phi}^{vn}_t-\hat{\Phi}^{vn}_s)]f(X^n)\}\to\bar{P}\{[e^{i<X_t,v>}-$$

$$-e^{i<X_s,v>}-\int_0^t e^{i<X_u,v>}\varphi^v_u(X_u)dN_u]f(X)\}.$$

Therefore $\bar{P}\in\mathfrak{I}(\nu, a, B, N)$. Theorem 5 is proved.

References

Bouc R., Pardoux E. (1983). Asymptotic analysis of PDEs with wideband driving noise. – Publ. Math. Appl. Marseille–Toulon, 83–4.

Dawson D. A. (1975). Stochastic evolution equations and related measure processes. J. Multivariate Anal. vol. 5, Nr. 1, p. 1–52.

Dawson D. A. (1981). Critical dynamics and fluctuations for a mean field model of cooperative behaviour. Carlton Math. Lecture Note, Nr. 33.

Dawson D. A. (1985). Asymptotic analysis of multilevel stochastic systems. Lecture Notes in Control and Inf. Sc. 69, p. 79–90.

Gallavotti G., Jona-Lasinio G. (1975). Limit theorems for multidimensional markovian processes. Comm. Math. Phys., v. 41, p. 301–307.

Gikhman I. I., Skorokhod A. V. (1982). Stochastic differential eqnations, Naukova dumka, Kiev.

Gyongy I., Krylov N. V. (1982). On stochastic equations with respect to semimartingales. II, Ito formula in Banach spaces. Stochastics. vol. 6, Nr. 3–4, p. 153–173.

Gorostiza L. (1983). High density limit theorems for infinite systems of unscaled branching Brownian motions. Ann. of Probab. vol. 11, Nr. 2, p. 374–392.

Gorostiza L. (1985). Space scaling limit theorems for infinite particle branching brownian motions with immigration. Lecture Notes in Control and Inf. Sc. v. 69, p. 91–99.

Grigelionis B., Mikulevičius R. (1981). On weak convergence of semimartingales. Lietuvos matem. rinkinys. v. 21, Nr. 4, p. 9–24.

Grigelionis B., Mikulevičius R. (1983). Stochastic evolution equations and densities of the conditional distributions. Lecture Note in Control and Inf. Sc. v. 49, p. 49–86.

Grigelionis B., Mikulevičius R. (1985). On the functional limit theorems. Stochastic space-time models and limit theorems. D. Reidel Publ. Co., p. 191–215.

Grigelionis B., Mikulevičius R. (1985). Weak solutions of stochastic evolution equations. Lecture Notes in Control and Inf. Sc. v. 69, p. 43–54.

Hitsuda M. (1983). Central limit theorem for a simple interacting diffusion model and S' – valued processes. Lecture Notes in Math., 1021, p. 238–242, Proc. IV USSR – Japan Symp. Probab. Th. and Math. Stat., Springer.

Holley R. A., Stroock D. W. (1976). A martingale approach to infinite systems of interacting processes. The Annals of Probab. v. 4, Nr. 2, p. 195–208.

Holley R., Stroock D. W. (1978). Generalized Ornstein – Uhlenbeck processes and infinite particle branching brownian motions. RIMS Kyoto Univ. vol. 14, p. 741–788.

Jacod J. (1979). Calcul stochastique et problemes de martingales. Lecture Notes in Math. v. 714, Springer Verlag, Berlin.

Jacod J., Memin J. (1981). Weak and strong solutions of stochastic differential equations: Existence and stability. Lecture Notes in Math. v. 851, Stochastic integrals, Springer, p. 169–212.

Joffe A., Metivier M. (1986). Weak convergence of sequences of semimartingales with applications to multitype branching processes. Adv. in Appl. Probab.

Kotelenez P. (1982). Law of large number and central limit theorem for chemical reactions with diffusion. Dr. dissertation, Univ. Bremen.

Kozlov S. M. (1978). Some questions of stochastic partial differential equations. Proc. Petrovskii Sem. vol. 4, p. 147–172.

Krylov N. V., Rozovskii B. L. (1979). On stochastic evolution equations. Modern Problems in Math. v. 14, VINITI, M., p. 71–146.

Lavrent'ev N. V. (1983). On weak convergence Hilbert-space valued semimartingales to stochasticaly continuous processes with conditionaly independent increments. Uspehi mat. nauk. v. 38, Nr. 3, p. 183–184.

Martin-Lof A. (1976). Limit theorems for motion of a Poisson system of independent markovian particles with high density. Z. Wahr. verw. Geb. B. 34, p. 205–223.

Metivier M. (1982). Semimartingales, W. de Gruyter, Berlin – New York.

Metivier M. (1984). Convergence faible et principe d'invariance pour des martingales a valeurs dans des espaces de Sobolev. Ann. Inst. H. Poincare. v. 20, Nr. 4, p. 329–348.

Mikulevičius R. (1985). On weak convergence of measures. Lietuvos matem. rinkinys. XXV, Nr. 1.

Mikulevičius R. (1985). On the convergence to the solution of martingale problem in the rigged Hilbert space. Abstracts of comm. of fourh intern. Vilnius conference on probab. theory and math. stat., p. 177–178.

Morrow G. J. (1983). A weak invariance principle for Hilbert space valued martingales. III. J. Math. v. 27, Nr. 4, p. 659–669.

Papanicolaou G., Stroock D., Varadhan S. R. S. (1977). Martingale approach to some limit theorems. In: Duke Turbulence Conference. Duke University.

Shiga T., Tanaka H. (1985). Central limit theorem for a system of markovian particles with mean field interactions. Z. Wahr. verw. Geb. B. 69, H. 3, p. 439–459.

Stroock D. W., Varadhan S. R. S. (1979). Multidimensional diffusion processes. Springer-Verlag, Berlin.

Tanaka H., Hitsuda M. (1981). Central limit theorem for a simple diffusion model of interacting particles. Hiroshima Math. J. v. 11, p. 415–423.

Viot M. (1976). Solutions faibles d'equations aux derivees partielles stochastiques non lineaires. These doct. sci. Univ. P. et. M. Curie, Paris.

Vishik M. I., Komech A. I., Fursikov A. V. (1979). Mathematical problems of hydromechanics. Uspechi Mat. nauk. v. 34, Nr. 5, p. 135–210.

Prob. Theory and Math. Stat., Vol.1, pp. 523-528
Prohorov et al. (eds)
1986 VNU Science Press

OPTIMAL STOPPING OF A MARKOV CHAIN WITH VECTOR-VALUED GAIN FUNCTION

Ulrich S. Gugerli

c/o SANDOZ LTD. Biostatistics Clinical Research
CH-4002 Basel Switzerland [1]

1. Introduction

Let $(x_k, \underline{\underline{F}}_k, P_x)_{k\in N, x\in E}$ be a (homogeneous, non-terminating) Markov chain with countable state space E, and $g: E \longrightarrow R^{n+1}$, $x \longmapsto g(x)=(g_0(x), g_1(x), \ldots, g_n(x))$, a vector-valued gain function satisfying

(1) $E_x(\sup_{j\geq 0} |g_i(x_j)|) < \infty$, $i=0,1,\ldots,n$.

Furthermore, let $E_x g(x_t)$ denote the vector expectation $(E_x g_0(x_t), E_x g_1(x_t), \ldots, E_x g_n(x_t))$ and T the set of all $(\underline{\underline{F}}_k)$- stopping times. A stopping time is understood to be P_x-a.s. finite for every $x\in E$.

The optimal stopping problem considered is to maximize the expected gain $E_x g(x_t)$ over $t\in T$, in what sense, however, has yet to be made precise. If n=0, there is no ambiguity. The solution of this scalar optimal stopping problem is well known (see Shiryayev[1]): the value function $s(x) := \sup_{t\in T} E_x g(x_t)$ is the smallest (P_x)-superharmonic function dominating g and satisfies Bellman's equation $s(x) = g(x) \vee E_x s(x_1)$, $x\in E$. The time t^ε of first entry into the set $\{x\in E \mid s(x)\leq g(x)+\varepsilon\}$, $\varepsilon\geq 0$, is an ε-optimal stopping time for any $\varepsilon>0$; if the state space E is finite, t^0 is an optimal stopping time.

[1] the author's former address was:
Department of Electrical Engineering Imperial College
London SW7 2BT UK

In general ($n>0$), however, the optimal stopping problem involves a vector maximization problem. A suitable optimality concept has, therefore, to be agreed upon. Let T^* denote the set of all stopping times randomized out of finitely many $(\underline{\underline{F}}_k)$-stopping times (see Siegmund[2] for randomization of stopping times). The set $O(x)$ of all optima with respect to the initial state $x \in E$ is, by definition, the set of all elements of the closed set $\overline{\{E_x g(x_t) \mid t \in T^*\}}$ which are maximal with regard to the coordinatewise partial ordering on R^{n+1} ($u \leqq v$ if and only if $u_i \leqq v_i$, $i=0,1,\ldots n$). The optimality concept thus generated is generally known as Pareto-optimality (Pareto[3]).

The objective of this paper is to characterize the sets $O(x)$ as well as corresponding ε-optimal stopping times.

2. Countable State Space

As is usual with vector maximization problems we consider weights $p=(\pi_1,\pi_2,\ldots,\pi_n)$ belonging to the (closed) canonical n-simplex $\Delta_n=\mathrm{conv}(e_0,e_1,\ldots,e_n) \subset R^n$, where e_0 is the null-vector and $(e_i)_{i=1}^n$ the canonical basis in R^n. Let $g(.|.): \Delta_n \times E \longrightarrow R$ be defined by

$$g(p|x) := (1-\sum_{i=1}^{n} \pi_i) g_0(x) + \sum_{j=1}^{n} \pi_j g_j(x). \text{ Then,}$$

$(x_k,\underline{\underline{F}}_k,P_x,g(p|x))_{k \in N, x \in E}$ constitutes a scalar optimal stopping problem the value function of which is denoted by $s(p|x)$. Set

$$S_p^{\varepsilon} := \{x \in E \mid s(p|x) \leqq g(p|x)+\varepsilon\}, \quad t_p^{\varepsilon} := \mathrm{INF}\{k \geqq 0 \mid x_k \in S_p^{\varepsilon}\}, \quad \varepsilon \geqq 0.$$

Making use of condition (1) it is not very hard to verify that for any fixed initial state $x \in E$ the function $s(.|x): \Delta_n \longrightarrow R$ is convex and continuous. Let $\partial s(p|x)$ denote the set of all subgradients of $s(.|x)$ at p and $\langle .,. \rangle$ the usual inner product on R^n. For any $x \in E$ and any $p \in \Delta_n$ the set $\partial s(p|x)$ is non-empty, convex and closed. Let $s'(p;q|x)$ denote the directional derivative of $s(.|x)$ at p in the direction q, $s'(p;q;r|x)$ the directional derivative of $s'(p;.|x)$ at q in the direction r, and so on.

Lemma: Suppose $x \in E$ and $p \in \Delta_n$. If p^* is an extreme point of $\partial s(p|x)$, the vector $u(p,p^*|x) := (s(p|x)+\langle p^*,e_i-p\rangle)_{i=0}^{n}$ belongs to $O(x)$. Moreover, there is a parameter $q \in \mathring{\Delta}_n$ such that t_q^{δ} is an ε-optimal stopping time in the sense that $\|u(p,p^*|x)-E_x g(x_{t_q}\delta)\| < \varepsilon$ for $\delta>0$ small enough.

We call $u(p,p^*|x)$ an 'R-value'. We have to confine ourselves to a rough outline of the proof. First, it has to be shown that $u(p,p^*|x)$ is ε-attainable. This is quite easy, if p belongs to the interior of Δ_n and $s(.|x)$ is differentiable at p. In the general case, using the machinery of convex analysis (see Rockafellar[4]) one can establish the existence of n parameters $p_i \in \Delta_n$, $i=1,2,\ldots,n$, such that, with $p_0:=p$, the following two conditions are satisfied:

(2) $(p_i)_{i=0}^{n}$ is affinely independent

(3) $\langle p^*,p_i-p_{i-1}\rangle = s'(p_0;p_1-p_0;\ldots;p_i-p_{i-1}|x)$, $i=1,2,\ldots,n$

The hypothesis that p^* is an extreme point of $\partial s(p|x)$ is crucial here. More convex analysis yields the existence of a parameter q such that q belongs to the interior of Δ_n, $s(.|x)$ is differentiable at q and $|(s(p_0|x)+\langle p^*,p_j\rangle) - (s(q|x)+\langle \mathrm{grad}\, s(q|x),p_j-q\rangle)| < \varepsilon$, for $j=0,1,\ldots,n$. For sufficiently small $\delta>0$, $E_x g(p_j|x_{t_q}\delta)$ belongs to ε-neighbourhood of the second term of the difference, $j=0,1,\ldots,n$. Note that t_q^{δ} belongs to T, i.e. is a non-randomized stopping time. An affine transformation now yields the desired result.

Suppose now that $v \in R^{n+1}$ is ε-attainable and such that $u(p,p^*|x) \leqq v$. In order to show that $u(p,p^*|x)$ is not improvable it suffices to show that the hyperplane in $R^n \times R$ passing through the points (e_i,v_i), $i=0,1,\ldots,n$, coincides with the hyperplane determined by the point $(p,s(p|x))$ and the linear form p^*.

As we can see from this sketch of the proof $u(p,p^*|x)$ is the result of the following stepwise optimization procedure with accumulating constraints: first, maximize $E_x g(p_0|x_t)$ over $t \in T$ without any constraints; second, maximize $E_x g(p_1|x_t)$ while retaining what has been achieved in the previous step, and so on. The procedure terminates after n+1 steps, ranking the parameters $(p_0, p_1, \ldots, p_n)$ according to their priority, hence the label 'R-value'.

Let $Z(\partial s(p|x))$, $p \in \Delta_n$, be the set of all subgradients $p^* \in \partial s(p|x)$ which satisfy the following condition:

(4) apart from p^*, there is no subgradient $q^* \in \partial s(p|x)$ such that $p^* - q^*$ belongs to the recession cone of $\partial s(p|x)$.

It is easy to see that $Z(\partial s(p|x))$ contains the set of all extreme points of $\partial s(p|x)$ and is itself contained in the convex hull of the set of all extreme points of $\partial s(p|x)$.

Theorem 1: $O(x) = \{u(p,p^*|x) \mid p \in \Delta_n,\ p^* \in Z(\partial s(p|x))\}$, $x \in E$

It follows from this characterization of $O(x)$ that R-values are 'extreme' optima in the sense that any optimum is a convex linear combination of not more than n+1 R-values. Corresponding ε-optimal stopping times may be obtained by randomizing the ε-optimal stopping times associated with the above R-values, the weights of the randomization coinciding with the weights of the convex linear combination.

3. Finite State Space

For reasons of brevity we only consider the case n=1, i.e. the case of a bivariate gain function. If the state space E is finite, optimal stopping times always exist, thus rendering redundant the notion of ε-optimality; the superscript in S_p^0 will be abandoned. The equivalence relation $p \sim q$ if and only if $S_p = S_q$ defined on $\Delta_1 \times \Delta_1$ generates a finite partition $0 =: q_0 < q_1 < \ldots < q_m < q_{m+1} := 1$ of the unit interval $\Delta_1 = [0,1]$.

Theorem 2: Let

$\hat{A} := A \cup \{x \in E \mid E_x s(0|x_1) < s(0|x)\}$, where A is the set of all absorbing states

$(x_k, \underline{F}_k, \hat{P}_x)_{k \in N, x \in E}$ the modified Markov chain with $\hat{A}$ made absorbing

$$\hat{g}_1(x) := \begin{cases} g_1(x), \text{ if } x \in S_0 \\ (\underset{y \in E}{\text{MIN}}\ g_1(y))-1, \text{ if } x \in E \setminus S_0 \end{cases}$$

$\hat{s}(1|x)$ the smallest $(\hat{P}_x)$-superharmonic function dominating $\hat{g}_1$

$S_{0+} := \{x \in E \mid \hat{s}(1|x) = \hat{g}_1(x)\}$,

$t_{0+} := \text{INF}\ \{k \geq 0 \mid x_k \in S_{0+}\}$, $\tau_{0+} := \text{INF}\ \{k > 0 \mid x_k \in S_{0+}\}$,

$d_i(x) := E_x g_i(x_{\tau_{0+}}) - g_i(x)$, $i=0,1$,

$$r(x) := \begin{cases} d_0(x)/[d_0(x)-d_1(x)], \text{ if } d_0(x)d_1(x)<0 , \\ 1 , \text{ otherwise} \end{cases}$$

If the state space E is finite, then:

(5) (a) $E_x g_0(x_{t_{0+}}) = s(0|x)$ and

(b) $E_x g_1(x_{t_{0+}}) = \text{MAXIMUM}$ under the constraint (a)

(6) $S_p = S_{0+}$ for all $p \in]0,q_1[$

(7) $q_1 = \underset{x \in E}{\text{MIN}}\ r(x)$

Solving the two scalar optimal stopping problems $(x_k, \underline{\underline{F}}_k, P_x, g_0(x))$ and $(x_k, \underline{\underline{F}}_k, \hat{P}_x, \hat{g}_1(x))$ yields the R-value with maximal first component, according to (5). Parameters $p \in \Delta_1$ do not enter these calculations. Assertion (6) then shows that for $p \in]0,q_1[$ the optimal stopping problems $(x_k, \underline{\underline{F}}_k, P_x, g(p|x))$ are all equivalent to the optimal stopping problem $(x_k, \underline{\underline{F}}_k, \hat{P}_x, \hat{g}_1(x))$.

From the Markov property and from the properties of the modified Markov chain we have:

(8) $E_x g_0(x_{\tau_{0+}}) = E_x s(0|x_1)$, $E_x g_1(x_{\tau_{0+}}) = E_x \hat{s}(1|x_1)$, $x \in E$

Using (7) we can now compute the first partition point q_1. Rescaling and iteration yield the complete partition $q_0 < q_1 < \ldots < q_m < q_{m+1}$ and thus all R-values, whence, by Theorem1, we get, for each $x\in E$, the set of optima $O(x)$ as well as corresponding optimal stopping times.

Moreover, we have the following

Corollary:

(9) $\hat{A} \subset S_{0+} \subset S_0$, $t_{0+} \geq t_0$ everywhere

(10) $S_{q_1} = S_{0+} \cup \{x\in E \mid r(x)=q_1\}$

A straight-forward generalization of the preceding theorem to arbitrary finite dimension n+1 is contained in Gugerli[5] as are the proofs of Theorems 1 and 2.

References:

[1] Shiryayev A.N. (1978) Optimal Stopping Rules Springer-Verlag NewYork Heidelberg Berlin

[2] Siegmund D. (1967) Some problems in the theory of optimal stopping rules Ann.Math.Stat. 38(6)

[3] Pareto V. (1927) Manuel d'Economie Politique Marcel Giard Paris

[4] Rockafellar R.T. (1972) Convex Analysis Princeton University Press Princeton

[5] Gugerli U.S. (1983) Optimales Stoppen von Markovketten mit vektorwertiger Auszahlungsfunktion (unpublished thesis) Diss. ETH 7426 ETHZ CH-8092 Zürich Switzerland

Prob. Theory and Math. Stat., Vol.1, pp. 529-534
Prohorov et al. (eds)
1986 VNU Science Press

SOME STRONG LAWS OF LARGE NUMBERS IN BANACH SPACES WITH REGULAR NORMS.

Bernard HEINKEL
Département de Mathématique
7, rue René Descartes
67084 Strasbourg - Cédex (France)

1. INTRODUCTION.

Let $(B, \| \|)$ be a real separable Banach space equipped with its Borel σ-field $\mathcal{B}$. A sequence (X_n) of independent $(B, \mathcal{B})$ valued random variables (r.v.) is said to satisfy the weak law of large numbers ($(X_n) \in$ WLLN) if and only if :

$(1/n) \|S_n\| = (1/n) \|X_1 + \ldots + X_n\| \to 0$ in probability ;

it is said to satisfy the strong law of large numbers ($(X_n) \in$ SLLN) if and only if :

$(1/n) \|S_n\| \to 0$ almost surely.

The problem of finding sufficient conditions for the SLLN (of course in the non i.i.d. setting) is one of the most famous of the theory of probability in Banach spaces. Many situations - mimicked from the scalar case - have been considered. Here we will be concerned with the following one : the norms of the r.v. X_n being bounded in the same manner as in Prohorov's (1959) scalar SLLN, how to extend this result to B ?

The first general extension of Prohorov's theorem is due to Kuelbs and Zinn (1979), and is as follows :

<u>Theorem 1.</u> Let (X_n) be a sequence of centered, independent r.v. with values in a real separable Banach space $(B, \| \|)$ such that :

(1) $\exists K > 0, \forall n \in \mathbb{N}, \quad \|X_n\| \le K(n/L_2 n)$ a.s. ,

where : $L_2 x = \text{Log Log} (\sup(e, x))$.

For every integer n, we denote by $\Gamma(n)$:

$\Gamma(n) = 2^{-2n} \sum_{j \in I(n)} E\|X_j\|^2$, where $I(n) = (2^n+1, \ldots, 2^{n+1})$;

we suppose that the following holds :

(2) $\forall\, \varepsilon > 0 \quad \sum_{n \geq 1} \exp(-\varepsilon / \Gamma(n)) < +\infty$.

Then : $(X_n) \in$ WLLN $\Leftrightarrow$ $(X_n) \in$ SLLN .

One remarks that on the one hand Theorem 1 has the disadvantage to make only norm assumptions on the r.v. X_n , but on the other hand it has the advantage to apply in every separable Banach space ; in that sense it is a good extension of Prohorov's theorem. When Goodman et al. (1981) solved the problem of the law of the iterated logarithm (LIL) in Hilbert spaces and Ledoux (1983) did the same in 2-uniformly smooth spaces, Theorem 1 began to appear too weak in that nice spaces, because it was somewhat shocking to have a knowledge of the LIL - which is a limiting form of Prohorov's SLLN - which was better than the one of the SLLN itself. This observation led me to give, in 2-uniformly smooth spaces, an extension of Prohorov's Theorem which is better than Theorem 1.

Before to state that result, I will recall the definition of a p-uniformly smooth Banach space.

Definition. A real separable Banach space $(B, \|\ \|)$ is p-uniformly smooth $(1 < p \leq 2)$ if its modulus of smoothness ρ :

$\forall\, t > 0$, $\rho(t) = \sup\left((1/2)(\|x+ty\| + \|x-ty\| - 2) , \|x\| = \|y\| = 1 \right)$

satisfies : $\rho(t) \leq Ct^p$, C being a positive constant.

The norm of such a Banach space is Fréchet differentiable, with derivative D, away from 0.

Moreover, if one defines $F : B \to B'$ by setting :

$\forall\, x \neq 0 \qquad F(x) = \|x\|^{p-1} D(x/\|x\|)$

and $\qquad F(0) = 0$,

one notices that the following properties hold (see Ledoux (1985)) :

(i) $\forall\, x \in B \quad \|F(x)\|_{B'} = \|x\|^{p-1}$;

(ii) $\exists\, C(p) > 0 : \forall\, (x, y) \in B^2, \ \left| \|x+y\|^p - \|x\|^p \right| \leq p\,|F(x)(y)| + C(p)\|y\|^p$.

2. PROHOROV'S SLLN IN P-UNIFORMLY SMOOTH SPACES.

Theorem 2. (Heinkel(1984a). Let (X_n) be a sequence of centered, independent r.v. with values in a real, separable, 2-uniformly smooth Banach space $(B, \| \|)$; suppose that the r.v. X_n satisfy Prohorov's boundedness assumption (1).
For every integer n one denotes by :

$$\Lambda(n) = 2^{-2n} \sum_{j \in I(n)} \sup \left(Ef^2(X_j), \quad \|f\|_{B'} \leq 1 \right)$$

and one supposes that :

$$(3) \qquad \forall \varepsilon > 0 \quad \sum_{n \geq 1} \exp(-\varepsilon / \Lambda(n)) < +\infty .$$

Then the following implication holds :

$$(4) \quad n^{-2} \sum_{1 \leq k \leq n} E\|X_k\|^2 \to 0 \quad \Rightarrow \quad (X_n) \in \text{SLLN}.$$

In fact condition (4) can be weakened in (4)' :

$$(4)' \quad n^{-2} \sum_{1 \leq k \leq n} \|X_k\|^2 \to 0 \ \text{in probability} \quad \text{(see Heinkel (1984b)).}$$

Notice that in a Hilbert space condition (4) is necessary for the WLLN in the Prohorov setting ; this condition (4) appears as a cotype 2 hypothesis and this shows the limit of application of Theorem 2. This hypothesis (4) is not surprising because the proof of Theorem 2 is directly based on property (ii) which says - in some sense - that B is close to a Hilbert space. Besides property (ii) the other ingredients of the proof of Theorem 2 are Yurinskii's (1974) inequality and an iterative use of exponential inequalities for real valued martingales.

A sequence (X_n) to which Theorem 2 applies but Theorem 1 doesn't is easy to construct in the space $B = l^2$. Consider an independent triangular array of independent real valued r.v. $(\lambda_j^n, j = 1, \ldots, [n^{1/6}])$, the notation [ ] denoting the integer part of a number. For each fixed n, the λ_j^n have all the same distribution : they are Cauchy r.v. truncated (symmetrically) at the level $n^{5/6} / L_3 n$ (where $L_3 x = \text{Log } L_2 x$) . Let's denote by (e_n) the canonical basis of l^2. The sequence :

$$X_n = \sum_{1 \le j \le [n^{1/6}]} \lambda_j^n e_j$$

fulfils the assumptions of Theorem 2, but not the ones of Theorem 1.

The state of knowledge of Prohorov's SLLN appeared relatively satisfactory till Ledoux (1985) solved the LIL problem in p-uniformly smooth spaces ($p \neq 2$). Then the problem of finding a better version of Theorem 1 came up. Fortunately it is also possible to prove a better Prohorov's SLLN in that case :

Theorem 3. (Heinkel 1985). Let (X_n) be a sequence of centered, independent r.v. with values in a real separable p-uniformly smooth ($1 < p \le 2$) Banach space $(B, \| \|)$, such that (1) holds. Let's suppose that the following conditions are satisfied :

a) $(X_n) \in$ WLLN .

b) $2^{-2np} \operatorname{Log} n \sum_{k \in I(n)} \|X_k\|^{2p} \to 0$ in probability .

c) $\forall\ \varepsilon > 0$, $\sum_{n \ge 1} \exp(-\varepsilon / \Lambda(n)) < +\infty$.

Then $(X_n) \in$ SLLN .

Remarks. 1) Other kinds of asymptotic behaviours can be considered for the sequence (S_n/n), for instance :

- $\sup \|S_n/n\| < +\infty$ a. s. ;

- for almost all ω the sequence $(S_n(\omega)/n)$ converges weakly to 0.

By using the same method as for the proof of Theorem 3 - and also some amarts techniques - sufficient conditions for such asymptotic behaviours can be obtained (see Heinkel (1985) for details).

2) It is worth to notice that Theorem 3 improves not only Theorem 1 but even Theorem 2. This is due to the technique of proof which is much less hilbertian ! The question of a possible improvement of Theorem 3 under an additional cotype assumption on B raises naturally. A very partial answer to this question is given by the following result :

Corollary. (Heinkel (1985)). Let $3/2 \le p \le 2$ and $2 \le q \le 2p - 1$. Consider (X_n) a sequence of independent, centered, r.v. with values in a p-uniformly smooth Banach space $(B, \|\ \|)$, of cotype q. Let's suppose that (1) holds and also :

$$\forall \varepsilon > 0, \quad \sum_{n \ge 1} \exp(-\varepsilon / \Lambda(n)) < +\infty .$$

Then : $(X_n) \in$ WLLN $\Leftrightarrow$ $(X_n) \in$ SLLN .

3. APPENDIX. A KOLMOGOROV-BRUNK SLLN IN P-UNIFORMLY SMOOTH SPACES.

It is natural to ask if the SLLN of Kolmogorov-Brunk type known previously in general Banach spaces can also be improved in p-uniformly smooth spaces. Of course this is possible. We will only state one result of that kind ; more can be found in Heinkel (1984c) and Heinkel (1985) :

Theorem 4. (Heinkel (1985)). Let (X_n) be a sequence of centered independent r.v. with values in a real separable p-uniformly smooth $(1 < p \le 2)$ Banach space $(B, \|\ \|)$ and such that :

a) $\sum_{n \ge 1} n^{-2p} E\|X_n\|^{2p} < +\infty$.

b) $\exists j \ge 1 : \sum_{n \ge 1} (\Lambda(n))^j < +\infty$.

Then : $(X_n) \in$ WLLN $\Leftrightarrow$ $(X_n) \in$ SLLN .

REFERENCES.

Goodman, V., Kuelbs, J., Zinn, J. (1981). Some results on the LIL in Banach spaces with applications to weighted empirical processes. Ann. Probability 9, 713-752.

Heinkel, B. (1984a). Une extension de la loi des grands nombres de Prohorov. Z. Wahrscheinlichkeitstheorie verw. Gebiete 67, 349-362.

Heinkel, B. (1984b). The non i.i.d. strong law of large numbers in 2-uniformly smooth Banach spaces. In : Probability Theory on Vector Spaces III - Lublin 1983 - . Lecture Notes in Math 1080. Szynal, D., and Weron, A. (Ed.). Springer, Berlin-Heidelberg-

New York, pp. 90-118.

Heinkel, B. (1984c). On Brunk's law of large numbers in some type 2 spaces. To appear in "Probability in Banach spaces 5" - Medford 1984 -. Lecture Notes in Math.

Heinkel, B. (1985). An application of a martingale inequality of Dubins and Freedman to the law of large numbers in Banach spaces. To appear in the proceedings of "Journées SMF de Calcul des Probabilités dans les espaces de Banach" - Strasbourg 1985 -. Lecture Notes in Math.

Kuelbs, J., and Zinn, J. (1979). Some stability results for vector valued random variables. Ann. Probability 7, 75-84.

Ledoux, M. (1983). Sur les théorèmes limites dans certains espaces de Banach lisses. In : Probability theory in Banach spaces 4 - Oberwolfach 1982 -. Lecture Notes in Math 990. Beck, A., and Jacobs, K. (Ed.). Springer, Berlin-Heidelberg-New York, pp. 150-169.

Ledoux, M. (1985). La loi du logarithme itéré dans les espaces de Banach uniformément convexes. C. R. Acad. Sc. Paris 300, Sér. I, n° 17, 613-616.

Prohorov, Yu. V. (1959). Some remarks on the strong law of large numbers. Theor. Probability Appl. 4, 204-208.

Yurinskii, V.V. (1974). Exponential bounds for large deviations. Theor. Probability Appl. 19, 154-155.

Prob. Theory and Math. Stat., Vol.1, pp. 535-541
Prohorov et al. (eds)
1986 VNU Science Press

OPTIMALITY IN ESTIMATION FOR STOCHASTIC PROCESSES UNDER BOTH FIXED AND LARGE SAMPLE CONDITIONS

C.C. Heyde
Department of Statistics, University of Melbourne, Parkville,
Vic. 3052, Australia

ABSTRACT. Most theory of inference for stochastic processes is based on asymptotics and implies the use of large samples. However, recent consideration has been given by Godambe (1985) to the choice of estimators from estimating equations to have a certain optimality property for fixed samples. In this paper it is shown that the estimation procedure advocated by Godambe also has substantial large sample justification. A discussion of the application of the methodology is also provided.

1. INTRODUCTION

The method of least squares has been of profound influence in Statistics, largely because there are strong justifications for its use based on both fixed sample considerations and on asymptotics. It is the objective of this paper to provide such a dual justification for an estimation procedure proposed by Godambe (1985) in a general discrete time stochastic process framework.

Let $\{X_1, X_2, \ldots, X_n\}$ be a sample drawn from some process whose distribution depends on a single parameter θ taking values in a open interval. Write $\mathcal{F}_k$ for the σ-field generated by $\{X_j, j \leq k\}$ and $E_k(\cdot)$ for the conditional expectation $E(\cdot | \mathcal{F}_k)$.

We shall confine attention to the class $\mathcal{G}$ of square integrable martingale estimating functions $G_n = G_n(X_1, X_2, \ldots, X_n)$. Thus, $G_n = \Sigma_{i=1}^n g_i$, say, with $E_{i-1} g_i = 0$ a.s. Estimators θ^* are found by solving the estimating equation $G_n(\theta^*) = 0$.

It should be noted that least squares, conditional least squares, maximum likelihood, quasi-likelihood and partial likelihood procedures all fit into the estimating equation framework. The use of estimating equations has been quite strongly advocated in the literature (e.g. Godambe (1960), (1985), Godambe and Thompson (1978), (1984)).

2. FIXED SAMPLE APPROACH

Let $L_n(\theta) = L_n(X_1, X_2, \ldots, X_n; \theta)$ denote the likelihood function and suppose that $L_n(\theta)$ is differentiable with respect to θ and $E(d \log L_n(\theta)/d\theta)^2 < \infty$. Suppose in addition that we confine attention to those G_n for which $dG_n/d\theta$ exists and $\int G_n L_n d\mu$ is differentiable under the integral sign, L_n being the density of $X_1, X_2, \ldots, X_n$ with respect to the measure μ.

Write $U_n = d \log L_n(\theta)/d\theta$ for the score function. It is well-known that, under minor regularity conditions, the score function provides a minimal sufficient partitioning of the sample space. In this sense the score function contains all the information in the sample about θ.

If the true distribution and hence the score function is unknown, it may be argued that it is best to choose from $\mathcal{G}$ an estimating function G_n^* which has maximum correlation with U_n. That is,

$$(1) \qquad \operatorname{corr}(G_n^*, U_n) \geq \operatorname{corr}(G_n, U_n), \quad G_n \in \mathcal{G}$$

for any allowable distribution.

The requirement (1) can, furthermore, be easily restated in a form which makes its implementation much more transparent. Under the regularity conditions that have been imposed,

$$EG_n U_n = -E \frac{dG_n}{d\theta}, \quad EG_n = EU_n = 0,$$

and hence (1) is equivalent to finding $G^* \in \mathcal{G}$ so that

$$(2) \qquad \left(E \frac{dG_n}{d\theta}\right)^2 / EG_n^2 \quad \text{is maximized}$$

(Godambe (1985)).

It is usually desirable to confine attention to some convenient subset $\mathcal{G}_1$ of $\mathcal{G}$ because U_n, the solution from $\mathcal{G}$ by definition, is ordinarily unknown, while some particular $\mathcal{G}_1$'s have such convenient properties that it is possible to immediately write down the optimal solution G^* within $\mathcal{G}_1$. One particularly important class which is especially easy to use is the class $\mathcal{G}_1(\{h_i\})$ of G_n of the form

$$G_n = \sum_{i=1}^{n} a_{i-1}(\theta) h_i(\theta)$$

where the h_i are fixed martingale differences and the weights a_{i-1}

are $\mathcal{F}_{i-1}$-measurable (predictable) for each i. The optimal $G_n^* = \Sigma_{i=1}^n a_{i-1}^* h_i \in \mathcal{G}_1(\{h_i\})$ satisfies

$$(3) \qquad a_{i-1}^* = -E_{i-1} \frac{dh_i}{d\theta} / c(\theta) E_{i-1} h_i^2 , \quad 1 \leq i \leq n ,$$

for some constant $c(\theta)$. A general discussion in which the result (3) is established as both a fixed and a large sample optimum will be given in Section 4.

3. ASYMPTOTIC APPROACH

Now $\{G_n = \Sigma_{i=1}^n g_i, \mathcal{F}_n, n \geq 1\}$ is a square integrable martingale and under quite broad conditions, for example as given in Chapter 3 of Hall and Heyde (1980),

$$(4) \qquad G_n / \left(\sum_{i=1}^n E_{i-1} g_i^2\right)^{\frac{1}{2}} \xrightarrow{d} N(0,1)$$

as $n \to \infty$.

Next, let θ^* be a solution of $G_n(\theta) = 0$ and use Taylor's expansion to obtain

$$(5) \qquad 0 = G_n(\theta^*) = G_n(\theta) + (\theta - \theta^*)\left(\frac{dG_n(\theta)}{d\theta}\right)_{\theta=\theta^+}$$

where $\theta^+ = \theta + \gamma(\theta^* - \theta)$ with $|\gamma| < 1$. Then, assuming that $dG_n/d\theta$ is uniformly continuous in θ in a suitable neighbourhood, (4) and (5) give

$$(6) \qquad (\theta - \theta^*) \frac{dG_n}{d\theta} / \left(\sum_{i=1}^n E_{i-1} g_i^2\right)^{\frac{1}{2}} \xrightarrow{d} N(0,1).$$

Now write

$$\frac{dG_n}{d\theta} = \sum_{i=1}^n \left(\frac{dg_i}{d\theta} - E_{i-1} \frac{dg_i}{d\theta}\right) + \sum_{i=1}^n E_{i-1} \frac{dg_i}{d\theta} ,$$

noting that the first term on the right hand side is a martingale while as $n \to \infty$,

$$\sum_{i=1}^n \left(\frac{dg_i}{d\theta} - E_{i-1} \frac{dg_i}{d\theta}\right) = o_p\left(\left|\sum_{i=1}^n E_{i-1} \frac{dg_i}{d\theta}\right|\right),$$

o_p denoting small order in probability, under modest conditions which it is not illuminating to spell out in detail. It is sufficient, for example, that $\{dG_n/d\theta, \mathcal{F}_n, n \geq 1\}$ is a supermartingale with $dG_n/d\theta \to -\infty$ a.s. as $n \to \infty$.

Thus, heuristic considerations which can be formalized under appropriate regularity conditions suggest that

$$(7) \quad (\theta - \theta^*) \sum_{i=1}^{n} E_{i-1} \frac{dg_i}{d\theta} \Big/ \Big(\sum_{i=1}^{n} E_{i-1} g_i^2 \Big)^{\frac{1}{2}} \xrightarrow{d} N(0,1).$$

Best (asymptotic) estimation is then achieved by choosing G so that the scale variable

$$(8) \quad S^2(\theta) = \Big(\sum_{i=1}^{n} E_{i-1} \frac{dg_i}{d\theta} \Big)^2 \Big/ \sum_{i=1}^{n} E_{i-1} g_i^2 \quad \text{is maximized}$$

since $S(\theta^*)(\theta - \theta^*)$ should be approximately normally distributed for large samples and maximum S^2 leads to approximate confidence intervals of minimum width.

4. FINDING THE OPTIMUM

Now let $\Sigma_{i=1}^{n} h_i \in \mathcal{G}$ and consider the class $\mathcal{G}_1(\{h_i\})$ of martingales of the form $\Sigma_{i=1}^{n} a_{i-1} h_i$ for which a_{i-1} is $\mathcal{F}_{i-1}$-measurable and $da_{i-1}/d\theta$ exists. It is easily checked that

$$\frac{\Big(E \sum_{i=1}^{n} a_{i-1} E_{i-1} \frac{dh_i}{d\theta} \Big)^2}{E \sum_{i=1}^{n} a_{i-1}^2 E_{i-1} h_i^2} \quad \text{and} \quad \frac{\Big(\sum_{i=1}^{n} a_{i-1} E_{i-1} \frac{dh_i}{d\theta} \Big)^2}{\sum_{i=1}^{n} a_{i-1}^2 E_{i-1} h_i^2}$$

are both maximized when (3) holds and hence (3) provides the optimum according to both fixed sample criterion (2) and asymptotic criterion (8) within the class $\mathcal{G}_1(\{h_i\})$. Notice that when (3) holds, $g_i = a_{i-1}^* h_i$ satisfies

$$(9) \quad -E_{i-1} \frac{dg_i}{d\theta} = c(\theta) E_{i-1} g_i^2$$

so that if attention is being focused on some $\mathcal{G}_1$, optimum estimating functions need only be sought amongst the subset of $\mathcal{G}_1$ for which (9) holds for the martingale differences g_i.

It should be remarked that if $\mathcal{G}_1$ includes the score function U_n, then U_n is the optimum estimating function according to both (2) and (8). The result for (2), which is obvious from (1), also follows immediately from an application of Schwarz' inequality to give

$$\frac{\left(E\frac{dG_n}{d\theta}\right)^2}{EG_n^2} = \frac{(EG_nU_n)^2}{EG_n^2} \leqslant EU_n^2 .$$

Similar considerations based on conditional expectations yield the result for (8). We find under the standard regularity conditions that

$$E_{i-1}\frac{dg_i}{d\theta} = -E_{i-1}u_ig_i$$

and hence

$$\frac{\left(\sum_{i=1}^{n} E_{i-1}\frac{dg_i}{d\theta}\right)^2}{\sum_{i=1}^{n} E_{i-1}g_i^2} \leqslant \frac{\left[\sum_{i=1}^{n}(E_{i-1}u_i^2)^{\frac{1}{2}}(E_{i-1}g_i^2)^{\frac{1}{2}}\right]^2}{\sum_{i=1}^{n} E_{i-1}g_i^2} \leqslant \sum_{i=1}^{n} E_{i-1}u_i^2$$

via two applications of Schwarz' inequality.

5. STRATEGY FOR APPLICATIONS

The simplest procedure for obtaining a set of basic martingale differences is to subtract conditional expectations. Thus we can use $h_i = X_i - E_{i-1}X_i$, where it is assumed that $E_{i-1}X_i$ is a function of the parameter θ. The optimal estimating equation can then be chosen from the class of martingales $\Sigma_{i=1}^{n} a_{i-1}h_i$, for nonnegative predictable weights a_{i-1}. For example, if $E_{i-1}h_i^2$ is constant, $1 \leqslant i \leqslant n$, then using (3) we obtain for the optimal estimating equation

$$(10) \qquad \sum_{i=1}^{n} \{X_i - E_{i-1}X_i\} \frac{d}{d\theta} E_{i-1}X_i = 0$$

which can also be obtained by minimizing $\Sigma_{i=1}^{n}(X_i - E_{i-1}X_i)^2$ with respect to θ. This procedure gives the so-called conditional least squares estimator (e.g. Klimko and Nelson (1978), Hall and Heyde (1980), Chapter 6, Section 3).

A less superficial strategy for applications is based on first computing the score function for some plausible underlying distribution, such as a convenient member of the exponential family. This score function can in general be interpreted in terms of a quasi-likelihood function, using terminology introduced by Wedderburn

(1974), and the martingale thereby generated should suggest a basic set of martingale differences. The optimal estimating function is then chosen within the class of martingales $\Sigma_{i=1}^{n} a_{i-1} h_i$, with nonnegative predictable weights a_{i-1}.

To illustrate this method, consider the estimation of the parameter θ in the first order autoregressive model $X_i = \theta X_{i-1} + \varepsilon_i$, where the ε_i are i.i.d. with zero mean and variance 1. Choosing a normal distribution for the ε_i gives $\Sigma_{i=1}^{n} X_{i-1}(X_i - \theta X_{i-1})$ for the derivative of the log of the likelihood and this suggests the choice $h_i = X_i - \theta X_{i-1}$, these being martingale differences whatever the underlying distribution. The best choice of weights according to (3) then gives the corresponding optimum estimating equation within this class of martingales as

$$\text{(11)} \qquad \sum_{i=1}^{n} X_{i-1}(X_i - \hat{\theta} X_{i-1}) = 0.$$

Now suppose that $|\theta| < 1$, so that the autoregression is stationary. It is instructive to illustrate the asymptotic performance of the estimator (11) in relation to that of the maximum likelihood estimator which would, as indicated previously, be used if the true distribution were known.

If the ε_i have the distribution of $t_\nu[(\nu-2)/\nu]^{\frac{1}{2}}$, $\nu > 2$, t_ν denoting the t distribution with ν degrees of freedom, and θ_{ML} denotes the maximum likelihood estimator of θ based on the sample $\{X_0, \ldots, X_n\}$, then it is easily checked that

$$n^{\frac{1}{2}}(\theta_{ML} - \theta) \xrightarrow{d} N\left(0, (1-\theta^2)\frac{(\nu-2)(\nu+3)}{\nu(\nu+1)}\right)$$

as $n \to \infty$, while for $\hat{\theta}$ given by (11), the corresponding result is

$$n^{\frac{1}{2}}(\hat{\theta} - \theta) \xrightarrow{d} N(0, 1-\theta^2).$$

Of course the asymptotic relative efficiency of $\hat{\theta}$ to θ_{ML} tends to unity as $\nu \to \infty$ as would be expected from the convergence of t_ν to N(0,1). Furthermore, the loss of efficiency will in most cases be a small price to pay for avoiding a specific distributional assumption about the ε_i. Other support for this point of view can be found in Godambe (1985).

6. CONCLUDING REMARKS

This paper has been concerned with optimality of martingale estimating equations for stochastic processes in discrete time. A similar formulation, based on local martingale estimating equations, can cope with continuous time as well. This is discussed in detail in Godambe and Heyde (1985); the fixed sample result has been treated in Thavaneswaran and Thompson (1984).

REFERENCES

Godambe, V.P. (1960). An optimum property of regular maximum likelihood estimation. Ann. Math. Statist., 31, 1208-1211.

Godambe, V.P. (1985). The foundations of finite sample estimation in stochastic processes. Biometrika, 72, 419-428.

Godambe, V.P. and Heyde, C.C. (1985). The general theory of optimality in estimation for stochastic processes. Preprint.

Godambe, V.P. and Thompson, M.E. (1978). Some aspects of the theory of estimating equations. J. Statist. Planning Inf., 2, 95-104.

Godambe, V.P. and Thompson, M.E. (1984). Robust estimation through estimating equations. Biometrika, 71, 115-125.

Hall, P. and Heyde, C.C. (1980). Martingale Limit Theory and its Application. Academic Press, New York.

Klimko, L.A. and Nelson, P.I. (1978). On conditional least squares estimation for stochastic processes. Ann. Statist., 6, 629-642.

Thavaneswaran, A. and Thompson, M.E. (1984). Optimal estimation for semimartingales. Preprint.

Wedderburn, R.W.M. (1974). Quasi-likelihood, generalized linear models and the Gauss-Newton method. Biometrika, 61, 439-447.

Prob. Theory and Math. Stat., Vol.1, pp. 543-545
Prohorov et al. (eds)
1986 VNU Science Press

ON URN SCHEMES IMBEDDED IN BIRTH PROCESSES

Lars Holst
Department of Mathematics, Uppsala University
Thunbergsv. 3, S-752 38 Uppsala
Sweden

ABSTRACT

Many different problems connected with urnschemes such as drawing with and without replacement and Pólya-sampling can be studied using imbedding in birth processes. This technique is illustrated by some waiting time problems for sampling with replacement and Pólya-sampling. Both exact and asymptotic results are given.

1. URNSCHEMES

An urn contains b balls of r different colours. Many different ways of drawing balls for such an urn have been studied since the early days of probability theory. The most wellknown examples are drawing at random with or without replacement. Another example is Pólya's urnscheme, i.e. draw a ball at random and replace it together with c balls of the same colour. Many problems connected with such drawing schemes can be studied in a unified way by imbedding in birth processes. In this paper we will give some examples on this, for more general results see Holst and Hüsler (1985) and Holst (1986).

2. IMBEDDING IN BIRTH PROCESSES.

Introduce r independent pure birth processes with birth intensities $\lambda_i = b + c\cdot i$, $i = 0,1,2,\ldots$. If at some time $i_1, i_2, \ldots, i_r$ events have occured in the processes, then the probability that the next event will occur in the k:th process is $(b + c\cdot i_k)/(rb + c\cdot(i_1+\ldots+ i_r))$. The same probability holds for getting a ball of the k:th colour in the draw $(i_1 +\ldots+ i_r+ 1)$ from the Pólya urn; $c = 0$ (-1) corresponds to drawing with (without) replacement. Every time an event occurs in the birth processes we can think of it as a drawing from a corresponding urn. The urnscheme is imbedded in the birth processes. The usefulness of this approach will be illustrated below.

3. A COLLECTOR'S PROBLEM

Draws at random with replacement from an urn containing r different balls. Continue until all balls have been obtained at least m times each. Let W_r denote the necessary number of draws. This drawing scheme can be imbedded in r birth processes each with intensity $\lambda_i = 1$, that is r independent Poisson processes. Let $X_1,\dots,X_r$ be the times in these processes when they get their m:th event. The X's are independent $\Gamma(m,1)$-distributed and the extreme order statistic $T_r = \max(X_1,\dots,X_r)$ is the waiting time until all the processes have got at least m events each. The super-position of the processes is a Poisson process with intensity r and inter-arrival times $Z_1/r, Z_2/r, \dots$ where the Z's are independent exponential random variables with mean 1. Thus we can write

$$T_r = \Sigma_1^{W_r} Z_j/r = W_r/r + \Sigma_1^{W_r}(Z_j - 1)/r$$

where W_r and the Z's are independent. From this relation it follows, for example, that $E(W_r) = r\,E(T_r)$. As $r\to\infty$ it holds that $E(T_r) = \ln r + (m-1)\ln\ln r - \ln(m-1)! + \gamma + o(1) = a_r + \gamma + o(1)$, where γ is Euler's constant, and furthermore $T_r - a_r \overset{d}{\to} \exp(-e^{-x})$. From this we find $T_r = W_r/r + o_p(1)$ and thus we have the convergence $W_r/r - a_r \to \exp(-e^{-x})$. For other results on similar problems see Holst (1986).

4. A PÓLYA URN PROBLEM

Draw at random from an urn containing r balls of different colours. Replace each ball together with one more ball of the same colour. Continue until all colours have been obtained at least once. Let W_r be the necessary number of draws. This situation can be imbedded in birth processes with intensities $\lambda_i = 1 + i$. The number of events X_t in the interval $(0,t)$ for such a process has the distribution $P(X_t = k) = e^{-t}(1-e^{-t})^k$. For a process having at least one event in $(0,t)$ let Y_t denote its number of events. We have $P(Y_t = k) = P(X_t | X_t \geq 1)$ and $E(Y_t) = e^t$. The waiting time Z until a process gets its first event satisfies $P(Z < t) = P(X_t = 0) = e^{-t}$, that is Z is exponential with mean 1. The time when all processes have got at least one event is $T_r = \max(Z_1,\dots,Z_r)$ where the Z's are independent copies of Z.

Given $T_r = t$ we have the representation $W_r = 1 + Y_{2t} + \ldots + Y_{rt}$ where the Y's are independent copies of Y_t. Hence by taking conditional expectations we find $E(W_r) = 1 + (r - 1)E(\exp(T_r)) = + \infty$! Using this approach we can get an explicit expression for the distribution of W_r and also prove that r^2/W_r converges in distribution to an exponential distribution with mean 1 as $r \to \infty$. We may remark that the mean waiting time for obtaining all but one colour is finite and equal to $(r - 1)^2$, but its second moment is infinite! For further results on such problems see Holst and Hüsler (1985).

REFERENCES

Holst,L., and Hüsler,J. (1985). Sequential urn schemes and birth processes. Adv. Appl. Prob. 17, 257 - 279.

Holst,L. (1986). On birthday, collectors', occupancy and other classical urn problems. Int. Statist. Rev. 54, April.

Prob. Theory and Math. Stat., Vol.1, pp. 547-552
Prohorov et al. (eds)
1986 VNU Science Press

LIMITING DISTRIBUTIONS AND MEAN-VALUES OF COMPLEX-VALUED MULTIPLICATIVE FUNCTIONS

Karl-Heinz Indlekofer
Füllekengrund 12, D-4799 Borchen-Dörenhagen

1. INTRODUCTION

An arithmetic function $f: \mathbb{N} \to \mathbb{C}$ is said to be multiplicative if for every pair of coprime positive integers m and n the relation $f(mn) = f(m)f(n)$ is satisfied. If in addition the relation $f(p) = f(p^k)$ holds for each prime power p^k we say that f is strongly multiplicative.

For each $n \in \mathbb{N}$ we define the distribution P_n of f by

$$P_n(B) := n^{-1}|\{m \le n: f(m) \in B\}| = n^{-1} \sum_{m \le n,\ f(m) \in B} 1$$

for all Borel sets $B \in \mathcal{B}(\mathbb{C})$. If as $n \to \infty$ the distributions P_n converge weakly to a probability measure P, $P_n \Rightarrow P$, we say that f has a limiting distribution P. We call P degenerate if $P(\{0\}) = 1$ and non-degenerate otherwise. Necessary and sufficient conditions for the existence of a non-degenerate limiting distribution P for real-valued and complex-valued multiplicative functions were found by Bakstys (1968), Galambos (1971) and Delange (1973) (cf. Laurinčikas (1974)), respectively.

In this paper we deal with the connections between the existence of the limiting distribution P of f and the asymptotic behaviour of the moments

$$\int_{\mathbb{R} \times \mathbb{R}} x\, dP_n(x) = n^{-1} \sum_{m \le n} f(m) =: M(f,n)$$

as $n \to \infty$. We say that f has a mean-value $M(f)$ if the limit $M(f) := \lim_{n\to\infty} M(f,n)$ exists. The case of real-valued multiplicative functions was handled by Indlekofer (1981). A typical result is the following:

Let f be a real-valued multiplicative function, which is uniformly summable. Then the mean-value $M(|f|)$ of f exists and is different from zero if and only if the limiting distribution P of f exists and is non-degenerate.

It turns out that also in the case of complex-valued multiplicative functions the space of uniformly summable functions is the appropriate space for describing the mentioned connections.

To ease notational difficulties we introduce the spaces $(\alpha \geq 1)$

$$L_\alpha := \{f: \mathbb{N} \to \mathbb{C},\ \|f\|_\alpha < \infty\}$$

of functions f with bounded seminorm

$$\|f\|_\alpha := \{\limsup_{n\to\infty} n^{-1} \sum_{m\leq n} |f(n)|^\alpha\}^{1/\alpha}$$

and the space L^* of uniformly summable functions, i.e. L^* is the set of all functions f such that

$$\lim_{K\to\infty} \sup_{n\in\mathbb{N}} n^{-1} \sum_{\substack{m<n \\ |f(m)|\geq K}} |f(m)| = 0.$$

Obviously $(\alpha > 1)$

$$L_\alpha \subsetneqq L^* \subsetneqq L_1 .$$

For further characterizations of L^* see Indlekofer (1980a).

2. RESULTS

For the proof of the following theorems one uses a Cramér-Wolddevice for $\mathbb{R}_{\geq 0} \times [0,2\pi)$ and properties of multiplicative functions $f \in L^*$ (see Indlekofer (1980b).

Theorem 1. Let $f \in L^*$ be multiplicative. Then the following assertions hold.

(i) $P_n \Rightarrow P$, where P is degenerate, if and only if the mean-value $M(|f|)$ exists and equals zero.

(ii) $P_n \Rightarrow P$, where P is non-degenerate, if and only if the following holds.

(a) the mean-value $M(|f|)$ exists and is different from zero;

(b) for each $m \in \mathbb{Z}$ the mean-value $M(|f| e^{im \arg f})$ exists.

The sharpness of this result will be illustrated by the following two examples.

Example 1. Define a multiplicative function f such that, for all primes p,

$$f(p^k) = \begin{cases} 1 & \text{if } k = 1 \\ p^k & \text{if } k \geq 2 \end{cases}.$$

Then the non-degenerate limiting distribution of f exists but

$$\sum_{m \leq n} f(m) > \sum_{m^2 \leq n} m^2 \gg n^{3/2}.$$

Therefore $M(|f|)$ does not exist.

Example 2. Let the multiplicative function f be defined by

$$f(p^k) = \begin{cases} 0 & \text{if } k \geq 1,\ k \neq 2 \\ p & \text{if } k = 2 \end{cases}$$

for all primes p. Then the limiting distribution of f is degenerate. On the other hand

$$\sum_{m \le n} f(m) = \sum_{m^2 \le n} \mu^2(m)m$$

and thus $M(f) = M(|f|) = 3\pi^{-2} \neq 0$.

For a given multiplicative function f we define a strongly multiplicative function f' by

$$(1) \qquad f'(p) = \begin{cases} 1 & \text{if } ||f(p)|-1| > \frac{1}{2} \\ f(p) & \text{if } ||f(p)| -1| \ge \frac{1}{2}. \end{cases}$$

Then we have

Theorem 2. Let $f\colon \mathbb{N} \to \mathbb{C}$ be multiplicative and f' be defined by (1). Then the following two assertions are equivalent.

(i) $P_n \Rightarrow P$, where P is non-degenerate.

(ii) (a) The mean-value $M(|f'|)$ exists and is different form zero.
(b) For each $m \in \mathbb{Z}$ the mean-value $M(|f'|e^{im \arg f'})$ exists.
(c) $f' \in L_2$ and $\sum_{\substack{p \\ f'(p) \neq f(p)}} p^{-1} < \infty$.

Remark 1. Concerning the assertion (ii), (c) the property $f' \in L_2$ implies $f' \in L_\alpha$ for all $\alpha > 2$ (see Indlekofer (1980a)).

Remark 2. The above results may be interpreted in the following way: Let f be a multiplicative function. Then the limiting distribution of f exists and is non-degenerate if and only if there exists a strongly multiplicative function $f^* \in L^*$ in the "neighbourhood" of f so that $M(|f^*|) \neq 0$ and the mean-values $M(|f^*|e^{im \arg f})$ exist for all $m \in \mathbb{Z}$. Here the neighbourhood of f is defined as the set $\{f^*\colon \mathbb{N} \to \mathbb{C}, \sum_{\substack{p \\ f^*(p) \neq f(p)}} p^{-1} < \infty\}$.

3. CONCLUDING REMARKS

A multiplicative function f is uniquely determined by its values $f(p^k)$ at prime powers p^k. Therefore the following results of the author give further equivalent assertions to Theorem 1 and Theorem 2, respectively.

Theorem 3. (Indlekofer (1980a). Let $f: \mathbb{N} \to \mathbb{C}$ be multiplicative and $\alpha \geq 1$. Then the following two assertions hold.

(i) If $f \in L^* \cap L_\alpha$ and if the mean-value $M(f)$ exists and is non-zero, then the series

$$\sum_p \frac{f(p)-1}{p}, \quad \sum_{\substack{p \\ |f(p)|\leq 3/2}} \frac{|f(p)-1|^2}{p}, \tag{2}$$

$$\sum_{\substack{p \\ |f(p)-1|\geq 1/2}} \frac{|f(p)|^\lambda}{p}, \quad \sum_p \sum_{k\geq 2} \frac{|f(p^k)|^\lambda}{p^k}$$

converge for all λ with $1 \leq \lambda \leq \alpha$ and, for each prime p,

$$1 + \sum_{k=1}^{\infty} \frac{f(p^k)}{p^k} = 0. \tag{3}$$

(ii) If the series (2) converge, then $f \in L^* \cap L_\alpha$ and the mean-values $M(f)$, $M(|f|^\lambda)$ exist for all λ with $1 \leq \lambda \leq \alpha$. If in addition (3) holds then $M(f) \neq 0$.

Theorem 4. (Indlekofer (1980b), Theorem 4). Let $f \in L^*$ be multiplicative and $\|f\|_1 > 0$. Then the mean-value $M(f)$ exists and equals zero if and only if one of the following conditions holds.

(a) For each $t \in \mathbb{R}$ the series

(4) $$\sum_{\substack{p \\ ||f(p)|-1|\leq 1/2}} \frac{1-\text{Re } f(p)\{|f(p)|p^{it}\}^{-1}}{p}$$

diverges.

(b) There exists a $t \in \mathbb{R}$ such that the series (4) converges and $\sum_{m=1}^{\infty} f(p^m)p^{-m}\, p^{-imt} + 1 = 0$ for some prime p.

REFERENCES

Bakstys, A. (1968). On the asymptotic distribution of multiplicative numbertheoretic functions. (Russian). Litovsk. Mat. Sb. 8, 5-20.
Delange, H. (1973). Sur la distribution des valeurs des fonctions multiplicatives complexes. Compt. Rend. Acad. Sci., Série A., 276, 161-164.
Galambos, J. (1971). On the distribution of strongly multiplicative functions. Bull. London Math. Soc. 3, 307-312.
Indlekofer, K.-H. (1980a). A mean-value theorem for multiplicative functions. Math. Z. 172, 255-271.
Indlekofer, K.-H. (1980b). Properties of uniformly summable multiplicative functions. Preprint. To appear in Periodica Math. Hung.
Indlekofer, K.-H. (1981). Limiting distributions and mean-values of multiplicative arithmetical functions. J. reine angew. Math. 328, 116-127.

Prob. Theory and Math. Stat., Vol.1, pp. 553-574
Prohorov et al. (eds)
1986 VNU Science Press

ASYMPTOTICALLY MINIMAX TESTING OF NONPARAMETRIC HYPOTHESES

Yu.I.Ingster

Leningrad Branch of Steklov Mathematical Institute, Leningrad, USSR

1. We shall consider the following problem of hypotheses testing. Let Φ be a set of probability distribution densities on the probability space $(\mathcal{X}, \mathcal{U}, P)$, $x_1,\ldots,x_N$ be the random sample distributed according to the unknown density $f \in \Phi$. Let $\Phi_0 \subset \Phi$ and null hypothesis H_0 be $f \in \Phi_0$. The alternative hypotheses of the following type will be considered. Let $d(f, \Phi_0)$ be a "measure of distance" for a density $f \in \Phi$ from the set Φ_0 and ρ_N be a positive value. We shall consider the set of densities

$$\Phi_N = \{ f \in \Phi : d(f, \Phi_0) \geqslant \rho_N \}. \tag{1}$$

The alternative hypothesis H_1 is $f \in \Phi_N$. It is necessary to choose hypothesis H_0 or H_1 using the sample $x_1,\ldots,x_N$.

This problem will be considered from the minimax point of view. Let

$$\gamma_N(\psi) = \gamma_N(\psi, \Phi_0, \Phi_N) \triangleq \sup_{f \in \Phi_0} M_{f,N}\psi + \sup_{f \in \Phi_N} M_{f,N}(1-\psi)$$

be the sum of maximum values of the first and second kind error probabilities for the test $\psi : (\mathcal{X}, \mathcal{U})^N \to ([0,1], \mathcal{B})$ where $M_{f,N}$ is the mean for the measure P_f^N on $(\mathcal{X}, \mathcal{U})^N$. This value characterizes the minimax properties of the test ψ. The value

$$\gamma_N = \gamma_N(\Phi_0, \Phi_N) \triangleq \inf_{\psi} \gamma_N(\psi, \Phi_0, \Phi_N)$$

characterizes the minimax discernability of hypotheses H_0 and H_1. It is clear that $0 \le \gamma_N \le 1$.

The difficulties of minimax hypotheses testing problem are well known [1]. Using minimax theorems [2] one may write the following equality

$$\gamma_N = 1 - \frac{1}{2} \, var\,(conv\{P_f^N, f \in \Phi_0\}, conv\{P_f^N, f \in \Phi_N\}), \quad (2)$$

where $var(A, B)$ is the variation distance between probability measure sets A and B, $conv\, A$ is the convex envelope of the set A, but calculations of γ_N or value of the same type using (2) or the less favourable distributions theorems [3, 4] are possible only in special cases.

It is possible to overcome these difficulties in the asymptotical version of the problem (asymptotic of the first order: calculation of γ_N and minimazation of $\gamma_N(\psi_N)$ with exactness to the value $o(1)$ if $N \to \infty$) for the case where $\Phi = \{f_\theta, \theta \in \Theta\}$ is a regular finite-parametrical family, $\Phi_0 = \{f_\theta, \theta \in \Theta_0\}$, Θ and Θ_0 are closed regions in R^n and R^m, $0 \le m < n$,

$$d(f, \Phi_0) = d(\theta, \Theta_0) = \inf_{\xi \in \Theta_0} d_\xi(\theta, \xi),$$

where d_ξ is the Fisher's metric of the family Φ in the point ξ ([5] for "local" problems of hypotheses testing, [6]).

The object of interest for us is the case of nonparametric set Φ. The version of problem will be considered which is analogous to asymptotic of the null order.

Let $N \to \infty$, ρ_N be the given positive sequence. Let us define sequencies of hypotheses H_0 and H_1 as asymptotically nondiscerning (A.N.) if $\gamma_N \to 1$, consistently nondiscerning (C.N.) if $\underline{\lim}\, \gamma_N > 0$ and consistently discerning (C.D.) if $\gamma_N \to 0$. The sequence of tests ψ_N such that $\gamma_N(\psi_N) \to 0$ let be defined as consistent in minimax sense. The case of the simple hypothesis $H_0: \mathcal{P}_0 = \{f_0\}$, $f_0 \in \mathcal{P}$ will only be considered, $d(f, \mathcal{P}_0) = d(f, f_0)$ be the metric in $\mathcal{P}$. Let us further assume that $f_0 \equiv 1$ and $P = P_{f_0}$.

The object of our interest is to investigate the dependence of discernability and nondiscernability conditions upon the properties of the set $\mathcal{P}$, the metric d and the asymptotic of the sequence ρ_N.

2. Let $\mathcal{P} = \{f_\theta, \theta \in \Theta\}$ be a regular (in the sense of [5-7]) finite-parametrical family, Θ be the compact in R^n, $f_0 = f_{\theta_0}$, $d(f, f_0) = d(\theta, \theta_0)$ be a metric in R^n which is equivalent to Euclidian metric. The conditions of discernability and nondiscernability are very simple in this case. Let $\rho_N^* = N^{-1/2}$. Then A.N. holds if $\rho_N = o(\rho_N^*)$, C.N. holds if $\rho_N = O(\rho_N^*)$ and C.D. holds if $\rho_N / \rho_N^* \to \infty$. In the last case, for example, the following sequence of tests is consistent in minimax sense:

$$\psi_N = \mathbb{1}_{\{\|\theta_N^* - \theta_0\| > T_N\}},$$

where θ_N^* is any uniform $\sqrt{N}$-consistent estimator of parameter (for example, maximum likelihood or Bayesian estimator) and the sequence T_N is such that

$$T_N = o(\rho_N^*), \quad T_N / \rho_N \to \infty. \tag{3}$$

The asymptotic of sequence ρ_N that characterizes discernability conditions in the abovementioned or similar sense we may call "the critical order" in the considered hypotheses testing problem. It can be seen that in all regular finite-parametrical problems "the critical order" is $\rho_N^* = N^{-1/2}$.

Similar situation holds for nonparametrical problem if the metric d is "sufficiently robust". Let for example $(\mathcal{X}, \mathcal{U}) = ([0,1], \mathcal{B})$, P be Lebesgue measure, $d(f, f_0) = \|F - F_0\|_p$, $1 \le p \le \infty$; where F is the distribution function for the density f, $F_0(t) = t$. Let $\mathcal{P}$ be a convex set of densities (this restriction may be replaced by a weaker condition). Then "the critical order" in this problem also is $\rho_N^* = N^{-1/2}$ and if $\rho_N^* / \rho_N \to \infty$ then the following sequence of tests is consistent in minimax sense:

$$\Psi_N = \mathbb{1}_{\{\|F_N - F_0\|_p > T_N\}}, \tag{4}$$

where F_N is the empirical distribution function, T_N is a sequence for which relation (3) holds.

For $p = \infty$ the relation (4) determines Kolmogorov's tests, for $p = 2$ it determines the ω^2-tests of von Mises-Smirnov. It is clear that this fact is related to the uniform $\sqrt{N}$-consistency of F_N as estimator of F in L_∞-metric.

A question arises: which metrics should be used for removing neighbourhoods of null hypothesis from the set of alternatives?

From the asymptotical theory of hypotheses testing point of view it seems most natural to choose metrics $d(f, f_0)$ describing the discernability of simple hypotheses corresponding to f and f_0, i.e. variation distance $\mathrm{var}(P_f^N, P_{f_0}^N)$. In the regular finite-parametrical case, $f_0 = f_{\theta_0}$, such metric is the Fisher's metric d_{θ_0}, in the general case such metric is Hellinger's metric or quadratic metric (in the least for sets of densities $\mathcal{P}$ such that $\sup_{f \in \mathcal{P}} \|f\|_\infty < \infty$):

$$d_2(f, f_0) \triangleq \|f - f_0\|_2 = M_{f_0}^{1/2} (f/f_0 - 1)^2;$$

relations $\mathrm{var}(P_{f_N}^N, P_{f_0}^N) \to 0$ or $\mathrm{var}(P_{f_N}^N, P_{f_0}^N) \to 2$ hold if and only if $\sqrt{N}\, d_2(f_N, f_0) \to 0$ or

$\sqrt{N}\, d_2(f_N, f_0) \to \infty$

Further a wider class of metrics d_p will be considered:

$$d_p(f, f_0) \overset{\triangle}{=} \| f - f_0 \|_p \ , \ 1 \le p \le \infty$$

The following theorem shows that discerning properties are distinctly different for this metrics.

Theorem 1. Let $(\mathcal{X}, \mathcal{U}) = ([0,1], \mathcal{B})$, P be Lebesgue measure, $d = d_p$. Then the equality $\gamma_N(f_0, \Phi_N) = 1$ holds for the following sets Φ, values p and ρ_N where Φ_N is determined in (1):

a) Φ is the set of all densities on $(\mathcal{X}, \mathcal{U}, P)$, $p \geqslant 2$, $\rho_N > 0$;

b) Φ is the set of densities f such that $\| f - 1 \|_\infty < a$, $a \le 1$; $0 < \rho_N \le a$, $p \geqslant 1$;

c) Φ is the starring set of densities $f_{i,t}(x) = 1 + t e_i(x)$ where e_i is the sequence of different elements of orthonormal Walsh basis in $L_2[0,1]$, $i = 1, 2, \dots$; $0 \le t \le a$, $a \le 1$; $0 < \rho_N \le a$, $p \geqslant 1$.

This theorem shows that for d_p-metrics the minimax problem of hypothesis testing without essential restrictions on the set Φ has no nontrivial solutions for every number N if the radiuses of removed null hypothesis neighbourhoods are not too large. The nondiscernability arises in particular if set Φ contains infinitely many densities f_i such that $(f_i - 1, f_j - 1) = \rho\, \delta_{ij}$, where $\rho > 0$, $\delta_{ij} = \{0$ if $i \ne j$; 1 if $i = j\}$ (see the theorem 4 further).

However nontrivial solutions for our problem can be obtained in the case of d_p-metric if some restrictions on the sets Φ are allowed.

Some general results about relations between the discernability and geometrical properties of sets Φ or Φ_N are stated in pp 4 and 5. Those properties are in some way similar to such characteristics of sets like interior radiuses and Kolmogorov's diameters. However at first let us consider the results for the special class of set of densities Φ : densities on the unit cube $\Delta_K = [0,1]^K \subset R^K$ with the given degree of smoothness. This class of sets is the object of interest because it permits to get conditions of discernability and nondiscernability in exact order as well as to compare the minimax properties of different classes of tests. In addition there are exact on the order results in the minimax estimation problem for the functional classes of densities of the same type and for the same metrics d_p and it is possible to compare minimax nonparametric problems of estimation and hypothesis testing.

3. Let $L>0$, $1 \le q \le \infty$, $\tau = m+\alpha$, where $m = 0,1,2,\dots, \alpha \in (0,1]$, K be a natural number. Let $H_K(\tau,L,q)$ be the set of densities f on $(\Delta_K, \mathcal{B}, P)$ where P is Lebesgue measure such that the following condition holds: all partial derivatives $f^{(\bar m)}$, $\bar m = (m_1,\dots,m_K), m_i \ge 0$, of the order $m = \sum_i m_i$ exist and the Hölder's condition of the order α with the constant L in L_q-metric if $q<\infty$ or the same Lipschitz's condition if $q=\infty$ holds: for every $h \in R^K$

$$\Big(\int_{\Delta_K^h} | f^{(\bar m)}(t+h) - f^{(\bar m)}(t)|^q dt \Big)^{1/q} < L\|h\|^\alpha, \; q<\infty;$$

$$\operatorname*{ess\,sup}_{t \in \Delta_K^h} | f^{(\bar m)}(t+h) - f^{(\bar m)}(t)| < L\|h\|^\alpha, \; q=\infty,$$

where $\Delta_K^h = \{ t \in \Delta_K : t+h \in \Delta_K \}$. Let $\tilde H_K(\tau,L,q)$ be the similar set of functions which are periodical with the period equal to 1 for every variable.

Let Φ be the set of densities $f \in H_K(\tau, L, q)$ or

$f \in \tilde{H}_K(r, L, q)$ such that $\|f\|_2 < C$ if $r \le 1$ and $q < 2$, $C > 1$. Let us consider the metric d_p in Φ and assume that if $p \le 2$ then $q \ge p$ and if $p > 2$ then $q = p$ and $r > K/2p$. Let

$$\rho_N^* = \begin{cases} N^{-2r/(4r+K)} & \text{if } 1 \le p \le 2, \\ N^{-r/(2r+K(1-p^{-1}))} & \text{if } 2 < p < \infty, \\ (\ln N/N)^{r/(2r+K)} & \text{if } p = \infty \end{cases} \tag{5}$$

<u>Theorem 2</u>.

1. Let $p < \infty$. Then the sequences of hypotheses which are corresponding to the density f_0 and the set Φ_N of the type (1) are A.N. if $\rho_N = o(\rho_N^*)$ and these are C.N. if $\rho_N = O(\rho_N^*)$. If $p = \infty$ then there exists such constant $\lambda_1 > 0$ that if $\overline{\lim}\, \rho_N/\rho_N^* < \lambda_1$ then the sequences of hypotheses are A.N.

2. If $p < \infty$ and $\rho_N/\rho_N^* \to \infty$ then the sequences of hypotheses are C.D. If $p = \infty$ then there exists such constant $\lambda_2 > 0$ that if $\underline{\lim}\, \rho_N/\rho_N^* > \lambda_1$ then the sequences of hypotheses are C.D.

3. If the conditions of p.2 hold then consistent in minimax sense sequences of tests may be constructed in following way. Let $m = m_N$ be the integer sequence, $m_N \asymp (\rho_N^*)^{-1/r}$, We devide every edge of the cube Δ_K in m equal intervals and consider the corresponding partition of the cube in $n = m^K$ small cubes $\delta_1, \dots, \delta_n$. Let K_i be the number of sample elements $x_\alpha \in \delta_i$, $y_i = K_i/N - 1/n$. Let

$$t_N = \begin{cases} \sum_i y_i^2 & \text{if } 1 \le p \le 2, \\ \sum_i |y_i|^p & \text{if } 2 < p < \infty, \\ \max_i |y_i| & \text{if } p = \infty. \end{cases} \tag{6}$$

Then there exists such sequence T_N that if the conditions of p.2 hold then the sequence of tests

$$\Psi_N = \mathbb{1}\{t_N > T_N\} \tag{7}$$

is consistent in minimax sense.

<u>Corollary.1</u>. Let $p<\infty$. Then the sequences of hypotheses are C.D. if and only if $\rho_N/\rho_N^* \to \infty$ where ρ_N^* is determined in (5).

The theorem 2 and corollary 1 are true if we substitute the set Φ by set $\Phi^{(\varepsilon)} = \{ f \in \Phi : \| f - 1 \|_\infty < \varepsilon \}$ for every $\varepsilon > 0$.

The proof of theorem 2 for the case of $k=1, p=q=2$ and natural r is stated in [8].

The theorem 2 showes that "the critical order" ρ_N^* in this problem as determined in (5) depends on p if $p>2$. The sense of term "critical order" is slightly different in the case $p=\infty$ because a border between the discernability and nondiscernability tends to be more sharp. If degree of smoothness $r \to \infty$ then $\rho_N^* \to$ $\to N^{-1/2}$ which corresponds to the finite-parametrical case.

The tests in p.3 of the theorem are the chi-square tests with increasing number of elements in the uniform partition of the cube Δ_k for the metrics d_ρ , $p \le 2$ and their generalizations (we call it χ^p -tests) for $p>2$.

It seems to be of interest that if $p \leqslant 2$ and $r < k/4$ then the number $n = m^k$ of elements in the cube partition is much larger then N but such chi-square tests are optimal in the sense of minimum order of asymptotic ρ_N necessary for the minimax consistence.

Let consider the minimax properties of some other classes of tests in this problem. Using the analogy of the finite-parametrical case and of the case of metric $\| F - F_0 \|_p$ one could attribute good properties in the minimax sense to the tests based on the minimax on the order estimators of densities in d_p -metrics:

$$\Psi_N = \mathbb{1}_{\{\|f_N - f_0\|_p > T_N\}}, \tag{8}$$

where f_N is the such estimator that

$$E_N(f_N, \Phi) \stackrel{\Delta}{=} \sup_{f \in \Phi} M_{f,N} \|f_N - f\|_p \asymp E_N(\Phi) \stackrel{\Delta}{=} \inf_{f_N} E_N(f_N, \Phi).$$

The problem of minimax density estimation for the functional sets of densities on R^K which are analogic to $H_K(r, L, p)$ is studied in [9, 10]. The methods and the results [10] may be extended on the functional sets $\Phi = \tilde{H}_K(r, L, p)$. In this case

$$E_N(\Phi) \asymp E_N(f_N, \Phi) \asymp \rho_N^* = \begin{cases} N^{-r/(2r+K)}, & 2 < p < \infty, \\ (\ln N/N)^{r/(2r+K)}, & p = \infty \end{cases} \tag{9}$$

where f_N is the estimator of type similar to "projecting" (see p.5). It is possible to show that with the estimators f_N of this type the tests (8) may be consistent only if $\lim \rho_N / \rho_N^* > C$ where ρ_N^* is determined in (9), C is a positive constant. We see that the minimax properties of the tests (8) for $2 < p < \infty$ are considerably worse than properties of χ^p-tests from the p.3 of theorem 2. This fact is connected with the difference in the requirements to the "projecting space" dimensions asymptotic in the minimax problems of hypotheses testing and estimation (see p.5).

Let us now compare minimax properties of chi-square tests and tests of the type (4) for $K = 1$. Let Φ be the set of all densities on $([0, 1], \mathcal{B}, P)$, $d(f, f_0) = \|F - F_0\|_p$. It was previously shown that the minimax consistency condition for the tests (4) and for the sets Φ_N of the type (1) is $\rho_N/\rho_N^* \to \infty$ where $\rho_N^* = N^{-1/2}$, where as for the chi-square tests with the uniform partition of $[0,1]$ on the $n = n_N$ intervals the minimum order in the similar condition is $\rho_N^* = N^{-2/5}$ and the optimum order of the number of elements in the partition is $n \asymp N^{2/5}$ (if $p=\infty$ it comes from theorem of Mann and Wald [11], see also [12], pp.583-585; the methods [12] are extended to the case $1 \le p \le \infty$). Thus in this problem the chi-square tests are considerably worse than the tests (4), in particular, Kolmogorov's tests for $p=\infty$ and ω^2-tests von Mises-Smirnov for $p=2$: the ratio of power degrees of corresponding "critical orderes" is equal to 4/5. This fact was discused for example in [12], p.p.615-616, [13]. It is related certainly to the choice of the metric d in Φ .

However in the case of the metric $d=d_2$ and functional sets Φ from the theorem 2 the situation is opposite.

Theorem 3. Let $\Phi = H_1(\tau, L, q)$, $q \ge 2$, $d = d_2$. Then the tests sequencies (4) for $p \ge 2$ and any sequence T_N will be consistent in minimax sense if and only if $\rho_N/\rho_N^* \to \infty$ where $\rho_N^* = N^{-\tau/(2\tau+2)}$.

Comparing the theorem 3 and the theorem 2 for $k=1$ and $p=2$ we see that the ratio of power degrees of "critical orders" ρ_N^* equals $\gamma(\tau) = (4\tau+1)/(4\tau+4)$; $\gamma(\tau) \to 1$ if $\tau \to \infty$ but $\gamma(1) = 5/8$ and $\gamma(\tau) \to 1/4$ if $\tau \to 0$.

The proof of the theorem 3 is based on the following property of the functional sets $H_1(\tau, L, q)$ for $q \ge 2$: if $p \ge 2$ and $\rho \to 0$ then

$$\inf_{f \in H_1(\tau, L, q):\, \|f - f_0\|_2 > \rho} \|F - F_0\|_p \asymp \rho^{1+1/\tau} .$$

4. Let now describe methods of proofs for nondiscernability theorems.

The equality (2) shows that for the proof of A.N.(or C.N.) it is sufficient to construct the sequence of density sets $\{f_1,\dots,f_M\}\subset \Phi_N$, $M=M(N)$ such that for the probability measures $P_0=P_{f_0}^N$ and $P_1=\frac{1}{M}\sum_i P_{f_i}^N$ on $(\mathcal{X},\mathcal{U})^N$ the following condition holds: $var(P_1,P_0)\to 0$ (or $\overline{\lim}\, var(P_1,P_0)<2$). We can construct such sequences in following way. Let $e_1,\dots,e_n$ be $P=P_{f_0}$-centred orthonormal statistics set in $L_2(\mathcal{X},\mathcal{U},P)$, $n=n(N)$ be integer sequence, $r=r(N)$ be positive sequence. We consider three types of density sets $\Phi(r,n,\alpha)=\{f_i^{(\alpha)}\}$, $f_i^{(\alpha)}=1+r\sum_j a_{ij}^{(\alpha)}e_j$, $i=1,\dots,M_\alpha$, $\alpha=1,2,3$:

1) $a_{ij}^{(1)}=1$ or -1, $M_1=2^n$;

2) $a_{ij}^{(2)}=0$ or 1, $\sum_j a_{ij}^{(2)}=K$ where $K=K(N)\asymp n^{1/2}$, $M_2=C_n^K$;

3) $a_{ij}^{(3)}=0$ or 1, $\sum_j a_{ij}^{(3)}=1$, $M_3=n$

Let

$$\xi^\alpha=\xi_N^\alpha=\begin{cases} Nr^2n^{1/2} & \text{if } \alpha=1,\\ Nr^2 & \text{if } \alpha=2,\\ n^{-1}\exp Nr^2 & \text{if } \alpha=3\end{cases} \tag{10}$$

<u>Theorem 4</u>. Let $\Phi_N\supset\Phi(r,n,\alpha)$ for any $\alpha=1,2,3$ and N sufficiently large. Then $\gamma_N(f_0,\Phi_N)\to 1$ if $\xi^\alpha=o(1)$ and $\underline{\lim}\,\gamma_N(f_0,\Phi_N)>0$ if $\xi^\alpha=O(1)$.

The proof of the theorem 4 is based on the estimation of quadratic distance $d_{P_0,2}(P_1,P_0)=M_{P_0}^{1/2}(dP_1/dP_0-1)^2$ which majorize the variation distance $var(P_1,P_0)$.

The value r in the theorem 4 is connected with d_2-

distance between $f_j^{(\alpha)}$ and f_0. For the sets $e_1,\dots,e_n$ of special type the nondiscernability conditions may be related with d_p-distance.

Let $\lambda > 1$ be a constant independent of N and the following conditions on the sets of centered statistics $e_1,\dots,e_n$ hold: there exist subsets $\tilde{\delta}_i \subset \delta_i \subset \subset \mathcal{X}$, $\delta_i \cap \delta_j = \emptyset$ for $i \neq j$, such that $e_i(x)=0$ if $x \bar{\in} \delta_i$, $P(\tilde{\delta}_i) \geq 1/n\lambda$, $P(\delta_i) \leq \lambda/n$,

$$\operatorname*{ess\,sup}_{x\in\delta_i} |e_i(x)| \,/\, \operatorname*{ess\,inf}_{x\in\tilde{\delta}_i} |e_i(x)| \leq \lambda .$$

Let $\Phi_\lambda(\rho, n, p)$ be the set of the type $\Phi(r, n, \alpha)$ where $\alpha=1$ if $1 \leq p \leq 2$, $\alpha=2$ if $2<p<\infty$ and $\alpha=3$ if $p=\infty$, the choice of the value r is such that $\min_i \| f_i^{(\alpha)} - f_0 \|_p = \rho$. It is clear that $\rho \asymp r n^{1/2}$ if $p \leq 2$ and $p=\infty$, $\rho \asymp r n^{1/2-1/2p}$ if $2<p<\infty$. Let

$$\xi_p = \xi_{p,N} = \begin{cases} \sqrt{N}\rho n^{-1/4} & \text{if } 1 \leq p \leq 2, \\ \sqrt{N}\rho n^{-1/2+1/2p} & \text{if } 2<p<\infty, \\ n^{-1}\exp(N\rho^2/n) & \text{if } p=\infty . \end{cases}$$

The following corollary from the theorem 4 can be deduced.

<u>Corollary 2</u>. Let $\Phi_N \supset \Phi_\lambda(\rho, n, p)$ for any $\lambda>1$, $\rho=\rho_N$, $n=n_N$, $1 \leq p \leq \infty$. Then $\gamma_N(f_0, \Phi_N) \to 1$ if $\xi_p = o(1)$ (for $p=\infty$, in particular, if $\overline{\lim}\, N\rho^2/n \ln n < 1$) and $\underline{\lim}\, \gamma_N(f_0, \Phi_N) > 0$ if $\xi_p = O(1)$.

We may deduce p.1 of theorem 2 from the corollary 2 if we consider the sets $e_1,\dots,e_n$, $n=m^K$ of the type $e_i(t_1,\dots,t_K) = n^{1/2}\varphi(mt_1-i_1)\times\dots\times$

$\times \varphi(mt_\kappa - i_\kappa)$, $0 \le i_\ell \le m-1$, where $\varphi(t)$ is an infinitely differentiable function on R^1, $supp\, \varphi \subset [0,1]$, $\int \varphi(t)dt = 0$, $\int \varphi^2(t)dt = 1$. We may choose the asymptotic $m = m_N$ as in p.3 of the theorem 2 such that $\Phi_\lambda(\rho_N, n_N, p) \subset \Phi_N$ for any $\lambda > 1$.

The theorem 4 and the corollary 2 provide the means for reducing probability problem (estimation of the variation distance $var(P^N_{f_0}, conv\{P^N_f, f \in \Phi_N\})$) to the geometrical problem: construction of sets $e_1, \dots, e_{n_N}$ of orthonormal centered statistics and corresponding sets $\Phi_\lambda(\rho_N, n_N, p)$ or $\Phi(r_N, n_N, \alpha)$ in Φ_N with the appropriate asymptotic of values ξ_p or $\xi^{(\alpha)}$.

For the set Ψ of densities on $(\mathcal{X}, \mathcal{U}, P)$ let $r_n(\Psi, \alpha)$ be the infimum of such values r that $\Phi(r, n, \alpha) \subset \Psi$ over all sets $e_1, \dots, e_n$ of orthonormal centered statistics. Let $\xi_{\alpha,N}(\Psi)$ be infimum over n values of the type (10) where $r = r_n(\Psi, \alpha)$. Then the theorem 4 may be reformulated in the following way.

<u>Theorem 4'</u>. If $\xi_{\alpha,N}(\Phi_N) = o(1)$ then $\gamma_N(f_0, \Phi_N) \to 1$ and if $\xi_{\alpha,N}(\Phi_N) = O(1)$ then $\underline{lim}\, \gamma_N(f_0, \Phi_N) > 0$.

For the functional sets Φ_N of the type (1) which are determined by the metric d and values ρ_N the nondiscernability conditions may be formulated in the terms of the following characteristics of the set Φ. Let $r_n(\Phi, \rho, d, \alpha)$ be the infimum of such values r that $\Phi(r, n, \alpha) \subset \Phi$ and $d(f, f_0) \ge \rho$ if $f \in \Phi(r, n, \alpha)$ over the all sets $e_1, \dots, e_n$ of orthonormal centered statistics. Let $\xi_{\alpha,N}(\Phi, \rho, d)$ be the infimum over n of the values of the type (10) where $r = r_n(\Phi, \rho, d, \alpha)$.

<u>Theorem 4"</u>. Let $\Phi_N = \{f \in \Phi : d(f, f_0) \ge \rho_N\}$. Then if $\xi_{\alpha,N}(\Phi, \rho_N, d) = o(1)$ then $\gamma_N(f_0, \Phi_N) \to$

$\to 1$ and if $\xi_{\alpha,N}(\Phi, \rho_N, d) = O(1)$ then $\underline{\lim}\, \gamma_N(f_0, \Phi_N) > 0$.

Geometrical characteristics of the sets Φ and Φ_N used in the theorems 4' and 4" are close in spirit to characteristics of the type of interior radiuses [14] but these are not quite usual. In the case of the quadratic metric d_2 we may get some properties of nondiscernability using usual characteristics. Let $R_n(\Phi, f_0)$ be n-dimensional interior radius of the set Φ in the point f_0 i.e. supremum of radiuses R of balls $\mathcal{D}_R^{(m)}(f_0)$ in d_2-metric in athenian m-dimensional spaces of centered statistics $L_{m,f_0} = \{f_0 + \sum_{i=1}^{m} \alpha_i e_i\}$, $n \geq m$: $\mathcal{D}_R^m(f_0) = \{f \in L_{m,f_0} : \| f - f_0 \|_2 \leq R\}$, such that $\mathcal{D}_R^m(f_0) \subset \Phi$ (the value of the same type are also colled Aleksandrov's diameters of the set Φ [14, 15]). Let $n(\Phi, f_0, \rho)$ be supremum of values n such that $R_n(\Phi, f_0) \geqslant \rho$ and let $\xi_N(\Phi, f_0, \rho) = \sqrt{N}\, \rho\, n^{-1/4}(\Phi, f_0, \rho)$.

Because all functions $f \in \Phi(n, \tau, 1)$ lie on the sphere $S_R^{n-1}(f_0)$ for $R = \sqrt{n}\, \tau$ we deduce the following corollary from the theorem 4 for $\alpha = 1$.

<u>Corollary 3.</u> Let $\Phi_N = \{f \in \Phi : \| f - f_0 \|_2 \geqslant \rho_N\}$. Then $\gamma_N(f_0, \Phi_N) \to 1$ if $\xi_N(\Phi, f_0, \rho_N) = o(1)$ and $\underline{\lim}\, \gamma_N(f_0, \Phi_N) > 0$ if $\xi_N(\Phi, f_0, \rho_N) = O(1)$. In partibulary let $n = n_N$ be such integer sequence that $n^{-1/4} R_n(\Phi, f_0)\sqrt{N} \asymp 1$. Then $\gamma_N(f_0, \Phi_N) \to 1$ if $\rho_N = o(R_{n_N}(\Phi, f_0))$.

The asymptotic of the interior radiuses is studied in the approximation theory [15] for many functional sets Φ. In particular for the sets Φ from the theorem 2 if $p = q = 2$, $\tau > 1/2$ then $R_n(\Phi, f_0) \asymp n^{-\tau/k}$ and under these restrictions the nondiscernability properties of the theorem 2 follow from the corollary 3.

It can also be seen how stronger are the results of the type given by the theorem 4 and corollary 2 than the results of the type given by corollary 3: although there are no balls in Φ or spheres in Φ_N with the requirement for nondiscernability of asymptotic of radiuses and dimensions those sets contain corresponding subsets $\Phi(r,n,\alpha)$.

5. Let describe the methods of consistent in minimax sense tests construction. First of all we consider χ^p-*tests*, $p \geq 2$.

Let $\Phi(C,p)$ be the set of such densities on $(\mathcal{X},\mathcal{U},P)$ that $\|f\|_p < C$. Let $\delta_1,\dots,\delta_n$ be P-uniform partition of $\mathcal{X}$ i.e. $P(\delta_i)=n^{-1}$. Let L_n be the space of statistics which are constant on the sets δ_i, Pr_{L_n} be the orthonormal projector on L_n:

$$Pr_{L_n} f = n \sum_i \mathbb{1}_{\delta_i} \int_{\delta_i} f\,dP .$$

<u>Theorem 5</u>. Let $\Phi_N \subset \Phi(C,p)$, $n=n_N$ be such integer sequence that $n=o(N)$ if $p>2$ and $n \ln n = o(N)$ if $p=\infty$. Then the χ^p-tests sequence of the type (6), (7) which has been generated by P-uniform partition sequence $\delta_1,\dots,\delta_n$ is consistent in minimax sense for any sequence T_N if the following relations hold:

$$\sqrt{N}\, n^{-1/2+1/2p} \inf_{f\in\Phi_N} \|Pr_{L_n}(f-f_0)\|_p \to \infty$$

if $2 \leq p < \infty$ and

$$\underline{\lim} \sqrt{\frac{N}{n \ln n}} \inf_{f\in\Phi_N} \|Pr_{L_n}(f-f_0)\|_\infty > \lambda$$

if $p=\infty$ where $\lambda=\lambda(C)$ is a positive constant.

The proof of the theorem 5 is based on the moments of statistics (6) estimation if $p<\infty$ and on the estimation of probability of large deviations for binomial distribution if $p=\infty$. The case $p=2$ was considered in [8].

The theorem 5 relates the minimax consistencety conditions with the properties of projectors of the functional sets $\Phi - f_0$ on the piecewise-constant function space L_n. For any functional sets Φ under some restrictions on the partitions the following inequality holds: if $f \in \Phi$ then

$$\| Pr_{L_n}(f-f_0)\|_p \geq b_1 \|f-f_0\|_p - b_2 \Gamma_n(\Phi) \qquad (11)$$

where $\Gamma_n(\Phi) \searrow 0$, $b_i = b_i(\Phi, p) > 0$, $i=1,2$.
In this case we can deduce the following corollary from the theorem 5.

Corollary 4. Let $\Phi_N = \{ f \in \Phi : \|f-f_0\|_p \geq \rho_N \}$, $\Phi \subset \Phi(C, p_1)$ where $p_1 = max(2, p)$. Let $n = n_N$ be integer sequence, $\delta_1, \dots, \delta_n$ be such P -uniform partition of $\mathcal{X}$ that for every n, $f \in \Phi$ inequality (11) holds and asymptotic of n_N is such that

$$\sqrt{N}\, n^{-1/2 + 1/2p_1} \Gamma_n(\Phi) \asymp 1 \quad \text{if } p < \infty,$$

$$\sqrt{N/n \ln n}\, \Gamma_n(\Phi) = 1 + o(1) \text{ if } p = \infty.$$

The the sequence of χ^{p_1} - tests is consistent in minimax sense for any sequence T_N if $\rho_N / \Gamma_{n_N}(\Phi) \to \infty$ for $p < \infty$ and if $\underline{\lim}\, \rho_N / \Gamma_{n_N}(\Phi) > \lambda_1$ for $p = \infty$ where $\lambda_1 = \lambda_1(C, b_1, b_2)$ is a positive constant.

The inequality (11) holds for the functional sets $H_K(r, L, q)$ from the theorem 2.

Theorem 6. Let $\delta_1, \dots, \delta_n$ be the partition of cube Δ_K on the $n = m^K$ parallelepiped with the edges of such length l_{ij} that $B^{-1} \leq m l_{ij} \leq B$ where $B > 1$. Then there exist such constants $b_i = b_i(K, r, L, q, B) > 0$, $i = 1, 2$ that for every $f \in H_K(r, L, q)$ the following inequality holds

$$\| Pr_{L_n}(f-f_0)\|_q \geq b_1 \| f-f_0\|_q - b_2\, n^{-r/K}.$$

We deduce p.3 of the theorem 2 from the corollary 4, theorem 6 and from the inclusions $\widetilde{H}_K(r, L, q) \subset \subset H_K(r, L, q) \subset H_K(r, L, p)$ if $q \geq p$.

If $p=2$ we can state the results which are analogus to the theorem 5 for a wider class of subspace L_n and for tests which are based on the corresponding quadratic forms.

Let $A>1$, $B>1$, $\Phi(A)$ be the set of densities on $(\mathcal{X}, \mathcal{U}, P)$ such that $A^{-1} \leq f(x) \leq A$, $x \in \mathcal{X}$. Let $n = n_N$ be the integer sequence, L_n be the sequence of n -dimensional spaces of statistics, $e_1, \ldots, e_n$ be such basis in L_n that

$$B^{-1} I \leq K \leq B I, \tag{12}$$

$$\mathcal{D}\left(\sum_i e_i^2\right) \leq B n \tag{13}$$

where K is the matrix formed of statistics $e_1, \ldots, e_n$ second P -moments, I is the unit $n \times n$ -matrix, $\mathcal{D}$ is the variance on the measure $P = P_{f_0}$.

The condition (13) holds for example if L_n is the space of functions which are constant on the elements of partition $\delta_1, \ldots, \delta_n$, $e_i = \sqrt{n}\, \mathbb{1}_{\delta_i}$ or if L_n is the space of trigonometrical polinoms on $(\Delta_K, \mathcal{B})$ of degree no more then m on the every variable, $n = = m^K$ or of the total degree no more then n, $e_1, \ldots, e_n$ is the standard basis from monomials. In this cases $\sum_i e_i^2 = const$: it follows from the equalities $\sum_i \mathbb{1}_{\delta_i} \equiv 1$ and $\sin^2 t + \cos^2 t \equiv 1$.

Let $\hat{e}_i(x) = e_i(x) - M_P e_i$, $y_{i,N} = N^{-1} \sum_{j=1}^{N} \hat{e}_i(x_j)$,

$$t_N = \sum_i y_{i,N}^2, \quad \Psi_N = \mathbb{1}_{\{t_N > T_N\}}. \tag{14}$$

Theorem 7. Let $\Phi_N \subset \Phi(A)$. Then the relation $\delta_N(\psi_N, f_0, \Phi_N) \to 0$ holds for the sequence of tests (14) and any sequence T_N if

$$\sqrt{N}\, n^{-1/2} \inf_{f \in \Phi_N} \| Pr_{L_n}(f - f_0)\|_2 \to \infty$$

where Pr_{L_n} is orthonormal projector on L_n .

The proof of the theorem 7 follows from the estimations obtained in [8].

In order to use the theorem 7 it is necessary to get inequality for projectors analogous to (11). We may get it on the base of exactness analysis of approximation for sets $\Phi - f_0$ by subspace L_n of considered type. Let

$$\delta(L_n, \Phi, f_0) \stackrel{\Delta}{=} \sup_{h \in \Phi - f_0} \inf_{g \in L_n} \| h - g\|_2 =$$

$$= \sup_{f \in \Phi} \| f - f_0 - Pr_{L_n}(f - f_0)\|_2 ,$$

$\delta_n(\Phi, f_0)$ and $\delta_{n,\beta}(\Phi, f_0)$ be infimums of the values $\delta(L_n, \Phi, f_0)$ over all n -dimensional spaces L_n and over spaces which contain such basis $e_1, \dots, e_n$ that conditions (12), (13) hold. The value $\delta_n(\Phi, f_0)$ is called n -dimensional Kolmogorov's diameter of the set $\Phi - f_0$ in d_2 -metric [14, 15]. The asymptotic of Kolmogorov's diameters was studid in approximation theory [15]. For the some functional sets infimum is reached on the order (i.e. $\delta_n(\Phi, f_0) \asymp \delta(L_n, \Phi, f_0)$) on the spaces L_n of trigonometrical polinoms (for example if $\Phi = \tilde{H}_K(r, L, 2)$ in this case $\delta_n(\Phi, f_0) \asymp n^{-r/K}$) i.e. $\delta_n(\Phi, f_0) \asymp \delta_{n,\beta}(\Phi, f_0)$ for this sets and any $\beta > 1$.

It is clear that inequality of type (11) holds

$$\| Pr_{L_n} (f-f_0) \|_2 \geqslant \| f-f_0 \|_2 - \delta(L_n, \Phi, f_0)$$

and from the theorem 7 we obtain

Corollary 5. Let $A>1$, $B>1$, $\Phi_N = \{ f \in \Phi : \| f - f_0 \|_2 \geqslant \rho_N \}$, $\Phi \subset \Phi(A)$. Let $n = n_N$ be such integer sequence that $\sqrt{N}\, n^{-1/4}\, \delta_{n,B}(\Phi, f_0) \asymp 1$. Then $\gamma_N(f_0, \Phi_N) \to 0$ if $\rho_N / \delta_{n_N, B}(\Phi, f_0) \to \infty$.

It is of some interest to compare the minimax problems of estimation and hypotheses testing. We consider only the quadratic metric $d = d_2$ in the problem of hypothesis testing and quality criteria in the estimation problem

$$E_N(\Phi) \triangleq \inf_{f_N} E_N(\Phi, f_N); \quad E_N^2(\Phi, f_N) = \sup_{f \in \Phi} M_{f,N} \| f - f_N \|_2^2$$

where $f_N = f_N(x; x_1, \ldots, x_N)$ is estimator of density. We consider the class $A_{N,n}$ of "projection estimators" f_N of the following type. Let L_n be the n -dimensional space of statistics, $e_1, \ldots, e_n$ be a orthonormal basis in L_n

$$f_N = \sum_i a_{i,N} e_i, \quad a_{i,N} = N^{-1} \sum_{j=1}^{N} e_i(x_j).$$

The simple computations show (see [14], § 25) that if $f \in \Phi(A)$ then

$$M_{f,N} \| f - f_N \|_2^2 \asymp \frac{n}{N} + \| f - Pr_{L_n} f \|_2^2,$$

$$\inf_{f_N \in A_{N,n}} E_N^2(\Phi, f_N) \triangleq E_{N,Pr}(\Phi, n) \asymp$$

$$\asymp \frac{n}{N} + \delta_n^2(\Phi)$$

where $\delta_n(\Phi)$ is the n -dimension Kolmogorov's diameter of the set Φ. Therefore for the asymptotic of optimal dimensions $n = n_N$ of space L_n the following relation must be satisfied:

$$\sqrt{N}\, n^{-1/2} \delta_n(\Phi) \asymp 1 \tag{15}$$

and

$$E_{N,p_\tau}(\Phi) \triangleq \inf_n E_{N,p_\tau}(\Phi, n) \asymp \delta_{n_N}(\Phi).$$

For any functional sets Φ the following relations hold

$$E_N(\Phi) \asymp E_{N,p_\tau}(\Phi),\ \delta_{n,B}(\Phi, f_0) \asymp \delta_n(\Phi, f_0) \asymp \delta_n(\Phi).$$

In this case comparing relation (15) and analogic relation in the corollary 5 we see the difference in the dimensions of"projection space" asymptotic which is related with the difference of power degries of n (1/4 and 1/2). It determines the difference in results for minimax nonparametric problems of estimation and hypothesis testing.

The described methods and results may be expanded to the case of composite and nonparametric null hypotheses Φ_0 about distribution densities (for $k=1$, $p=2$ and integer τ see [8, 16]) and to other statistical models: the hypotheses about signals in the Gaussian white noise [17] and about spectral densities of the Gaussian process [18].

REFERENCES

1. Линник Ю.В. Приближенно минимаксное обнаружение векторного сигнала при гауссовской помехе. - Теория ве-

роятностей и её примен., 1966, т.II, с.561-578.
2. Fan Ky. Minimax theorems. - Proc.Nat.Acad.Sci.USA, 1953, 39, N 1, p.42-47.
3. Wald A. Statistical decision functions Wiley, New York, 1950.
4. Lehmann E.L. Testing statistical hypotheses. Wiley, New York, and Chapman & Hall, London, 1959.
5. Боровков А.А. Математическая статистика. М.: Наука, 1984, 472 с.
6. Ингстер Ю.И. Асимптотически оптимальные тесты проверки сложных конечнопараметрических гипотез. - Теория вероятностей и её примен., 1985, т.30, № 2, с.289-308.
7. Ибрагимов И.А., Хасьминский Р.З. Асимптотическая теория оценивания. М.: Наука, 1979, 528 с.
8. Ингстер Ю.И. Асимптотически минимаксная проверка непараметрических гипотез о плотности распределения независимой выборки. - Записки научн.семинаров ЛОМИ, 1984, т.136, с.74-96.
9. Bretagnolle J. Huber C. Estimation de densities: risque minimax. - Z.Wahrscheinlichkeittheorie verw. Gebiete, 1979, v.47, p.119-137.
10. Ибрагимов И.А., Хасьминский Р.З. Об оценке плотности распределения. - Записки научн.семинаров ЛОМИ, 1980, т.98, с.61-85.
11. Mann H.B., Wald A. On the choice of the number of intervals on the application of the chi-square test. - Ann.Math.Statist., 1942, N 13, p.306.
12. Кендалл М., Стьюарт А. Статистические выводы и связи. М.: Наука, 1973, 899 с.
13. Kac M., Kiefer J., Wolfowitz J. On test of normality and other test of goodness of fit based on distance methods. - Ann.Math.Statist., 1955, v.26, p.189.
14. Ченцов Н.Н. Статистические решающие правила и оптимальные выводы. М.: Наука, 1972, 520 с.
15. Тихомиров В.М. Некоторые вопросы теории приближений.

М.: Изд-во МГУ, 1976.

16. Ингстер Ю.И. Асимптотически минимаксная проверка непараметрических гипотез о типе распределения независимой выборки. - Теория вероятностей и её примен., 1983, т.28, № 4, с.799-800.
17. Ингстер Ю.И. О минимаксном непараметрическом обнаружении сигнала в гауссовском белом шуме. - Проблемы передачи информации, 1982, т.28, № 2, с.61-73.
18. Ингстер Ю.И. Асимптотически минимаксная проверка непараметрических гипотез о спектральной плотности.- Теория вероятностей и её примен., 1984, т.29, № 4, с.809-810.

Bei Fragen zur Produktsicherheit wenden Sie sich bitte an:
If you have any questions regarding product safety,
please contact:

Walter de Gruyter GmbH
Genthiner Straße 13
10785 Berlin
productsafety@degruyterbrill.com